DESIGN OF MODERN STEEL RAILWAY BRIDGES

DESIGN OF MODERN STEEL RAILWAY BRIDGES

JOHN F. UNSWORTH

CRC Press is an imprint of the
Taylor & Francis Group, an **informa** business

CRC Press
Taylor & Francis Group
6000 Broken Sound Parkway NW, Suite 300
Boca Raton, FL 33487-2742

Printed in the United States of America on acid-free paper
10 9 8 7 6 5 4 3 2 1

International Standard Book Number: 978-1-4200-8217-3 (Hardback)

Library of Congress Cataloging-in-Publication Data

Unsworth, John F.
Design of modern steel railway bridges / John F. Unsworth.
p. cm.
Includes bibliographical references and index.
ISBN 978-1-4200-8217-3 (hardcover : alk. paper)
1. Railroad bridges--Design and construction. 2. Iron and steel bridges. I. Title.

TG445.U57 2010
624.2--dc22 2009047373

Visit the Taylor & Francis Web site at
http://www.taylorandfrancis.com

and the CRC Press Web site at
http://www.crcpress.com

To my extraordinary wife, Elizabeth, without whose support and patience this book could not have been started or completed.

Contents

Preface

It is estimated that, in terms of length, just over 50% of the approximately 80,000 bridges in the North American freight railroad bridge inventory have steel superstructures. These bridges are critical components of the railroad infrastructure and, therefore, essential elements of an effective and competitive national transportation system. Many of these railway bridges are over 80 years old* and have experienced substantial increases in both the magnitude and frequency of freight railroad live load. The assessment (inspection, condition rating, strength rating, and fatigue life cycle analysis), maintenance (repair and retrofitting), and rehabilitation (strengthening) of existing railway bridges are fundamental aspects of a sustainable, safe, and reliable national railroad transportation infrastructure. However, in many cases, due to functional and/or structural obsolescence (age [fatigue], condition, and/or strength), the replacement of steel railway superstructures is required.

In response, this book is an attempt to provide a focus on the design of new steel superstructures for modern railway bridges. However, while the focus is on replacement superstructures, many of the principles and methods outlined will also be useful in the maintenance and rehabilitation of existing steel railway bridges. This book is intended to supplement existing structural steel design books, manuals, handbooks, guides, specifications, and technical reports currently used by railway bridge design engineers. In particular, the book complements the recommended practices of Chapter 15—Steel Structures in the American Railway Engineering and Maintenance-of-way Association (AREMA) Manual for Railway Engineering (MRE). The recommended practices of the MRE are updated by an active committee of railway bridge owners, engineers, consultants, suppliers, academics, and researchers. The reader is recommended to consult the most recent version of the AREMA MRE as a basis for steel railway superstructure design. This book references AREMA (2008), the MRE edition current at the time of writing. Nevertheless, the majority of the information contained herein is fundamental and will remain valid through many editions of the MRE.

It is hoped that this book will serve as a practical reference for experienced bridge engineers and researchers, and a learning tool for students and engineers newly engaged in the design of steel railway bridges. The book is divided into nine chapters. The first three chapters provide introductory and general information as a foundation

* Estimated as the typical design life of a steel railway superstructure.

for the subsequent six chapters examining the detailed analysis and design of modern steel railway superstructures.

Modern structural engineering has its roots in the history and development of steel railway bridges. Chapter 1 provides a brief history of iron and steel railway bridges. The chapter concludes with the evolution and advancement of structural mechanics and design practice precipitated by steel railway bridge development.

Chapter 2 considers the engineering properties of structural steel typically used in modern steel railway bridge design and fabrication. The chapter focuses on the significance of these properties in steel railway superstructure design.

Chapter 3 presents information regarding the planning and preliminary design of steel railway bridges. The planning of railway bridges considering economic, business, regulatory, hydraulic, clearance, and geotechnical criteria is outlined. Following a general discussion of the first three of these criteria, simple methods of hydraulic analysis are covered before a general discussion of scour evaluation for ordinary railway bridges. Planning deliberations conclude with a discussion of the horizontal and vertical geometries of the general bridge arrangement. This material is intended to provide guidance regarding the scope and direction of planning issues in advance of the preliminary design. Preliminary design concerns, such as aesthetics, form, framing, and deck type, are discussed in terms of typical modern steel railway superstructures. The subjects of bearings, walkways, fabrication, and erection for ordinary steel railway superstructures are also briefly considered.

The remaining six chapters deal with the development of loads, the structural analysis, and the detailed design of modern steel railway bridge superstructures.

Chapter 4 outlines the loads and forces on railway superstructures. Many of these loads and forces are specific to railway bridges and others are characteristic of bridges in general. The design live load and related dynamic effects are particular to railroad traffic. Longitudinal, centrifugal, and some lateral forces are also railroad traffic specific. The theoretical and experimental development of modern steel railway bridge design live, impact (dynamic), longitudinal and lateral loads or forces is succinctly covered. This precedes a discussion of load distribution, and the wind and seismic forces on ordinary steel railway superstructures.

Railway live loads are heavy and dimensionally complex moving loads. Modern structural analyses of moving loads are often effectively performed by digital computer software. However, an intuitive and analytical understanding of moving load effects is a necessary tool for the railway bridge design engineer to correctly interpret computer analyses and conduct simple evaluations manually. Therefore, the criteria for the maximum effects from moving loads and their use in developing design live loads are presented in Chapter 5. The effects of moving loads are outlined in terms of qualitative and quantitative influence lines for beam, girder, truss, and arch spans. The moving load discussion ends with the equivalent uniform load concept, and charts, tables, and equations available for the analysis of simply supported railway spans. The chapter contains many examples intended to illustrate various principles of moving load structural analysis. Chapter 5 also outlines lateral load analytical methods for superstructure bracing. The chapter ends with a general discussion of strength, stability, serviceability, and fatigue design criteria as a foundation for subsequent chapters concerning the detailed design of superstructure members.

The next three chapters concern the design of members in modern steel railway superstructures. Chapters 6 and 7 describe the detailed design of axial and flexural members, respectively, and Chapter 8 investigates combinations of forces on steel railway superstructures. The book concludes with Chapter 9 concerning connection design.

Trusses containing axial members are prevalent for relatively long-span railway bridge superstructures. Axial tension and compression member design are outlined in Chapter 6. Built-up member requirements, for compression members in particular, are also considered.

Beam and girder spans comprise the majority of small- and medium-span steel railway bridge superstructures. Chapter 7 examines flexural members of noncomposite and composite design. The detailed design of plate girder flange, stiffened web, and stiffener plate elements is considered based on yield, fracture, fatigue and stability criteria.

Chapter 8 is concerned with members subjected to the combination of stresses that may occur in steel railway superstructures from biaxial bending, unsymmetrical bending, and combined axial and bending forces. The chapter presents a discussion of simplified analyses and the development of interaction equations suitable for use in routine design work.

The final chapter, Chapter 9, provides information concerning the design of connections for axial and flexural members in steel railway superstructures. The chapter discusses weld and bolt processes, installation and types prior to outlining typical welded and bolted joint types used in modern steel railway superstructures. Welded and bolted connections that transmit axial shear, combined axial tension and shear, and eccentric shear are examined.

This book is an endeavor to provide fundamental information on the design of ordinary modern steel railway superstructures and does not purport to be a definitive text on the subject. Other books, manuals, handbooks, codes, guides, specifications, and technical reports/papers are essential for the safe and reliable design of less conventional or more complex superstructures. Some of those resources, that were available to the author, were used in the preparation of the information herein. In all cases, it is hoped that proper attribution has been made. The author gratefully appreciates any corrections that are drawn to his attention.

John F. Unsworth
Calgary, Alberta, Canada

Acknowledgments

I must respectfully acknowledge the efforts of my father, who provided opportunities for an early interest in science and engineering (e.g., by presenting me with the 56th edition of the *CRC Handbook of Chemistry and Physics*) and my mother, whose unconditional support in all matters has been truly appreciated. My wife, Elizabeth, also deserves special recognition for everything she does, and the kind and thoughtful way in which she does it. She, my daughters, Tiffany and Genevieve, and granddaughter, Johanna, furnish my greatest joys in life. In addition, the guidance and friendship provided by many esteemed colleagues, in particular Dr. R. A. P. Sweeney and W. G. Byers are greatly appreciated.

Author

John F. Unsworth is a professional engineer (P Eng). Since his completion of a bachelor of engineering degree in civil engineering in 1981 and a master of engineering degree in structural engineering in 1987, he has held professional engineering and management positions concerning track, bridge, and structures maintenance, design, and construction at the Canadian Pacific Railway. He is currently the vice president of Structures of the American Railway Engineering and Maintenance-of-way Association (AREMA) and has served as chairman of AREMA Committee 15—Steel Structures. In addition, he is the current Chair of the Association of America Railroads (AAR) Bridge Research Advisory Group and is a member of the National Academy of Sciences Transportation Research Board (TRB) Steel Bridges Committee. He is also a member of the Canadian Society for Civil Engineering (CSCE) and International Association of Bridge and Structural Engineers (IABSE). He is a licensed professional engineer in six Canadian Provinces. He has written papers and presented them at AREMA Annual Technical Conferences, the International Conference on Arch Bridges, TRB Annual Meetings, the CSCE Bridge Conference, and the International Bridge Conference (IBC). He has also contributed to the fourth edition of the *Structural Steel Designer's Handbook* and the *International Heavy Haul Association (IHHA) Best Practices* books.

1 History and Development of Steel Railway Bridges

1.1 INTRODUCTION

The need for reliable transportation systems evolved with the industrial revolution. By the early nineteenth century, it was necessary to transport materials, finished goods, and people over greater distances in shorter times. These needs, in conjunction with the development of steam power,* heralded the birth of the railroad. The steam locomotive with a trailing train of passenger or freight cars became a principal means of transportation. In turn, the railroad industry became the primary catalyst in the evolution of materials and engineering mechanics in the latter half of the nineteenth century.

The railroad revolutionized the nineteenth century. Railroad transportation commenced in England on the Stockton to Darlington Railway in 1823 and the Liverpool and Manchester Railway in 1830. The first commercial railroad in the United States was the Baltimore and Ohio (B&O) Railroad, which was chartered in 1827.

Construction of the associated railroad infrastructure required that a great many wood, masonry, and metal bridges be built. Bridges were required for live loads that had not been previously encountered by bridge builders.† The first railroad bridge in the United States was a wooden arch-stiffened truss built by the B&O in 1830. Further railroad expansion‡ and rapidly increasing locomotive weights, particularly in the United States following the Civil War, provoked a strong demand for longer and stronger railway bridges. In response, a great many metal girder, arch, truss, and suspension bridges were built to accommodate railroad expansion, which was

---

* Nicolas Cugnot is credited with production of the first steam-powered vehicle in 1769. Small steam-powered industrial carts and trams were manufactured in England in the early years of the nineteenth century and George Stephenson built the first steam locomotive, the "Rocket," for use on the Liverpool and Manchester Railway in 1829.

† Before early locomotives, bridges carried primarily pedestrian, equestrian, and light cart traffic. Railroad locomotive axle loads were about 11,000 lb on the B&O Railroad in 1835.

‡ For example, in the 1840s charters to hundreds of railway companies were issued by the British government.

occurring simultaneously in the United States and England following the British industrial revolution.

In the United States, there was an intense race among emerging railroad companies to expand west. Crossing the Mississippi River became the greatest challenge to railroad growth. The first railway bridge across the Mississippi River was completed in 1856 by the Chicago, Rock Island, and Pacific Railroad.* The efforts of the B&O Railroad company to expand its business and cross the Mississippi River at St. Louis, Missouri, commencing in 1839† and finally realized in 1874, proved to be a milestone in steel railway bridge design and construction. Although the St. Louis Bridge never served the volume of railway traffic anticipated in 1869 at the start of construction, its engineering involved many innovations that provided the foundation for long-span railway bridge design for many years following its completion in 1874.

The need for longer and stronger railway bridges precipitated a materials evolution from wood and masonry to cast and wrought iron, and eventually to steel. Many advances and innovations in construction technology and engineering mechanics can also be attributed to the development of the railroads and their need for more robust bridges of greater span.

1.2 IRON RAILWAY BRIDGES

1.2.1 Cast Iron Construction

A large demand for railway bridges was generated as railroads in England and the United States prospered and expanded. Masonry and timber were the principal materials of early railway bridge construction, but new materials were required to span the greater distances and carry the heavier loads associated with railroad expansion. Cast iron had been used in 1779 for the construction of the first metal bridge, a 100 ft arch span over the Severn River at Coalbrookedale, England. The first bridge to use cast iron in the United States was the 80 ft arch, built in 1839, at Brownsville, Pennsylvania. Cast iron arches were also some of the first metal railway bridges constructed and their use expanded with the rapidly developing railroad industry.‡ Table 1.1 indicates some notable cast iron arch railway bridges constructed between 1847 and 1861.

The oldest cast iron railway bridge in existence is the 47 ft trough girder at Merthyr Tydfil in South Wales. It was built in 1793 to carry an industrial rail tram. The first iron railway bridge for use by the general public on a chartered railroad was built in 1823 by George Stephenson on the Stockton to Darlington Railway (Figure 1.1).

---

* The bridge was constructed by the Rock Island Bridge Company after U.S. railroads received approval to construct bridges across navigable waterways. The landmark Supreme Court case that enabled the bridge construction also provided national exposure to the Rock Island Bridge Company solicitor, Abraham Lincoln.

† In 1849, Charles Ellet, who designed the ill-fated suspension bridge at Wheeling, West Virginia, was the first engineer to develop preliminary plans for a railway suspension bridge to cross the Mississippi at St. Louis. Costs were considered prohibitive, as were subsequent suspension bridge proposals by J.A. Roebling, and the project never commenced.

‡ Cast iron bridge connections were made with bolts because the brittle cast iron would crack under pressures exerted by rivets as they shrank from cooling.

TABLE 1.1
Notable Iron and Steel Arch Railway Bridges Constructed between 1847–1916

Location	Railroad	Engineer	Year	Material	Hinges	Span (ft)
Hirsk, UK	Leeds and Thirsk	—	1847	Cast iron	0	—
Newcastle, UK	Northeastern	R. Stephenson	1849	Cast iron	0	125
Oltwn, Switzerland	Swiss Central	Etzel and Riggenbach	1853	Wrought iron	0	103
Paris, France	Paris—Aire	—	1854	Wrought iron	2	148
Victoria, Bewdley, UK	—	J. Fowler	1861	Cast iron	—	—
Albert, UK	—	J. Fowler	1861	Cast iron	—	—
Coblenz, Germany	—	—	1864	Wrought iron	2	—
Albert, Glasgow, Scotland	—	Bell and Miller	1870	Wrought iron	—	—
St. Louis, MO	Various	J. Eads	1874	Cast steel	0	520
Garabit, France	—	G. Eiffel	1884	Wrought iron	2	540
Paderno, Italy	—	—	1889	Iron	—	492
Stony Creek, BC	Canadian Pacific	H.E. Vautelet	1893	Steel	3	336
Keefers, Salmon River, BC	Canadian Pacific	H.E. Vautelet	1893	Steel	3	270
Surprise Creek, BC	Canadian Pacific	H.E. Vautelet	1897	Steel	3	290
Grunenthal, Germany	—	—	1892	Steel	2	513
Levensau, Germany	—	—	1894	Steel	2	536
Mungsten, Prussia	—	A. Rieppel	1896	Steel	0	558
Niagara Gorge (2), NY	—	—	1897	Steel	2	550
Viaur Viaduct, France	—	—	1898	Steel	0	721
Worms, Germany	—	Schneider and Frintzen	1899	Steel	—	217
Yukon, Canada	Whitepass and Yukon	—	—	Steel	0	240
Passy Viaduct, France	Western Railway of Paris	—	—	Steel	—	281
Rio Grande, Costa Rica	Narrow gage	—	1902	Steel	2	448
Birmingham, AL	Cleveland and Southwestern Traction	—	1902	Steel	—	—
Mainz, Germany	—	—	1904	Steel	—	—
Paris, France	Metropolitan	—	1905	Steel	—	460
Song-Ma, China	Indo-China	—	—	Steel	3	532
Iron Mountain, MI	Iron ore	—	—	Steel	3	—
Zambesi, Rhodesia	—	G.A. Hobson	1905	Steel	—	500
Thermopylae, Greece	—	P. Bodin	1906	Steel	3	262
Nami-Ti Gorge, China	Yunnan	—	1909	Steel	3	180
Hell Gate, NY	Pennsylvania	G. Lindenthal	1916	Steel	2	978

FIGURE 1.1 Gaunless River Bridge of the Stockton and Darlington Railway built in 1823 at West Auckland, England. (Chris Lloyd, *The Northern Echo*, Darlington.)

The bridge consisted of 12.5 ft long lenticular spans* in a trestle arrangement. This early trestle was a precursor to the many trestles that would be constructed by railroads to enable almost level crossings of wide and/or deep valleys. Table 1.2 summarizes some notable cast iron railway trestles constructed between 1823 and 1860.

George Stephenson's son, Robert, and Isambard Kingdom Brunel were British railway engineers who understood cast iron material behavior and the detrimental effects on arches created by moving railroad loads. They successfully built cast iron arch bridges that were designed to act in compression. However, the relatively level grades required for train operations (due to the limited tractive effort available to early locomotives) and use of heavier locomotives also provided motivation for the extensive use of cast iron girder and truss spans for railway bridges.

Commencing about 1830, Robert Stephenson built both cast iron arch and girder railway bridges in England. Cast iron plate girders were also built in the United States by the B&O Railroad in 1846, the Pennsylvania Railroad in 1853, and the Boston and Albany Railroad in 1860. The B&O Railroad constructed the first cast iron girder trestles in the United States in 1853. One of the first cast iron railway viaducts in Europe was constructed in 1857 for the Newport to Hereford Railway line at Crumlin, England. Nevertheless, while many cast iron arches and girders were built in England and the United States, American railroads favored the use of composite trusses of wood and iron.

American railroad trusses constructed after 1840 often had cast iron, wrought iron, and timber members. In particular, Howe trusses with wood or cast iron compression members and wrought iron tension members were used widely in early American railroad bridge construction.

* Also referred to as Pauli spans.

TABLE 1.2
Notable Iron and Steel Viaduct Railway Bridges Constructed between 1823–1909

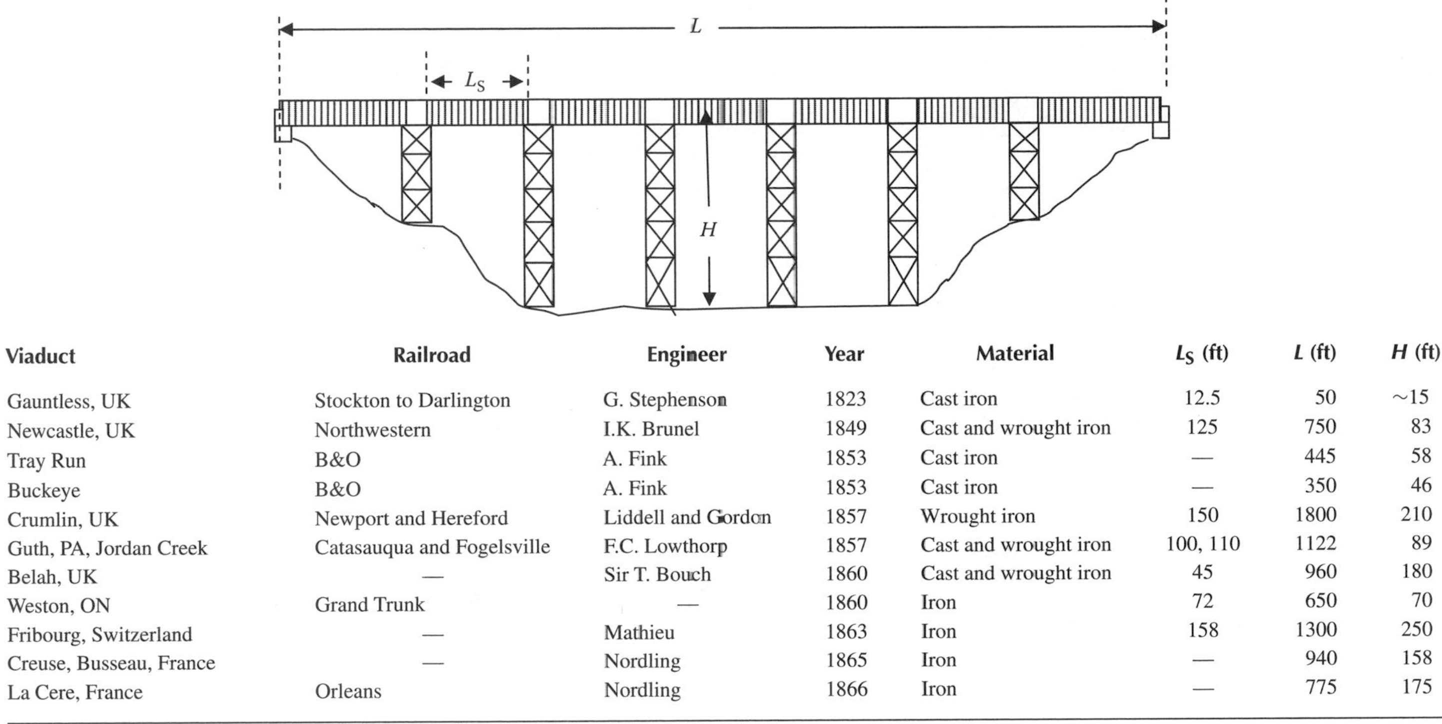

Viaduct	Railroad	Engineer	Year	Material	L_S (ft)	L (ft)	H (ft)
Gauntless, UK	Stockton to Darlington	G. Stephenson	1823	Cast iron	12.5	50	~15
Newcastle, UK	Northwestern	I.K. Brunel	1849	Cast and wrought iron	125	750	83
Tray Run	B&O	A. Fink	1853	Cast iron	—	445	58
Buckeye	B&O	A. Fink	1853	Cast iron	—	350	46
Crumlin, UK	Newport and Hereford	Liddell and Gordon	1857	Wrought iron	150	1800	210
Guth, PA, Jordan Creek	Catasauqua and Fogelsville	F.C. Lowthorp	1857	Cast and wrought iron	100, 110	1122	89
Belah, UK	—	Sir T. Bouch	1860	Cast and wrought iron	45	960	180
Weston, ON	Grand Trunk	—	1860	Iron	72	650	70
Fribourg, Switzerland	—	Mathieu	1863	Iron	158	1300	250
Creuse, Busseau, France	—	Nordling	1865	Iron	—	940	158
La Cere, France	Orleans	Nordling	1866	Iron	—	775	175

continued

TABLE 1.2 (continued)
Notable Iron and Steel Viaduct Railway Bridges Constructed between 1823–1909

Viaduct	Railroad	Engineer	Year	Material	L_S (ft)	L (ft)	H (ft)
Assenheim, Germany	—	—	~1866	Iron	—	—	—
Angelroda, Germany	—	—	~1866	Iron	100	300	—
Bullock Pen	Cincinnati and Louisville	F.H. Smith	1868	Iron	—	470	60
Lyon Brook, NY	New York, Oswego, and Midland	—	1869	Wrought iron	30	820	162
Rapallo Viaduct	New Haven, Middletown, and Willimantic	—	1869	Iron	30	1380	60
St. Charles Bridge over the Mississippi River	—	—	1871	—	—	—	—
La Bouble, France	Commentary-Gannat	Nordling	1871	Wrought iron	160	1300	216
Bellon Viaduct, France	Commentary-Gannat	Nordling	1871	Steel	131	—	160
Verragus, Peru	Lima and Oroya	C.H. Latrobe	1872	Wrought iron	110, 125	575	256
Olter, France	Commentary-Gannat	Nordling	1873	Steel	—	—	—
St. Gall, France	Commentary-Gannat	Nordling	1873	Steel	—	—	—
Horse Shoe Run	Cincinnati Southern	L.F.G. Bouscaren	~1873	Wrought iron	—	900	89
Cumberland	Cincinnati Southern	L.F.G. Bouscaren	~1873	Wrought iron	—	—	100
Tray Run (2)	B&O	—	1875	Steel	—	—	58
Fishing Creek	Cincinnati Southern	L.F.G. Bouscaren	1876	Wrought iron	—	—	79
McKees Branch	Cincinnati Southern	L.F.G. Bouscaren	1878	Wrought iron	—	—	128
Portage, NY	Erie	G.S. Morison and O. Chanute	1875	Wrought iron	50, 100	818	203
Staithes, UK	Whitby and Loftus	J. Dixon	1880	—	—	690	150
Oak Orchard, Rochester, NY	Rome, Watertown, and Western	—	~1881	Steel	30	690	80
Kinzua (1), PA	New York, Lake Erie, and Western	G.S. Morison, O. Chanute, T.C. Clarke and A. Bonzano	1882	Wrought iron	—	2053	302
Rosedale, Toronto, ON	Ontario and Quebec	—	1882	—	30, 60	—	—

Dowery Dell, UK	Midland	Sir T. Bouch	~1882	—	—	—	—
Marent Gulch, MT	Northern Pacific	—	1884	Steel	116	800	200
Loa, Bolivia	Antofagasta	—	1885–1890	—	—	800	336
Malleco, Chile	—	A. Lasterria	1885–1890	—	—	1200	310
Souleuvre, France	—	—	1885–1890	—	—	1200	247
Moldeau, Germany	—	—	1885–1890	—	—	886	214
Schwarzenburg, Germany	—	—	1889	Steel	—	—	—
Panther Creek, PA	Wilkes-Barre and Eastern	—	1893	Steel	—	1650	154
Pecos, CA	—	—	1894	Steel	—	2180	320
Grasshopper Creek	Chicago and Eastern Illinois	—	1899	Steel	—	—	—
Lyon Brook (2), NY	New York, Ontario, and Western	—	1894	Steel	30	820	162
Kinzua (2), PA	New York, Lake Erie, and Western	C.R. Grimm	1900	Steel	—	2052	302
Gokteik, Burma	Burma	Sir A. Rendel	1900	Steel	—	2260	320
Boone, IA	Chicago and Northwestern	G.S. Morison	1901	Steel	45, 75, 300	2685	185
Portage, NY (2)	Erie	—	1903	Steel	50, 100	818	203
Richland Creek, IN	—	—	1906	Steel	40, 75	—	158
Moodna Creek	Erie	—	1907	Steel	40, 80	3200	182
Colfax, CA	—	—	1908	Steel	—	810	190
Makatote, New Zealand	—	—	1908	Steel	—	860	300
Cap Rouge, QC	Transcontinental	—	1908	Steel	40, 60	—	173
Battle River, AB	Grand Trunk Pacific	—	1909	Steel	—	~2700	184
Lethbridge, AB	Canadian Pacific	Monsarrat and Schneider	1909	Steel	67, 100	5328	314

The failure of a cast iron girder railway bridge in 1847* stimulated an interest in wrought iron among British railway engineers.† British engineers were also concerned with the effect of railway locomotive impact on cast iron railway bridges. In addition, many were beginning to understand that, while strong, cast iron was brittle and prone to sudden failure. Concurrently, American engineers were becoming alarmed by cast iron railway bridge failures, and some even promoted the exclusive use of masonry or timber for railway bridge construction. For example, following the collapse of an iron truss bridge in 1850 on the Erie Railroad, some American railroads dismantled their iron trusses and replaced them with wood trusses. However, the practice of constructing railway bridges of iron was never discontinued on the B&O Railroad.

European and American engineers realized that a more ductile material was required to resist the tensile forces developed by heavy railroad locomotive loads. Wrought iron‡ provided this increase in material ductility and was integrated into the construction of many railway bridges after 1850. The use of cast iron for railway bridge construction in Europe ceased in about 1867. One of the last major railway bridges in Europe to be constructed in cast iron was Gustave Eiffel's 1600 ft long Garonne River Bridge built in 1860. However, cast iron continued to be used (primarily as compression members) in the United States, even in some long-span bridges, for more than a decade after its demise in Europe.§

1.2.2 Wrought Iron Construction

Early short- and medium-span railway bridges in the United States were usually constructed from girders or propriety trusses (e.g., the Bollman, Whipple, Howe, Pratt, and Warren trusses shown in Figure 1.2). The trusses typically had cast iron or wood compression members and wrought iron tension members.** United States patents were granted for small- and medium-span iron railway trusses after 1840 and they became widely used by American railroads.

The wooden Howe truss with wrought iron vertical members (patented in 1840) was popular on American railroads up to the 1860s and used on some railroads upto the turn of the century.†† The principal attraction of the Howe truss was the use of wrought iron rods, which did not permit the truss joints to come apart when diagonal members were in tension from railway loading. However, the Howe truss form is

* This was Stephenson's cast iron girder bridge over the River Dee on the London–Chester–Holyhead Railroad. In fact, Stephenson had recognized the brittle nature of cast iron before many of his peers and reinforced his cast iron railway bridge girders with wrought iron rods. Nevertheless, failures ensued with increasing railway loads.

† Hodgekinson, Fairbairn, and Stephenson had also performed experiments with cast and wrought iron bridge elements between 1840 and 1846. The results of those experiments led to a general acceptance of wrought iron for railway bridge construction among British engineers.

‡ Wrought iron has a much lower carbon content than cast iron and is typically worked into a fibrous material with elongated strands of slag inclusions.

§ J.H. Linville was a proponent of all-wrought-iron truss construction in the early 1860s.

** Wrought iron bridge construction provided the opportunity for using riveted connections instead of bolts. The riveted connections were stronger due to the clamping forces induced by the cooling rivets.

†† During construction of the railroad between St. Petersburg and Moscow, Russia (ca. 1842), American Howe truss design drawings were used for many bridges.

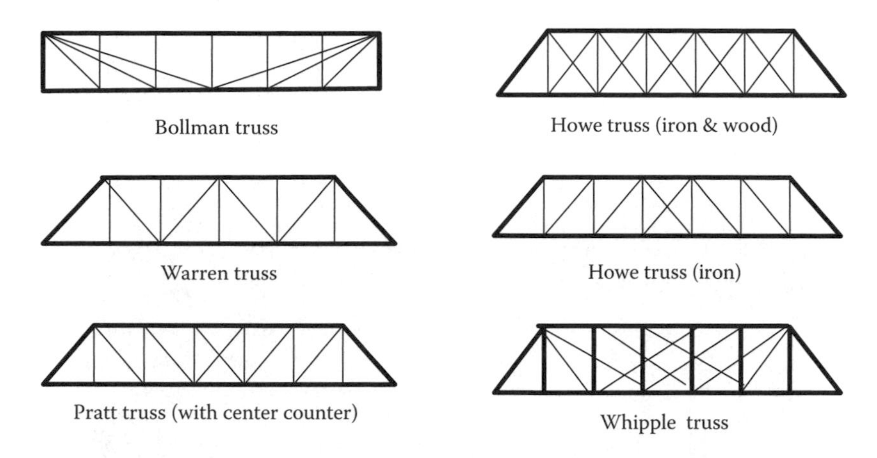

FIGURE 1.2 Truss forms used by railroads in the United States.

statically indeterminate and, therefore, many were built on early American railroads without the benefit of applied scientific analysis.

The first railway bridge in the United States constructed entirely in iron was a Howe truss with cast iron compression and wrought iron tension members built by the Philadelphia and Reading Railroad in 1845 at Manayunk, Pennsylvania. Following this, iron truss bridges became increasingly popular as American railroads continued their rapid expansion. Iron Howe trusses were also constructed by the Boston and Albany Railroad in 1847 near Pittsfield, Massachusetts, and on the Harlem and Erie Railroad in 1850. Early examples of Pratt truss use were the Pennsylvania Railroad's cast and wrought iron arch-stiffened Pratt truss bridges of the 1850s. An iron railway bowstring truss, also utilizing cast iron compression and wrought iron tension members, was designed by Squire Whipple* for the Rensselaer and Saratoga Railway in 1852. Fink and Bollman, both engineers employed by the B&O Railroad, used their own patented cast and wrought iron trusses extensively between 1840 and 1875.† Noteworthy, iron trusses were also built by the North Pennsylvania Railroad in 1856 (a Whipple truss) and the Catasauqua and Fogelsville Railroad in 1857. The Erie Railroad pioneered the use of iron Post truss bridges in 1865 and they remained a standard of construction on the B&O Railroad for the next 15 years.

However, due to failures in the 30 years after 1840 occurring predominantly in cast iron bridge members, the use of cast iron ceased and wrought iron was used exclusively for railway girders and trusses. Isambard Kingdom Brunel used thin-walled wrought iron plate girders in his designs for short and medium railway spans on the Great Western Railway in England during the 1850s. Between 1855 and 1859, Brunel also designed and constructed many noteworthy wrought iron lattice girder, arch, and suspension bridges for British railways. In particular, the Royal Albert Railway Bridge across the Tamar River, completed at Saltash in 1859, is a significant example of a

* In 1847, Whipple published *A Treatise on Bridge Building*, the first book on scientific or mathematical truss analysis.

† The first all-iron trusses on the B&O were designed by Fink in 1853.

FIGURE 1.3 The Royal Albert Bridge built in 1859 over Tamar River at Saltash, England. (Courtesy of Owen Dunn, June 2005.)

Brunel wrought iron railway bridge using large lenticular trusses (Figure 1.3). Other important railway bridges built by Brunel on the Great Western Railway were the Wharnecliffe Viaduct, Maidenhead, and Box Tunnel bridges. Table 1.3 lists some notable wrought iron truss railway bridges constructed between 1845 and 1877.

The English engineer William Fairbairn constructed a tubular wrought iron through girder bridge on the Blackburn and Bolton Railway in 1846. Later, in partnership with Fairbairn, Robert Stephenson designed and built the innovative and famous wrought iron tubular railway bridges for the London–Chester–Holyhead Railroad at Conway in 1848 and at Menai Straits (the Britannia Bridge) in 1849. The Conway Bridge is a simple tubular girder span of 412 ft and the Britannia Bridge consists of four continuous tubular girder spans of 230, 460, 460, and 230 ft (Figure 1.4). Spans of up to 460 ft were mandated for navigation purposes, making this the largest wrought iron bridge constructed. It was also one of the first uses of continuity to reduce dead load bending moments in a bridge. Arch bridges were also proposed by Stephenson* and Brunel.† However, arch bridges were rejected due to concerns about interference with navigation and the wrought iron tubular girder spans were built in order to obtain the stiffness required for wind and train loadings. The construction of the Conway and Britannia tubular iron plate girder bridges also provided the opportunity for further investigations into issues of plate stability, riveted joint construction, lateral wind

* Stephenson had studied the operating issues associated with some suspension railway bridges, notably the railway suspension bridge built at Tees in 1830, and decided that suspension bridges were not appropriate for railway loadings. He proposed an arch bridge.

† In order to avoid the use of falsework in the channel, Brunel outlined the first use of the cantilever construction method in conjunction with his proposal for a railway arch bridge across Menai Straits.

TABLE 1.3
Notable Iron and Steel Simple Truss Span Railway Bridges Constructed between 1823–1907

Location	Railroad	Engineer	Year Completed	Type	Material	L (ft)
West Auckland, UK	Stockton to Darlington	G. Stephenson	1823	Lenticular	Cast iron	12.5
Ireland	Dublin and Drogheda	G. Smart	1824	Lattice	Cast iron	84
Manayunk, PA	Philadelphia and Reading	R. Osborne	1845	Howe	Cast and wrought iron	34
Pittsfield, MA	Boston and Albany	—	1847	Howe	Cast and wrought iron	30
Windsor, UK	Great Western	I.K. Brunel	1849	Bowstring	Iron	187
Newcastle, UK	Northwestern	I.K. Brunel	1849	Bowstring	Cast and wrought iron	125
—	Harlem and Erie	—	1850	Howe	Iron	—
Various	Pennsylvania	H. Haupt	1850s	Pratt with cast iron arch	Iron	—
Harper's Ferry	B&O	W. Bollman	1852	Bollman	Cast and wrought iron	124
Fairmont, WV	B&O	A. Fink	1852	Fink	Cast and wrought iron	205
—	Rennselaer and Saratoga	S. Whipple	1852	Whipple	Iron	—
Newark Dyke, UK	Great Northern	C. Wild	1853	Warren	Cast and wrought iron	259
—	North Pennsylvania	—	1856	Whipple	Iron	—
Guth, PA, Jordan Creek	Catasauqua and Fogelsville	F.C. Lowthorp	1857	—	Cast and wrought iron	110
Phillipsburg, NJ	Lehigh Valley	J.W. Murphy	1859	Whipple (pin-connected)	Iron	165
Plymouth, UK	Cornish (Great Western)	I.K. Brunel	1859	Lenticular	Wrought iron	455
Frankfort, Germany	—	—	1859	Lenticular	Iron	345
Various	New York Central	H. Carroll	1859	Lattice	Wrought iron	90
Kehl River, Germany	Baden State	Keller	1860	Lattice	Iron	197

continued

TABLE 1.3 (continued)
Notable Iron and Steel Simple Truss Span Railway Bridges Constructed during 1823–1907

Location	Railroad	Engineer	Year Completed	Type	Material	L (ft)
Schuylkill River	Pennsylvania	J.H. Linville	1861	Whipple	Cast and wrought iron	192
Steubenville, OH	Pennsylvania	J.H. Linville	1863	Murphy-Whipple	Cast and wrought iron	320
Mauch Chunk, PA	Lehigh Valley	J.W. Murphy	1863	—	Wrought iron	—
Liverpool, UK	London and Northwestern	W. Baker	1863	—	Iron	305
Blackfriar's Bridge, UK	—	Kennard	1864	Lattice	Iron	—
Orival, France	Western	—	~1865	Lattice	Iron	167
Various	B&O	S.S. Post	1865	Post	Iron	—
Lockport, IL	Chicago and Alton	S.S. Post	~1865	Post	Cast and wrought iron	—
Schuylkill River	Connecting Railway of Philadelphia	J.H. Linville	1865	Linville	Wrought iron	—
Dubuque, IA	Chicago, Burlington, and Quincy	J.H. Linville	1868	Linville	Wrought iron	250
Quincy, IA	Chicago, Burlington, and Quincy	T.C. Clarke	1868	—	Cast and wrought iron	250
Kansas City (Hannibal) (1), MO	Chicago, Burlington, and Quincy	J.H. Linville and O. Chanute	1869	—	Iron	234
Louisville, KY	B&O	A. Fink	1869	Subdivided Warren and Fink	Wrought iron	390
Parkersburg and Benwood, WV	B&O	J.H. Linville	1870	Bollman	Iron	348
St. Louis, MO	North Missouri	C. Shaler Smith	1871	—	Iron	250
Atcheson	Various	—	1875	Whipple	Iron	260
Cincinnati, OH	Cincinnati Southern	J.H. Linville and L.F.G. Bouscaren	1876	Linville	Wrought iron	515

Tay River (1), Scotland	—	Sir T. Bouch	1877	Lattice	Wrought iron	—
Glasgow, MO	Chicago and Alton	—	1879	Whipple	Steel	—
Bismark, ND	—	G.S. Morison and C.C. Schneider	1882	Whipple	Steel	—
Tay River (2), Scotland	—	—	1887	—	Steel	—
Sioux City, IA	—	—	1888	—	Steel	400
Cincinnati, OH	—	W.H. Burr	1888	—	Steel	550
Benares, India	—	—	1888	Lattice	Steel	356
Hawkesbury, Australia	—	—	1889	—	Steel	416
Henderson Bridge	Louisville and Nashville	—	~1889	Subdivided Warren	Steel	525
Cairo, IL	Illinois Central	—	1889	Whipple	Steel	518
Ceredo RR Bridge	—	Doane and Thomson	~1890	—	Steel	521
Merchant's Bridge, St. Louis	—	G.S. Morison	1890	Petit	Steel	517
Kansas City (Hannibal) (2), MO	—	—	1891	—	Steel	—
Louisville, KY	—	—	1893	Petit	Steel	550
Nebraska City, NB	—	G.S. Morison	1895	Whipple	Steel	400
Sioux City, IA	—	—	1896	—	Steel	490
Montreal, QC	Grand Trunk	—	1897	—	Steel	348
Kansas City, MO	Kansas City Southern	J.A.L. Waddell	1900	Pratt	Steel	—
Rumford, ON	Canadian Pacific	—	1907	Subdivided Warren	Steel	412

FIGURE 1.4 The Britannia Bridge built in 1849 across the Menai Straits, Wales. (Postcard from the private collection of Jochem Hollestelle.)

pressure, and thermal effects. Fairbairn's empirical work on fatigue strength and plate stability during the design of the Conway and Britannia bridges is particularly significant.*

A small 55 ft long simple span tubular wrought iron plate girder bridge was built in the United States by the B&O Railroad in 1847. However, the only large tubular railway bridge constructed in North America was the Victoria Bridge built in 1859 for the Grand Trunk Railway over the St. Lawrence River at Montreal† (Figure 1.5). The Victoria Bridge was the longest bridge in the world upon its completion.‡ The bridge was replaced with steel trusses in 1898 due to rivet failures associated with increasing locomotive weights and ventilation problems detrimental to passengers traveling across the 9144 ft river crossing with almost 6600 ft of tubular girders. Table 1.4 indicates some notable continuous span railway bridges constructed after 1850.

These tubular bridges provided the stiffness desired by their designers but proved to be costly. Suspension bridges were more economical but many British engineers were hesitant to use flexible suspension bridges for long-span railroad crossings.§ Sir Benjamin Baker's 1867 articles on long-span bridges also promoted the use of

* Also, later in 1864, Fairbairn studied iron plate and box girder bridge models under a cyclical loading representative of railway traffic. These investigations assisted in the widespread adoption of wrought iron, in lieu of cast iron, for railway bridge construction in the latter quarter of the nineteenth century.

† The Victoria Bridge over the St. Lawrence at Montreal was also designed by Stephenson.

‡ The longest span in the Victoria Bridge was 330 ft.

§ The first railway suspension bridge built over the Tees River in England in 1830 (with a 300 ft span) had performed poorly by deflecting in a very flexible manner that even hindered the operation of trains. It was replaced by cast iron and steel girders, respectively, in 1842 and 1906. The Basse–Chaine suspension bridge in France collapsed in 1850, as did the suspension bridge at Wheeling, West Virginia, in 1854, illustrating the susceptibility of flexible suspension bridges to failure under wind load conditions.

FIGURE 1.5 The Victoria Bridge under construction (completed in 1859) across the St. Lawrence River, Montreal, Canada. (William Notman, Library and Archives Canada.)

more rigid bridges for railway construction. Furthermore, Baker had earlier recommended cantilever trusses for long-span railway bridges.* Also in 1867, Heinrich Gerber constructed the first cantilever bridge in Hanover, Germany, and some short-span cantilever arch and truss bridges were built in New England and New Brunswick between 1867 and 1870.

Nevertheless, railway suspension bridges were built in the United States in the last quarter of the nineteenth century. Unlike the aversion for suspension bridges that was prevalent among British railway engineers, American engineers were using iron suspension bridges for long spans carrying relatively heavy freight railroad traffic. Modern suspension bridge engineering essentially commenced with the construction of the 820 ft span railway suspension bridge over the Niagara Gorge in 1854. This bridge, designed by John A. Roebling, was used by the Grand Trunk Railway and successor railroads for over 40 years. Roebling had realized the need for greater rigidity in suspension bridge design after the failure of the Wheeling† and other suspension bridges. As a consequence, his Niagara Gorge suspension bridge was the first to incorporate stiffening trusses into the design (Figure 1.6). Rehabilitation works were required in 1881 and 1887, but it was replaced with a steel spandrel braced hinged arch bridge in 1897 due to capacity requirements for heavier railway loads. The railway suspension bridge constructed in 1840 over the Saone River in France

* Baker's 1862 book *Long-Span Railway Bridges* and A. Ritter's calculations of the same year outlined the benefits of cantilever bridge design.

† The 1010 ft wire rope suspension bridge over the Ohio River at Wheeling, West Virginia collapsed due to wind loads in 1854, just five years after completion of construction.

TABLE 1.4
Notable Continuous Span Railway Bridges Constructed between 1850–1929

Location	Railroad	Engineer	Year	Type	Largest Span (ft)
Torksey, UK	—	J. Fowler	1850	Three span continuous tubular girder	130
Britannia Bridge, Menai Straits, UK	London–Chester–Holyhead	R. Stephenson	1850	Four span continuous tubular	460
Montreal, QC	Grand Trunk	R. Stephenson	1860	Twenty-five span continuous tubular	330
Montreal, QC	Canadian Pacific	C. Shaler Smith	1886	Four span continuous trusses	408
Sciotoville, OH	Chesapeake and Ohio	G. Lindenthal and D.B. Steinman	1917	Two span continuous truss	775
Allegheny River	Bessemer and Lake Erie	—	1918	Three span continuous truss	520
Nelson River	Bessemer and Lake Erie	—	1918	Three span continuous truss	400
Cincinnati, OH	C.N.O. and T.P.	—	1922	Three span continuous truss	516
Cincinnati, OH	Cincinnati and Ohio	—	1929	Three span continuous truss	675

FIGURE 1.6 The railway suspension bridge built in 1854 across the Niagara Gorge between New York, USA, and Ontario, Canada. (Niagara Falls Public Library.)

was also replaced only four years after completion due to poor performance under live loads.* The railway suspension bridge constructed in 1860 at Vienna, Austria, was also prematurely replaced with an iron arch bridge in 1884 after concerns over the flexibility of the suspended span. The early demise of these and other suspension bridges generated new concerns among some American engineers over the lack of rigidity of cable-supported bridges under steam locomotive and moving train loads.

The first all-wrought-iron bridge in the United States, a lattice truss, was completed in 1859 by the New York Central Railroad.† In the same year, the Lehigh Valley Railroad built the first pin-connected truss. In 1861, the Pennsylvania Railroad pioneered the use of forged eyebars in a pin-connected truss over the Schuylkill River. After this many American railway bridges were constructed with pinned connections, while European practice still favored the use of riveted construction. Riveted construction was considered superior but pin-connected construction enabled the economical and

* The suspension bridge was replaced by a stone masonry bridge.

† The New York Central Railroad also initiated the use of iron stringers (as opposed to wooden) in railway trusses in the 1860s.

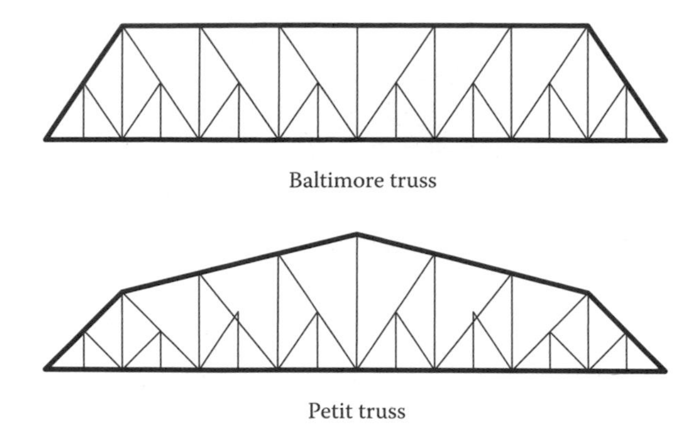

FIGURE 1.7 Baltimore trusses (the inclined chord truss is also called Petit truss).

rapid erection of railway bridges in remote areas of the United States. The principal exception was the New York Central Railroad, which used riveted construction exclusively for its iron railway bridges.

In 1863, the Pennsylvania Railroad successfully crossed the Ohio River using a 320 ft iron truss span. The railroad used the relatively rigid Whipple truss for such long spans. This construction encouraged greater use of longer span iron trusses to carry heavy freight railroad traffic in the United States. Another notable wrought iron railway truss was the 390 ft span built by the B&O Railroad at Louisville, Kentucky, in 1869.

In the 1870s the Pratt truss (patented in 1844) became prevalent for short- and medium-span railway bridges in the United States. Pratt trusses are statically determinate and their form is well suited for use in iron bridges. Whipple, Warren, and Post trusses were also used by U.S. railroads in the 1870s. The Bollman truss bridge, patented in 1852 and used by the B&O and other railroads until 1873, was an example of the innovative* use of wrought iron in American railway bridge construction. For longer wrought iron railway bridge spans, the Baltimore or Petit truss was often used (Figure 1.7).† The first use of a Baltimore truss (a Pratt truss with subdivided panels) was on the Pennsylvania Railroad in 1871.

Large railway viaduct bridges were also constructed in wrought iron. The 216 ft high and 1300 ft long Viaduc de la Bouble was built in France in 1871. In 1882 the Erie Railroad completed construction of the 300 ft high and over 2000 ft long wrought iron Kinzua Viaduct in Pennsylvania (Figure 1.8). Also in France, Gustave Eiffel designed the wrought iron Garabit Viaduct, which opened to railroad traffic in 1884 (Figure 1.9).

A large number of iron railway bridges built after 1840 in the United States and England failed under train loads. It was estimated that about one-fourth of railway

* Bollman trusses used wrought iron tension members and cast iron compression members. The redundant nature of the truss form reduced the possibility of catastrophic failure.

† The Petit truss was used extensively by American railroad companies.

FIGURE 1.8 Kinzua Viaduct 1882, Pennsylvania. (Historic American Engineering Record.)

bridges in the American railroad inventory were failing annually between 1875 and 1888. Most of these failures were related to fatigue and fracture, and the buckling instability of compression members (notably top chords of trusses). Although most of the failures were occurring in cast iron truss members and girders, by 1850

FIGURE 1.9 The Garabit Viaduct built in 1884 over the Tuyere River, France. (Courtesy of GFDL J. Thurion, July 2005.)

many American engineers had lost confidence in even wrought iron girder, truss, and suspension railway bridge construction.*

At this time railway construction was not well advanced in Germany, and these failures interested Karl Culmann during the construction of some major bridges for the Royal Bavarian Railroad. He proposed that American engineers should use lower allowable stresses to reduce the fatigue failures of iron truss railway bridges and he recognized the issue of top chord compressive instability. Culmann also proposed the use of stiffening trusses for railroad suspension bridges after learning of concerns expressed by American bridge engineers with respect to their flexibility under moving live loads.

A railroad Howe truss collapsed under a train at Tariffville, Connecticut, in 1867 and a similar event occurred in 1877 at Chattsworth, Illinois. However, the most significant railway bridge failure, due to the considerable loss of life associated with the incident, was the collapse of the cast iron Howe deck truss span on the Lake Shore and Michigan Southern Railroad at Ashtabula, Ohio, in 1876 (Figures 1.10a and b). The Ashtabula bridge failure provided further evidence that cast iron was not appropriate for heavy railway loading conditions and caused American railroad companies to abandon the use of cast iron elements for bridges.† This was, apparently, a wise decision as modern forensic analysis indicates that the likely cause of the Ashtabula failure was a combination of fatigue and brittle fracture initiated at a cast iron flaw.

FIGURE 1.10a The Ashtabula Bridge, Ohio before the 1876 collapse. (Ashtabula Railway Historical Foundation.)

---

* For example, following the collapse of an iron bridge in 1850, all metal bridges on the Boston and Albany Railroad were replaced with timber bridges.

† With the exception of cast iron bearing blocks at the ends of truss compression members.

FIGURE 1.10b The Ashtabula Bridge, Ohio after the 1876 collapse. (Ashtabula Railway Historical Foundation.)

In addition, the collapse of the Tay Railway Bridge in 1879, only 18 months after completion, promoted a renewed interest in wind loads applied to bridges (Figures 1.11a and b). The Tay bridge collapse also reinforced the belief, held by many engineers, that light and relatively flexible structures are not appropriate for railway bridges.

These bridge failures shook the foundations of bridge engineering practice and created an impetus for research into new methods (for design and construction) and materials to ensure the safety and reliability of railway bridges. The investigation and specification of wind loads for bridges also emerged from research conducted following these railway bridge collapses. Furthermore, in both Europe and the United States, a new emphasis on truss analysis and elastic stability was developing in response to railway bridge failures.

FIGURE 1.11a The Tay River Bridge, England before the 1879 collapse.

FIGURE 1.11b The Tay River Bridge, England after the 1879 collapse.

A revitalized interest in the cantilever construction method occurred, particularly in connection with the erection of arch bridges. Early investigations by Stephenson, Brunel, and Eads had illustrated that the erection of long arch spans using the cantilever method* was feasible and precluded the requirement for falsework as temporary support for the arch. The cantilevered arms were joined to provide fixed or two-hinged arch action† or connected allowing translation of members to provide a statically determinate structure. The cantilever construction method was also proposed for long-span truss erection where the structure is made statically determinate after erection by retrofitting to allow appropriate members to translate. This creates a span suspended between two adjacent cantilever arms that are anchored by spans adjacent to the support pier, which provides a statically determinate structure.‡ Alternatively, the cantilever arms may progress only partially across the main span and be joined by a suspended span erected between the arms.§ Other benefits of cantilever construction are smaller piers (due to a single line of support bearings) and an economy of material for properly proportioned cantilever arms, anchor spans, and suspended spans.

Iron trusses continued to be built in conjunction with the rapid railroad expansion of the 1860s. However, in the second half of the nineteenth century, steel started to replace iron in the construction of railway bridges.** For example, the iron Kinzua Viaduct of 1882 was replaced with a similar structure of steel only 18 years

* Often using guyed towers and cable stays as erection proceeds.

† Depending on whether fixed or pinned arch support conditions were used.

‡ Statically indeterminate structures are susceptible to stresses caused by thermal changes and support settlements. Therefore, statically indeterminate cantilever bridges must incorporate expansion devices and be founded on unyielding foundations to ensure safe and reliable behavior.

§ This was the method used in the 1917 reconstruction of the Quebec Bridge.

** In 1895, steel completely replaced wrought iron for the production of manufactured structural shapes.

after construction due to concerns about the strength of wrought iron bridges under increasing railroad loads.

1.3 STEEL RAILWAY BRIDGES

Steel is stronger and lighter than wrought iron, but was expensive to produce in the early nineteenth century. Bessemer developed the steel-making process in 1856 and Siemens further advanced the industry with open-hearth steel making in 1867. These advances enabled the economical production of steel. These steel-making developments, in conjunction with the demand for railway bridges following the American Civil War, provided the stimulus for the use of steel in the construction of railway bridges in the United States. In the latter part of the nineteenth century, North American and European engineers favored steel arches and cantilever trusses for long-span railway bridges, which, due to their rigidity, were considered to better resist the effects of dynamic impact, vibration, and concentrated moving railway loads.

The first use of steel in a railway bridge* was during the 1869 to 1874 construction of the two 500 ft flanking spans and 520 ft central span of the St. Louis Bridge (now named the Eads Bridge after its builder, James Eads†) carrying heavy railroad locomotives across the Mississippi River at St. Louis, Missouri. Eads did not favor the use of a suspension bridge for railway loads‡ and proposed a cast steel arch bridge. Eads' concern for stiffness for railway loads is illustrated by the trusses built between the railway deck and the main steel arches of the St. Louis Bridge (Figure 1.12). The Eads Bridge features not only the earliest use of steel but also other innovations in American railway bridge design and construction. The construction incorporated the initial use of the pneumatic caisson method§ and the first use of the cantilever method of bridge construction in the United States.** It was also the first arch span over 500 ft and incorporated the earliest use of hollow tubular chord members.†† The extensive innovations associated with this bridge caused considerable skepticism among the public. In response, before it was opened, Eads tested the bridge using 14 of the heaviest locomotives available. It is also interesting to note that the construction of the Eads Bridge almost depleted the resources of the newly developed American steel-making industry.

The initial growth of the American steel industry was closely related to the need for steel railway bridges, particularly those of long span. The American railroads' demand

---

* The first use of steel in any bridge was in the 1828 construction of a suspension bridge in Vienna, Austria, where open-hearth steel suspension chains were incorporated into the bridge.

† Eads was assisted in design by Charles Pfeiffer and in construction by Theodore Cooper.

‡ A suspension bridge was proposed by John Roebling in 1864.

§ This method of pier construction was also used by Brunel in the construction of the Royal Albert Bridge at Saltash, England, in 1859.

** The cantilever method was proposed in 1800 by Thomas Telford for a cast iron bridge crossing the Thames at London and in 1846 by Robert Stephenson for construction of an iron arch railway bridge in order to avoid falsework in the busy channel of the Menai Straits. Eads had to use principles developed in the seventeenth century by Galileo to describe the principles of cantilever construction of arches to skeptics of the method.

†† The tubular arch chords used steel with 1.5–2% chromium content providing for a relatively high ultimate stress of about 100 ksi.

FIGURE 1.12 The St. Louis (Eads) Bridge built across the Mississippi River in 1874 at St. Louis, MO. (Historic American Engineering Record.)

for longer spans and their use of increasingly heavier locomotives and freight cars caused Andrew Carnegie* and others to invest considerable resources toward the development of improved steels of higher strength and ductility. The first exclusively steel railway bridge (comprising Whipple trusses) was built by the Chicago and Alton Railway in 1879 at Glasgow, Missouri.

Despite concerns about suspension bridge flexibility under train and wind loads, some American bridge engineers continued to design and construct steel suspension railway bridges. The famous Brooklyn Bridge, when completed in 1883, carried two railway lines. However, lingering concerns with suspension bridge performance and increasing locomotive weights precipitated the general demise of this relatively flexible type of railway bridge construction.

The structural and construction efficacy of cantilever-type bridges for carrying heavy train loads led to the erection of many long-span steel railway bridges of trussed cantilever design after 1876. The Cincinnati Southern Railway constructed the first cantilever, or Gerber† type, steel truss railway bridge in the United States over the Kentucky River in 1877.‡ In 1883 the Michigan Central and Canada South Railway completed the construction of a counterbalanced cantilever deck truss bridge§

* Andrew Carnegie worked for the Pennsylvania Railroad prior to starting the Keystone Bridge Company (with J.H. Linville) and eventually going into the steelmaking business.

† This type of bridge design and construction is attributed to the German engineer Heinrich Gerber who patented and constructed the first cantilever-type bridge in 1867.

‡ At the location of an uncompleted suspension bridge by John Roebling.

§ This was the first use of cantilever construction using a suspended span.

FIGURE 1.13 Fraser River Bridge built in 1884, British Columbia, Canada. (From Canadian Pacific Archives NS.11416, photograph by J.A. Brock. With permission.)

across the Niagara Gorge parallel to Roebling's railway suspension bridge. Shortly afterward, in 1884, the Canadian Pacific Railway crossed the Fraser River in British Columbia with the first balanced cantilever steel deck truss (Figure 1.13). Cantilever bridges became customary for long-span railway bridge construction as they provided the rigidity required to resist dynamic train loads, may be made statically determinate, and require no main span (composed of cantilever arms and suspended span) falsework to erect. Table 1.5 summarizes some notable cantilever railway bridges constructed after 1876.

Theodore Cooper promoted the exclusive use of steel for railway bridge design and construction in his 1880 paper to the American Society of Civil Engineers (ASCE) titled "The Use of Steel for Railway Bridges." Following this almost all railway bridges, and by 1895 all other bridges, in the United States were constructed of steel. Structural steel shape production was well developed for the bridge construction market by 1890.*

The British government lifted its ban on the use of steel in railway bridge construction in 1877. More than a decade later Benjamin Baker reviewed precedent cantilever bridges in North America (in particular, those on the Canadian Pacific Railway) and

* By 1895, structural shapes were no longer made with iron, and steel was used exclusively.

TABLE 1.5
Notable Steel Cantilever Railway Bridges Constructed between 1876–1917

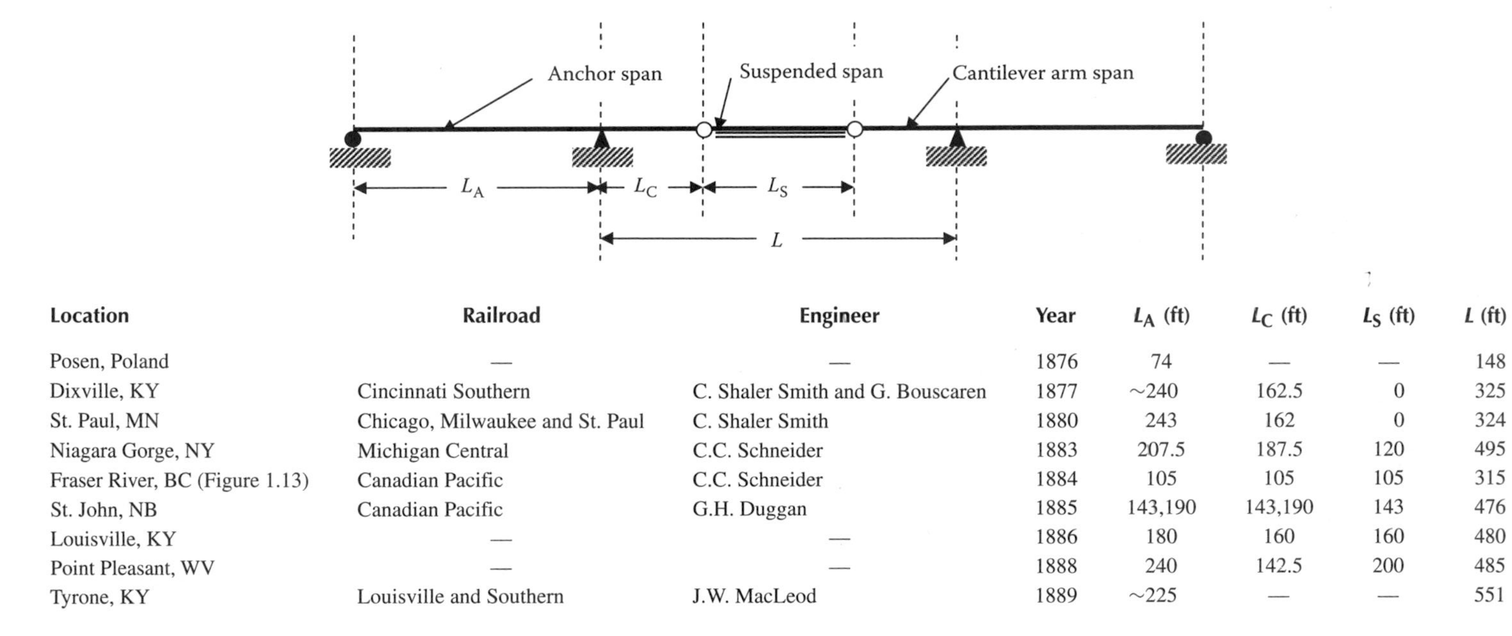

Location	Railroad	Engineer	Year	L_A (ft)	L_C (ft)	L_S (ft)	L (ft)
Posen, Poland	—	—	1876	74	—	—	148
Dixville, KY	Cincinnati Southern	C. Shaler Smith and G. Bouscaren	1877	~240	162.5	0	325
St. Paul, MN	Chicago, Milwaukee and St. Paul	C. Shaler Smith	1880	243	162	0	324
Niagara Gorge, NY	Michigan Central	C.C. Schneider	1883	207.5	187.5	120	495
Fraser River, BC (Figure 1.13)	Canadian Pacific	C.C. Schneider	1884	105	105	105	315
St. John, NB	Canadian Pacific	G.H. Duggan	1885	143,190	143,190	143	476
Louisville, KY	—	—	1886	180	160	160	480
Point Pleasant, WV	—	—	1888	240	142.5	200	485
Tyrone, KY	Louisville and Southern	J.W. MacLeod	1889	~225	—	—	551

Poughkeepsie, NY	Central New England	—	1889	262.5	~160	~228	548
Hooghly, India	East India	Sir B. Leslie	1890	—	—	—	—
Firth of Forth, Scotland (Figure 1.14)	North British	Sir B. Baker and Sir J. Fowler	1890	680	680	350	1710
Pecos River	Southern Pacific	A. Bonzano	1891	—	—	—	~200
Red Rock, CO	—	J.A.L. Waddell	1892	165	165	330	660
Callao, Peru	Lima and Oroya	L.L. Buck	~1892	—	—	—	265
Cernavoda, Romania	—	—	~1892	233.5	164	295	623
Memphis, TN	—	G.S. Morison	1892	226 and 310	170	450	790.5
Ottawa, ON	Canadian Pacific	G.H. Duggan	1900	247	123.5	308	555
Loch Etive, Scotland	—	Sir J.W. Barry	1903	139.5	146	232	524
Pittsburgh, PA	Wabash	—	1904	346	226	360	812
Mingo Junction, OH	Wabash	—	1904	298	—	—	700
Thebes, IL	—	A. Noble and R. Modjeski	1905	260.5 (1/2 of span)	152.5	366	671
Blackwell's Island (Queensboro), NY	City of New York (light rail)	G. Lindenthal	1907	469.5 and 630	591	0	1182
Khushalgarth, India	—	Rendel and Robertson	1908	—	—	—	—
Westerburg, Prussia	Prussian State	—	1908	—	—	110	—
Daumer Bridge, China	Yunnan	—	1909	123	90	168	348
Beaver, PA	Pittsburgh and Lake Erie	—	1910	320	242	285	769
Quebec, QC (Figure 1.15)	Canadian Government	T. Cooper and G.H. Duggan	1917	515	580	640	1800

FIGURE 1.14 The Forth Railway Bridge built over the Firth of Forth in 1890, Scotland. (Courtesy of GFDL Andrew Bell, January 2005.)

proposed a cantilever truss for the Firth of Forth railway bridge crossing in Scotland.* It was a monumental undertaking completed in 1890 (Figure 1.14). It is an example of steel truss cantilever-type railway bridge construction on a grand scale with cantilever arms of 680 ft supporting a 350 ft suspended span. Baker used the relatively new Bessemer steel in the bridge even though it was an untested material for such large structures and some engineers thought it susceptible to cracking. The bridge is very stiff and the $3\frac{1}{2}$ in deflection, measured by designer Baker under the heaviest locomotives available on the North British Railway, compared well with his estimate of 4 in. The bridge was further tested under extreme wind conditions with two long heavy coal trains and the cantilever tip deflection was <7 in.

The Forth Railway Bridge used a large quantity of steel and was costly. This prompted engineers such as Theodore Cooper (who had worked with Eads on the St. Louis Bridge) to consider cantilever construction with different span types using relatively smaller members. Two such statically determinate railway bridges were the 671 ft main span bridge crossing the Mississippi at Thebes, Illinois, and the 1800 ft main span Quebec Bridge. The Thebes bridge, constructed in 1905, consists of five pin-connected through truss spans, of which two spans are 521 ft fixed double anchor spans (anchoring four 152.5 ft cantilever arms) and three contain 366 ft suspended spans. The Quebec Bridge, an example of economical long-span steel cantilevered truss construction for railroad loads, was completed in 1917 after two construction failures (Figures 1.15a and b). The initial 1907 failure was likely due to calculation error in determining dead load compressive stresses in the bottom chord members

* Before this, Baker may not have known of the work of engineers C. Shaler Smith or C.C. Schneider who had already constructed cantilever railway bridges in the United States.

during construction as the cantilever arms were increased in length. The bridge was redesigned* and a new material, nickel steel,† was used in the reconstruction. In 1916, the suspended span truss fell while being hoisted into place. It was quickly rebuilt and the Quebec Bridge was opened to railway traffic in 1917 (Figure 1.15c). Another major cantilever-type bridge was not to be constructed until after 1930. It remains the longest span cantilever bridge in the world.

Continuous spans were often used for long-span steel railway bridge construction in Europe but seldom in North America due to the practice of avoiding statically indeterminate railway bridge structures. The first long-span continuous steel truss railway bridge was built by the Canadian Pacific Railway over the St. Lawrence River at Montreal in 1887 (Figure 1.16).‡ The 408 ft main spans were erected by the cantilever method without falsework. The Viaur Viaduct, built in 1898, was the first major steel railway bridge in France.§

FIGURE 1.15a The 1907 Quebec Bridge collapse, Canada. (Carleton University Civil Engineering Exhibits.)

* The original designer was Theodore Cooper. Following the collapse a design was submitted by H.E. Vautelet, but the redesign of the bridge was carried out by G.H. Duggan under the review of C.C. Schneider, R. Modjeski, and C.N. Monsarrat.

† Alloy nickel steel was first used in 1909 on the Blackwell's Island (now Queensboro) Bridge in New York. Nickel steel was also used extensively by J.A.L. Waddell for long-span railway bridge designs. A.N. Talbot conducted tests of nickel steel connections for the Quebec Bridge reconstruction.

‡ These spans were replaced in 1912 due to concern over performance under heavier train loads.

§ This cantilever truss arch bridge is unusual in that it incorporates no suspended span, thereby rendering the structure statically indeterminate. Many engineers believe that the design was inappropriate for railroad loading.

FIGURE 1.15b The 1916 Quebec Bridge collapse, Canada. (A.A. Chesterfield, Library and Archives Canada.)

FIGURE 1.15c The Quebec Bridge completed in 1917 across the St. Lawrence River at Quebec City, Canada. (Carleton University Civil Engineering Exhibits.)

Many iron and steel railway bridges were replaced in the first decades of the twentieth century due to the development of substantially more powerful and heavier locomotives.* Riveting was used extensively in Europe but only became a standard of American long-span steel railway bridge fabrication after about 1915† with construction of the Hell Gate and Sciotoville bridges. Hell Gate is a 978 ft two-hinged steel

* Locomotive weights were typically about 40 ton in 1860, 70 ton in 1880, 100 ton in 1890, 125 ton in 1900, and 150 ton in 1910.

† Riveting was used on smaller spans earlier in the twentieth century.

FIGURE 1.16 The St. Lawrence Bridge built in 1886 at Montreal, Canada. (From Canadian Pacific Archives NS.1151, photograph by J.W. Heckman. With permission.)

trussed arch bridge in New York. It was built to carry four heavily loaded railroad tracks of the New England Connecting Railroad and Pennsylvania Railroad when it was completed in 1916 (Figure 1.17). It is the largest arch bridge in the world and was erected without the use of falsework. It was also the first major bridge to use high carbon steel members in its construction.* The Chesapeake and Ohio Railroad completed construction of two 775 ft span continuous steel trusses across the Ohio River at Sciotoville, Ohio, in 1917. This bridge remains the largest continuous span bridge in the world.

It has been estimated that in 1910 there were 80,000 iron and steel bridges† with a cumulative length of 1400 miles on about 190,000 miles of track. Railroads were the catalyst for material and construction technology innovation in the latter half of the nineteenth century as the transition from wood and masonry to iron and steel bridges occurred in conjunction with construction methods that minimized interference with rail and other traffic.‡ The art and science of bridge engineering was emerging from

* Primarily, due to the high cost of alloy steel.

† The majority being steel by the beginning of the twentieth century.

‡ For example, in order to not interfere with railway traffic, the tubular spans of the Victoria Bridge at Montreal were replaced by extension of substructures and erecting steel trusses around the exterior of the tubular girders.

FIGURE 1.17 The Hell Gate Bridge built across the East River in 1916, New York. (Library of Congress from Detroit Publishing Co.)

theoretical and experimental mechanical investigations prompted, to a great extent, by the need for rational and scientific bridge design in a rapidly developing and expanding railroad infrastructure.

1.4 DEVELOPMENT OF RAILWAY BRIDGE ENGINEERING

1.4.1 Strength of Materials and Structural Mechanics

The early work of Robert Hooke (1678) concerning the elastic force and deformation relation, of Jacob Bernoulli (1705) regarding the shape of deflection curves, of Leonard Euler (1759) and C.A. Coulomb (1773) about elastic stability of compression members,* and of Louis M.H. Navier (1826) on the subject of the theory of elasticity laid the foundation for the rational analysis of structures. France led the world in the development of elasticity theory and mechanics of materials in the eighteenth century and produced well-educated engineers, many of whom became leaders in American railway bridge engineering practice.† Railroad expansion continued at a considerable

* Between 1885 and 1889, F. Engesser, a German railway bridge engineer, further developed compression member stability analysis for general use by engineers.

† Charles Ellet (1830), Ralph Modjeski (1855), L.F.G. Bouscaren, Chief Engineer of the Cincinnati Southern Railroad (1873), and H.E. Vautelet, Bridge Engineer of the Canadian Pacific Railway (ca. 1876), were graduates of early French engineering schools.

pace for another 80 years following inception in the 1820s. During that period, due to continually increasing locomotive loads, it was not uncommon for railway bridges to be replaced at 10–15 year intervals. The associated demand for stronger and longer steel bridges, coupled with failures that were occurring, compelled engineers in the middle of the nineteenth century to engage in the development of a scientific approach to the design of iron and steel railway bridges.

American railway bridge engineering practice was primarily experiential and based on the use of proven truss forms with improved tensile member materials. Many early Town, Long, Howe, and Pratt railway trusses were constructed without the benefit of a thorough and rational understanding of forces in the members. The many failures of railway bridge trusses between 1850 and 1870 attest to this. This empirical practice had served the burgeoning railroad industry until heavier loads and longer span bridges, in conjunction with an increased focus on public safety, made a rational and scientific approach to the design of railway bridges necessary. In particular, American engineers developed a great interest in truss analysis because of the extensive use of iron trusses on U.S. railroads. In response, Squire Whipple published the first rational treatment of statically determinate truss analysis (the method of joints) in 1847.

The rapid growth of engineering mechanics theory in Europe in the mid-nineteenth century also encouraged French and German engineers to design iron and steel railway bridges using scientific methods. At this juncture, European engineers were also interested in the problems of truss analysis and elastic stability. B.P.E. Clapyron developed the three-moment equation in 1849 and used it in an 1857 postanalysis of the Britannia Bridge.* Concurrently, British railway bridge engineers were engaged in metals and bridge model testing for strength and stability. Following Whipple, two European railway bridge engineers, D.J. Jourawski† and Karl Culmann, provided significant contributions to the theory of truss analysis for iron and steel railway bridges. Karl Culmann, an engineer of the Royal Bavarian Railway, was a strong and early proponent of the mathematical analysis of trusses. He presented, in 1851, an analysis of the Howe and other proprietary trusses‡ commonly used in the United States. The Warren truss was developed in 1846,§ and by 1850 W.B. Blood had developed a method of analysis of triangular trusses. Investigations, conducted primarily in England in the 1850s, into the effects of moving loads and speed were beginning. Fairbairn considered the effects of moving loads on determinate trusses as early as 1857.

J.W. Schwedler, a German engineer, presented the fundamental theory of bending moments and shear forces in beams and girders in 1862. Earlier he had made a substantial contribution to truss analysis by introducing the method of sections. Also in 1862, A. Ritter improved truss analysis by simplifying the method of sections

* The design of the Britannia Bridge was based on simple span analysis, even though Fairbairn and Stephenson had a good understanding of continuity effects on bending. The spans were erected simply supported, and then sequentially jacked up at the appropriate piers and connected with riveted plates to attain continuous spans.

† Jourawski was critical of Stephenson's use of vertical plate stiffeners in the Britannia Bridge.

‡ Culmann also analyzed Long, Town, and Burr trusses using approximate methods for these statically indeterminate forms.

§ The Warren truss was first used in a railway bridge in 1853 on the Great Northern Railway in England.

through development of the equilibrium equation at the intersection of two truss members. James Clerk Maxwell* and Culmann† both published graphical methods for truss analysis. Culmann also developed an analysis for the continuous beams and girders that were often used in the 1850s by railroads. Later, in 1866, he published a general description of the cantilever bridge design method.‡ In subsequent years, Culmann also developed moving load analysis and beam flexure theories that were almost universally adopted by railroad companies in the United States and Europe. Bridge engineers were also given the powerful tool of influence lines for moving load analysis, which was developed by E. Winkler in 1867.

The effects of moving loads, impact (from track irregularities and locomotive hammer blow), pitching, nosing, and rocking of locomotives continued to be of interest to railway bridge engineers and encouraged considerable testing and theoretical investigation. Heavier and more frequent railway loadings were also creating an awareness of, and initiating research into, fatigue (notably by A. Wohler for the German railways).

North American engineers recognized the need for rational and scientific bridge design, and J.A.L. Waddell published comprehensive books on steel railway bridge design in 1898 and 1916. Furthermore, Waddell and others promoted independent bridge design in lieu of the usual proprietary bridge design and procurement practice of the American railroad companies. The Erie Railroad was the first to establish this practice and only purchased fabricated bridges from their own scientific designs, which soon became the usual practice of all American railroads.

1.4.2 Railway Bridge Design Specifications

Almost 40 bridges (about 50% of them iron) were collapsing annually in the United States during the 1870s. This was alarming as the failing bridges comprised about 25% of the entire American bridge inventory of the time. In particular, between 1876 and 1886 almost 200 bridges collapsed in the United States.

Most of these bridges were built by bridge companies without the benefit of independent engineering design. As could be expected, some bridge companies had good specifications for design and construction but others did not. Therefore, without independent engineering design, railroad company officials required a good knowledge of bridge engineering to ensure public safety. This was not always the case, as demonstrated by the Ashtabula collapse where it was learned in the subsequent inquiry that the proprietary bridge design had been approved by a railroad company executive without bridge design experience.§ Many other proprietary railway bridges were also failing, primarily due to a lack of rigidity and lateral stability. American engineers

* Truss graphical analysis methods were developed and improved by J.C. Maxwell and O. Mohr between 1864 and 1874. Maxwell and W.J.M. Rankine were also among the first to develop theories for steel suspension bridge cables, lattice girders, bending force, shear force, deflection, and compression member stability.

† Culmann published an extensive description of graphical truss analysis in 1866.

‡ Sir Benjamin Baker also outlined the principles of cantilever bridge design in 1867.

§ There were also material quality issues with the cast iron compression blocks, which were not discovered as the testing arranged by the Lake Shore and Michigan Southern Railroad Co. was inadequate.

were proposing the development and implementation of railroad company specifications that all bridge fabricators would build in accordance with, to preclude further failures. Developments in the fields of materials and structural mechanics had supplied the tools for rational and scientific bridge design that provided the basis on which to establish specifications for iron and steel railway bridges.

The first specification for iron railway bridges was made by the Clarke, Reeves and Company (later the Phoenix Bridge Co.) in 1871. This was followed in 1873 by G.S. Morison's "Specifications for Iron Bridges" for the Erie Railroad (formerly the New York, Lake Erie and Western Railroad). L.F.G. Bouscaren of the Cincinnati Southern Railroad published the first specifications with concentrated wheel loads in 1875.* Following this, in 1878, the Erie Railroad produced a specification (at least partially written by Theodore Cooper) with concentrated wheel loads that specifically referenced steam locomotive loads.

By 1876 the practice of bridge design by consulting engineers working on behalf of the railroads became more prevalent in conjunction with the expanding railroad business. In particular, Cooper's publications concerning railway loads, design specifications, and construction were significant contributions in the development of a rational basis for the design of steel railway bridges. Cooper produced specifications for iron and steel railway bridges in 1884, intended for use by all railroad companies. By 1890 Cooper provided his first specification for steel railway bridges. This portended the development of general specifications for steel railway bridges by the American Railway Engineering and Maintenance-of-way Association (AREMA) in 1905. This latter specification has been continuously updated and is the current recommended practice on which most North American railroad company design requirements are based. Other significant milestones in the development of general specifications for iron and steel railway bridges were

- 1867 St. Louis Bridge Co. specifications for Eads' steel arch†
- 1873 Chicago and Atchison Railroad Co.
- 1877 Chicago, Milwaukee and St. Paul Railway Co. (C. Shaler Smith)
- 1877 Lake Shore and Michigan Southern Railway (C. Hilton)
- 1877 Western Union Railroad Co.
- 1880 Quebec Government Railways
- 1880 New York, Pennsylvania, and Ohio Railroad
- 1895 B&O Railroad

The large magnitude dynamic loads imposed on bridges by railroad traffic created a need for scientific design in order to ensure safe, reliable, and economical‡ construction. Railroad and consulting engineers engaged in iron and steel railway

* However, it appears that the first use of concentrated wheel loads for bridge design was by the New York Central Railroad in 1862.

† This was not a general specification but was the first use of specification documents in the design and construction of railway bridges in the United States. The specification also included the first requirements for the inspection of material.

‡ This can be a critical consideration as most railway bridge construction projects are privately funded by railroad companies.

bridge design were the leaders in the development of structural engineering practice.* Evidence of this leadership was the publication, in 1905, of the first general structural design specification for steel bridges in the United States by AREMA.

1.4.3 Modern Steel Railway Bridge Design

The basic forms of ordinary steel railway superstructures have not substantially changed since the turn of the twentieth century. Steel arch, girder, and truss forms remain commonplace. However, considerable improvements in materials, construction technology, structural analysis and design, and fabrication technology occurred during the twentieth century.

The strength, ductility, toughness, corrosion resistance, and weldability of structural steel have improved substantially since the middle of the twentieth century. These material enhancements, combined with a greater understanding of hydraulics, geotechnical, and construction engineering, have enabled the design of economical, reliable, and safe modern railway bridges.

Modern structural analysis has also enabled considerable progress regarding the safety and economics of modern railway superstructures. Vast advancements in the theory of elasticity and structural mechanics were made in the nineteenth century as a result of railroad expansion. Today, the steel railway bridge engineer can take advantage of modern numerical methods, such as the matrix displacement (or stiffness) method, to solve difficult and complex structures. These methods of modern structural analysis may be efficiently applied using digital computers and have evolved into multipurpose finite element programs capable of linear, nonlinear, static, dynamic (including seismic), stability, fracture mechanics and other analyses. Furthermore, modern methods of structural design that facilitate the efficient and safe design of modern structures have followed from research.

Advances in manufacturing and fabrication technologies have permitted plates, sections, and members of large and complex dimensions to be fabricated and erected using superior fastening techniques such as welding and high-strength bolting. Modern fabrication with computer-controlled machines has produced economical, expedient, and reliable steel railway superstructures.

REFERENCES

Akesson, B., 2008, *Understanding Bridge Collapses*, Taylor & Francis, London, UK.
Baker, B., 1862, *Long-Span Railway Bridges*, Reprint from Original, BiblioBazaar, Charleston, SC.
Bennett, R. and Skinner, T., 1996, *Bridge Failures, Recent and Past Lessons for the Future*, American Railway Bridge and Building Association, Homewood, IL.
Billington, D.P., 1985, *The Tower and the Bridge*, Princeton University Press, Princeton.
Chatterjee, S., 1991, *The Design of Modern Steel Bridges*, BSP Professional Books, Oxford.

* The advanced state of steel design and construction knowledge possessed by railway bridge engineers made them a greatly sought after resource by architects from about 1880 to 1900 during the rebuilding of Chicago after the Great Fire.

Clark, J.G., 1939, *Specifications for Iron and Steel Railroad Bridges Prior to 1905*, Published by Author, Urbana, IL.

Cooper, T., 1889, *American Railroad Bridges*, Engineering News, New York.

Gasparini, D.A. and Fields, M., 1993, Collapse of Ashtabula Bridge on December 29, 1876, *Journal of Performance of Constructed Facilities, ASCE*, 7(2), 109–125.

Ghosh, U.K., 2006, *Design and Construction of Steel Bridges*, Taylor & Francis, London, UK.

Griggs, F.E., 2002, Kentucky High River Bridge, *Journal of Bridge Engineering, ASCE*, 7(2), 73–84.

Griggs, F.E., 2006, Evolution of the continuous truss bridge, *Journal of Bridge Engineering, ASCE*, 12(1), 105–119.

Johnson, A., 2008, *CPR High Level Bridge at Lethbridge*, Occasional Paper No. 46, Lethbridge Historical Society, Lethbridge, Alberta, Canada.

Kuzmanovic, B.O., 1977, History of the theory of bridge structures, *Journal of the Structural Division, ASCE*, 103(ST5), 1095–1111.

Marianos, W.N., 2008, G.S. Morison and the development of bridge engineering, *Journal of Bridge Engineering, ASCE*, 13(3), 291–298.

Middleton, W.D., 2001, *The Bridge at Quebec*, Indiana University Press, Bloomington, IN.

Petroski, H., 1996, *Engineers of Dreams*, Random House, New York.

Plowden, D., 2002, *Bridges: The Spans of North America*, W. W. Norton & Co., New York.

Ryall, M.J., Parke, G.A.R., and Harding, J.E., 2000, *Manual of Bridge Engineering*, Thomas Telford, London.

Timoshenko, S.P., 1983, *History of Strength of Materials,* Dover Publications, New York.

Troitsky, M.S., 1994, *Planning and Design of Bridges,* Wiley, New York.

Tyrrell, H.G., 1911, *History of Bridge Engineering*, H.G. Tyrell, Chicago, IL.

Unsworth, J.F., 2001, Evaluation of the Load Capacity of a Rehabilitated Steel Arch Railway Bridge, *Proceedings of 3rd International Arch Bridges Conference*, Presses de L'ecole Nationale des Ponts et Chaussees, Paris, France.

Waddell, J.A.L., 1898, *De Pontibus*, Wiley, New York.

Waddell, J.A.L., 1916, *Bridge Engineering—Volume 1*, Wiley, New York.

Waddell, J.A.L., 1916, *Bridge Engineering—Volume 2*, Wiley, New York.

Whipple, S., 1873, *Treatise on Bridge Building*, Reprint from Original 2nd Edition, University of Michigan, Ann Arbor, MI.

2 Steel for Modern Railway Bridges

2.1 INTRODUCTION

Modern steel is composed of iron with small amounts of carbon, manganese, and traces of other alloy elements added to enhance physical properties. Carbon is the principal element controlling the mechanical properties of steel. The strength of steel may be increased by increasing the carbon content, but at the expense of ductility and weldability. Steel also contains deleterious elements, such as sulfur and phosphorous, that are present in the iron ore.

Steel material development in the latter part of the twentieth century has been remarkable. Chemical and physical metallurgical treatments have enabled improvements to many steel properties. Mild carbon and high-strength low-alloy (HSLA) steels have been used for many years in railway bridge design and fabrication. Recent research and development related to high-performance steel (HPS) metallurgy has provided modern structural steels with even further enhancements to physical properties.

The important physical properties of modern structural bridge steels are

- Strength
- Ductility
- Fracture toughness
- Corrosion resistance
- Weldability

2.2 ENGINEERING PROPERTIES OF STEEL

2.2.1 STRENGTH

Strength may be defined in terms of tensile yield stress, F_y, which is the point where plastic behavior commences at almost constant stress (unrestricted plastic flow). Strength or resistance may also be characterized in terms of the ultimate tensile stress, F_U, which is attained after yielding and significant plastic behavior. An increase in strength is associated with plastic behavior (due to strain hardening) until the ultimate tensile stress is attained (Figure 2.1). The most significant properties of steel that are

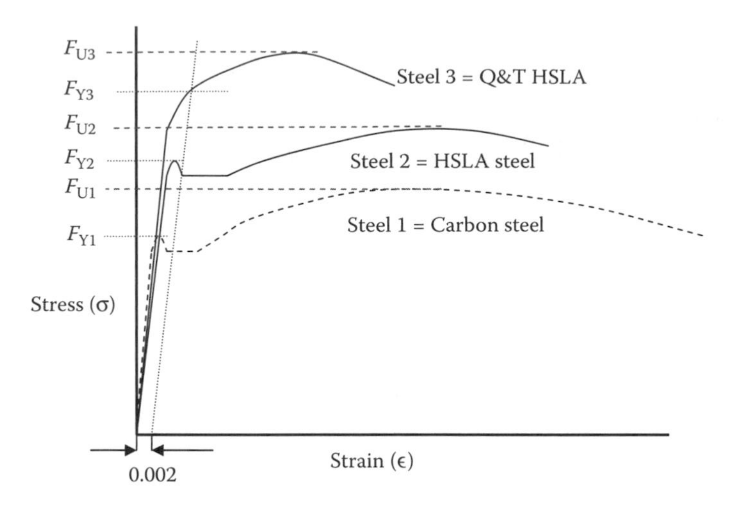

FIGURE 2.1 Idealized tensile stress–strain behavior of typical bridge structural steels.

exhibited by stress–strain curves are the elastic modulus (linear slope of the initial portion of the curve up to yield stress), the existence of yielding, and plastic behavior, with some unrestricted flow and strain hardening, until the ultimate stress is attained.

Yield stress in tension can be measured by simple tensile tests (ASTM, 2000). Yield stress in compression is generally assumed to be equal to that in tension.* Yield stress in shear may be established from theoretical considerations of the yield criteria. Various yield criteria have been proposed, but most are in conflict with experimental evidence that yield stress is not influenced by hydrostatic (or octahedral normal) stress. Two theories, the Tresca and von Mises yield criteria, meet the necessary requirement of being pressure independent. The von Mises criterion is most suitable for ductile materials with similar compression and tensile strength, and also accounts for the influence of intermediate principal stress (Chen and Han, 1988; Chatterjee, 1991). It has also been shown by experiment that the von Mises criterion best represents the yield behavior of most metals (Chakrabarty, 2006).

The von Mises yield criterion is based on the octahedral shear stress, τ_h, attaining a critical value, τ_{hY}, at yielding. The octahedral shear stress, τ_h, in terms of principal stresses, $\sigma_1, \sigma_2, \sigma_3$ is

$$\tau_h = \frac{1}{3}\sqrt{(\sigma_1 - \sigma_2)^2 + (\sigma_1 - \sigma_3)^2 + (\sigma_2 - \sigma_3)^2}. \tag{2.1}$$

Yielding in uniaxial tension will occur when $\sigma_1 = \sigma_Y$ and $\sigma_2 = \sigma_3 = 0$. Substitution of these values into Equation 2.1 provides

$$\tau_{hY} = \frac{\sqrt{2}}{3}\sigma_Y \tag{2.2}$$

* It is actually about 5% higher that the tensile yield stress.

or the criterion, that at yielding,

$$\sigma_Y = \frac{1}{\sqrt{2}}\sqrt{(\sigma_1 - \sigma_2)^2 + (\sigma_1 - \sigma_3)^2 + (\sigma_2 - \sigma_3)^2}, \tag{2.3}$$

where σ_Y is the yield stress from the uniaxial tensile test.

It can also be shown that the octahedral shear stress at yield is (Hill, 1989)

$$\tau_{hY} = \sqrt{\frac{2}{3}}\tau_Y, \tag{2.4}$$

which when substituted into Equation 2.2 provides

$$\tau_Y = \frac{\sigma_Y}{\sqrt{3}}, \tag{2.5}$$

where τ_Y is the yield stress in pure shear. Therefore, a theoretical relationship is established between yield stress in shear and tension.

Example 2.1

Determine the allowable shear stress for use in design, f_v, if the allowable axial tensile stress, f_t, is specified as $0.55F_y$ and $0.60F_y$ (F_y is the axial tensile yield stress).

For $f_t = 0.55F_y$; f_v = allowable shear stress $= 0.55F_y/\sqrt{3} = 0.32F_y$.

For $f_t = 0.60F_y$; f_v = allowable shear stress $= 0.60F_y/\sqrt{3} = 0.35F_y$.

AREMA (2008) recommends the allowable shear stress for structural steel to be $0.35F_y$.

2.2.2 Ductility

Ductility is the ability of steel to withstand large strains after yielding and prior to fracture. Ductility is necessary in railway bridges and many civil engineering structures to provide advance warning of overstress conditions and potential failure. Ductility also enables the redistribution of stresses when a member yields in redundant systems, in continuous members, and at locations of stress concentration (i.e., holes and discontinuities). Adequate ductility also assists in the prevention of lamellar tearing in thick elements.* Ductility is measured by simple tensile tests and specified as a minimum percentage elongation over a given gage length (usually 8 in.). Only ductile steels are used in modern railway bridge fabrication.

2.2.3 Fracture Resistance

Brittle fracture occurs as cleavage failure with little associated plastic deformation. Once initiated, brittle fracture cracks can propagate at very high rates as elastic strain

* Such as the relatively thick flange plates typically required for railway loads on long-span girders.

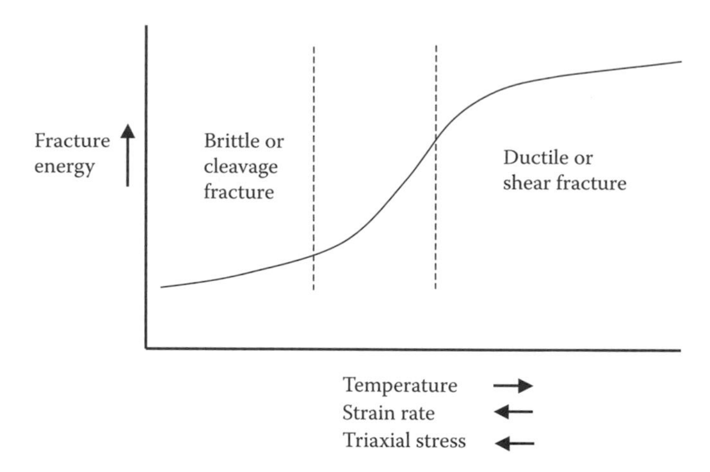

FIGURE 2.2 Fracture toughness transition behavior of typical bridge structural steel.

energy is released (Fisher, 1984; Barsom and Rolfe, 1987). In steel railway bridges, this fracture can be initiated below the yield stress.

Design- and fabrication-induced cracks, notches, discontinuities, or defects can create stress concentrations that may initiate brittle fracture in components in tension. Welding can also create hardened heat-affected zones (HAZ), hydrogen-induced embrittlement, and high residual tensile stresses near welds. All of these may be of concern with respect to brittle fracture. Rolled sections might contain rolling inclusions and defects that may also initiate brittle fracture. Other factors that affect brittle fracture resistance are galvanizing (hot-dip), poor heat treatments, and the presence of nonmetallic alloy elements. Brittle fracture most often occurs from material effects in cold service temperatures, high load rates, and/or triaxial stress states (Figure 2.2).

Normal railway bridge strain rate application is relatively slow (in comparison to, e.g., machinery components or testing machines). Brittle fracture can, however, be caused by high strain rates associated with large impact forces from live loads.* Triaxial stress distributions and high stress concentrations can be avoided by good welding and detailing practice. Thick elements are often more susceptible to brittle fracture due to the triaxial stress state. Normalizing, a supplemental heat treatment, can be beneficial in improving material toughness through grain size reduction in thick elements (Brockenbrough, 2006). Adequate material toughness for the coldest service temperature likely to be experienced by the bridge (generally a few degrees cooler than the coldest ambient temperature) is critically important.

Temperature changes the ductile to brittle behavior of steel. A notch ductility measure, the Charpy V-Notch (CVN) test, is used to ensure adequate material toughness against brittle fracture at intended service temperatures. A fracture control plan (FCP) should ensure that weld metals have at least the same notch ductility as the specified base metal and some specifications indicate even greater notch toughness requirements for welds in fracture critical members (FCM). CVN testing is performed to

* Caused by poor wheel and/or rail conditions or collision.

establish notch ductility or material toughness based on energy absorbed at different test temperatures. CVN testing is done at a rapid load rate, so adjustments are made to the specified test temperature to account for the greater ductility associated with the slower strain rate application of railway traffic. For design purposes, temperature service zones are established with a specified minimum energy absorption at a specified test temperature for various steel types and grades. CVN requirements are often specified independently for FCM and non-FCM. Tables 2.1 and 2.2 show the specified CVN test requirements for steel railway bridges recommended by AREMA (2008).

2.2.4 Weldability

If the carbon content of steel is $<0.30\%$, it is generally weldable. Higher-strength steels, where increased strength is attained through increased carbon and manganese content, will become hard and difficult to weld. The addition of other alloy elements to increase strength (Cr, Mo, and V) and weathering resistance (Ni and Cu) will also reduce the weldability of steel.

The weldability of steel is estimated from a carbon equivalency equation, given as

$$\mathrm{CE} = \mathrm{C} + \frac{\mathrm{Mn} + \mathrm{Si}}{6} + \frac{\mathrm{Ni} + \mathrm{Cu}}{15} + \frac{\mathrm{Cr} + \mathrm{Mo} + \mathrm{V}}{5}, \tag{2.6}$$

where C, Mn, Si, Ni, Cu, Cr, Mo, and V are the percentage of elemental carbon, manganese, silicon, nickel, copper, chromium, molybdenum, and vanadium in the steel, respectively. Carbon equivalence (CE) of about 0.5% or greater generally indicates that special weld treatments may be required.

Weld cracking generally results from resistance to weld shrinkage upon cooling. Thicker elements are more difficult to weld. Preheat and interpass temperature control, in conjunction with the use of low hydrogen electrodes, will prevent welding-induced hardening and cracking.

Modern structural steels have been developed with excellent weldability.* The increase in weldability enables limited preheat requirements and postweld treatments (translating into fabrication savings), and may eliminate hydrogen-induced weld cracking.

2.2.5 Weather Resistance

Atmospheric corrosion-resistant (weathering) steel chemistry (chromium, copper, nickel, and molybdenum alloys) is such that a thin iron oxide film forms upon initial wetting cycles and prevents the further ingress of moisture. This type of corrosion protection works well where there are alternate wetting and drying cycles. It may not be appropriate in locations where deicing chemicals and salts are prevalent, in marine environments, or where there is a high level of sulfur content in the atmosphere.

Weldability is slightly compromised because CE is raised through the addition of alloy elements for weathering resistance. However, these steels have about 4 times

* For example, HPS for bridges such as ASTM A709 HPS 50W, 70W, and 100W.

TABLE 2.1
Fracture Toughness Requirements for FCM[a]

ASTM Designation	Thickness in (mm)	Minimum Test Value Energy ft-lb(J)	Minimum Average Energy. ft-lb(J) and Test Temperature		
			Zone 1	Zone 2	Zone 3
A36/A36M[b]	To 4(100) incl.	20(27)	25(34) @ 70°F(21°C)	25(34) @ 40°F(4°C)	25(34) @ 10°F(−12°C)
A709/A709M, Grade 36F (Grade 250F)[b,c]					
A992/A992M	To 2(50) incl.	20(27)	25(34) @ 70°F(21°C)	25(34) @ 40°F(4°C)	25(34) @ 10°F(−12°C)
A709/A709M, Grade 50SF (Grade 345SF)[b,c]	Over 2(50) to 4(100) incl.	24(33)	30(41) @ 70°F(21°C)	30(41) @ 40°F(4°C)	30(41) @ 10°F(−12°C)
A572/A572M, Grade 50 (Grade 345)[b,d]					
A709/A709M, Grade 50SF(Grade 345SF)[b,c,d]					
A588/A588M[b,d]					
A709/A709M, Grade 50WF(Grade 345WF)[b,c,d]					
A709/A709M, Grade HPS 50WF (Grade HPS 345WF)[c,d]	To 4(100) incl.	24(33)	30(41) @ 10°F(−12°C)	30(41) @ 10°F(−12°C)	30(41) @ 10°F(−12°C)

A709/A709M, Grade HPS 70WF (Grade HPS 485WF)[c,e]	To 4(100) incl.	28(38)	35(48) @ −10°F(−23°C)	35(48) @ −10°F(−23°C)	35(48) @ −10°F(−23°C)
Minimum service temperature[f]			0°F(−18°C)	−30°F(−34°C)	−60°F(−51°C)

Source: From AREMA, 2008, *Manual for Railway Engineering*, Chapter 15, Lanham, MD. With permission.

[a] Impact teats shall be CVN impact testing "P" plate frequency, in accordance with ASTM Designation A673/A673M except for plates of A709/A709M Grades 36F (250F), 50F (345F), 50WF (345WF), HPS 50WF (HPS 345WF), and HPS 70WF (HPS) 485WF and their equivalents in which case specimens shall be selected as follows:

1. As-rolled plates shall be sampled at each end of each plate-as-rolled.
2. Normalized plates shall be sampled at one end of each plate-as-heat treated.
3. Quenched and tempered plates shall be sampled at each end of each plate-as-heat-treated.

[b] Steel backing for groove welds joining steels with a minimum specified yield strength of 50 ksi (345 MPa) or loss may be base metal conforming to ASTM A36/A36M, A709/A709M, A588/Af88M, and/or A572/A572M, at the contractor's option provided the backing material is furnished as bar stock rolled to a size not exceeding 2/8 in (10 mm) $1\frac{1}{4}$ in (32 mm). The bar stock so furnished need not conform to the CVN impact test requirements of this table.

[c] The suffix "F" is an ASTM A709 (A709M designation for fracture critical material requiring impact testing, with supplemental requirement S84 applying. A numeral 1, 2, or 3 shall be added to the F marking to indicate the applicable service temperature zone.

[d] If the yield point of the material exceeds 65 ksi (450 MPa), the test temperature for the minimum average energy and minimum test value energy required shall be reduced by 15°F (8°C) for each increment or fraction of 10 ksi (70 MPa) above 65 ksi (450 MPa). The yield point is the value given on the certified "Mill Test Report."

[e] If the yield strength of the material exceeds 85 ksi (585 MPa), the test temperature for the minimum average energy and minimum test value energy required shall be reduced by 15°F (8°C) for each increment of 10 ksi (70 MPa) above 85 ksi (585 MPa). The yield strength is the value given on the certified "Mill Test Report."

[f] Minimum service temperature of 0°F (−18°C) corresponds to Zone 1, −30°F (−34°C) to Zone 2, and −60°F (−51°C) to Zone 3 referred to in Part 9, Commentary, Article 9.1.2.1.

TABLE 2.2
Fracture Toughness Requirements for Non-FCM[a,b]

ASTM Designation	Thickness in (mm)	Minimum Average Energy. ft-lb(J) and Test Temperature		
		Zone 1	Zone 2	Zone 3
A36/A36M	To 6(150) incl.	15(20) @ 70°F(21°C)	15(20) @ 40°F(4°C)	15(20) @ 10°F(−12°C)
A709/A709M, Grade 36T (Grade 250T)[c]	To 4(100) incl.	15(20) @ 70°F(21°C)	15(20) @ 40°F(4°C)	15(20) @ 10°F(−12°C)
A992/A992M	To 2(50) incl.	15(20) @ 70°F(21°C)	15(20) @ 40°F(4°C)	15(20) @ 10°F(−12°C)
A709/A709M, Grade 50ST (Grade 345ST)[c]	Over 2(50) to 4(100) incl.	20(27) @ 70°F(21°C)	20(27) @ 40°F(4°C)	20(27) @ 10°F(−12°C)
A588/A588M[b]				
A572/A572M, Grade 42 (Grade 290)[d]				
A572/A572M, Grade 50 (Grade 345)[d]				
A709/A709M, Grade 50T (Grade 345T)[c,d]				
A709/A709M, Grade 50WT (Grade 345WT)[c,d]				
A572/A572M, Grade 42 (Grade 290)[c]	Over 4(100) to 6(150) incl.	20(27) @ 70°F(21°C)	20(27) @ 40°F(4°C)	20(27) @ 10°F(−12°C)
A588/A588M[c]	Over 4(100) to 5(125) incl.	20(27) @ 70°F(21°C)	20(27) @ 40°F(4°C)	20(27) @ 10°F(−12°C)
A709/A709M, Grade HPS 50WT (Grade HPS 345WT)[c,d]	To 4(100) incl.	25(34) @ −10°F(−23°C)	25(34) @ −10°F(−23°C)	25(34) @ −10°F(−23°C)
A709/A709M, Grade HPS 70WT (Grade HPS 485WT)[c,e]				
Minimum service temperature[f]		0°F (−18°C)	−30°F (−34°C)	−60°F (−51°C)

Source: From AREMA, 2008, *Manual for Railway Engineering*, Chapter 15, Lanham, MD. With permission.

[a] Impact test requirements for structural steel FCM are specified in Tables 15-1-14.

[b] Impact teats shall be in accordance with the CVN tests as governed by ASTM Specification A673/A673M with frequency of testing H for all grades except for A709/A709M, Grade HPS 70WT (Grade HPS 485WT), which shall be frequency of testing P.

[c] The suffix T is an ASTM A709/A709M, designation for nonfracture critical material requiring impact testing, with Supplemental Requirement S83 applying. A numeral 1, 2, or 3 should be added to the T marking to indicate the applicable service temperature zone.

[d] If the yield point of the material exceeds 65 ksi (450 MPa), the test temperature for the minimum average energy required shall be reduced by 15°F (8°C) for each increment or fraction of 10 ksi (70 MPa) above 65 ksi (450 MPa).

[e] If the yield strength of the material exceeds 85 ksi (585 MPa), the test temperature for the minimum average energy required shall be reduced by 15°F (8°C) for each increment or fraction of 10 ksi (70 MPa) above 85 ksi (585 MPa).

[f] Minimum service temperature of 0°F(−18°C) corresponds to Zone 1, −30°F(−34°C) to Zone 2, and −60°F(−51°C) to Zone 3 referred to in Article 9.1.2.1.

the resistance to atmospheric corrosion as carbon steels (Kulak and Grondin, 2002), which makes their use in bridges economical from a life cycle perspective. Weathering resistance can be estimated by alloy content equations given in ASTM G101.* An index of 6.0 or higher is required for typical bridge weathering steels.

Nonweathering steels can be protected with paint or sacrificial coatings (hot-dip or spray applied zinc or aluminum). Shop applied three-coat paint systems are currently used by many North American railroads. Two, and even single, coat painting systems are being assessed by the steel coatings industry and bridge owners. An effective modern three-coat paint system consists of a zinc-rich primer, epoxy intermediate coat, and polyurethane top coat. For aesthetic purposes, steel with zinc or aluminum sacrificial coatings can be top coated with epoxy or acrylic paints.

2.3 TYPES OF STRUCTURAL STEEL

2.3.1 Carbon Steels

Modern carbon steel contains only manganese, copper, and silicon alloys. Mild carbon steel has a carbon content of 0.15–0.29%, and a maximum of 1.65% manganese (Mn), 0.60% copper (Cu), and 0.60% silicon (Si). Mild carbon steel is not of high strength but is very weldable and exhibits well-defined upper and lower yield stresses (Steel 1 in Figure 2.1). Shapes and plates of ASTM A36 and A709 Grade 36 are mild carbon steels used in railway bridge fabrication.

2.3.2 High-Strength Low-Alloy Steels

Carbon content must be limited to preclude negative effects on ductility, toughness, and weldability. Therefore, it is not desirable to increase strength by increasing carbon content and manipulation of the steel chemistry needs to be considered. HSLA steels have increased strength attained through the addition of many alloys.

Alloy elements can significantly change steel phase transformations and properties (Jastrewski, 1977). The addition of small amounts of chromium, columbium, copper, manganese, molybdenum, nickel, silicon, phosphorous, vanadium, and zirconium in specified quantities results in improved mechanical properties. The total amount of these alloys is <5% in HSLA steels. These steels typically have a well-defined yield stress in the 44–60 ksi range (Steel 2 in Figure 2.1). Shapes and plates of ASTM A572, A588, and A992 (rolled shapes only) and A709 Grade 50, 50S, and 50W are HSLA steels used in railway bridges.

A572 Grade 42, 50, and 55 steels are used for bolted or welded construction. Higher-strength A572 steel (Grades 60 and 65) is used for bolted construction only, due to reduced weldability. A572, A588, and A992 steels are not material toughness graded at the mills and often require supplemental CVN testing to ensure adequate toughness, particularly for service in cold climates. A588 and A709 Grade 50W steels are atmospheric corrosion-resistant (weathering) steels. ASTM A709 Grade 50, 50S, and 50W steel is mill certified with a specific toughness in terms of the minimum

* Other equations, such as the Townsend equation, have also been proposed and may be of greater accuracy.

CVN impact energy absorbed at a given test temperature (e.g., designations 50T2 indicating non-FCM Zone 2 and 50WF3 indicating FCM Zone 3 toughness criteria).

Further increases in strength, ductility, toughness, and weathering resistance through steel chemistry alteration have been made in recent years. HSLA steels with 70 ksi yield stress have been manufactured with niobium, vanadium, nickel, copper, and molybdenum alloy elements. These alloys stabilize either austenite or ferrite so that martensite formation and hardening does not occur, as it would for higher-strength steel attained by heat treatment. A concise description of the effects of various alloy and deleterious elements on steel properties is given in Brockenbrough (2006).

2.3.3 Heat-Treated Low-Alloy Steels

Higher-strength steel plate (with yield stress in excess of 70 ksi) is produced by heat treating HSLA steels. A disadvantage of higher-strength steels is a decrease in ductility. Heat treatment restores loss of ductility through quenching and tempering processes. The quenching of steel increases strength and hardness with the formation of martensite. Tempering improves ductility and toughness through temperature relief of the high internal stresses caused by martensite formation. Even HPS steel with 100 ksi yield stress has been quench and temper heat treated to provide good ductility, weldability, and CVN toughness (Chatterjee, 1991). However, after quenching, tempering, and controlled cooling these steels will not exhibit a well-defined yield stress (Steel 3 in Figure 2.1). In such cases, the yield stress is determined at the 0.2% offset from the elastic stress–strain relation.

Use of these steels may result in considerable weight reductions and precipitate fabrication, shipping, handling, and erection cost savings. High-strength steel can also allow for design of shallower superstructures. ASTM A514, A852, and A709 Grade 70W and 100W are quench and tempered low-alloy steel plates. However, none of these steels are typically used in ordinary railway bridges.

2.3.4 High-Performance Steels

HPS plates have been developed in response to the need for enhanced toughness, weldability, and weathering resistance of high-strength steels. HPS 70W and 100W steels are produced by a combination of chemistry manipulation and quench and temper operations or, for longer plates, thermo-mechanical controlled processing (TMCP). The first HPS steels were produced with a yield stress of 70 ksi. However, HPS with 50 ksi yield stress soon followed due to the weldability, toughness, and atmospheric corrosion resistance property improvements of HPS. HPS 50W is produced with the same chemistry as HPS 70W using conventional hot or controlled rolling techniques. HPS plates with 100 ksi yield stress are also available. HPS 100 W is considered an improvement of A514 steel plates (Lwin et al., 2005).

Weldability is increased by lowering the carbon content (e.g., below 0.11% for HPS 70W), therefore, benefiting the CE (Equation 2.6). This weldability increase results in the elimination of preheat requirements for thin members and limited preheat requirements for thicker members. Also, postweld treatments are reduced and hydrogen-induced cracking at welds eliminated (provided correct measures are taken to eliminate hydrogen from moisture, contaminants, and electrodes). Welding of HPS steels using low-hydrogen electrodes is done by submerged arc welding (SAW) or shielded metal arc welding (SMAW) processes (see Chapter 9).

Toughness is significantly increased through reductions in sulfur content (0.006% max.) and control of inclusions (by calcium treatment of steel). The fracture toughness of HPS is, therefore, much improved with the ductile to brittle transition occurring at lower temperatures (the curve shifts to the left in Figure 2.2). Higher toughness also translates into greater crack tolerance for fatigue crack detection and repair procedure development. HPS steels meet or exceed the CVN toughness requirements specified for the coldest climates (Zone 3 in AREMA, 2008).

The weathering properties of HPS are based on quench and tempered ASTM A709 Grade 70W and 100W steels. Chromium, copper, nickel, and molybdenum are alloyed for improved weathering resistance. Improved weathering resistant steels are under development that might provide good service in even moderate chloride environments.

Hybrid* applications of HPS steels with HSLA steels have proven technically and economically successful on a number of highway bridges (Lwin, 2002) and may be appropriate for some railway bridge projects.

2.4 STRUCTURAL STEEL FOR RAILWAY BRIDGES

There is no increase in stiffness associated with higher-strength steels (deflections, vibrations, and elastic stability are proportional to the modulus of elasticity and not strength). Also, because fatigue strength depends on applied stress range and detail (see Chapter 5), there is no increase in fatigue resistance for higher-strength steels. Therefore, the material savings associated with the use of higher-strength steels (with greater than 50 ksi yield stress) may not be available because deflection and fatigue criteria often govern critical aspects of ordinary steel railway superstructure design. The steel bridge designer must carefully consider all design limit states (strength, serviceability, fatigue, and fracture), fabrication, and material cost aspects when selecting the materials for railway bridge projects.

2.4.1 Material Properties

The following material properties are valid for steel used in railway bridge construction:

- Density, $\gamma = 490$ lb/ft^3
- Modulus of elasticity (Young's modulus), $E = 29 \times 10^6$ psi $= 29{,}000$ ksi
- Coefficient of thermal expansion, $\alpha = 6.5 \times 10^{-6}$ per °F
- Poisson's ratio, $\upsilon = 0.3$ (lateral to longitudinal strain ratio under load)
- In accordance with the theory of elasticity, shear modulus, $G = E/[2(1 + \upsilon)]$ (11.2×10^6 psi)

* An example is the use of HPS steels for tension flanges in simple and continuous girders.

TABLE 2.3
Structural Steel for Railway Bridges[a]

ASTM Designation	F_y (Min. Yield Point) pel	F_U (Ultimate Strength) pel	Thickness Limitation: For Plates and Bare, Inches	Thickness Limitation: Applicable to Shapes
A36	36,000	58,000 min 80,000 max	To 6 incl.	All
A709, Grade 36	36,000	58,000 min 80,000 max	To 4 incl.	All
A588[b] A709, Grade 50W[b] A709, Grade HPS 50W[b]	50,000	70,000 min	To 4 incl.	All
A588[b]	46,000	70,000 min	Over 4–5 incl.	All
A588[b]	42,000	63,000 min	Over 5–8 incl.	None
A992 A709, Grade 50S	50,000	65,000 min	None	All
A572, Grade 50 A709, Grade 50	50,000	65,000 min	To 4 incl.	All
A572, Grade 42	42,000	60,000 min	To 6 incl.	All
A709, Grade HPS 70W[b]	70,000	85,000 min 110,000 max	To 4 incl.	None

Source: From AREMA, 2008, *Manual for Railway Engineering*, Chapter 15, Lanham, MD. With permission.

[a] These data are current as of January 2002.

[b] A588 and A709, Grade 50W, Grade HPS 50W, and Grade HPS 70W have atmospheric corrosion resistance in most environments substantially better than that of carbon steels with or without copper addition. In many applications these steels can be used unpainted.

2.4.2 Structural Steels Specified for Railway Bridges

Modern structural bridge steels provide good ductility, weldability, and weathering resistance. The AREMA (2008) recommendations do not include heat-treated low-alloy steels. The only steel with a yield stress >50 ksi currently recommended is A709 HPS 70W. Also, as seen in Table 2.3, AREMA (2008) recommends the use of weathering steels such as ASTM A588 and A709. Nonweathering steels such ASTM A36 and A572 are also indicated for use. Since A572 Grades 42 and 50 are recommended for welded and bolted construction, with higher grades used for bolted construction only, the AREMA (2008) recommendations for structural steel do not include A572 grades higher than Grade 50.

REFERENCES

American Railway Engineering and Maintenance-of-way Association (AREMA), 2008, Steel Structures, in *Manual for Railway Engineering*, Chapter 15, Lanham, MD.

American Society of Testing and Materials (ASTM), 2000, *Standards Vol. 01.04, A36, A572, A588, A709, A992; Vol. 03.01, E8; and Vol. 03.02, G101*, 2000 Annual Book of ASTM Standards West Conshohocken, PA.

Barsom, J.M. and Rolfe, S.T., 1987, *Fracture and Fatigue Control in Structures*, 2nd Edition, Prentice-Hall, New Jersey.

Brockenbrough, R.L., 2006, Properties of structural steels and effects of steelmaking, in *Structural Steel Designer's Handbook*, 4th Edition, R.L. Brockenbrough and F.S. Merritt, Eds, McGraw-Hill, New York.

Chakrabarty, J., 2006, *Theory of Plasticity*, 3rd Edition, Elsevier, Oxford, UK.

Chatterjee, S., 1991, *The Design of Modern Steel Bridges*, BSP Professional Books, Oxford, UK.

Chen, W.F. and Han, D.J., 1988, *Plasticity for Structural Engineers*, Springer-Verlag, New York.

Fisher, J.W., 1984, *Fatigue and Fracture in Steel Bridges*, Wiley, New York.

Hill, R., 1989, *The Mathematical Theory of Plasticity*, Oxford University Press, Oxford.

Jastrewski, Z.D., 1977, *The Nature and Properties of Engineering Materials*, 2nd Edition, Wiley, New York.

Kulak, G.L. and Grondin, G.Y., 2002, *Limit States Design in Structural Steel*, Canadian Institute of Steel Construction, Toronto.

Lwin, M.M., 2002, *High Performance Steel Designer's Guide*, FHWA Western Resource Center, U.S. Department of Transportation, San Francisco, CA.

Lwin, M.M., Wilson, A.D., and Mistry, V.C., 2005, *Use and Application of High-Performance Steels for Steel Structures—High-Performance Steels in the Untied States*, International Association for Bridge and Structural Engineering, Zurich, Switzerland.

3 Planning and Preliminary Design of Modern Railway Bridges

3.1 INTRODUCTION

The primary purpose of railway bridges is to safely and reliably carry freight and passenger train traffic within the railroad operating environment. It is estimated that, in terms of length, between 50% and 55% of the approximately 80,000 bridges (with an estimated total length of almost 1800 miles) in the North American freight railroad bridge inventory are composed of steel spans (Unsworth, 2003; FRA, 2008).* Structural and/or functional obsolescence precipitates the regular rehabilitation and/or replacement of steel railway bridges. Many of the steel bridges in the North American freight railroad bridge inventory are over 80 years old and may require replacement due to the effects of age, increases in freight equipment weight,† and the amplified frequency of the application of train loads.‡ Bridge replacement requires careful planning with consideration of site conditions and transportation requirements in the modern freight railroad operating environment.

Site conditions relating to hydraulic or roadway clearances, as well as the geotechnical and physical environment (during and after construction), are important concerns during planning and preliminary design. Railroad and other transportation entity operating practices also need careful deliberation. Interruption to traffic flow in rail, highway, or marine transportation corridors and safety (construction and public) are

* In 1910, it was estimated that there were about 80,000 metal bridges with a cumulative length of 1400 miles on 190,000 miles of track (see Chapter 1). In 2008, it was estimated that there were about 77,000 bridges of all materials with a cumulative length of 1760 miles on 191,000 miles of track. In 2008 the cumulative length of steel bridges was estimated as 935 miles.

† In 1910, locomotives typically weighed about 300,000 lb (see Chapter 1). Over the next few decades the weight of some heavy service locomotives increased by over 50%. The weight of typical locomotives currently used on North American railroads approaches 450,000 lb.

‡ Trains with loads causing many cycles of stress ranges that might accumulate significant fatigue damage did not occur until the latter half of the twentieth century when typical train car weights increased from 177,000 lb to over 263,000 lb on a regular basis.

of paramount concern. The planning phase should yield information concerning optimum crossing geometry, layout, and anticipated construction methodologies. This information is required for the selection of span lengths, types, and materials for preliminary design. Preliminary design concepts are often the basis of regulatory reviews, permit applications, and budget cost estimates. Therefore, planning and preliminary design can be critical to successful project implementation and, particularly for large or complex bridges, warrants due deliberation. Detailed design of the bridge for fabrication and construction can proceed following preliminary design.

3.2 PLANNING OF RAILWAY BRIDGES

Planning of railway bridges involves the careful consideration and balancing of multifaceted, and often competing, construction economics, business, public, and technical requirements.

3.2.1 Bridge Crossing Economics

In general, other issues not superseding, the bridge crossing should be close to perpendicular to the narrowest point of the river or flood plain. The economics of a bridge crossing depends on the relative costs of foundations, substructures, and superstructures.* Estimates related to the cost of foundation and substructure construction are often less reliable than those for superstructures. Superstructure fabrication cost estimates are often more dependable than erection estimates due to the inherently greater uncertainty and risk associated with field construction. Excluding, in particular, public and technical (hydraulic and geotechnical) constraints from the cost-estimating procedure enables the economical span length, l, to be established based on very simple principles. Considering a fairly uniform bridge with similar substructures and multiple equal length spans, the total estimated cost, CE, of a bridge crossing may be expressed as

$$\mathrm{CE} = n_s C_{\mathrm{sup}} w_s l + (n_s - 1) C_{\mathrm{pier}} + 2C_{\mathrm{abt}}, \tag{3.1}$$

where n_s is the number of spans and is equal to L/l, where L is the length of the bridge; C_{sup} is the estimated cost of steel per unit weight (purchase, fabricate, and erect); w_s is the weight of span elements that depend on span length, l (e.g., the bridge deck of a span or the floor system of a through span is independent of span length and excluded from w_s); C_{pier} is the average cost of one pier (materials, foundation, and construction); and C_{abt} is the average cost of one abutment (materials, foundation, and construction).

If $w_s = \alpha l + \beta$, where α and β are constants independent of span length and dependent only on span type and design live load, Equation 3.1 may be expressed as

$$\mathrm{CE} = C_{\mathrm{sup}} L(\alpha l + \beta) + C_{\mathrm{pier}} L\left(\frac{L - l}{lL}\right) + 2C_{\mathrm{abt}}, \tag{3.1a}$$

* A rule of thumb for economical relatively uniform multispan bridges is that the cost of superstructure (fabrication and erection) equals the cost of foundation and substructure construction (Byers, 2009).

which may be differentiated in terms of span length, l, to determine an expression for the minimum total estimated cost, CE, as

$$\frac{\mathrm{dCE}}{\mathrm{d}l} = C_{\mathrm{sup}}\alpha L - \frac{C_{\mathrm{pier}}L}{l^2} = 0. \quad (3.2)$$

Rearrangement of Equation 3.2 provides the economical span length, l, as

$$l = \sqrt{\frac{C_{\mathrm{pier}}}{C_{\mathrm{sup}}\alpha}}. \quad (3.3)$$

However, while Equation 3.3 provides a simple estimate of economical span length, the final general arrangement in terms of span lengths, l, may depend on other business, public, and technical requirements.

3.2.2 Railroad Operating Requirements

Most new freight railway bridges are constructed on existing routes on the same alignment. Construction methods that minimize the interference to normal rail, road, and marine traffic enable simple erection and are cost-effective must be carefully considered during the planning process. Often, in order to minimize interruption to railroad traffic, techniques such as sliding spans into position from falsework, floated erection of spans from river barges, span installation with movable derricks and gantries, construction on adjacent alignment,* and the use of large cranes must be developed (Unsworth and Brown, 2006). These methodologies may add cost to the reconstruction project that are acceptable in lieu of the costs associated with extended interruption to railway or marine traffic.

New rail lines are generally constructed in accordance with the requirements established by public agencies† and railroad business access. It is not often that bridge crossings are selected solely on the basis of localized bridge economics planning principles‡ as outlined in Section 3.2.1. Therefore, site reconnaissance (surveying and mapping) and route selection are performed on the basis of business, technical, and public considerations.

The railroad operating environment presents specific challenges for bridge design, maintenance, rehabilitation, and construction. The design of steel railway bridges involves the following issues related to railroad operations:

- The magnitude, frequency, and dynamics of railroad live loads
- Other loads specific to railroad operations

* Either the new bridge is constructed on an adjacent alignment or a temporary bridge is built on an adjacent alignment (shoo-fly) in order to not interrupt the flow of rail traffic. This may not always be feasible due to cost, site conditions, and/or route alignment constraints.

† Generally, the requirements relate to environmental, fish and wildlife, land ownership, and cultural considerations and/or regulations.

‡ The exception might be very long bridges.

- The location of the bridge (in relation to preliminary design of bridge type, constructability, and maintainability)
- Analysis and design criteria specific to railway bridges

3.2.3 Site Conditions (Public and Technical Requirements of Bridge Crossings)

Site conditions are of critical importance in the determination of location, form, type, length, and estimated cost of railway bridges. Existing records and drawings of previous construction are of considerable value during planning of railway bridges being reconstructed on the same, or nearby, alignment. In terms of the cost and constructability of bridges being built on a new alignment, the preferred bridge crossing is generally the shortest or shallowest crossing.

3.2.3.1 Regulatory Requirements

Bridge location, form, and length are often governed by existing route location, preestablished design route locations, and/or regulation. Regulatory requirements relating to railway bridge location, construction, and environmental mitigation vary by geographic location and jurisdiction. Environmental protection (vegetation, fish, and wildlife) and cultural considerations are often critical components of the bridge planning phase. Land ownership and use regulations also warrant careful review with respect to potential bridge crossing locations. Railway bridge construction project managers and engineers must be well versed in the jurisdictional permitting requirements for bridge crossings. Regulatory concerns regarding bridge location and construction that may affect bridge form must be communicated to the railway bridge designer during the planning phase.

3.2.3.2 Hydrology and Hydraulics of the Bridge Crossing

Hydrological and hydraulic assessments are vital to establishing the required bridge opening at river crossings, and ensuing form, type, length, and estimated cost. The bridge opening must safely pass the appropriate return frequency (probability of occurrence) water discharge,* ice and debris† past constrictions and obstructions created by the bridge crossing substructures. The basic form of the bridge may depend on whether the flood plain is stable. Stable flood plains may be spanned with shorter spans unless shifting channel locations require the use of longer spans. Hydraulic studies must also consider the potential for scour at substructures.

3.2.3.2.1 Bridge Crossing Discharge Hydraulics

A study of area hydrology and river hydraulics will provide information concerning the existing average channel velocity, V_u, at the required return frequency discharge, Q. If there are no piers to obstruct the river crossing and no constriction of the channel

---

* Developed from hydrological evaluations for the river crossing.

† In general, piers should not be skewed to river flow to avoid impact from ice, debris, or vessels.

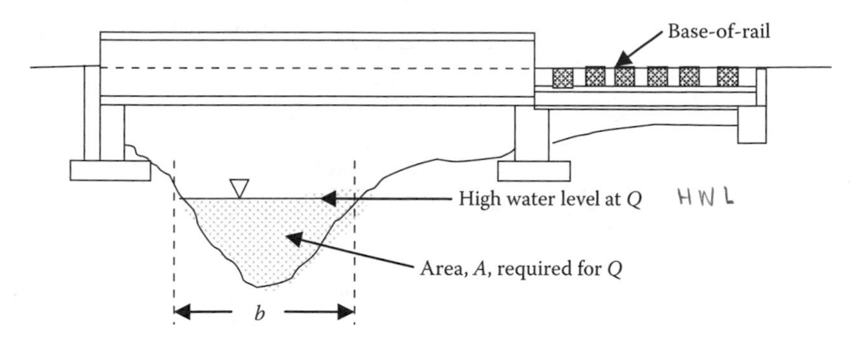

FIGURE 3.1 River crossing profile without constriction or obstruction.

(Figure 3.1), the required area of the crossing is simply established as

$$A = \frac{Q}{V_u}. \tag{3.4}$$

3.2.3.2.1.1 Constricted Discharge Hydraulics Where abutments constrict the channel (Figure 3.2), the flow may become rapidly varied and exhibit a drop in water surface elevation (hydraulic jump) as a result of the increased velocity. Four types of constriction openings have been defined (Hamill, 1999) as follows:

- Type 1: Vertical abutments with and without wings walls with vertical embankments
- Type 2: Vertical abutments with sloped embankments
- Type 3: Sloped abutments with sloped embankments
- Type 4: Vertical abutments with wings walls and sloped embankments (typical of many railroad embankments at bridge crossings)

The hydraulic design should strive for subcritical flow ($F < 1.0$ with a stable water surface profile). Discharge flows that exceed subcritical at, or even immediately downstream of, the bridge may also be acceptable with adequate scour protection. Supercritical flow ($F > 1.0$) is undesirable and may create an increase in water surface elevation at the bridge crossing. However, supercritical flow may be unavoidable for river crossings with steep slopes [generally >0.5–1% (Transportation Association of Canada (TAC), 2004)], and careful hydraulic design is required. In the case of constrictions, the required area of the crossing is established as

$$A = \frac{Q}{C_c V_u}, \tag{3.5}$$

where A is the minimum channel area required under the bridge; Q is the required or design return frequency discharge (e.g., 1:100); V_u is the average velocity of the existing (upstream) channel for discharge, Q; and C_c is the coefficient of contraction

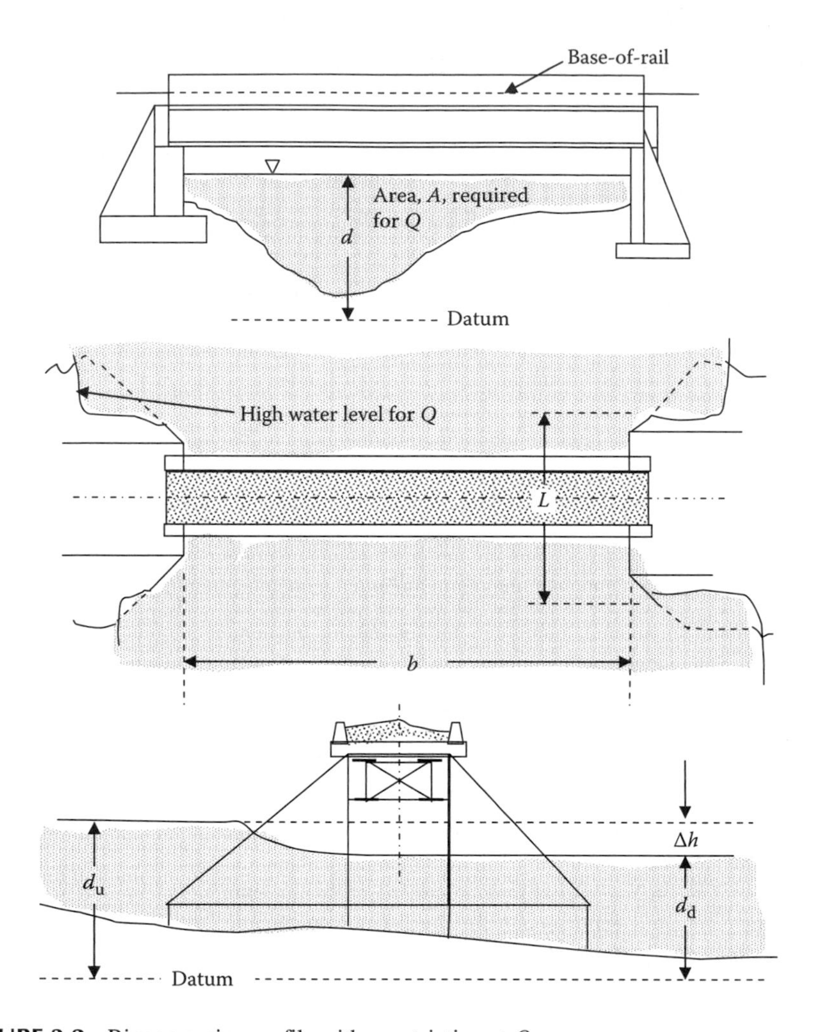

FIGURE 3.2 River crossing profile with constriction at Q.

of the new channel cross section at the constriction. It can be shown that the coefficient of contraction, C_c, depends on the following:

- Contraction ratio
- Constriction edge geometry and angularity
- Submerged depth of abutment
- Slope of abutment face
- Eccentricity of the constriction relative to normal stream flow
- Froude number, which is

$$F = \frac{Q}{A\sqrt{gd}}, \tag{3.6}$$

where g is the acceleration due to gravity and d is the effective depth of the channel under the bridge and is equal to A/b, where b is the net or effective width of the bridge opening.

The theoretical determination of C_c is difficult and numerical values are established experimentally. Published values of C_c for various constriction geometries are available in the literature on open channel hydraulics (Chow, 1959).

One commonly used hydraulic analysis* is the U.S. Geological Survey (USGS) method, which is based on extensive research. It determines a base coefficient of discharge, C', for four opening types in terms of the opening ratio, N, and constriction length ratio, L/b. The coefficient of discharge, C', is further modified by adjustment factors based on the opening ratio, Froude number, F, and detail abutment geometry to obtain the discharge coefficient, C. In terms of the USGS method, the coefficient of contraction is

$$C_c = \frac{C}{V_u}\sqrt{2g\left(\Delta h + \frac{\alpha_u V_u^2}{2g} - h_f\right)}, \tag{3.7}$$

where C is the USGC discharge coefficient, which depends on N, L, b, F, and other empirical adjustment factors based on skew angle of the crossing, the conveyance, K, details the geometry and flow depth at the constriction; N is the bridge opening ratio and is equal to Q_c/Q, where Q_c is the undisturbed flow that can pass the bridge constriction and Q is the flow in the not constricted channel; L is the length of channel at the constricted bridge crossing; $K = AR^{2/3}/n$; $\Delta h = d_u - d_d$, where d_u is the depth of the channel upstream of the bridge and d_d is the depth of the channel downstream of the bridge; h_f is the friction loss upstream and through the constricted opening, which is

$$h_f = L_u\left(\frac{Q^2}{K_u K_d}\right) + L\left(\frac{Q}{K_d}\right)^2, \tag{3.8}$$

where L_u is the length of upstream reach (from the uniform flow to the beginning of constriction); K_u is the upstream conveyance and is equal to $A_u R_u^{2/3}/n_u$; K_d is the downstream conveyance and is equal to $A_d R_d^{2/3}/n_d$; A_u and A_d are the upstream and downstream channel cross-sectional areas, respectively; R_u and R_d are the upstream and downstream hydraulic radii, respectively and are equal to area, A_u or A_d, divided by the channel wetted perimeter; n_u and n_d are the upstream and downstream Manning's roughness coefficients, respectively.

3.2.3.2.1.2 Obstructed Discharge Hydraulics Due to the large live loads, long railway bridges are often composed of many relatively short spans where the topography allows such construction. In these cases, many piers are required that may create an obstruction to the flow and consideration of the contraction effects due to obstruction is also necessary (Equation 3.5 with C_c being the coefficient of contraction of the new channel cross section at the obstruction). The flow past an obstruction is

* Other methods such as the U.S. Bureau of Public Roads, Biery and Delleur, and the U.K. Hydraulic Research methods are also used.

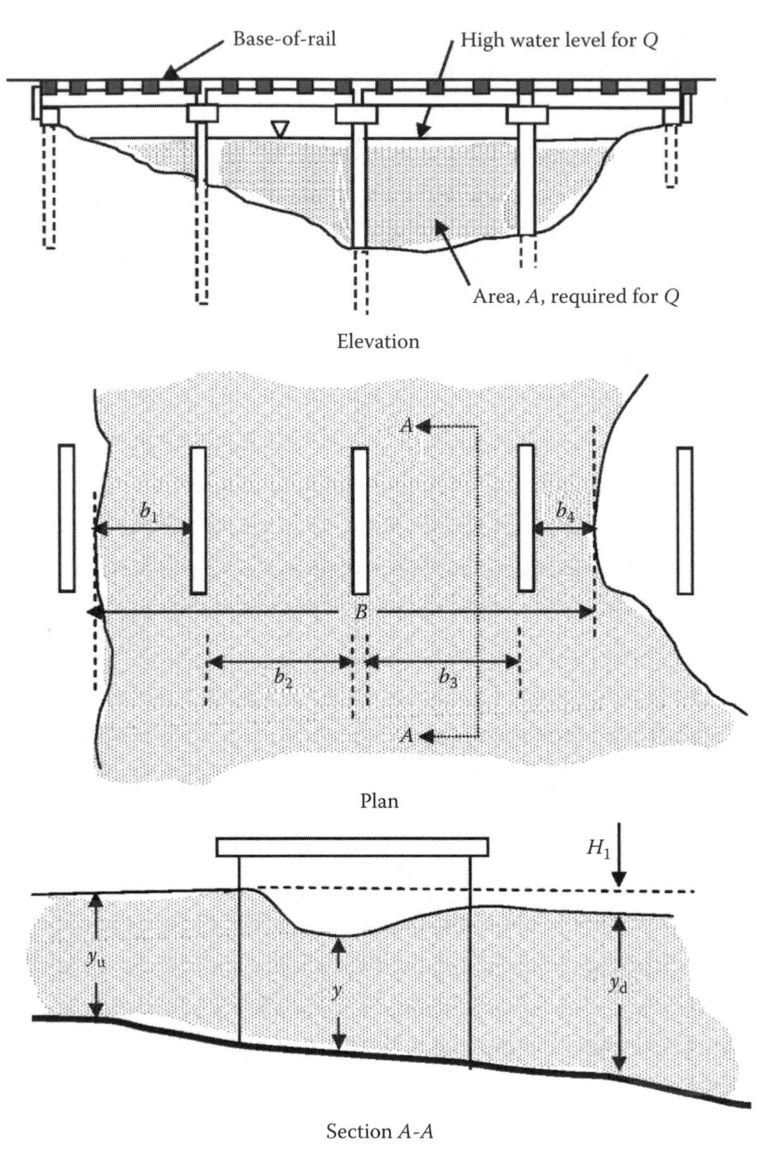

FIGURE 3.3 River crossing profile with obstructions at Q.

similar to the flow past a constriction but with more openings (Figure 3.3). The degree of contraction is usually less for obstructions than constrictions. Published values of C_c for various obstruction geometries are available in the literature on open channel hydraulics (Yarnell, 1934; Chow, 1959).

The flow about an obstruction consisting of bridge piers was extensively investigated (Nagler, 1918) and Equation 3.9, which is similar in form to Equation 3.5, was derived from the results:

$$A = by = \frac{Q}{K_{\mathrm{N}}\sqrt{2gH_1 + \beta V_{\mathrm{u}}^2}}, \tag{3.9}$$

where b is the effective width at the obstruction and is equal to $b_1 + b_2 + b_3 + b_4$ (Figure 3.3); y is the depth at the pier or obstruction and is equal to $y_{\mathrm{d}} - \phi V_{\mathrm{d}}^2/2g$, where y_{d} is the depth downstream of the pier or obstruction, V_{d} is the average velocity of the downstream channel for discharge, Q; ϕ is the adjustment factor (it has been evaluated from experiments that ϕ is generally about 0.3); K_{N} is the coefficient of discharge, which depends on the geometry of pier or obstruction and the bridge opening ratio, $N = b/B$; H_1 is the downstream afflux; β is the correction for upstream velocity, V_{u}, head (from experiments and depends on bridge opening ratio, N) (for $N < 0.6$, $\beta \sim 2$).

However, for subcritical flow, the Federal Highway Administration (FHWA, 1990) recommends the use of energy equation (Schneider et al., 1977) or momentum balance methods (TAC, 2004) when pier drag is a relatively small proportion of the friction loss. When pier drag forces constitute the predominate friction loss through the contraction, the momentum balance or Yarnell equation methods are applicable. The momentum balance method yields more accurate results when pier drag becomes more significant.

The Yarnell equation is based on further experiments (summarized by Yarnell, 1934) with relatively large piers (typical of railway bridge substructures) that were performed to develop equations for the afflux for use with Equation 3.10 (the d'Aubuisson equation, which is applicable to subcritical flow conditions only):

$$A_{\mathrm{d}} = by_{\mathrm{d}} = \frac{Q}{K_{\mathrm{A}}\sqrt{2gH_1 + V_{\mathrm{u}}^2}}, \tag{3.10}$$

where K_{A} is the coefficient from Yarnell's experiments, which depends on the geometry of pier or obstruction and the bridge opening ratio, $N = b/B$, where B is the width of the channel without obstruction.

The afflux depends on whether the flow is subcritical or supercritical (Hamill, 1999). For subcritical flow conditions:

$$H_1 = Ky_{\mathrm{d}}F_{\mathrm{d}}^2\left(K + 5F_{\mathrm{d}}^2 - 0.6\right)\left[(1 - N) + 15(1 - N)^4\right], \tag{3.11}$$

where K is Yarnell's pier shape coefficient (between 0.90 and 1.25 depending on pier geometry) and F_{d} is the normal depth Froude number and is given by

$$F_{\mathrm{d}} = \frac{Q}{A\sqrt{gy_{\mathrm{d}}}} \leq 1.0.$$

The normal depth, y_{d}, is readily calculated from the usual open channel hydraulics methods.

For supercritical flow conditions (which will cause downstream hydraulic jump), the analysis is more complex and design charts have been made to assist in establishing the discharge past obstructions (Yarnell, 1934).

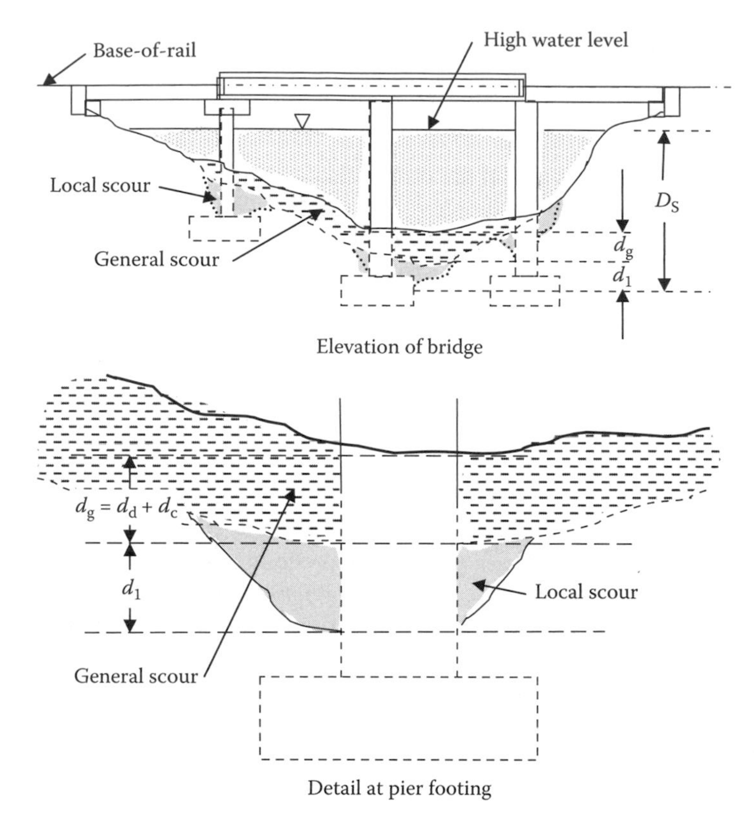

FIGURE 3.4 Scour at bridge crossings due to constrictions and obstructions.

3.2.3.2.2 Scour at Bridge Crossings

Once the required bridge opening and general geometry of the crossing is established, scour conditions at constrictions and obstructions must be investigated in order to sustain overall stability of the bridge foundations and substructure (Figure 3.4). Scour can occur when the streambed is composed of cohesive or cohesionless materials. However, scour generally occurs at a much higher rate for cohesionless materials, which will be the focus of the present discussion.

General scour occurs due to streambed degradation at the constriction or obstruction of the waterway opening caused by the bridge substructures (abutments and piers). Contraction scour (due to opening constriction and/or obstruction) and local scour (due to obstruction) are also components of total scour that may occur under both live-bed and clear-water (the upstream streambed is at rest and there is no sediment in water) conditions. General scour may also occur due to degradation, or adjustment of the river bed elevation, due to overall hydraulic changes not specifically related to the bridge crossing. This component of the general scour may occur with live-bed scour (the streambed material upstream of the bridge is moving) conditions. For preliminary design scour, degradation can be estimated by assuming that the cross-sectional area of the degradation scour is equal to the loss of cross-sectional discharge area due to the submerged substructures. The depth of the degradation scour, d_d, can then be estimated based on the width of the channel bed. Accurate assessment methods for general scour, d_g, are available (TAC, 2004; Richardson and Davis, 1995).

3.2.3.2.2.1 Contraction Scour During live-bed scour the contraction scour depths are affected by deposits of sediment from upstream. Scour will cease when the rate of sediment deposit equals the rate of loss by contraction scour. During clear-water conditions sediment is not transported into the contraction scour depth increase (creating channel bed depressions or holes). Scour will equilibrate and cease when the velocity reduction caused by the increased area becomes less than that required for contraction scour. Equation 3.12 provides an estimate of the approach streambed velocity at which live-bed scour will initiate, V_s, as (Laursen, 1963)

$$V_s = 6y_u^{1/6}D_{50}^{1/3}, \tag{3.12}$$

where y_u is the depth of channel upstream of bridge crossing, D_{50} is the streambed material median diameter at which 50% by weight are smaller than that specified (the size that governs the beginning of erosion in well-graded materials).

Contraction scour depth, d_c, for cohesionless materials under live-bed conditions can be estimated as (Laursen, 1962)

$$d_c = y_u\left[\left(\frac{Q_c}{Q}\right)^{6/7}\left(\frac{B}{b}\right)^{k1}\left(\frac{n_c}{n_u}\right)^{k2} - 1\right], \tag{3.13}$$

where Q_c is the discharge at contracted channel (at bridge crossing); b is the net or effective width of bridge opening; B is the width of the channel without obstruction or constriction; n_c is Manning's surface roughness coefficient at the contracted channel; n_u is Manning's surface roughness coefficient at the upstream channel; $k1$ and $k2$ are exponents that depend on $\sqrt{gy_uS_u/V_{D_{50}}}$, where S_u is the upstream energy slope (often taken as streambed slope); $V_{D_{50}}$ is the median fall velocity of flow based on D_{50} median particle size.

Contraction scour depth, d_c, for cohesionless materials under clear-water conditions can be estimated as (Laursen, 1962; Richardson and Davis, 1995)

$$d_c = y_u\left[\left(\frac{B}{b}\right)^{6/7}\left(\frac{V_u^2}{42(y_u)^{1/3}(D_{50})^{2/3}}\right)^{3/7} - 1\right]. \tag{3.14}$$

3.2.3.2.2.2 Local Scour Local scour occurs at substructures as a result of vortex flows induced by the localized disturbance to flow caused by the obstruction. The determination of local scour depths is complex but there are published values relating local and general scour depths that are useful for preliminary scour evaluations. Procedures for establishing local scour relationships for abutments and piers are available (TAC, 2004; Richardson and Davis, 1995). For most of the modern bridges, local scour can generally be precluded, particularly at abutments, by the use of properly designed revetments and scour protection. Local scour depth for cohesionless materials at piers can be estimated as

$$d_{\mathrm{l}} = 2yK_{1\mathrm{P}}K_{2\mathrm{P}}K_{3\mathrm{P}}\left(\frac{b_{\mathrm{pier}}}{y}\right)^{0.65}F^{0.43}. \tag{3.15}$$

Local scour depth for cohesionless materials at abutments can be estimated as (Richardson and Davis, 1995)

$$d_{\mathrm{l}} = 4y\left(\frac{K_{1A}}{0.55}\right)K_{2\mathrm{A}}F^{0.33}, \tag{3.16}$$

where $K_{1\mathrm{P}}$ is the pier nose geometry adjustment factor (from experimental values), $K_{2\mathrm{P}}$ is the angle of the flow adjustment factor (from experimental values), $K_{3\mathrm{P}}$ is the bed configuration (dune presence) adjustment factor (from experimental values), $K_{1\mathrm{A}}$ is the abutment type (vertical, sloped, wingwalls) adjustment factor (from experimental values), $K_{2\mathrm{A}}$ is the abutment skew adjustment factor (from experimental values), y is the depth at pier or abutment, and F is the Froude number calculated at pier or abutment.

3.2.3.2.2.3 Total Scour The total scour depth, d_{t}, is then estimated as

$$d_{\mathrm{t}} = d_{\mathrm{g}} + d_{\mathrm{l}} = d_{\mathrm{d}} + d_{\mathrm{c}} + d_{\mathrm{l}}, \tag{3.17}$$

where d_{g} is the general scour, d_{d} is the degradation depth, d_{c} is the contraction scour (under live-bed or clear-water conditions), and d_{l} is the local scour at substructure (abutment or pier).

In some cases, substructure depth must also be designed anticipating extreme natural scour and channel degradation events. Therefore, for ordinary railway bridges, spread footings or the base of pile caps are often located such that the underside of the footing or cap is 5–6 ft below the estimated total scour depth, d_{t}. Also, it is often beneficial to consider the use of fewer long piles than a greater number of short piles when the risk of foundation scour is relatively great.

3.2.3.3 Highway, Railway, and Marine Clearances

Railway bridges crossing over transportation corridors must provide adequate horizontal and vertical clearance to ensure the safe passage of traffic under the bridge. Railway and highway clearances are prescribed by States and navigable waterway clearance requirements are the responsibility of the United States Coast Guard (USCG). Provision for changes in elevation of the under-crossing (i.e., a highway or track rise) and widening should be considered during planning and preliminary design.

The minimum railway bridge clearance envelope recommended by AREMA (2008) is generally 23 ft from the top of the rail and 9 ft each side of the track centerline. AREMA (2008, Chapters 15 and 28) outline detailed clearance requirements for railway bridges. These dimensions must be revised to properly accommodate track curvature.

3.2.3.4 Geotechnical Conditions

Geotechnical site conditions are often critical with respect to the location, form, constructability, and cost of railway bridges. Soil borings should generally be taken at or near each proposed substructure location. Soil samples are submitted for laboratory testing and/or tested *in situ* to determine soil properties required for foundation design such as permeability, compressibility, and shear strength.

For the purposes of railway bridge design, the subsurface investigation should yield a report making specific foundation design recommendations. The geotechnical investigation should encompass:

- Foundation type and depth (spread footings, driven piles, drilled shafts, etc.)
- Construction effects on adjacent structures (pile driving, jetting, and drilling)
- Foundation settlement*
- Foundation scour analyses and protection design
- Foundation cost

Driven steel pipe, steel HP sections, and precast concrete piles are cost-effective bridge foundations. Although typically more costly than driven piles, concrete piles may also be installed by boring when required by the site conditions. Geotechnical investigations for driven pile foundations should include the following:

- Recommended pile types based on design and installation criteria
- Consideration of soil friction and/or end bearing
- Pile tip elevation estimation
- Allowable pile loads
- Recommended test pile requirements and methods

Investigations for spread footing foundations should include the following:

- Footing elevation (scour or frost protection)
- Allowable soil bearing pressure
- Groundwater elevation
- Stability (overturning, sliding)
- Bedding materials and compaction

The design and construction of drilled shaft foundations should be based on geotechnical investigations and recommendations relating to:

- Friction and end bearing conditions (straight shaft, belled base)
- Construction requirements (support of hole)

---

* Generally simply supported spans and ballasted deck bridges tolerate greater settlements. However, tolerable settlements may depend on longitudinal and lateral track geometry regulations, which are often specified as maximum permissible variations in rail profile and cross level.

- Allowable side shear and base bearing stresses
- Downdrag and uplift conditions

Also, specific dynamic soil investigations may be required in areas of high seismic activity to determine soil strength, foundation settlement, and stability under earthquake motions.

In some cases it is possible that, due to geotechnical conditions, foundations must be relocated. This will result in significant changes in the proposed bridge arrangement and should be carefully and comparatively cost estimated. A geotechnical engineer experienced in shallow and deep bridge foundation design and construction should be engaged to manage geotechnical site investigations and provide recommendations for foundation design.

3.2.4 Geometry of the Track and Bridge

Railway horizontal alignments consist of simple curves (Figure 3.5) and a tangent track connected by transition or spiral curves. Track profile, or vertical alignment, is composed of constant grades connected by parabolic curves. Many high-density rail lines have grades of $<1\%$ and restrict curvature to safely operate at higher train speeds.

3.2.4.1 Horizontal Geometry of the Bridge

If curved tracks traverse a bridge, the consequences for steel superstructure design are effects due to the:

- Centrifugal force created as the train traverses the bridge at speed, V (force effect)

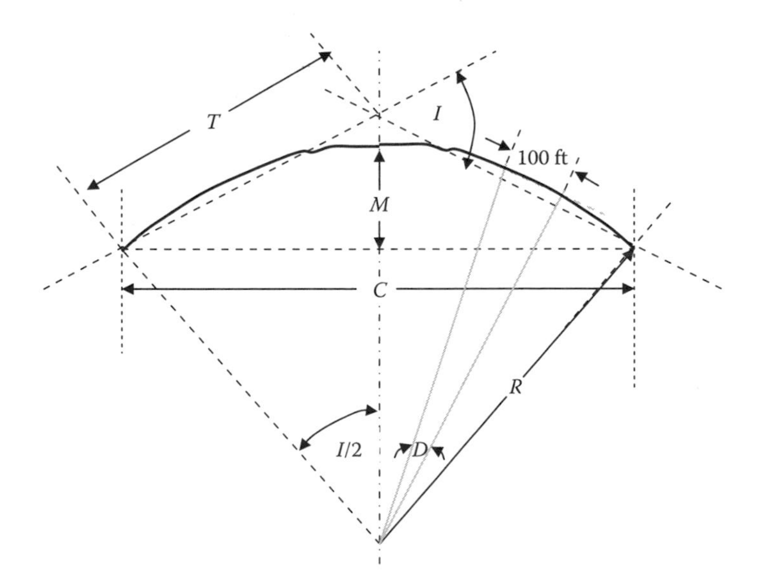

FIGURE 3.5 Simple curve geometry.

- Offset or eccentricity of the track alignment with respect to the centerline of the span or centerline of supporting members (geometric effect)
- Offset or eccentricity of the center of gravity of the live load as it traverses the superelevated curved track (geometric effect)

The centrifugal force is horizontal and transferred to supporting members as a vertical force couple. The magnitude of the force depends on the track curvature and live load speed and is applied at the center of gravity of the live load (see Chapter 4). Track alignment and superelevation also affect vertical live load forces (including impact) in supporting members based on geometrical eccentricity of the live load.

3.2.4.1.1 Route (Track) Geometrics

Train speed is governed by the relationship between curvature and superelevation. Railway bridge designers must have an accurate understanding of route geometrics in order to develop the horizontal geometry of the bridge, determine centrifugal forces and ensure adequate horizontal and vertical clearances in through superstructures. The central angle subtended by a 100 ft chord in a simple curve, or the degree of curvature, D, is used to describe the curvature of North American railroad tracks. The radius, R, ft, and other simple curve data are then

$$R = \frac{360(100)}{2\pi D} = \frac{5729.6}{D}, \tag{3.18}$$

$$I \approx \frac{L_c D}{100}, \tag{3.19}$$

$$T = R \tan \frac{I}{2}, \tag{3.20}$$

$$C = 2R \sin \frac{I}{2}, \tag{3.21}$$

$$M = R\left(1 - \cos \frac{I}{2}\right) \approx \frac{C^2}{8R}, \tag{3.22}$$

where L_c is the length of the curve, ft ($L_c \gg 100$ ft for a typical railway track), I is the intersection angle, T is the tangent distance, C is the chord length, and M is the mid-ordinate of the curve. The track is superelevated to accommodate the centrifugal forces that occur as the train traverses through curved track (Figure 3.6). For equilibrium, with weight equally distributed to both wheels, the superelevation, e, is (Hay, 1977)

$$e = \frac{\text{CF}(d)}{W}. \tag{3.23}$$

Also, since

$$\text{CF} = \frac{mV^2}{R} \tag{3.24}$$

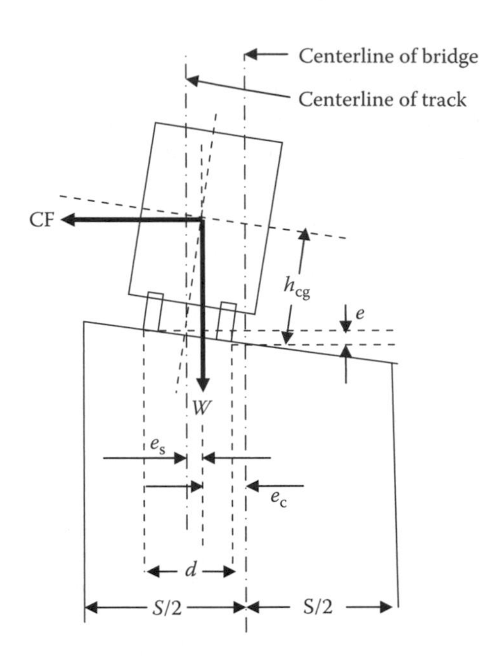

FIGURE 3.6 Railway track superelevation.

the superelevation can be expressed as

$$e = \frac{dV^2}{gR}, \tag{3.25}$$

where d is the horizontal projection of track contact point distance, which may be taken as 4.9 ft for North American standard gage tracks; CF is the centrifugal force; W is the weight of the train; $m = W/g$ is the mass of the train (g is the acceleration due to gravity); V is the speed of the train.

Substitution of $R = 5730/D$, $d = 4.9$ ft, and $g = 32.17$ ft/s^2 into Equation 3.25 yields

$$e \approx 0.0007DV^2, \tag{3.26}$$

where e is the equilibrium superelevation, inches, and D is the degree of the curve.

Transition curves are required between tangent and curved tracks to gradually vary the change in the lateral train direction. The cubic parabola is used by many freight railroads as a transition from the tangent track to an offset simple curve. The length of the transition curve is based on the rate of change of superelevation. Safe rates of superelevation "run-in" are prescribed by regulatory authorities and railroad companies. For example, with a rate of change of superelevation that is equal to 1.25 in/sec, the length of the transition curve, L_s, is

$$L_s = 1.17\,\mathrm{eV}, \tag{3.27}$$

where e is the equilibrium superelevation, inches, and V is the speed of the train, mph.

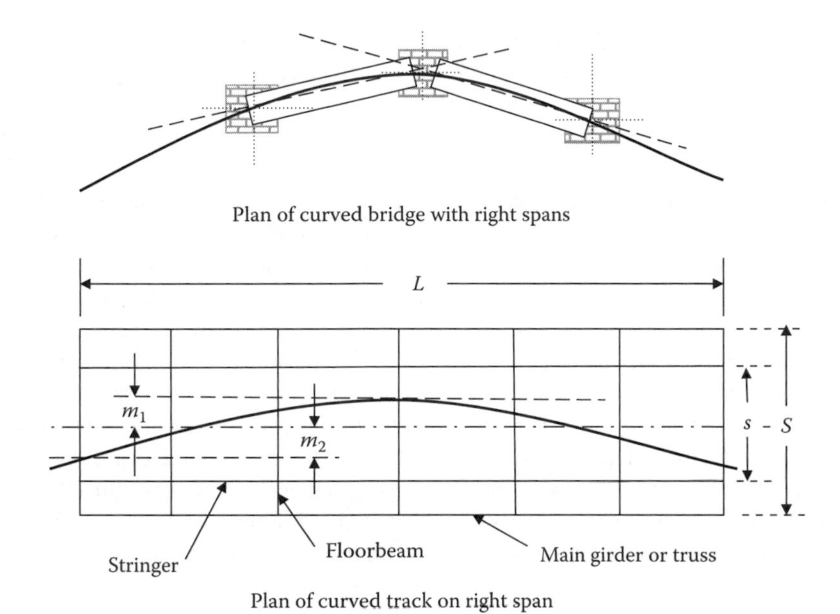

FIGURE 3.7 Horizontally curved bridges using right spans.

3.2.4.1.2 Bridge Geometrics

Track curvature can be accommodated by laying out bridges using right or curved spans. Right spans must be laid out on a chord to form the curved track alignment. The individual right spans must be designed for the resulting eccentricities* as the curved track traverses a square span (Figure 3.7). The stringer spacing, s, may be adjusted to equalize eccentricities in the center and end panels ($m_1 = m_2$); and in some cases, such as sharp curves, it may be economical to offset the stringers equally in each panel. However, fabrication effort and costs must be carefully considered prior to designing offset floor systems. It is common practice on freight railroads to use m_1 between $M/6$ and $M/2$,† where $M = m_1 + m_2$ is given by Equation 3.22.

The superelevated and curved track creates horizontal eccentricities based on the horizontal curve geometry (track curvature effect), e_c, and vertical superelevation (track shift effect), e_s. These eccentricities must be considered when determining the lateral distribution of live load forces (including dynamic effects) to members (stringers, floorbeams, and main girders or trusses). The shift effect eccentricity, e_s, is (Figure 3.6)

$$e_s = \frac{h_{cg}e}{d}. \quad (3.28)$$

* For example, through girder or truss spacing and design forces are increased due to eccentricity of the track (curvature effect) and superelevation (shift effect).

† An eccentricity, $m_1 = M/3$, is often used, which provides for equal shear at the ends of the longitudinal members.

Expressed as a percentage, this will affect the magnitude of forces to supporting members each side of the track in the following proportion:

$$\frac{2e_s}{s}(100), \tag{3.29}$$

where h_{cg} is the distance from the center of gravity of the rail car to the base of the track [AREMA (2008) recommends 8 ft for the distance from the center of gravity of the car to the top of the rail]; e is the superelevation of the track; d is the horizontal projection of track gage distance (generally taken as 4.9 ft); s is the width of the span (distance between the center of gravity of longitudinal members supporting the curved track).

The eccentricity due to track curvature, e_c, is determined by considering an equivalent uniform live load, W_{LL}, along the curved track across the square span length, L. The curvature effects on shear force and bending moment depend on the lateral shift of the curved track with respect to the centerline of the span and the degree of curvature. These effects are often negligible for short spans or shallow curvature (Waddell, 1916). However, if necessary, they can be determined in terms of the main member shear force and bending moment for the tangent track across the span as (Figure 3.7)

$$V_o = \frac{W_{LL}L}{4}\left(1 + \frac{2m_1}{S} - \frac{L^2}{12RS}\right), \tag{3.30}$$

$$V_i = \frac{W_{LL}L}{4}\left(1 - \frac{2m_1}{S} + \frac{L^2}{12RS}\right), \tag{3.31}$$

$$M_o = \frac{W_{LL}L^2}{16}\left(1 + \frac{2m_1}{S} - \frac{L^2}{24RS}\right), \tag{3.32}$$

$$M_i = \frac{W_{LL}L^2}{16}\left(1 - \frac{2m_1}{S} + \frac{L^2}{24RS}\right), \tag{3.33}$$

where V_o is the shear force on the outer girder, V_i is the shear force on the inner girder, M_o is the bending moment on the outer girder, M_i is the bending moment on the inner girder, W_{LL} is the equivalent uniform live load per track, L is the length of the span, S is the distance between the center of gravity of supporting longitudinal members, m_1 is the offset of the track centerline to the bridge centerline at the center of the span, and R is the radius of the track curve.

For the condition of equal shear (from Equations 3.22, 3.30, and 3.31) at the ends of girders, each side of the track

$$m_1 = \frac{L^2}{24R} = \frac{M}{3}. \tag{3.34}$$

For the condition of equal moment (from Equations 3.22, 3.32, and 3.33) at the centers of girders, each side of the track

$$m_1 = \frac{L^2}{48R} = \frac{M}{6}. \tag{3.35}$$

Example 3.1 outlines the calculation of the geometrical (shift and curvature) effects on vertical live load forces on right superstructures.

Example 3.1

A ballasted steel through plate girder railway bridge is to be designed with a 6° curvature track across its 70 ft span. The railroad has specified a 5 in. superelevation based on operating speeds and conditions. The track tie depth and rail height are taken as 7 in. each and the girders are spaced at 16 ft. Determine the geometrical effects of the curvature on the design live load shear and bending moment for each girder.

The effect of the offset of the live load center of gravity is (Equation 3.28) $e_S = [8 + (14/12)](5)/(4.9) = 9.35$ in. (from the centerline of the track).

The outside girder forces will be reduced by, and the inside girder forces increased by, $[2(9.35)/(16)(12)](100) = 9.74\%$ (Equation 3.29).

The curve mid-ordinate over a 70 ft span (Equation 3.22) is

$$M = \frac{(70)^2 6}{8(5730)} = 0.64\text{ ft} = 7.7\text{ in.}$$

Design for equal shear at the girder ends and use a track offset at the centerline of

$$m_1 = \frac{7.7}{3} = 2.56,\ \text{use } 2.5\text{ in.}$$

The effect of the curvature alignment (Equations 3.30 and 3.31) on girder shear forces is

$$\left(1 \pm \frac{2m_1}{S} \mp \frac{L^2}{12RS}\right) = \left(1 \pm \frac{2m_1}{S} \mp \frac{L^2 D}{12(5730)S}\right)$$

$$= \left(1 \pm \frac{2(2.5)}{(16)(12)} \mp \frac{(70)^2 6}{12(5730)16}\right) = (1 \pm 0.026 \mp 0.027) = 0.$$

The effect of the curvature alignment (Equations 3.32 and 3.33) on girder flexural forces is

$$\left(1 \pm \frac{2m_1}{S} \mp \frac{L^2}{24RS}\right) = \left(1 \pm \frac{2m_1}{S} \mp \frac{L^2 D}{24(5730)S}\right)$$

$$= \left(1 \pm \frac{2(2.5)}{(16)(12)} \mp \frac{(70)^2 6}{24(5730)16}\right) = (1 \pm 0.026 \mp 0.013).$$

The outside girder and inside girder forces are multiplied by $(1 + 0.026 - 0.013) = 1.013$ and $(1 - 0.026 + 0.013) = 0.987$, respectively, to account for the offset of curved track alignment.

Therefore, the following is determined for the shear, V_{LL+I}, and bending moment, M_{LL+I}, live load forces:

$$V_{out} = V_{LL+I}(1 - 0.097) = 0.903V_{LL+I},$$
$$V_{in} = V_{LL+I}(1 + 0.097) = 1.097V_{LL+I},$$
$$M_{out} = M_{LL+I}(1.013 - 0.097) = 0.916M_{LL+I},$$
$$M_{in} = M_{LL+I}(0.987 + 0.097) = 1.084M_{LL+I}.$$

It should be noted that these shear and bending moment forces do not include the effects of the centrifugal force. Section 4.3.2.3 and Example 4.10 in Chapter 4 outline the calculation for centrifugal forces.

The bridge deck must be superelevated as indicated in Section 3.2.4.1.1. The required superelevation is easily accommodated in ballasted deck bridges (Figure 3.8) but may also be developed in open deck bridges by tie dapping, shimming, or varying the elevation of the supporting superstructure or bearings (Figure 3.9). Varying the elevation of the deck supporting members may be problematic from a structural behavior, fabrication, and maintenance perspective and, generally, is not recommended.

Curved spans must be designed for flexural and torsional effects. Dynamic behavior under moving loads is particularly complex for curved girders as flexural and torsional vibrations may be coupled. Even the static design of curved girder railway bridges requires careful consideration of torsional and distortional* warping stresses and shear lag considerations. These analyses are complex and often carried out with general purpose finite element or grillage computer programs.† Curved girders are best suited to continuous span construction and, therefore, not often used for freight railroad bridges. Continuous construction is relatively rare for ordinary steel freight railway bridges due to remote location erection requirements (field splicing, falsework, and

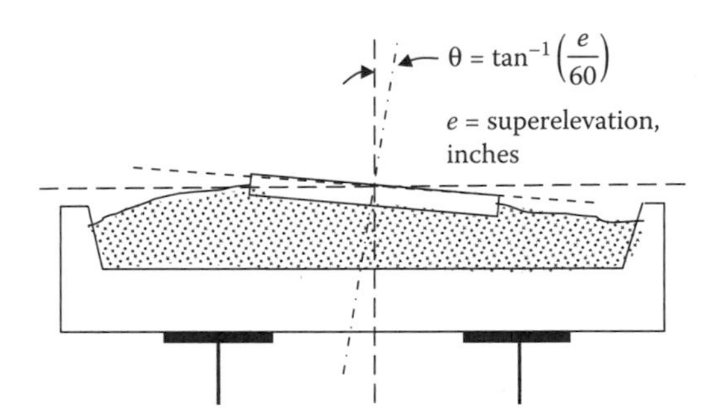

FIGURE 3.8 Track superelevation on ballasted deck bridges.

---

* In the case of box girders.

† The FHWA has also performed extensive research on steel curved girders at the Turner-Fairbanks laboratory. A synthesis of the research is available from the FHWA.

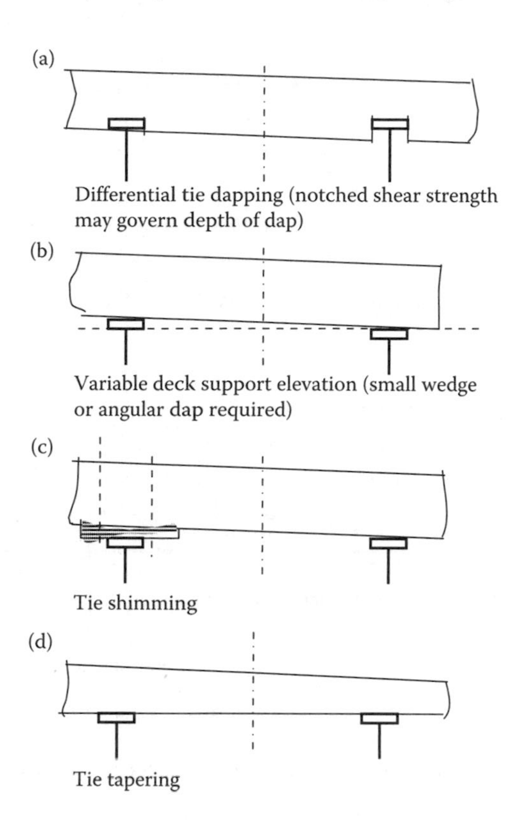

FIGURE 3.9 Track superelevation on open deck bridges.

large cranes) and to preclude uplift that may occur due to the large railway live load to superstructure dead load ratio. Nevertheless, curved girders are often effectively utilized for light transit applications.

3.2.4.1.3 Skewed Bridges

Skewed bridges have been considered as a necessary inconvenience to an abomination (Waddell, 1916) by bridge designers. There are many salient design and construction reasons for avoiding skewed bridge construction. Torsional moments and unequal distribution of live load occur with larger skew angles and compromise performance. Also, skewed spans generally require more material than square spans and include details that increase fabrication cost. However, on occasion, and particularly in congested urban areas or where large skew crossings exist, skewed construction may be unavoidable.

Many railroads have specific design requirements regarding skew angle and type of construction for skewed railway bridges. Skew connections and bent plates may be prohibited requiring that the track support at the ends of skewed spans be perpendicular to the track. This can be accommodated many ways depending on bearing and span types. Figure 3.10 shows examples of two types of floor systems in skewed spans over a pier.

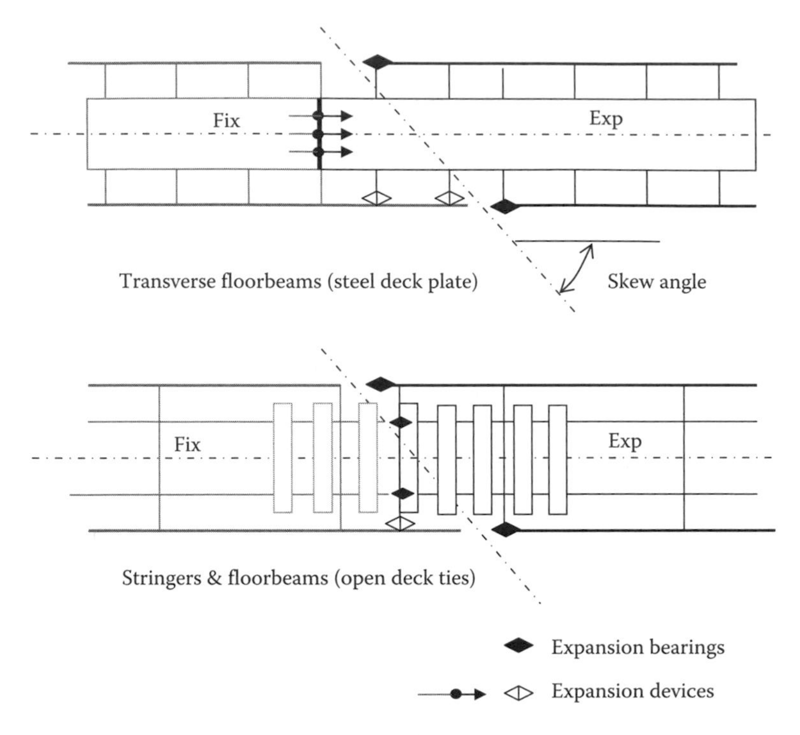

FIGURE 3.10 Square track support at skewed ends of steel railway spans over piers.

3.2.4.2 Vertical Geometry of the Bridge

In addition to a ground profile survey at the crossing, the dimensions generally required to develop a preliminary general arrangement of a railway bridge crossing are shown in Figure 3.11. These basic dimensions provide the information for preliminary design of the superstructure elements, where L is the length of the bridge and is given by $L = \sum_{i=1}^{\text{ns}} L_i$, where i is the span number, ns is the number of spans and L_i is the length of spans; W_i is the width of spans; H_i is the height (depth) of spans; H_{ci} is the construction depth of spans and is equal to H_i for deck type spans; H_w is the distance from the base-of-rail to water level; g is the grade of the bridge.

The length of spans is generally governed by site conditions, such as hydraulic or geotechnical considerations, or transportation corridor clearances (railroad, highway, or marine). Width is controlled by the number of tracks and the applicable railway company and regulatory clearances.

3.3 PRELIMINARY DESIGN OF STEEL RAILWAY BRIDGES

3.3.1 Bridge Aesthetics

Bridge aesthetics may be considered in terms of the structure itself and/or its integration into the environment. Bridge aesthetics is of particular importance in urban or accessible natural environments. Perception of beauty varies extensively among

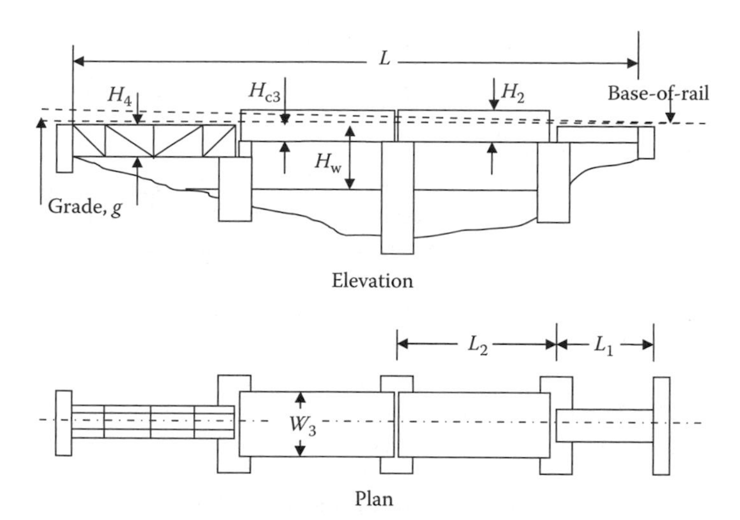

FIGURE 3.11 Basic dimensions of a railway bridge crossing.

persons. However, there are some basic tenets of aesthetic bridge design that appear to be generic in nature (Leonhardt, 1984; Billington, 1985; Taly, 1998; Bernard-Gely and Calgaro, 1994).

Harmony is often of primary importance to the public who generally desire bridges to integrate and be compatible with their environment.* Therefore, the bridge should be of materials and form suitable to achieve cultural and environmental congruence.

The bridge should also be expressive of function† and materials. In this manner the bridge will be a visual expression of the engineering mechanics and mathematics involved in achieving safety, efficiency, and economy. However, the economical proportioning of bridges does not necessarily produce aesthetic structures and other issues, in addition to harmony and expression of function, also warrant careful deliberation.

Proportion and scale are important. The dimensional relationships and relative size of components, elements, and/or parts of steel railway structures may affect public perception and support for the project. Slenderness, simplicity, and open space‡ often contribute toward attaining public acceptance of railway structures constructed in both urban and rural environments. Ornamentation that conceals function should generally be avoided.

The arrangement, rhythm, repetition, and order of members and/or parts of the structure are also essential deliberations of aesthetic bridge design. Light, shade, color, and surface treatments are further means of aesthetic improvement in structures within urban or accessible rural environments.

* The requirements related to environmental compatibility will vary depending upon whether the bridge is to be constructed in an urban or rural environment.

† Sullivan's famous "form follows function" statement on architecture applies well to bridges, which are often most aesthetically pleasing when designed primarily for economy and strength.

‡ Structures that look enclosed or cluttered are often unacceptable from an aesthetic perspective.

3.3.2 Steel Railway Bridge Superstructures

Railway bridges transmit loads to substructures through decks, superstructures, and bearings. The superstructure carries loads and forces with members that resist axial, shear, and/or flexural forces.

The steel superstructure forms typically used in freight railway bridge construction are beams, trusses, and arches. These superstructures have the rigidity required to safely and reliably carry modern heavy dynamic railroad live loads and the lightness required for transportation to, and erection at, remote locations. Beam, truss, and arch bridges can be constructed as deck or through structures depending on the geometry of the crossing and clearance requirements (Figure 3.12). Steel frame and suspended structures (i.e., suspension and cable-stayed bridges) are less common but sometimes used in lighter passenger rail bridge applications. Simple span construction is prevalent on North American freight railroads for performance,* rapid erection,† and maintenance considerations. AREMA (2008) recommends simple span types, based on length, for typical modern steel railway bridges as follows:

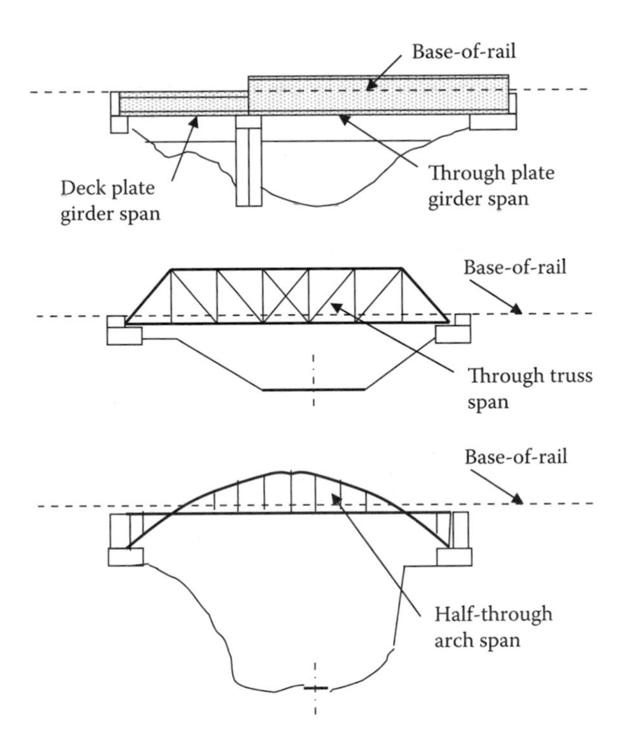

FIGURE 3.12 Basic forms of steel railway bridges—beams, trusses, and arches.

* The high railway live load to steel superstructure dead load ratio often precludes the use of continuous spans due to uplift considerations. AREMA (2008) recommends that dead load reactions exceed live load reactions by 50% to avoid uplift.

† Simple span construction is generally preferred by railroads due to relative ease of erection in comparison to continuous spans or spans requiring field splicing.

- Rolled or welded beams for spans up to about 50 ft in length (cover plates may increase strength to reduce span and/or depth of construction) (often used in floor systems of through plate girder and truss spans)
- Bolted or welded plate girders for spans between 50 and 150 ft
- Bolted or welded trusses for spans between 150 and 400 ft*

Steel freight railway bridge girder spans can be economically designed with a minimum depth to span ratio of about 1/15. Typically, depth to span ratios in the range of 1/10 to 1/12 are appropriate for modern short- and medium-span steel girder freight railway bridges. Beam, truss, and arch railway bridges can be constructed with open or closed (i.e., ballasted) decks.

3.3.2.1 Bridge Decks for Steel Railway Bridges

3.3.2.1.1 Railway Track on Bridge Decks

Rails with elastic fasteners seated on steel tie plates fastened to wooden ties or embedded in prestressed concrete ties are typical of modern North American railroad tracks. Steel ties have also been used and preclude the need for steel tie plates.† On ballasted deck bridges the wood, steel, or concrete ties are bedded in compacted granular rock ballast for drainage and track stability. Concrete ties may require damping devices, such as rubber pads applied to the bottom of ties with adhesive (Akhtar et al., 2006), on ballasted decks and are generally discouraged by railroads for use in open deck applications. Wooden ties are used in both open and closed or ballasted deck construction.

3.3.2.1.2 Open Bridge Decks

Open decks using wooden ties‡ are still used in many instances on modern railway bridges. Open deck bridges are often used in situations where new superstructures are being erected on existing substructures where it is necessary to reduce dead weight to preclude substructure overloading and foundation creep. On open bridge decks the ties are directly supported on steel structural elements (i.e., stringers, beams, girders) (Figures 3.13a and b). Dead load is relatively small, but dynamic amplification of live load may be increased because the track modulus is discontinuous. Bridge tie sizes can be large for supporting elements spaced far apart and careful consideration needs to be given to the deck fastening systems. Most railroads have open bridge deck standards based on the design criteria recommended by AREMA (2008, Chapter 7—Timber Structures).

Open bridge decks are often the least costly deck system and are free draining. However, they generally require more maintenance during the deck service life. Continuous welded rail (CWR) on long span bridges can create differential movements causing damage to, and skewing of, open bridge decks (see Chapter 4).

* Based on the upper limit of span length that AREMA (2008, Chapter 15) recommendations consider.

† Steel ties may also allow for a reduction in ballast depth on bridges.

‡ Steel, concrete, and composite ties may also be used but, due to their relatively large stiffness, may require a detailed analysis of structural behavior. Wooden ties may be used in accordance with the recommendations outlined in AREMA (2008, Chapter 7—Timber Structures).

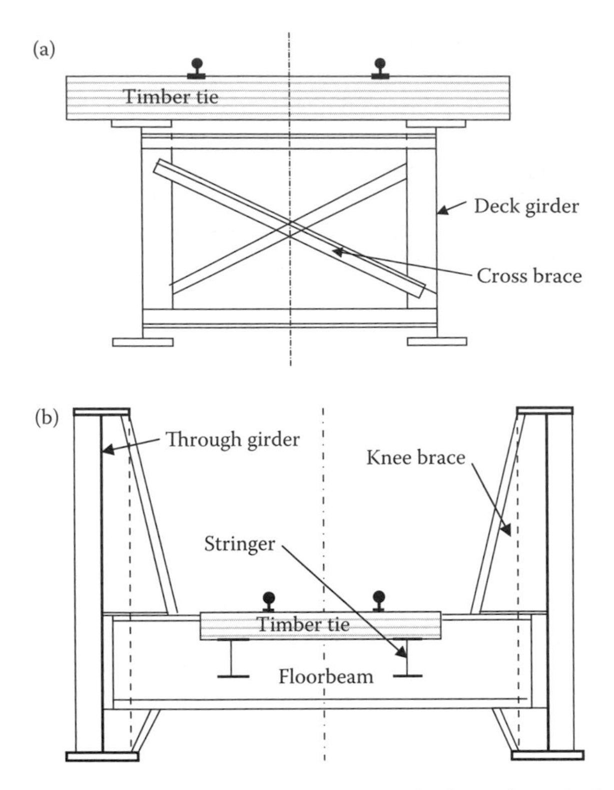

FIGURE 3.13 (a) Open deck plate girder (DPG) span. (b) Open through plate girder (TPG) span.

3.3.2.1.3 Ballasted Bridge Decks

Closed or ballasted steel plate and concrete slab deck bridges are common in new railway bridge construction. On ballasted or closed bridge decks, track ties are laid in stone ballast that is supported by steel or concrete decks (Figures 3.14a and b). The deck design may be composite or noncomposite. Dead load can be considerable, but dynamic effects are reduced and train ride quality is improved due to a relatively constant track modulus.

Steel plate decks are usually of isotropic design as orthotropic decks are often not economical for ordinary superstructures due to fabrication, welding, and fatigue design requirements. However, steel orthotropic plate modular deck construction is an effective means of rapid reconstruction of decks on existing steel railway bridges. Cast-in-place reinforced concrete and reinforced or prestressed precast concrete construction can also be used for deck slabs. Composite steel and concrete construction is structurally efficient (see Chapter 7), but may not be feasible due to site constraints (i.e., need for falsework and site concrete supply). Noncomposite precast concrete deck systems may be considered when site and installation time constraints exist in the particular railroad operating environment. However, precast concrete decks may be made composite with steel superstructures by casting recesses for shear connection

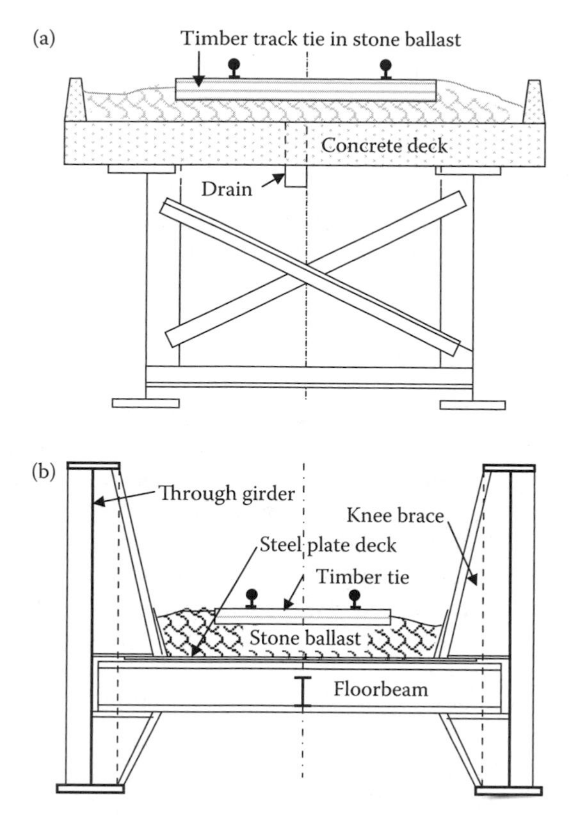

FIGURE 3.14 (a) Ballasted deck plate girder (BDPG) span. (b) Ballasted through plate girder (BTPG) span.

devices and grouting after installation. AREMA (2008) recommends a minimum deck thickness of 1/2 and 6 in. for steel plate and concrete slab decks, respectively.

Ballasted decks generally require less maintenance and are often used due to curved track geometry or when the bridge crosses over a roadway or sensitive waterway. Ballasted deck structures allow for easier track elevation changes, but drainage must be carefully considered. Drainage of the deck is often accomplished by sloping the deck surface to scuppers or through drains. In some cases, the through drains are connected to conduits to carry water to the ends of spans. In particular, deck drainage at the ends of spans using expansion plates under the ballast between decks must be carefully considered. Most railroads have standards for minimum ballast depth and waterproofing requirements. AREMA (2008, Chapter 8) contains information on recommended deck waterproofing systems.

3.3.2.1.4 Direct Fixation Decks

Tracks may be fastened directly to the deck or superstructure where live loads are light and dynamic forces effectively damped. Direct fixation decks are most often used in passenger rail service with rails firmly fastened to steel or concrete decks.

Dead load and structure depth are reduced, but dynamic forces can be large. Direct fixation decks are generally not used in freight railway bridges and require careful design and detailing (Sorgenfrei and Marianos, 2000).

3.3.2.2 Bridge Framing Details

Open deck through plate girder, through truss, and some deck truss spans have floor systems composed of longitudinal stringers and transverse floorbeams. Ballasted deck through plate girder spans generally have the concrete or steel plate decks supported on closely spaced transverse floorbeams framing into the main girder or truss (Figure 3.14b). In some cases, stringers with less closely spaced transverse floorbeams are used.*

Stringers should be placed parallel to the longitudinal axis of the bridge, and transverse floorbeams should be perpendicular to main girders or trusses. Stringers are usually framed into the floorbeams and have intermediate cross frames or diaphragms. Floorbeams should frame into the main girders or trusses such that lateral bracing may be connected to both the floorbeam and main member. The end connections of stringers and floorbeams should generally be made with two angles† designed to ensure flexibility of the connection in accordance with the structural analysis used. Due to cyclical live load stresses, welded end connections should not be used on the flexing leg of connections (see Chapter 9). Freight railway bridge spans should have end floorbeams, or other members, designed to permit lifting and jacking of the superstructure without producing stresses in excess of 50% of the basic allowable stresses (see Chapter 4). Multiple beams, girders, and stringers should be arranged to equally distribute live load to all the members.

Redundancy, particularly for FCM, is an important consideration in modern steel railway bridge design. Although costly from a fabrication perspective, internal redundancy can be achieved by the use of bolted built-up members. Structural redundancy can be achieved through establishing alternate load paths with additional members. For example, an open deck steel truss designer may decide to use two rolled beam stringers per rail instead of one stringer per rail. The nonredundant system of a single stringer per rail will require higher material toughness, and more stringent welding procedures and inspection if a built-up member is required.

3.3.2.3 Bridge Bearings

Freight railway spans that are 50 ft or greater in length should have fixed and expansion bearings that accommodate rotation due to live load and other span deflections.‡ All spans should also have provision for expansion to accommodate horizontal

---

* For example, when ballasted decks are used on through truss spans.

† Where floorbeams frame into girder webs at transverse web stiffener locations, it is often acceptable to use an angle connection on one side of the floorbeam and on the other side to directly connect the floorbeam web to the outstanding leg or plate of the girder web stiffener. This requires careful coping (or blocking) of the top and bottom flanges of the floorbeam.

‡ For example, rotations due to bridge skew, curvature, camber, construction misalignments and loads, support settlements, and thermal effects.

movements due to temperature or other longitudinal effects.* In addition to these translations and rotations, the bearings must also transmit vertical, lateral horizontal, and in the case of fixed bearings, longitudinal horizontal forces. Vertical forces are transferred through bearing plates directly to the substructure. Uplift forces may exist that require anchor bolts and many designers consider a nominal uplift force for the design of bearings, in any case. Horizontal forces are usually resisted by guide or key arrangements in bearing elements that transmit the horizontal forces to the substructure through anchor bolts.

Elastomeric bearings may be used at the ends of short spans of usual form† to accommodate expansion and rotation. However, for longer spans end rotation is permitted using spherical discs, curved bearing plates, or hinges, and expansion is enabled by low-friction sliding plates, rockers, or roller devices. Multirotational bearings may be required for long, skewed, curved, complex framed, and/or multiple track bridges or for those bridges where substructure settlement may occur. Constrained elastomeric (pot) bearings have been used with success in many applications. However, they are not recommended for steel railway superstructure support due to experience with bearing component damage from the high-magnitude cyclical railway live loads.

Typical fixed bearing components used on North American freight railroad steel bridges that transmit vertical and horizontal forces while allowing for rotation between superstructure and substructure are

- Flat steel plates—this type of bearing component has limited application due to inability for rotation and should not be used in spans >50 ft and in any span without careful deliberation concerning long-term performance.
- Disc bearings—this spherical segment bearing component allows rotation in any direction (e.g., longitudinal rotations combined with horizontal rotations due to skew and/or radial rotations due to curvature).
- Fixed hinged bearings—this type of hinged bearing uses a pin and pedestal arrangement to resist vertical and horizontal forces and enable rotation at the pin.
- Elastomeric bearings—these plain or steel reinforced rubber, neoprene, or polyurethane bearing pads allow rotation through elastic compression of the elastomer. The design of elastomeric bearing pads is a balance between the required stiffness of the pad to carry vertical loads and that needed to allow rotation by elastic compression.

Typical expansion bearing components used on North American freight railroad steel bridges that transmit vertical forces while allowing for rotation and translation between superstructure and substructure are

- Flat steel plates—this type of bearing component has limited application due to a deficient ability for translation unless maintained with lubrication.

---

* For example, translations due to braking and traction forces, construction misalignments and loads, support settlements, and thermal effects (particularly concerning CWR as outlined in Chapter 4).

† For example, elastomeric bearings might not be appropriate for spans greater than about 50 ft or for heavily skewed bridges.

- Bronze, copper-alloy, or polytetrafluoroethylene (PTFE) flat, cylindrical, and spherical sliding plates—these bearing components enable translation on low-friction surfaces. Bronze and copper-alloy sliding elements are self-lubricating by providing graphite or other solid lubricants in multiple closely spaced trepanned recesses. PTFE sliding plates should mate with stainless steel or other corrosion-resistant surfaces and contain self-lubricating dimples containing a silicone grease lubricant.
- Roller bearings—these bearing elements allow translation through rotation of single or multiple cylindrical rollers.
- Linked bearings—this type of bearing uses a double pin and link arrangement between pedestals to allow for horizontal translation.
- Expansion hinged bearings—this type of hinged bearing uses a pin and rocker (segmental roller) arrangement with the pin allowing rotation and the rocker permitting translation.
- Elastomeric bearings—these plain or steel reinforced rubber, neoprene, or polyurethane bearing pads allow translation through shear deformation of the elastomer.

In addition to steel span expansion bearings, the bearings at the bases of columns in steel bents and viaduct towers should be designed to allow for expansion and contraction of the tower or bent bracing system.

There are many proprietary types of fixed and expansion bearings available to the steel railway bridge engineer. Most are similar, or combinations of, the basic elements described above (Stanton et al., 1999; Ramberger, 2002). Bearings of mixed element types are not recommended (e.g., elastomeric fixed bearings with PTFE sliding bearings). Detailed recommendations on types, design, and fabrication of fixed and expansion bearings for steel freight railway bridge spans are found in AREMA (2008, Chapter 15). Due to the large vertical cyclical loads and exposed environment of most railway bridges, bearing designs should generally produce simple, robust, and functional bearings that are readily replaced by jacking of the superstructure.

3.3.3 Bridge Stability

Girders and trusses should be spaced to prevent overturning instability created by wind and equipment-based lateral loads (centrifugal, wheel/rail interface, and train rocking). AREMA (2008) recommends that the spacing should be greater than 1/20 of the span length for through spans and greater than 1/15 of the span length for deck spans of freight railway bridges. The spacing between the center of pairs of beams, stringers, or girders should not be less than 6.5 ft.

3.3.4 Pedestrian Walkways

Most railroad companies have policies, based on Federal and State regulations, regarding walkway and guardrail requirements for bridges. Width of walkways is often prescribed by the railroad company, but should not be less than 2 ft. Guardrail height is generally prescribed as a minimum of 3.5 ft by regulatory authorities but greater heights might be required at some crossings. Handrails and posts consisting of tubular, pipe, or angle sections are often used for railway bridge guardrails where safety without the need to consider aesthetics is acceptable. The designer should consult with the railroad company and applicable regulations concerning specific safety appliances that are required.

3.3.5 General Design Criteria

In North America, elastic structural analysis is used for freight railway steel bridge design based on the allowable stress design (ASD) methods of AREMA (2008).* The AREMA (2008) design criteria outline has recommended practices relating to materials, type of construction, loads, strength, serviceability, and fatigue design of steel railway bridges.

Dynamic amplification of live load (commonly referred to as impact) is very large in freight railway structures (see Chapter 4). Serviceability criteria (vibrations, deflections) and fatigue are important aspects of steel railway bridge design. Railway equipment, such as long unit trains (some with up to 150 cars), can create a significant number of stress cycles on busy rail lines, particularly on bridge members with relatively small influence lines (see Chapter 5). Railroads may limit span deflections based on operating conditions. Welded connections and other fatigue-prone details should be avoided in the high-magnitude cyclical live load stress range regime of freight railroad bridges (see Chapters 5, 6, 7, and 9).

AREMA (2008) recommends a performance-based approach to seismic design. Steel freight railway bridges have performed well in seismic events due to the type of construction usually employed. Typically, steel freight railway bridges have relatively light superstructures, stiff substructures, large bridge seat dimensions, and substantial bracing and anchor bolts (used to resist the considerable longitudinal and lateral forces associated with train operations). In general, steel railway bridges have suffered little damage or displacement during many recent earthquake events (Byers, 2006).

3.3.6 Fabrication Considerations

The steel fabrication process commences with shop drawings produced by the fabricator from engineering design drawings. The approved shop drawings are then used for cutting, drilling, punching, bolting, bending, welding, surface finishing, and assembly processes in the shop. Tolerances from dimensions on engineering drawings concerning straightness, length, cross section, connection geometry, clearances, and surface

* Recommended practices for the design of railway bridges are developed and maintained by the AREMA. Recommended practices for the design of fixed railway bridges are outlined in Part 1—Design and those for the design of movable railway bridges are outlined in Part 6—Movable Bridges, in Chapter 15—Steel Structures, of the AREMA MRE. Chapter 15—Steel Structures provides detailed recommendations for the design of steel railway bridges for spans up to 400 ft in length, standard gage track (56.5″), and North American freight and passenger equipment at speeds up to 79 and 90 mph, respectively. The recommendations may be used for longer span bridges with supplemental requirements. Many railroad companies establish steel railway bridge design criteria based on, and incorporating portions of these, recommended practices.

contact must be adhered to during fabrication.* Design and shop drawings should indicate all FCM since these members require specific material (see Chapter 2) and fabrication considerations. Fabricators must make joints and connections with high-strength steel bolts in accordance with ASTM A325 or A490†—Standard Specifications for Structural Bolts or welds in accordance with ANSI/AASHTO/AWS D1.5—Bridge Welding Code. Steel freight railway bridges are generally designed with slip critical connections and pretensioning is required for bolt installation (see Chapter 9). Bolts should be installed with a minimum tension‡ by turn-of-nut, tension-control bolt, direct-tension-indicator, or calibrated wrench tensioning.

Welding procedures, preparation, workmanship, qualification, and inspection requirements for steel railway bridges (dynamically loaded structures) should conform to the ANSI/AASHTO/AWS D1.5—Bridge Welding Code. In particular, for FCM, additional provisions concerning welding processes, procedures, and inspection merit careful attention during fabrication.§ Welding procedures typically used for steel railway bridges are SMAW, SAW, and flux cored arc welding (FCAW) (see Chapter 9). Railroad companies often prescribe limitations concerning acceptable welding procedures for superstructure fabrication.

Steel railway bridge fabrication, particularly for FCM, should be accompanied by testing of materials, fastenings, and welding. Material mill certifications should be reviewed to confirm material properties such as ductility, strength, fracture toughness, corrosion resistance, and weldability. Bolted joints and connections should be inspected by turn, tension, and torque tests to substantiate adequate joint strength. Quality assurance inspection of welding procedures, equipment, welder qualification, and nondestructive testing (NDT) is also required to validate the fabrication. NDT of welds is performed by magnetic particle testing (MPT), ultrasonic testing (UT), and/or radiographic testing (RT) by qualified personnel. Railroad companies often have specific criteria regarding the testing of fillet, complete joint penetration (CJP), or partial joint penetration (PJP) welds.

Steel bridges fabricated with modern atmospheric corrosion resistant (weathering) steels are often not coated, with the exception of specific areas that may be galvanized, metallized, and/or painted for localized corrosion protection.** Where required, modern multiple coat painting systems are used for steel railway bridge protection and many railroads have developed their own cleaning and painting guidelines or specifications. Many modern steel railway bridges are protected with a three-coat system consisting of a zinc rich primer, epoxy intermediate coat, and polyurethane topcoat.

Recommendations related to the fabrication of steel freight railway bridges are included in AREMA (2008).†† Engineers should consult with experienced fabricators

---

* These tolerances are outlined in AREMA (2008, Chapter 15, Part 3).

† ASTM A490 bolts are sometimes discouraged or prohibited by bridge owners due to brittleness concerns.

‡ For example, AREMA (2008) recommends a minimum tension force of 39,000 lb for 7/8 in diameter A325 bolts.

§ FCPs are usually specified to ensure that FCM fabrication is performed in accordance with the additional requirements indicated by AREMA (2008) and ANSI/AASHTO/AWS D1.5.

** For example, at bearing areas or top flanges of open deck spans.

†† Recommended practices for the fabrication of steel railway bridges are outlined in AREMA (2008, Chapter 15, Part 3).

early in the design process concerning rolled section and plate size availability from steel mills or supply companies. After the completion of fabrication, marking, loading, and shipping is arranged. Large steel railway spans may also require experienced engineers to provide loading and shipping arrangements that ensure safety of the fabricated structure, railway operations, and public.

3.3.7 Erection Considerations

Erection of steel railway bridges may be by steel fabricator, general contractor, or railroad construction forces. Steel span erection procedures and drawings should be in conformance with the engineering design drawings, specifications, special provisions, shop drawings, camber diagrams, match marking diagrams, fastener material bills, and all other information concerning erection planning requirements. Erection procedures should always be made with due consideration of safety and transportation interruption. The stresses due to erection loads in members and connections may exceed usual allowable stresses by 25% in steel freight railway bridges. This may be increased to 33% greater than usual allowable stresses for load combinations including erection and wind loads (see Chapter 4). Field joints should be made in order to connect the members without exceeding the calculated erection stresses until the complete connection is made. Recommendations related to the erection of steel freight railway bridges are also included in AREMA (2008).*

3.3.8 Detailed Design of the Bridge

Detailed design may proceed following all deliberations related to the railroad operating environment, site conditions, geometrics, aesthetics, superstructure type, deck type, preliminary design of framing systems, fabrication, and erection are completed to an appropriate level.† Detailed design will proceed from load development through structural analysis and design of members and connections to prepare structural steel design drawings and specifications for fabrication and erection of the superstructure (see Chapters 4 through 9).

REFERENCES

Akhtar, M.N., Otter, D., and Doe, B., 2006, *Stress-State Reduction in Concrete Bridges Using Under-Tie Rubber Pads and Wood Ties*, Transportation Technology Center, Inc. (TTCI), Association of American Railroads (AAR), Pueblo, Colorado.

American Railway Engineering and Maintenance-of-way Association (AREMA), 2008, *Manual for Railway Engineering*, Lanham, Maryland.

American Society of Testing and Materials (ASTM), 2000, *Standards A325 and A490*, 2000 Annual Book of ASTM Standards, West Conshohocken, PA.

American Welding Society (AWS), 2005, *Bridge Welding Code*, ANSI/AASHTO/AWS D1.5, Miami, FL.

* Recommended practice for the erection of steel railway bridges is outlined in AREMA Chapter 15, Part 4.

† The level of planning and preliminary design effort is related to the scope, magnitude, and complexity of the proposed bridge.

Bernard-Gély, A. and Calgaro, J.-A., 1994, *Conception des Ponts*, Presses de l'Ecole Nationale des Ponts et Chaussées, Paris, France.

Billington, D.P., 1985, *The Tower and the Bridge, The New Art of Structural Engineering*, Princeton University Press, Princeton, New Jersey.

Byers, W.G., 2006, Railway Bridge Performance in Earthquakes that Damaged Railroads, *Proceedings of the 7th International Conference on Short and Medium Span Bridges, Canadian Society for Civil Engineering* (CSCE), Montreal, QC.

Byers, W.G., 2009, Overview of bridges and structures for heavy haul operations, in *International Heavy Haul Association (IHHA) Best Practices*, Chapter 7, R. & F. Scott, North Richland Hills, TX.

Chow, V.T., 1959, *Open Channel Hydraulics*, McGraw-Hill, New York.

Federal Highway Administration (FHWA), 1990, *User Manual for WSPRO*, Publication IP-89-027, Washington, DC.

Federal Railroad Administration (FRA), 2008, *Railroad Bridge Working Group Report to the Railroad Safety Advisory Committee*, Washington, DC.

Hamill, L., 1999, *Bridge Hydraulics*, E & FN Spon, London, UK.

Hay, W.W., 1977, *Introduction to Transportation Engineering*, Wiley, New York.

Laursen, E.M., 1962, *Scour at Bridge Crossings*, *Transactions of the ASCE*, 127, Part 1, New York.

Laursen, E.M., 1963, *Analysis of Relief Bridge Scour*, *Journal of Hydraulics Division, ASCE*, 89(HY3), New York.

Leonhardt, F., 1984, *Bridges*, MIT Press, Cambridge, MA.

Nagler, F.A., 1918, Obstruction of bridge piers to the flow of water, *Transactions of American Society of Civil Engineers (ASCE)*, 82.

Ramberger, G., 2002, *Structural Bearings and Expansion Joints for Bridges*, Structural Engineering Document 6, International Association for Bridge and Structural Engineering (IABSE), Zurich, Switzerland.

Richardson, E.V. and Davis, S.R., 1995, *Evaluating Scour at Bridges*, Hydraulic Engineering Circular 18, FHWA, McLean, VA.

Schneider, V.R., Board , J.W., Colson, B.E., Lee, F.N., and Druffel, L., 1977, *Computation of back-water and discharge at width constrictions of heavily vegetated floodplains*, Water Resources Investigations 76-129, U.S. Geological Survey.

Sorgenfrei, D.F. and Marianos, W.N., 2000, Railroad bridges, in *Bridge Engineering Handbook*, Chapter 23, W.F. Chen and L. Duan, Eds, CRC Press, Boca Raton, FL.

Stanton, J.F., Roeder, C.W., and Campbell, T.I., 1999, *High-Load Multi-Rotational Bridge Bearings*, NCHRP Report 432, National Academy Press, Washington, DC.

Taly, N., 1998, *Design of Modern Highway Bridges,* McGraw-Hill, New York.

Transportation Association of Canada (TAC), 2004, *Guide to Bridge Hydraulics*, Thomas Telford, London, UK.

Unsworth, J.F., 2003, *Heavy Axle Load Effects on Fatigue Life of Steel Bridges*, TRR 1825, Transportation Research Board, Washington, DC.

Unsworth, J.F. and Brown, C.H., 2006, *Rapid Replacement of a Movable Steel Railway Bridge*, TRR 1976, Transportation Research Board, Washington, DC.

Waddell, J.H., 1916, *Bridge Engineering,* Vol. 1, Wiley, New York.

Yarnell, D.L., 1934, *Bridge Piers as Channel Obstructions*, Technical Bulletin 442, U.S. Department of Agriculture, Washington, DC.

4 Loads and Forces on Steel Railway Bridges

4.1 INTRODUCTION

The loads and forces on steel railway bridge superstructures are gravity, longitudinal, or lateral in nature.

Gravity loads, comprising dead, live, and impact loads, are the principal loads to be considered for steel railway bridge design. Live load impact (dynamic effect) is included due to the relatively rapid application of railway live loads. However, longitudinal forces (from live load or thermal forces) and lateral forces (from live load, wind forces, centrifugal forces, or seismic activity) also warrant careful consideration in steel railway bridge design. An excellent resource for the review of load effects on structures, in general, is ASCE (2005).

Railway bridges are subjected to specific forces related to railroad moving loads. These are live load impact from vertical and rocking effects, longitudinal forces from acceleration or deceleration of railroad equipment, lateral forces caused by irregularities at the wheel-to-rail interface (commonly referred to as "truck hunting" or "nosing"), and centrifugal forces due to track curvature.

4.2 DEAD LOADS

Superstructure dead load consists of the weight of the superstructure itself, track, deck (open or ballasted), utilities (conduits, pipes, and cables), walkways (some engineers also include walkway live load as a component of superstructure dead load), permanent formwork, snow, ice, and anticipated future dead loads (e.g., larger deck ties, increases in ballast depth, and additional utilities). However, snow and ice loads are generally excluded from consideration due to their relatively low magnitude. For ordinary steel railway bridges, dead load is often a small proportion of the total superstructure load (steel railway bridges typically have a relatively high live load to dead load ratio).

Curbs, parapets, and sidewalks may be poured after the deck slab in reinforced concrete construction. This superimposed dead load may be distributed according to superstructure geometry (e.g., by tributary widths). However, it is common practice to equally distribute superimposed dead loads to all members supporting the hardened deck slab. This is appropriate for most superstructure geometries, but may require

TABLE 4.1
Dead Loads on Steel Railway Bridges

Item	Dead Load
Track (rails and fastenings)	200 lb/ft
Steel	490 lb/ft^3
Reinforced and prestressed concrete	150 lb/ft^3
Plain (unreinforced) concrete	145 lb/ft^3
Timber	35–60 lb/ft^3
Sand and gravel, compacted (railroad ballast)	120 lb/ft^3
Sand and gravel, loose	100 lb/ft^3
Permanent formwork (including concrete in valleys)	15 lb/ft^2
Waterproofing on decks	10 lb/ft^2

Source: From American Railway Engineering and Maintenance-of-Way Association (AREMA), 2008, *Manual for Railway Engineering*, Chapter 15. Lanham, MD. With permission.

refinement in multibeam spans where exterior beams may be subjected to a greater proportion of any superimposed dead load.

At the commencement of design, dead load must be estimated from experience or review of similar superstructure designs. This estimated design load must be reviewed against the actual dead load calculated after final design of the superstructure. Small differences between the estimated and actual dead load are not important, provided the dead load is a reasonably small component of the total design load. Steel railway bridge engineers will often include an allowance of 10–15% of estimated steel superstructure weight to account for bolts, gusset plates, stiffeners, and other appurtenant steel components. Dead loads typically used for ordinary steel railway bridge design are shown in Table 4.1.

Temporary construction dead loads and the transfer of dead load during shored or unshored construction of steel and concrete composite deck spans should also be considered during design (see Chapter 7).

4.3 RAILWAY LIVE LOADS

Railroad locomotives and equipment (box and flat cars, commodity gondolas, and hopper and tank cars) vary greatly with respect to weight, number of axles, and axle spacing.

Modern freight locomotives have two three-axle sets with a spacing between axles of between 6.42 and 6.83 ft, and a spacing between axle sets (commonly refereed to as "truck spacing") of between 45.62 and 54.63 ft. These modern generation locomotives weigh up to 435,000 lb. There are, however, many four- and six-axle locomotives of weight between about 250,000 and 400,000 lb, and with lengths between 50 and 80 ft operating on the railroad infrastructure.

Axle spacing is typically 5–5.83 ft for North American four-axle freight car equipment. Truck spacing may vary from 17 to 66 ft (Dick, 2002). Gross car weights up to

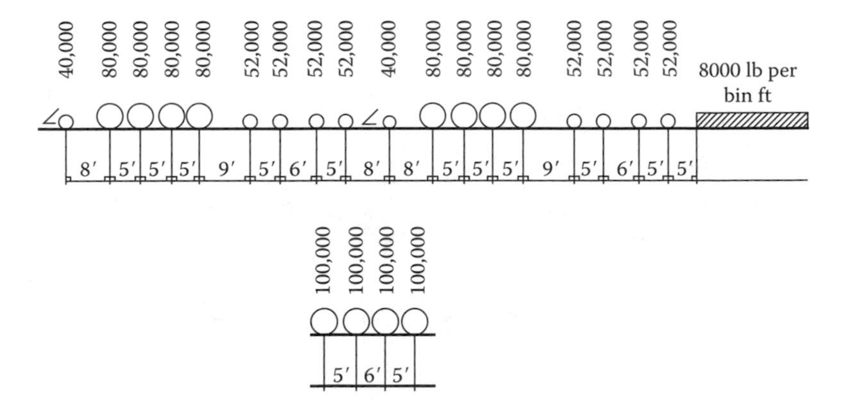

FIGURE 4.1 Cooper's E80 and alternate live load–axle loads. (From American Railway Engineering and Maintenance-of-Way Association (AREMA), 2009, *Manual for Railway Engineering*, Chapter 15, Lanham, MD. With permission.)

286,000 lb are common on North American railroads and some railroad lines carry 315,000 lb cars.

This variability in railroad equipment weight and geometry requires a representative live load model for design that provides a safe and reliable estimate of railroad operating equipment characteristics within the design life of the bridge.

4.3.1 Static Freight Train Live Load

The railway bridge design live load recommended in AREMA (2008) is Cooper's E80 load. This design load is based on two Consolidation-type steam locomotives with trailing cars represented by a uniformly distributed load (Cooper, 1894). The maximum locomotive axle load is 80,000 lb and freight equipment is represented by a uniform load of 8000 lb per ft of track. An alternate live load, consisting of four 100,000 lb axles, is also recommended in order to represent the stress range effects of adjacent heavy rail cars on short spans. These design live loads are shown in Figure 4.1.

This design load appears antiquated, particularly with respect to the use of steam locomotive geometry. However, it is a good representation of the load effects of modern freight traffic as illustrated in Figure 4.2. Figure 4.2 is plotted from a moving load analysis (see Chapter 5) of shear and flexure* on simple spans for continuous and uniform strings of various heavy freight equipment vehicles.† The unbalanced loads (indicated as UB with 25% of the total railcar load shifted to adjacent axle sets) for the car weight and configurations investigated in Figure 4.2 exceed the Cooper's

* It should be noted that the Equivalent Cooper's E loads shown in Figure 4.2 are for mid-span flexure. Effects may be even more severe at or near span quarter points for typical railway freight loads (Dick, 2006).

† In this case, six-axle 432,000 lb locomotives, and four-axle 315,000 lb freight cars with balanced and unbalanced (UB) loads.

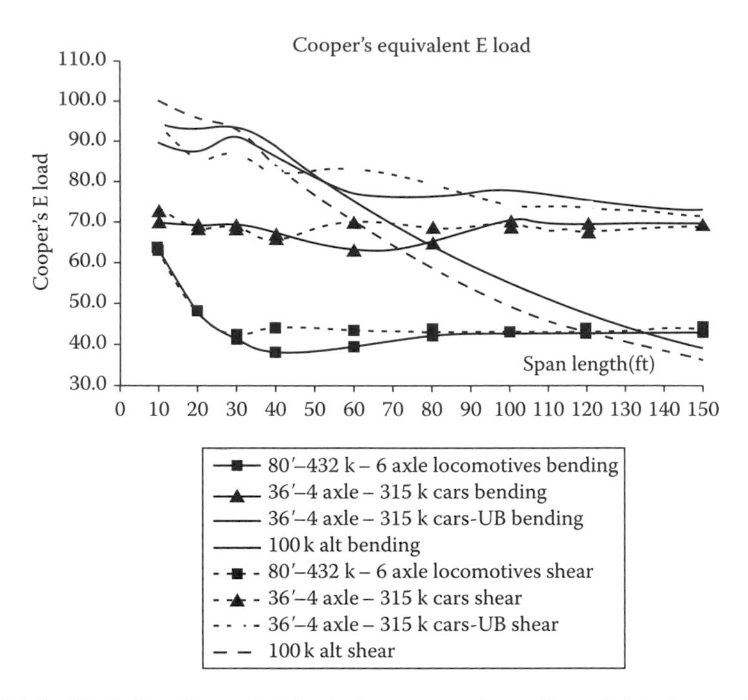

FIGURE 4.2 Equivalent Cooper's E loads for some modern railway freight locomotives and equipment on simply supported bridge spans up to 150 ft in length.

E80 design load for spans less than 80 ft in length.* Figure 4.2 also indicates that the alternate live load is appropriate for short-span design where the effects of short heavy axle cars with unbalanced loads can be considerable. The alternate live load governs superstructure design for simply supported spans up to about 55 ft in length. On short simply supported spans less than 30 ft in length, the alternate live load is equivalent to about Cooper's E94 in flexure and has even greater effects in shear.

Some railroad companies may vary from the Cooper's E80 bridge design load based on their operating practice. It is usual that the magnitude of the axle loads is changed (e.g., 72 or 90 kips), but the axle spacing is unaltered. Therefore, different bridge designs can be readily compared.

However, the flexural cyclical stress ranges created by Cooper's E80 design load do not necessarily accurately reflect the cyclical stress ranges created by modern railway freight equipment. Figures 4.3a and b show the variation in mid-span bending moment when 25 and 60 ft simply supported spans, respectively, are traversed by various train configurations. The Cooper's E80 live loads appear to be conservative representations of the design stress range magnitude for both the 25 and 60 ft spans.†

* These cars are very short and heavy, and not typical of those routinely used on North American railroads. They are, however, representative of equipment currently used on some specific routes and, in terms of weight, the potential direction for future freight equipment.

† This is accounted for with adjustments to the number of equivalent constant amplitude cycles based on the ratio of typical train loads to Cooper's E80 load (see Chapter 5).

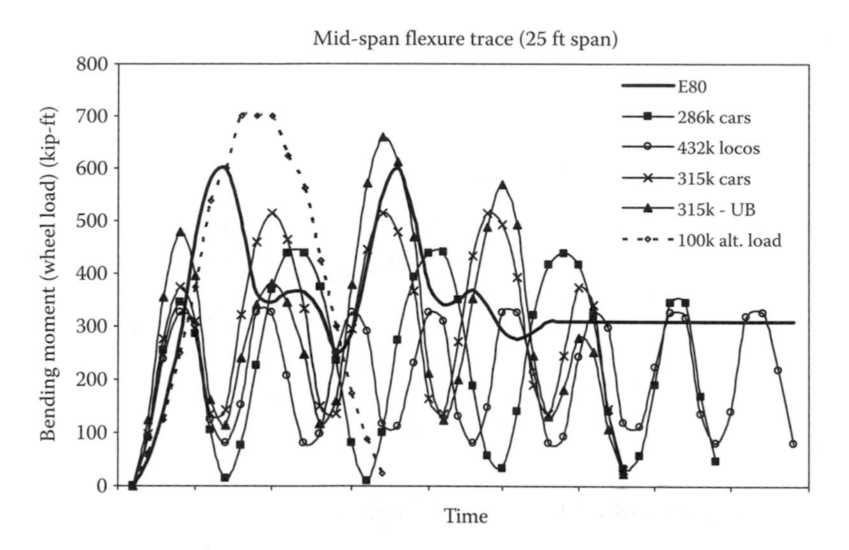

FIGURE 4.3a Mid-span bending moments for Cooper's E80 load, and modern railway freight locomotives and cars, on a 25 ft simply supported bridge span.

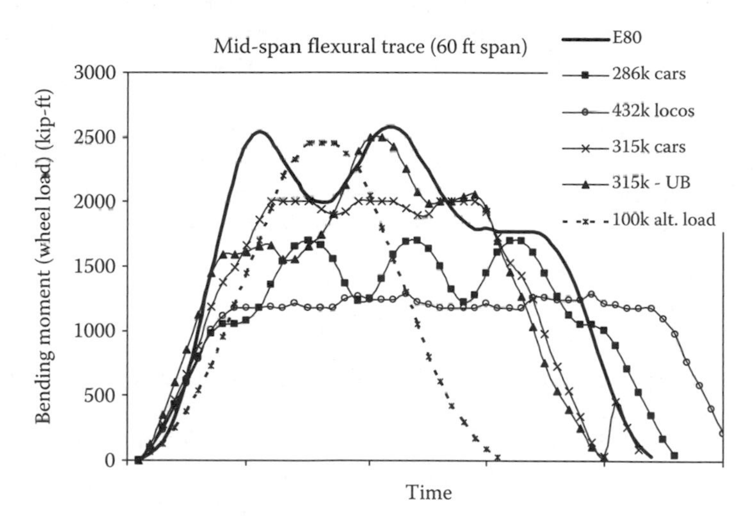

FIGURE 4.3b Mid-span bending moments for Cooper's E80 load, and modern railway freight locomotives and cars, on a 60 ft simply supported bridge span.

The alternate live load appears to better represent the cyclical behavior of the various train configurations, particularly on short spans. Therefore, the allowable fatigue stress ranges, S_{Rfat}, recommended for design in AREMA (2008) are based on equivalent constant amplitude stress cycles from projected variable amplitude stress cycles due to typical railroad traffic.* In order to establish the appropriate number of stress cycles

* Alternatively, the number of equivalent constant amplitude cycles could be developed for a loaded length traversed by a recommended fatigue design load vehicle.

for design, the projected variable amplitude stress cycles are developed for an 80 year life with the number of stress cycles based on the length of member influence lines (see Chapter 5).

4.3.2 Dynamic Freight Train Live Load

A train traversing a railway bridge creates actions in longitudinal, lateral, and vertical directions. Longitudinal forces and pitching rotations (rotations around an axis perpendicular to the longitudinal axis of the bridge) are caused by applied train braking and traction forces. Lateral forces are caused by wheel and truck yawing or "hunting." Lateral centrifugal forces are also created on curved track bridges. Rocking (rotations around an axis parallel to the longitudinal axis of the bridge) and vertical dynamic forces are created by structure–track–vehicle conditions and interactions.

4.3.2.1 Rocking and Vertical Dynamic Forces

Lateral rocking of moving vehicles will provide amplification of vertical wheel loads. This amplification will increase stresses in members supporting the track, and AREMA (2008) includes this load effect as a component of the impact load. Superstructure–vehicle interaction also creates a vertical dynamic amplification of the moving loads. This dynamic amplification results in vibrations that also increase stresses in members supporting the track.

The unloaded simply supported beam fundamental frequency, ω_1, of Equation 4.1 provides a basic indicator of superstructure vertical dynamic response, and can be used to establish superstructure stiffness requirements for this serviceability criterion. The fundamental frequency of free vibration of an unloaded simply supported beam is

$$\omega_1 = \frac{\pi^2}{L^2}\sqrt{\frac{EI}{m}}, \tag{4.1}$$

where L is the span length, EI is the flexural rigidity of the span, and m is the mass per unit length of the span.

The mathematical determination of a dynamic load allowance, or impact load (Equation 4.2), even for simply supported steel railway bridge superstructures is complex. AREMA (2008) provides an empirical impact factor based only on length (which appears reasonable based on Equation 4.1) in order to provide deterministic values for vertical impact design purposes. The dynamic load effect is

$$\mathrm{LE_D} = I_\mathrm{F}[\mathrm{LE_S}], \tag{4.2}$$

where $\mathrm{LE_D}$ is the dynamic load effect and is equal to impact load (or dynamic response for a linear elastic system), $\mathrm{LE_S}$ is the maximum static load effect (or maximum static response for a linear elastic system), and I_F is the impact factor for the simply supported bridge span.

Therefore, for steel railway bridges, the impact factor comprises the effects due to vehicle rocking, RE, and the vertical effects due to superstructure–vehicle interaction, I_V, or

$$I_F = RE + I_V. \tag{4.3}$$

4.3.2.1.1 Rocking Effects

Railroad freight equipment will rock or sway in a lateral direction due to wind forces, rail profile variances, and equipment spring stiffness differences. Rocking due to rail and equipment conditions will affect the magnitude of equipment axle loads and is considered as a dynamic increment of axle load by AREMA (2008). Rocking effects are independent of train speed (AREA, 1949; Ruble, 1955).

The rocking effect, RE, is determined for each member supporting the track as a percentage of the vertical live load. The applied rocking effect, as recommended in AREMA (2008), is the force couple created by an upward force on one rail and downward force on the other rail of 20% of the design wheel load, or $0.20W$, where W is the wheel load (1/2 of axle load). The calculation of RE for an open deck multibeam deck span is shown in Examples 4.1 and 4.2.

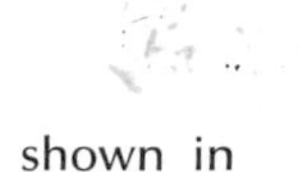

Example 4.1

A double track open deck steel multibeam railway bridge is shown in Figure E4.1. Determine the rocking effect, RE, component of the AREMA impact load.

The applied rocking force is a force couple, $R_A = 0.20W(5.0) = W$, as shown in Figure E4.2. If the vertical live load is equally distributed to three longitudinal beams (AREMA allows this provided that beams are equally spaced and adequately laterally braced), the applied rocking forces are resisted by a force couple with an arm equal to the distance between the centers of resisting members each side of the track centerline.

The resisting force couple (Figure E4.2) is $R_R = F_R\ (6.00)$. Since $R_A = R_R$, $F_R = 0.167(W)$ and the rocking effect, RE, expressed as a percentage of vertical live load, W, is $RE = [F_R(100)/W] = 16.7\%$.

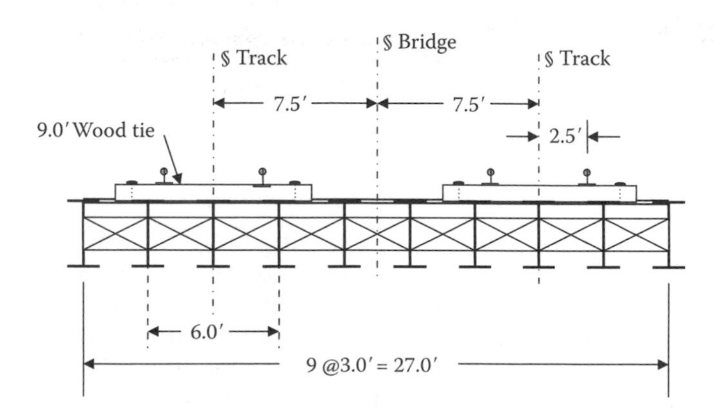

FIGURE E4.1

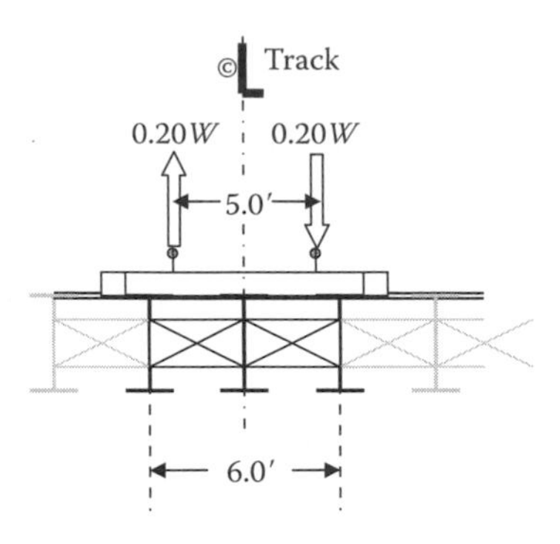

FIGURE E4.2

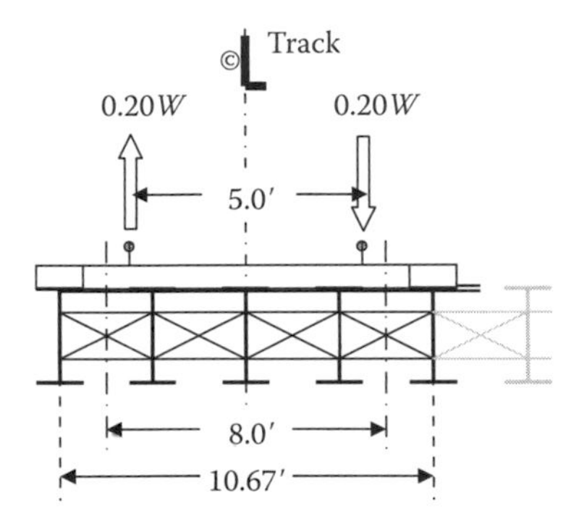

FIGURE E4.3

Example 4.2

A double track open deck steel multibeam railway bridge is similar to that shown in Figure E4.1. Determine the rocking effect, RE, component of the AREMA impact load if the beams are spaced at 2 ft-8 in centers.

The resisting force couple (Figure E4.3) is $R_R = F_R$ (8.00). Since $R_A = R_R$, $F_R = 0.125(W)$ and the rocking effect, RE, expressed as a percentage of vertical live load, W, is RE $= [F_R(100)/W] = 12.5\%$.

4.3.2.1.2 Vertical Effects on Simply Supported Spans

Superstructure vibration is induced by the moving load (locomotives and cars) suspension systems as the loads traverse a railway bridge with surface irregularities. When the length of the bridge span is large relative to the locomotive and car axle spacing, dynamic locomotive and car loads may be approximated by a concentrated and continuous moving load, respectively. When the span length is not large relative to the locomotive and car axle spacing, dynamic train loads may be approximated by concentrated moving loads.

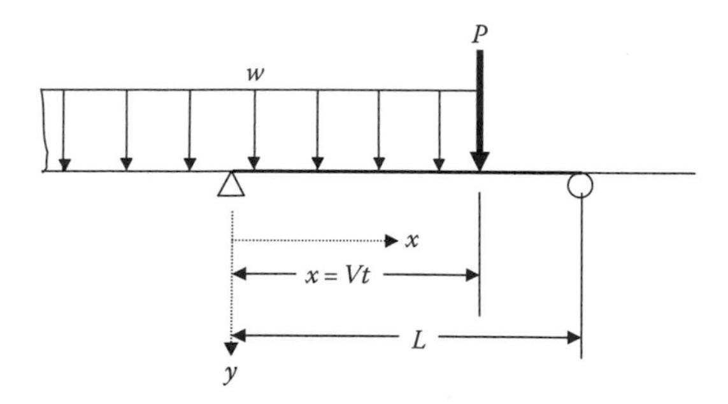

FIGURE 4.4 Moving train model on a simple beam.

Therefore, for short spans,* the forced vibration effects from freight equipment can be determined by superposition of the effects of concentrated moving loads. It is generally necessary to consider two cases for long-span bridge design:

1. The locomotives and trailing train on the bridge (Figure 4.4).
2. The train only on the bridge (both completely loaded and partially loaded).

Case 1 causes the greatest dynamic effects on long spans and can be represented by superposition of the effects of a moving constant concentrated force, P, and constant continuous load, w. Accordingly, because they are of practical importance in understanding the dynamics of steel railway bridges, and may be superimposed in linear elastic analysis, three loading cases will be discussed:

- A moving uniform continuous load
- A moving constant magnitude concentrated load
- A moving harmonically varying concentrated load that might represent steam locomotive loads on short and medium bridge spans.

The vertical dynamic effect, I_V, at a location, x, in a span at time, t, can be determined as the ratio of the dynamic displacement, $\bar{y}(x,t)$, to the static displacement, $y(x,t)$. The partial differential equation of motion† can be established for a simply supported span of constant mass and stiffness from force equilibrium as‡

$$EI\frac{\partial^4\bar{y}(x,t)}{\partial x^4} + m\frac{\partial^2\bar{y}(x,t)}{\partial t^2} + c\frac{\partial\bar{y}(x,t)}{\partial t} = p(x,t), \tag{4.4}$$

where $\bar{y}(x,t)$ is the superstructure vertical dynamic deflection at distance x and time t, EI is the flexural stiffness of the superstructure, I is the vertical moment of inertia

* Assuming linear elastic behavior.

† Assuming small deformations, Hooke's law, Navier's hypothesis, and the St. Venant principle apply. Also, this equation assumes that the internal (strain velocity) damping is negligible in comparison to the external (transverse velocity) damping of the steel beam.

‡ Also known as the Bernoulli–Euler equation.

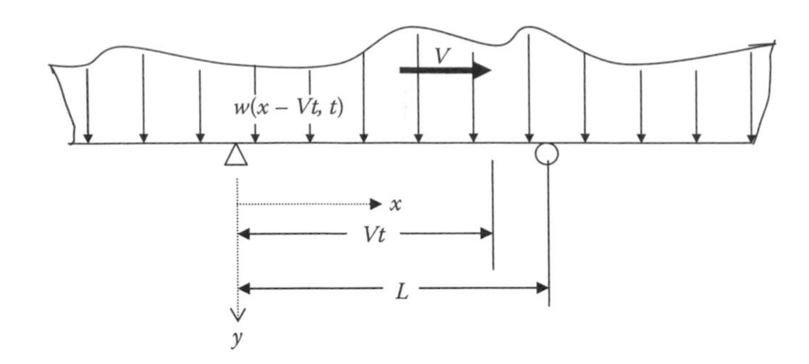

FIGURE 4.5 Moving continuous load on a simple beam.

(with respect to the horizontal axis), m is the mass of the superstructure per unit length, c is the equivalent coefficient of viscous damping of the superstructure and is equal to $2m\omega_c$, ω_c is the viscous damping frequency, and $p(x,t)$ is the dynamic load on the bridge at distance x and time t.

Mass develops inertial forces in direct proportion, and in opposite direction, to its acceleration in accordance with d'Alembert's principle. These inertial forces must be included in the analysis of relatively light steel railway bridges* traversed by large locomotive and train weights.

For a moving continuous load (Figure 4.5), the load on the bridge may be expressed as

$$p(x,t) = w(\xi,t) - m_{\mathrm{w}}(\xi)\frac{d^2\overline{y}(x,t)}{\mathrm{d}t^2}, \tag{4.5}$$

where $w(\xi,t)$ is the magnitude of uniform load at distance $\xi = x - Vt$ and time t; $m_w(\xi) = w(\xi,t)/g$ is the mass of uniform load at distance $\xi = x - Vt$ and time t; $V =$ is the constant velocity of load; and g is the acceleration due to gravity.

Since, due to the inertial effects of the stationary continuous mass, the load, $p(x,t)$, depends on the superstructure response, $\overline{y}(x,t)$, it is necessary to determine the derivative expression in Equation 4.5. The derivative, at $\xi = 0$ ($x = Vt$) with constant train velocity, V, can be expanded as (Fryba, 1996)

$$\frac{\mathrm{d}^2\overline{y}(Vt,t)}{\mathrm{d}t^2} = V^2\frac{\partial^2\overline{y}(Vt,t)}{\partial x^2} + 2V\frac{\partial^2\overline{y}(Vt,t)}{\partial x\partial t} + \frac{\partial^2\overline{y}(Vt,t)}{\partial t^2}. \tag{4.6}$$

For a uniform continuous moving load, $w(\xi, t) = w$, simply supported boundary conditions (common for steel railway bridges), initial conditions of zero displacement, and velocity, Equation 4.4 (with Equations 4.5 and 4.6) can be written as

$$EI\frac{\partial^4\overline{y}(x,t)}{\partial x^4} + m_{\mathrm{w}}V^2\frac{\partial^2\overline{y}(x,t)}{\partial x^2} + (m + m_{\mathrm{w}})\frac{\partial^2\overline{y}(x,t)}{\partial t^2} + c\frac{\partial\overline{y}(x,t)}{\partial t} = w \tag{4.7}$$

---

* Steel railway bridges typically have a very large live load to dead load ratio.

by neglecting the second term of Equation 4.6.* The solution of Equation 4.7 may be achieved by Fourier integral series transformation† as (Fryba, 1972)

$$\bar{y}(x,t) = \frac{5wL^4}{384EI} \sum_{i=1,3,5,\ldots}^{\infty} \frac{1}{i^5\left(1-(\alpha^2\bar{m}/i^2)\right)} \sin\frac{i\pi x}{L}, \tag{4.8}$$

where

$$\alpha = \frac{\pi V}{\omega_1 L} = \frac{LV}{\pi\sqrt{(EI/m)}}$$

and

$$\overline{m} = \frac{m_{\mathrm{w}}}{m}.$$

The case of the uniform continuous moving load only partially on the span is also of practical importance for long-span railway bridges. For uniform continuous loads arriving at (on the span over the distance $x = Vt$) and departing from (off the span over the distance $x = Vt$) lightly damped spans at low speeds, the first vibration mode ($i = 1$) is given by Equations 4.9a and 4.9b, respectively (Fryba, 1972).

$$\bar{y}(x,t) \approx \frac{5wL^4}{768EI}\left(1-\cos\frac{\pi Vt}{L}\right)\sin\frac{\pi x}{L}, \tag{4.9a}$$

$$\bar{y}(x,t) \approx \frac{5wL^4}{768EI}\left(1+\cos\frac{\pi Vt}{L}\right)\sin\frac{\pi x}{L}. \tag{4.9b}$$

However, in the development of Equations 4.8, 4.9a, and 4.9b, the dynamic effects of the uniform load (vehicle suspension system dynamics)‡ and the effects of surface irregularities are neglected. Surface irregularities are of considerable importance in the determination of railway live load dynamic effects (Byers, 1970).§ An advanced analytical and testing evaluation** of railway live load vertical dynamic effects, I_{V}, will provide an accurate assessment of impact in long-span railway bridges.

For a moving constant concentrated force (Figure 4.6)

$$p(x,t) = \delta(\xi)P, \tag{4.10}$$

---

* This assumes the superstructure is considered relatively torsionally stiff, which is generally the case for properly braced steel railway spans. AREMA (2008) provides recommendations for lateral bracing of steel railway spans. In many practical situations the effects of the second term and third term of Equation 4.6 may be neglected.

† The Fourier integral transform (Kreyszig, 1972) is $y(x,t) = \sum_{i=1}^{\infty} b_n \sin\frac{i\pi x}{L}$, where $b_n = \frac{2}{L}\int_0^L y(x,t)\sin\frac{i\pi x}{L}\,\mathrm{d}x$ and $i = 1, 2, 3, \ldots$.

‡ Long spans with a low natural frequency may not excite vehicle spring movements.

§ Surface irregularities such as flat wheels, rail joints, poor track geometry and even lesser aberrations such as rail undulation from bending between ties can excite vehicle and superstructure vibrations. Flat wheels and rail joints are of particular concern on short span bridges with high natural frequency.

** Advanced analysis by Finite Element Analysis (FEA) may be supplemented with experiments to determine dynamic response in long span and complex steel railway superstructures.

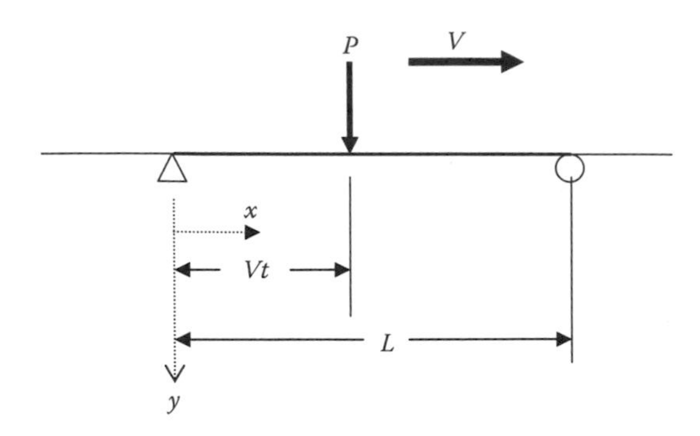

FIGURE 4.6 Moving concentrated load on a simple beam.

where $\delta(\xi)$ is the Dirac delta function, which mathematically describes a constant velocity unit concentrated force at $\xi = x - Vt$, considering the force $p(x, t)$ as a unit impulse force (Tse et al., 1978); P is the concentrated force and is equal to $F - m_{\mathrm{v}}(\mathrm{d}^2\bar{y}(Vt, t)/\mathrm{d}t^2)$ (d'Alembert's principle of inertial effects); $F = F(t) + m_{\mathrm{v}}g$, where $F(t)$ is the dynamic forces from concentrated moving load, such as forces from locomotive suspension system dynamics, m_{v} is the mass of the concentrated force, and g is the acceleration due to gravity.

For simply supported boundary conditions with initial conditions of zero displacement and velocity, Equation 4.4, for a moving constant concentrated force, can be written as

$$EI\frac{\partial^4\bar{y}(x,t)}{\partial x^4} + m\frac{\partial^2\bar{y}(x,t)}{\partial t^2} + c\frac{\partial\bar{y}(x,t)}{\partial t} = \delta(\xi)P. \tag{4.11}$$

For bridges carrying freight rail traffic, we can assume a relatively slow vehicle speed, V, and neglect the dynamic vehicle suspension load, $F(t) = 0$, which means that $P = m_{\mathrm{v}}[g - (\mathrm{d}^2\bar{y}(Vt, t)/\mathrm{d}t^2)]$. In this case, using $i = 1$ [since for simply supported spans it is generally sufficient to consider only the fundamental mode of vibration (Veletsos and Huang, 1970)], the solution of Equation 4.11 may be achieved by transformation techniques as (Fryba, 1996)

$$\bar{y}(x,t) = \frac{2\bar{F}L^3}{\pi^4 EI}\sum_{i=1}^{\infty}\frac{1}{i^2\left[i^2(i^2-(\omega/\omega_1)^2)^2 + 4(\omega/\omega_1)^2(\omega_c/\omega_1)^2\right]}$$

$$\left[\begin{array}{l} i^2\left[i^2-(\omega/\omega_1)^2\right]\sin i\omega t - \dfrac{i(\omega/\omega_1)\left[i^2(i^2-(\omega/\omega_1)^2)-2(\omega_c/\omega_1)^2\right]}{\sqrt{i^4-(\omega_c/\omega_1)^2}} \\ \mathrm{e}^{-\omega_c t}\sin\sqrt{\omega_i^2-\omega_c^2}t - 2i(\omega/\omega_1)(\omega_c/\omega_1)\Big(\cos i\omega t - \mathrm{e}^{-\omega_c t} \\ \quad\times\cos\sqrt{\omega_i^2-\omega_c^2}t\Big) \end{array}\right]\sin\frac{i\pi x}{L}, \tag{4.12a}$$

where $\bar{F} = m_{\mathrm{v}}g$ [when $F(t) = 0$], $\omega = (\pi V/L)$ (forcing frequency of $p(xt)$), and ω_1 is the first or fundamental frequency of span (resonance occurs when $\bar{\omega} = \bar{\omega}_1$).

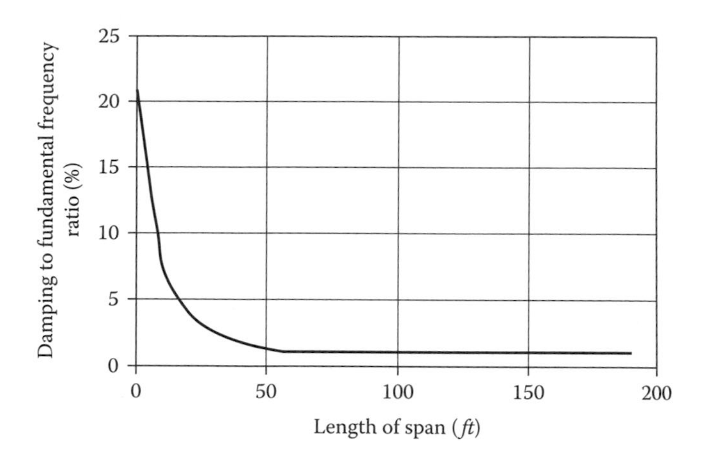

FIGURE 4.7 Empirical values of viscous damping frequency to fundamental frequency for steel railway bridges. (After Fryba, L., 1996, *Dynamics of Railway Bridges*, Thomas Telford, London, UK.)

Furthermore, for structures with light damping, where ω_c is much less than 1 (which is generally the case for steel railway bridges as illustrated in Figure 4.7), Equation 4.12a with the concentrated moving force, P, at mid-span is

$$\bar{y}(x,t) = \frac{2FL^3}{\pi^4 EI\left(1-(\omega/\omega_1)^2\right)}\left(\sin\omega t - \left(\frac{\omega}{\omega_1}\right)e^{-\omega_c t}\sin\omega_1 t\right)\sin\frac{\pi x}{L}. \quad (4.12b)$$

The solution of Equation 4.11 is also greatly simplified for simply supported spans with light damping by neglecting the damping ($c = 2m\omega_c = 0$) and assuming a generalized single degree of freedom system with a sinusoidal shape function of $\sin(\pi x/L)$ (Clough and Penzien, 1975; Chopra, 2004). The forced vibration solution for mid-span ($x = L/2$) deflection may then be expressed as

$$\bar{y}(L/2,t) = \frac{2P}{mL\left[\omega_1^2 - (\pi V/L)^2\right]}\left(\sin\frac{\pi Vt}{L} - \frac{\pi V}{\omega_1 L}\sin\omega_1 t\right). \quad (4.13)$$

Equation 4.13 indicates effectively static behavior for very short or stiff spans with a high natural frequency. However, in the development of Equation 4.13, the inertia effects of the stationary mass, dynamic effects of the load (vehicle suspension system dynamics),* damping, and the effects of surface irregularities are neglected. Therefore, Equation 4.13 will not provide satisfactory results for typical railway spans,† as shown in Example 4.3.

Example 4.3

Neglecting inertia effects, load dynamics, damping, and higher vibration modes and assuming a sinusoidal shape function, determine the maximum

* Vehicle dynamic loads may be particularly important in medium span superstructures.

† It may be appropriate to use equation (4.13) as a preliminary design tool for long or complex bridges.

dynamic mid-span deflection and associated impact factor for a 45 ft long ballasted deck through plate girder span traversed by a single concentrated load of 400 kips moving at 70 mph (103 ft/s) on a smooth surface. The span weighs 150,000 lb (total dead load including ballasted deck) and has a moment of inertia of 96,000 in.^{4}.

$$m = \frac{150{,}000}{32.17(45)} = 103.6\ \text{lb-s}^2/\text{ft}^2,$$

$$\omega_1 = \frac{\pi^2}{(45)^2}\sqrt{\frac{(29E6)(96{,}000)}{(103.6)(144)}} = 66.58\ \text{rad/s},$$

$$T_1 = \frac{2\pi}{\omega_1} = 0.094\ \text{s},$$

$$\frac{\pi V}{L} = \frac{\pi(103)}{45} = 7.19\ \text{rad/s}.$$

The duration of the forcing function pulse, $L/V = 45/(103) = 0.44$ s, is greater than one-half the span period, $T_1/2$; therefore, the maximum response will occur when the moving load is on the span. The maximum forced vibration mid-span deflection from Equation 4.13 is $\overline{y}(L/2, t) = 0.47(\sin 7.19t - 0.108 \sin 66.58t)$ inches, which is plotted in Figure E4.4, from which the maximum dynamic mid-span deflection is 0.50 in. The maximum static mid-span deflection is

$$\frac{PL^3}{48EI} = 0.47\ \text{inches}.$$

The impact factor is, then, $0.50/0.47 = 1.06$ (i.e., increase static forces by 6% to account for dynamic effects).

This impact factor would not be appropriate for design considering the assumptions made during the development of the equation of motion and its subsequent solutions. This is why impact measurements taken on actual bridges show appreciably greater impacts for the reasons discussed later in this section.

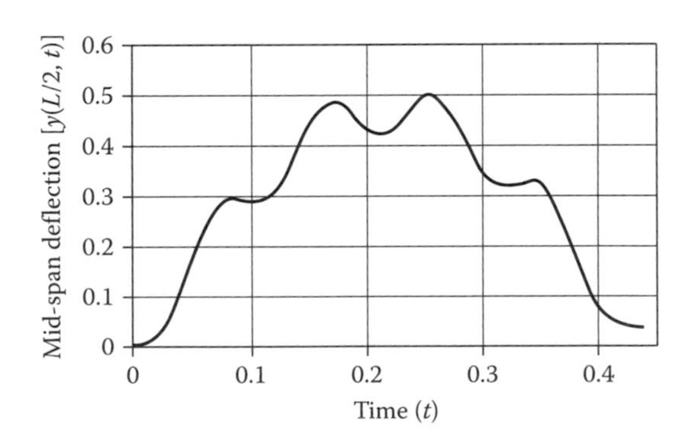

FIGURE E4.4

For the moving harmonically varying concentrated force shown in Figure 4.6,

$$p(x,t) = \delta(\xi)P \sin \omega_F t, \tag{4.14}$$

where ω_F is the frequency of the harmonic force, P.

The steady state solution for maximum* dynamic deflection of relatively long-span steel railway bridges with light damping, considering the relatively slow speed of heavy freight traffic, is (Fryba, 1972)

$$\bar{y}(x,t) = \frac{PL^3}{96EI}\omega_1\left(\frac{\cos \omega_1 t}{(\pi V/L)^2 + (c/2m)^2}\right)\left[\left(\frac{\pi V}{L}\right)\left[\left(\cos\left(\frac{\pi V}{L}\right)t - e^{-(c/2m)t}\right)\right]\right.$$
$$\left. - \frac{c}{2m}\sin\left(\frac{\pi V}{L}\right)t\right]\sin\frac{\pi x}{L}. \tag{4.15}$$

This solution for mid-span deflection would be applicable for a live load with harmonically varying frequency, such as a steam locomotive.

Dynamic analyses can be performed with the moving locomotives and trains idealized as multi-degree of freedom vehicles with wheels modeled as unsprung masses, bodies modeled as sprung masses, and with body and wheels connected by linear springs with parallel viscous dampers. Track irregularities can be estimated by equations† and a variable stiffness elastic layer can be used to account for open deck or ballasted tie conditions. The analytical solution is onerous and generally accomplished by the differential equations using numerical methods, such as the Runge–Kutta method (Carnahan et al., 1969). Using spectral analysis techniques, a closed form solution of Equation 4.4, including damping, dynamic vehicle load effect, and surface roughness, has been accomplished for the variation of dynamic deflection due to live load (Lin, 2006). This is valuable information concerning the statistical behavior of railway bridge vibrations, but does not provide a definitive mathematical solution for dynamic load allowance.

As indicated earlier, the natural frequency, ω_n, of the bridge span is a useful dynamic property that depends on the stiffness and mass of the span. The undamped natural frequency of various beam spans may be calculated using free vibration analysis [$c = 0$ and $p(x,t) = 0$] and some approximations for vibration modes, i, are shown in Table 4.2.

However, for short- and medium-span steel railway bridges, free vibration calculations that yield the natural frequency of the span must be made considering the inertial effects of the locomotive and trailing car weights. Some approximations for the unloaded fundamental ($n = 1$) frequency, ω_1 for railway bridges, developed from statistical analysis of measurements‡ on European bridges, are shown in Table 4.3 with L given in ft. These equations are also plotted in Figure 4.8 with a typical estimate for highway bridges.§ It can be observed that the fundamental natural frequency

* Occurs where forcing frequency equals fundamental frequency, $\omega_F = \omega_1$ (resonance).

† A harmonically varying equation is often used to facilitate solution of the differential equations.

‡ Based on 95% reliability.

§ An approximation for the unloaded fundamental frequency, ω_1, for highway bridges is 2060/L rad/s (L in ft) (Heywood, Roberts and Boully, 2001).

TABLE 4.2
Undamped Natural Frequencies of Various Beams

Beam	ω_{ni} (rad/s)
	$\dfrac{i^2\pi^2}{L^2}\sqrt{\dfrac{EI}{m}}$
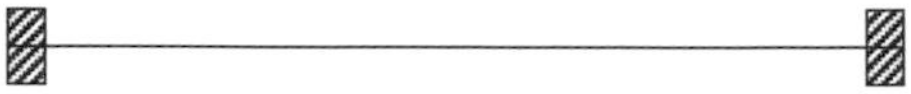	$\dfrac{(4i^2+4i+1)\pi^2}{4L^2}\sqrt{\dfrac{EI}{m}}$
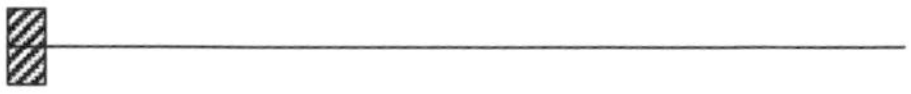	$\dfrac{(4i^2-4i+1)\pi^2}{4L^2}\sqrt{\dfrac{EI}{m}}$
	$\dfrac{(16i^2+8i+1)\pi^2}{16L^2}\sqrt{\dfrac{EI}{m}}$

for the span of Example 4.3, $\omega_1 = 66.58$ rad/s $= 10.6$ Hz, is slightly greater than the estimate of 9.4 Hz for railway ballasted girder spans in Figure 4.8.

Due to the inertial effects of the relatively large railway live load on steel spans, the loaded simply supported beam natural frequencies are required in the dynamic analysis of steel railway superstructures. Approximate equations for the loaded simply supported beam fundamental frequency, ω_{L1}, have been proposed (Fryba, 1972) as

$$\omega_{L1} = \omega_1\sqrt{\frac{1}{1+(2P/mgL)+(m_w/2mg)}} \quad \text{for moving uniform continuous loads} \tag{4.16}$$

and

$$\omega_{L1} = \omega_1\sqrt{\frac{1}{1+(2P/mgL)}} \quad \text{for moving constant concentrated loads.} \tag{4.17}$$

A similar equation for the loaded simply supported beam fundamental frequency, ω_{L1}, was proposed for a moving harmonically varying concentrated force

TABLE 4.3
Unloaded Fundamental Frequencies of Steel Railway Bridges (Empirical Equations from Fryba, 1996)

Superstructure	Unloaded Fundamental Frequency, f_1 (Hz)
Steel truss spans	$1135(L)^{-1.1}$
Ballasted girder spans	$135(L)^{-0.7}$
Open deck girder spans	$680/L$

Note: L = length in feet.

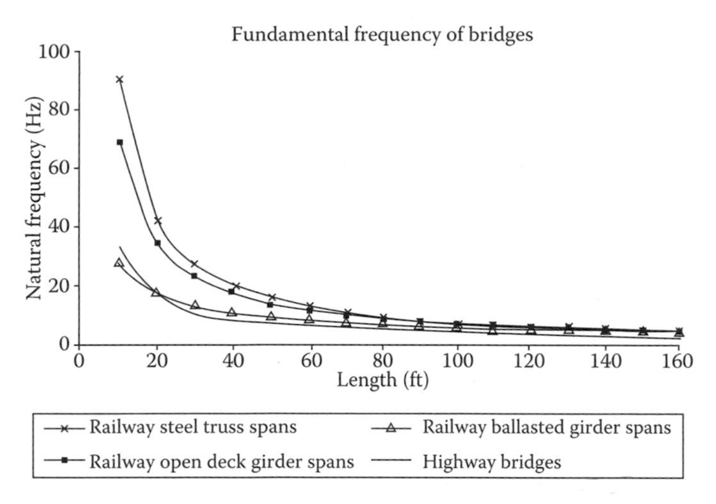

FIGURE 4.8 Unloaded fundamental frequencies of various steel bridge types.

(Inglis, 1928) as

$$\omega_{L1} = \omega_1 \sqrt{\frac{1}{1 + (2P/mgL)\sin^2(\pi x/L)}}. \tag{4.18}$$

Example 4.4

Estimate the loaded natural frequency of the bridge in Example 4.3.
Unloaded natural frequency $= \omega_1 = 66.58$ rad/s $= 10.6$ Hz.

Loaded natural frequency $= \omega_{L1}$

$$= 66.58\sqrt{\frac{1}{1 + [2(400{,}000)/(103.6)(32.17)(45)]}}$$

$$= 26.45 \text{ rad/s} = 4.2 \text{ Hz}.$$

It is evident that many of the parameters affecting the dynamic behavior of a steel railway bridge are complex and stochastic in nature. Deterministic solutions are difficult, even with many simplifying assumptions. Modern dynamic finite element analysis (FEA) methods and software enable incremental, mode superposition, frequency domain, and response spectra analysis of structures. FEA is of particular use in the dynamic analysis of long-span, continuous, and complex superstructures.* However, the dynamic effects of moving concentrated live loads on ordinary railway

* Up to three modes of vibration should be considered for continuous and cantilever bridges (Veletsos and Huang, 1970).

bridges are best developed for routine bridge design using empirical data.* The parameters that affect the dynamic behavior of steel railway bridges are (Byers, 1970; Yang et al., 1995; Taly, 1998; Heywood et al., 2001; Uppal et al., 2003)

- Dynamic characteristics of the live load (mass, vehicle suspension stiffness, natural frequencies, and damping).
- Train speed (a significant parameter).
- Train handling (causing pitching acceleration).
- Dynamic characteristics of the bridge (mass, stiffness, natural frequencies, and damping).
- Span length and continuity (increased impact due to higher natural frequencies of short-span bridges).
- Deck and track geometry irregularities on the bridge (surface roughness) (a significant parameter).
- Track geometry irregularities approaching the bridge.
- Rail joints and flat or out-of-round wheel conditions (a significant parameter of particular importance for short spans).
- Bridge supports (alignment and elevation).
- Bridge layout (member arrangement, skewed, and curved).
- Probability of attaining the maximum dynamic effect concurrently with maximum load.

Many of these parameters are nondeterministic and difficult to assess. Therefore, as with highway bridge design procedures, ordinary steel railway bridges are designed for dynamic allowance based on empirical equations developed from service load testing. AREMA (2008) provides deterministic values for design impact that are considered large enough, with an estimated probability of exceeding 1% or less for an 80 year service life, based on in-service railway bridge testing (Ruble, 1955; AREMA, 2008). The AREMA (2008)-recommended impact due to vertical effects for simply supported open deck steel bridges is shown in Figure 4.9 and Example 4.5. The impact load for ballasted deck steel bridges may be reduced to 90% of the total impact load determined for open deck steel bridges (AREA, 1966).

Example 4.5

The double track open deck steel multibeam railway bridge shown in Figure E4.5 is composed of two 45 ft simple spans.

The impact due to vertical effects, I_V, on a 45 ft span is 36.2% (Figure 4.9).

A statistical investigation of steel railway bridge impact (Byers, 1970) revealed that the test data (AREA, 1960) followed a normal frequency distribution with mean values and standard deviation increasing with increasing speed and decreasing with

* Although not often used in modern railway bridge design, impact equations for steam locomotives are provided in AREMA (2008) in addition to those recommended for modern diesel and diesel–electric powered locomotives.

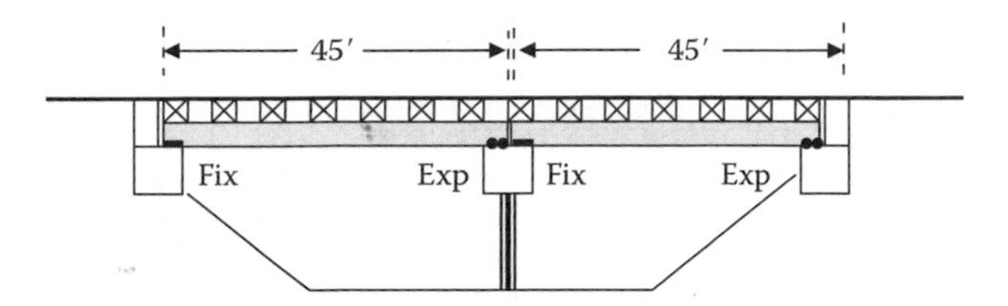

FIGURE E4.5

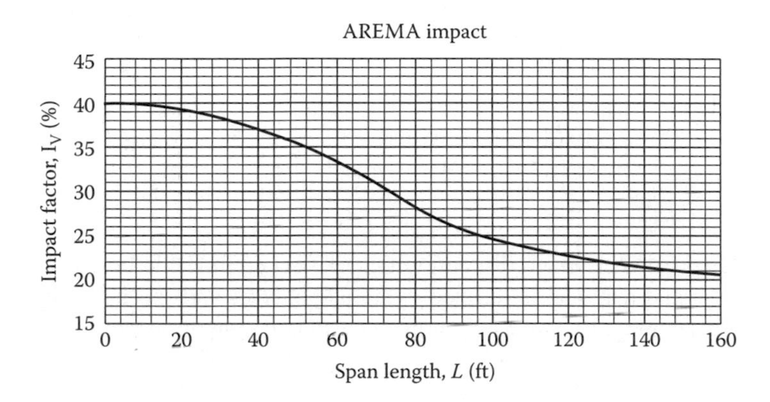

FIGURE 4.9 AREMA design impact for simply supported spans due to vertical effects as a percentage of live load for modern railroad equipment (diesel locomotives and modern freight cars).

increasing span length. The study also indicated that track irregularity effects may be a relatively large component of the total impact and short-span impacts are more sensitive to speed effects than those of longer spans.

Also, based on this same statistical investigation of the test data, the mean dynamic amplification (impact) values are presented in AREMA (2008) for fatigue design based on Cooper's E80 stress ranges. Fatigue is member-detail-sensitive and the criteria are given in Table 4.4 for various members as a percentage of the design impact load. However, these reductions should not be used in fatigue design for members less than 80 ft in length where poor track or wheel conditions exist.

4.3.2.1.3 Design Impact Load

The total impact load is the sum of the impact load due to rocking and vertical effects as shown in Equation 4.3.

Modern bridge codes have attempted to formulate dynamic load allowance as a function of fundamental frequency. However, the great number of random parameters generally leads research and development in the direction of simplification for ordinary bridge design. Therefore, many modern highway bridge design codes typically represent dynamic load allowance or impact as a simple function of length or specify a constant value within span ranges. The AREMA (2008) recommendations provide simple equations based on span type and length. Impact for direct fixation of track

TABLE 4.4
Mean Impact Loads for Fatigue Design

Member	Percentage of Total Impact Load
Beams (stringers, floorbeams) and girders	35
Members with loaded lengths less than or equal to 10 ft and no load sharing capabilities	65
Truss members (except hangers)	65
Hangers in through trusses	40

Source: From American Railway Engineering and Maintenance-of-Way Association (AREMA), 2008, *Manual for Railway Engineering*, Chapter 15. Lanham, MD. With permission.

to the bridge, or where track discontinuities exist (i.e., movable bridge joints), can be very large and may require refined dynamic analyses and special design considerations for damping. Example 4.6 outlines the calculation of impact for an ordinary simple span steel railway bridge.

Example 4.6

The governing Cooper's E80 or alternate live load maximum dynamic live load bending moment is required for each track of the open deck steel multibeam simple span railway bridge shown in Figures E4.1 and E4.5.

The maximum bending moment, shear forces, and pier reaction for each track of a 45 ft span due to Cooper's E80 and alternate live load (Figure 4.1) are given in Table E4.1 (see Chapter 5).

The appropriate values for determination of the maximum dynamic live load bending moment are

- The maximum static bending moment = 3420.0 ft-kips (alternate live load governs in Table E4.1).
- The rocking effect RE = 16.67% (Example 4.1).
- The vertical impact factor I_V = 36.2% (Example 4.5).
- The mean impact percentage for fatigue design = 35% (Table 4.4).

TABLE E4.1
Static Force from Moving Load

Maximum E80 bending moment	3202.4 ft-kips
Maximum E80 shear force	326.8 kips
Maximum E80 pier reaction	474.5 kips
Maximum alternate live load bending moment	3420.0 ft-kips
Maximum alternate live load shear force	328.9 kips

Calculation of the maximum dynamic live load bending moments for strength and fatigue design is as follows:

- The maximum bending moment impact for strength design = (0.167 + 0.362)(3420.0) = 1809.2 ft-kips.
- The mean bending moment range impact for fatigue design = [0.35(0.167 + 0.362)](3420.0) = 633.2 ft-kips.
- The maximum dynamic live load bending moment for strength design = 3420.0 + 1809.2 = 5229.2 ft-kips.
- The mean dynamic live load bending moment range for fatigue design = 3420.0 + 633.2 = 4053.2 ft-kips.

4.3.2.2 Longitudinal Forces due to Traction and Braking

Longitudinal forces, due to train braking (acting at the center of gravity of the live load) and locomotive tractive effort (acting at the freight equipment drawbars or couplers), are considerable for modern railway freight equipment. Longitudinal forces from railway live loads exhibit the following characteristics (Otter et al., 2000):

- Tractive effort and dynamic braking forces are greatest when accelerating/decelerating at low train speeds.
- Span length does not affect the relative magnitude of braking forces, due to the distributed nature of emergency train braking systems.
- Traction forces from locomotives may affect a smaller length of the bridge.
- Participation of the rails is relatively small (particularly when the bridge and approaches are loaded) due to the relatively stiff elastic fastenings used in modern bridge deck construction.
- The ability of the approach embankments to resist longitudinal forces is reduced when the bridge and approaches are loaded.
- Grade-related traction is relatively insignificant for modern high adhesion locomotives.

The locomotive and car wheels may be modeled as accelerating or decelerating rolling* masses that do not slide (complete adhesion†) as they traverse the bridge superstructure. The forces created by the vertical, horizontal, and rotational translation of the rolling mass are shown in Figures 4.10a–c. The longitudinal traction forces applied to the superstructure may be determined by superposition of the vertical, horizontal, and rotational effects of the rolling mass for linear elastic structures.

Neglecting axle bearing and wheel rim friction,‡ the force equilibrium relating to the vertical effects of rolling motion, considering complete adhesion (no sliding), provides (Figure 4.10a)

$$W - m_F \frac{d^2 y(t)}{dt^2} - R_V(t) = 0. \quad (4.19)$$

* Rolling is the superposition of translation and rotation (Beer and Johnston, 1976).

† Nonuniform speed (acceleration for starting and deceleration for braking) and adhesion must exist between the wheel and rail interface to start and stop trains.

‡ Axle bearing and wheel rim friction are very small in comparison to rolling friction.

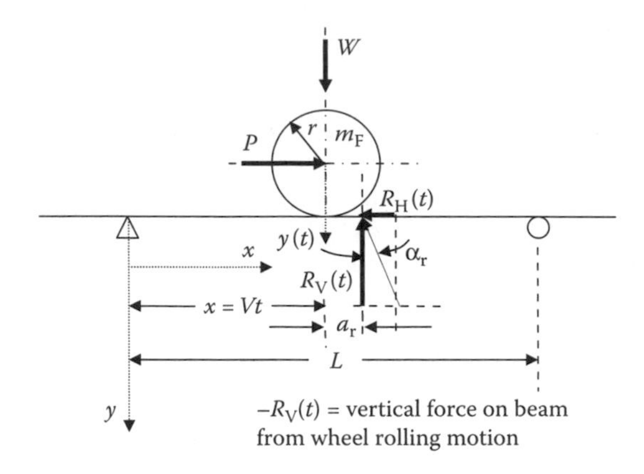

FIGURE 4.10a Vertical effects of concentrated rolling mass on a simply supported span.

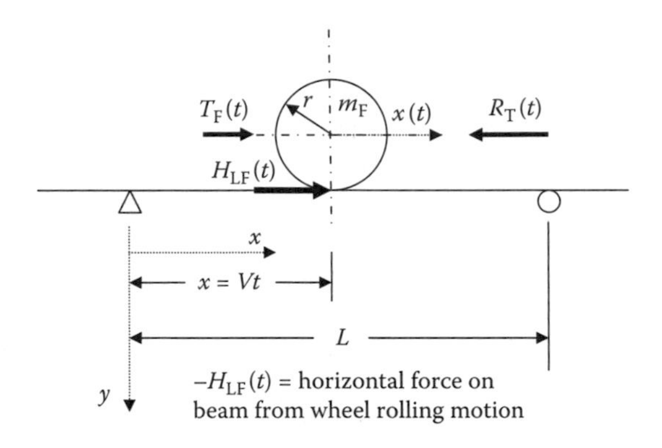

FIGURE 4.10b Horizontal effects of concentrated rolling mass on a simply supported span.

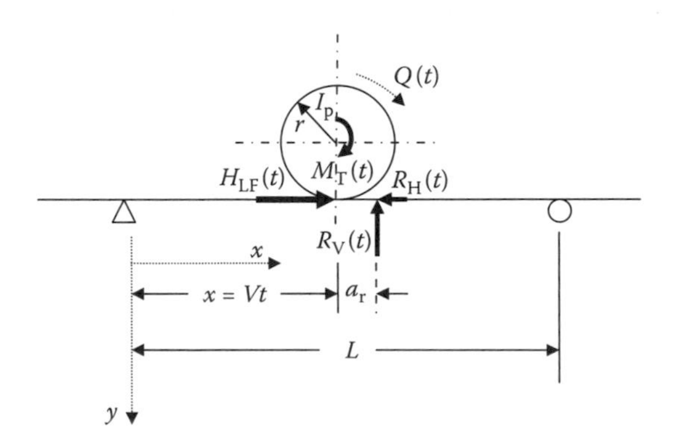

FIGURE 4.10c Rotational effects of concentrated rolling mass on a simply supported span.

The horizontal reaction at the wheel axle, P, is

$$P = R_{\mathrm{H}}(t) = \frac{Wa_r}{r}, \tag{4.20}$$

where $W = m_{\mathrm{F}}g$ is the weight of the concentrated force, m_{F} is the mass of the concentrated force, r is the wheel radius, and $R_{\mathrm{V}}(t)$ and $R_{\mathrm{H}}(t)$ are the vertical and horizontal components, respectively, of the reaction force due to rolling friction. The resultant reaction force, $R(t)$, is located at a horizontal distance, a_{r}, from the wheel centroid as a result of rolling friction (McLean and Nelson, 1962). The distance a_{r} is often referred to as the coefficient of rolling resistance. Rolling friction is small at constant train speed and greater at nonuniform train speeds. The horizontal component of the reaction, $R_{\mathrm{H}}(t)$, is generally small because the applied vertical forces greatly exceed applied horizontal forces. Neglecting axle bearing and wheel rim friction again, the force equilibrium relating to the horizontal effects of rolling motion, considering complete adhesion, yields (Figure 4.10b)

$$H_{\mathrm{LF}}(t) - R_{\mathrm{T}}(t) + T_{\mathrm{F}}(t) - m_{\mathrm{F}}\frac{\mathrm{d}^2x(t)}{\mathrm{d}t^2} = 0, \tag{4.21}$$

where $H_{\mathrm{LF}}(t)$ is the longitudinal force transferred to rails and deck/superstructure and $R_{\mathrm{T}}(t)$ is the resistance to horizontal movement (primarily air resistance or vehicle drag forces since axle bearing and wheel flange friction is considered negligible). $R_{\mathrm{T}}(t)$ is generally relativity small in comparison to other horizontal forces and it is not too conservative to neglect this force. $T_{\mathrm{F}}(t)$ is the locomotive traction force and is equal to $(M_{\mathrm{T}}(t)c'/r)$, where $M_{\mathrm{T}}(t)$ is the driving torque applied to wheel, and c' is a constant depending on locomotive engine characteristics and gear ratio.

Therefore, Equation 4.21 may be simplified to

$$H_{\mathrm{LF}}(t) + T_{\mathrm{F}}(t) - m_{\mathrm{F}}\frac{\mathrm{d}^2x(t)}{\mathrm{d}t^2} = 0. \tag{4.21a}$$

Also, neglecting axle bearing and wheel rim friction, the force equilibrium relating to the rotational effects of rolling motion, considering complete adhesion, provides (Figure 4.10c)

$$-rH_{\mathrm{LF}}(t) + M_{\mathrm{T}}(t) - a_{\mathrm{r}}R_{\mathrm{V}}(t) + rR_{\mathrm{H}}(t) - I_{\mathrm{p}}\frac{\mathrm{d}^2\theta(t)}{\mathrm{d}t^2} = 0, \tag{4.22}$$

where I_{p} is the rotational moment of the inertia of mass.

Since the distance, a_{r}, is small, the moment from rolling friction, $a_{\mathrm{r}}R_{\mathrm{V}}(t)$, may be neglected. In addition, because $R_{\mathrm{H}}(t)$ is relatively small, Equation 4.22 may be simplified to

$$-rH_{\mathrm{LF}}(t) + M_{\mathrm{T}}(t) - I_{\mathrm{p}}\frac{\mathrm{d}^2\theta(t)}{\mathrm{d}t^2} = 0. \tag{4.23}$$

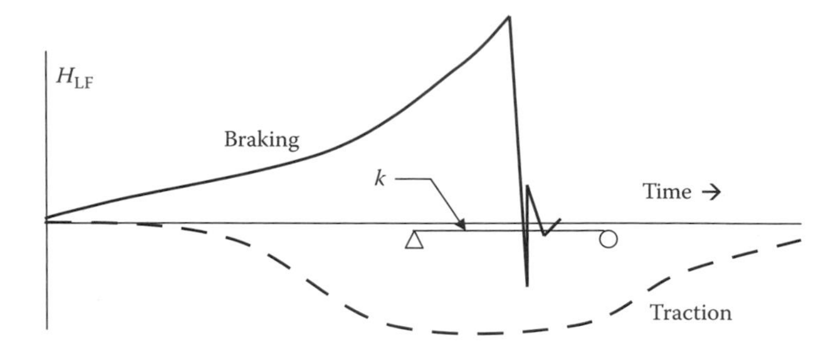

FIGURE 4.11 Time history of braking and traction forces (at fixed bearing) from railroad equipment.

For the condition of no slippage (complete adhesion), $\theta(t) = (x(t)/r)$. Substitution of $(d^2\theta(t)/dt^2) = (d^2x(t)/r dt^2)$ into Equation 4.23 yields

$$\frac{d^2x(t)}{dt^2} = \frac{I_p}{r}(M_T(t) - rH_{LF}(t)). \tag{4.24}$$

Substitution of Equation 4.24 into Equation 4.21a provides

$$H_{LF}(t) = \frac{1}{1 - m_F I_p}\left(\frac{m_F I_p}{r}(M_T(t) + T_F(t))\right) \leq \mu R_V(t), \tag{4.25}$$

where μ is the coefficient of adhesion between locomotive wheels and rail without slippage (can be as high as 0.35 for modern locomotives with software-controlled wheel slip).

Equation 4.25 allows the numerical solution for longitudinal force, $H_{LF}(t)$, which remains, however, too arduous for ordinary design. The longitudinal forces described by Equation 4.25 (including the effects of axle bearing, wheel rim friction, air resistance, rolling friction, and other effects) have been observed and recorded by field testing in both Europe and the United States. The longitudinal forces exhibit almost static behavior since maximum traction and braking forces occur at low speeds when starting and at the end of braking, respectively (Figure 4.11). Therefore, a static analysis can be performed with $H_{LF} = \mu R_V = \text{LF} = \mu W$.

For a static longitudinal analysis, the bridge may be modeled as a series of longitudinal elastic bars (with independent longitudinal and flexural deformations) on horizontal elastic foundations simply modeled* as equivalent horizontal springs with stiffness, k_i. The static longitudinal equilibrium equations for a system of i bars (spans and rails) on elastic foundations (elastic horizontal stiffness of bridge deck†

* Other models that incorporate different longitudinal restraint at the rail-to-deck and deck-to-superstructure may be used to provide greater accuracy.

† Particularly appropriate for modern elastic rail fastening systems.

and approach track) is (see Figure E4.6)

$$-E_iA_i\frac{d^2\overline{x_i}(x)}{dx^2} + k_i\overline{x}_i(x) = q_i(x), \tag{4.26}$$

where E_iA_i is the axial stiffness of the member (rail or span).

The resulting system of equations may be solved for the longitudinal displacements, $\overline{x}_i(x)$, and forces, $N_i(x) = E_iA_i(d\overline{x_i}(x)/dx)$, in the bars. The solution may be obtained using transformation methods (Fryba, 1996) and the appropriate boundary conditions (e.g., Table E4.2 in Example 4.7).

The longitudinal traction and braking forces transferred to the bearings may be determined from equilibrium following computation of the rail and span axial forces, $N_i(x)$. However, as seen in Example 4.7, even the simplest bridge models will involve considerable calculation.

Example 4.7

Develop the equations of longitudinal forces and boundary conditions for the open deck steel railway bridge shown in Figure E4.6.

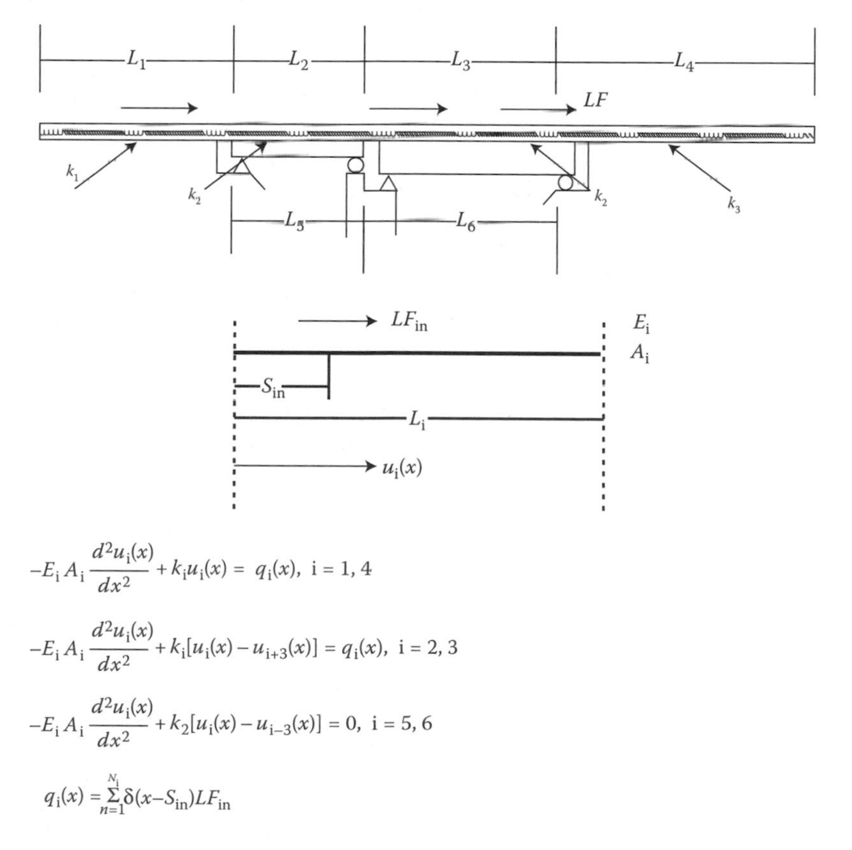

FIGURE E4.6

TABLE E4.2

Boundary Conditions	Force and Displacement Conditions
Rails ($i = 1, 2, 3, 4$)	$N_1(0) = N_4(L_4) = 0$
	$N_1(L_1) - \mathrm{LF}_{L1} = N_2(0)$
	$N_2(L_2) - \mathrm{LF}_{L2} = N_3(0)$
	$N_3(L_3) - \mathrm{LF}_{L3} = N_4(0)$
	$u_1(L_1) = u_2(0)$
	$u_2(L_2) = u_3(0)$
	$u_3(L_3) = u_4(0)$
Span ($i = 5, 6$)	$N_5(L_5) = N_6(L_6) = 0$
	$u_5(0) = u_6(0) = 0$
Particular	
Expansion joints at end of bridge	$L_1 = L_4 = 0$
CWR across bridge	$L_1 = L_4$
No longitudinal rail restraint (free rails)	$k_2 = 0$
Rails fixed (direct fixation to deck)	k_2

The equations of longitudinal forces and boundary conditions are shown in Figure E4.6 and Table E4.2, respectively.

Extensive testing and analytical work has been performed (see references Foutch et al., 1996, 1997; LoPresti et al., 1998; LoPresti and Otter, 1998; Otter et al., 1996, 1997, 1999, 2000; Tobias Otter and LoPresti, 1998; Tobias et al., 1999; Uppal et al., 2001) to overcome the theoretical model complexities and numerical modeling efforts. This work has established relationships for braking and traction dependent on the length of the portion of the bridge under consideration. Testing in the United States has provided longitudinal forces for Cooper's E80 design live load that are shown in Figure 4.12 and Equations 4.27 and 4.28. It appears that, for loaded lengths less than about 350 ft, longitudinal force due to traction governs. However, locomotive traction occurs over a relatively small length and braking forces on a loaded length consisting of the entire bridge may exceed the tractive effort (see Examples 4.9 and 4.10).* The force due to traction governs for short- and medium-length bridges.

$$\mathrm{LF}_B = 45 + 1.2L, \tag{4.27}$$

$$\mathrm{LF}_T = 25\sqrt{L}, \tag{4.28}$$

where LF_B is the longitudinal force due to train braking (kips), LF_T is the longitudinal force due to locomotive traction (kips), and L is the length of the portion of the bridge under consideration (ft).

However, while an estimate of the magnitude of the applied longitudinal traction and braking forces appropriate for design is readily available, the distribution of

* As illustrated by Figure 4.13 showing the ratio of the longitudinal force transmitted to the bearings, H_B, to the applied longitudinal force for bridges with continuous welded rail and steel bearings (based on European tests reported by Fryba, 1996).

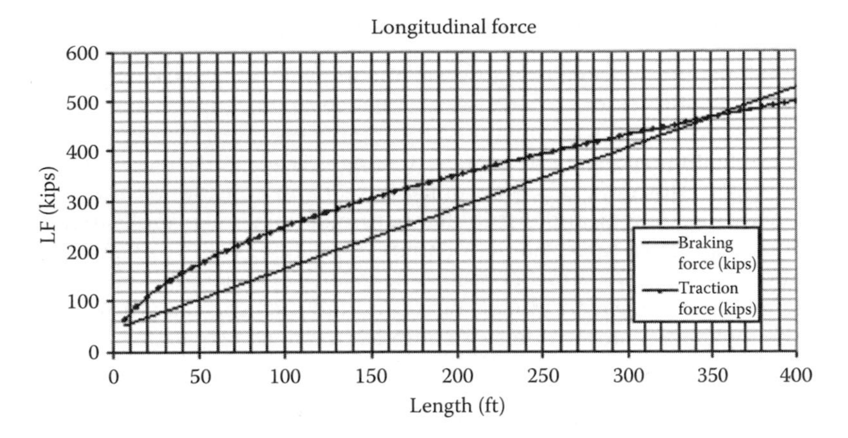

FIGURE 4.12 AREMA design longitudinal forces.

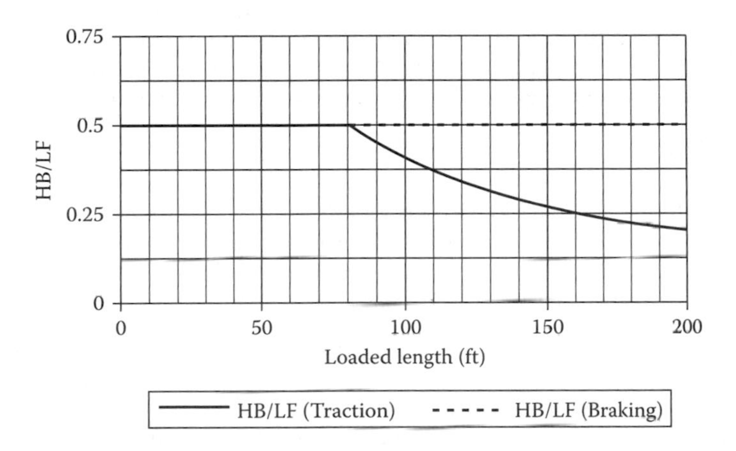

FIGURE 4.13 Bearing forces from European testing. (After Fryba, L., 1996, *Dynamics of Railway Bridges*, Thomas Telford, London, UK.)

longitudinal forces for the design of span bracing, bearings, substructures, and foundations needs careful consideration. The distribution and path of longitudinal forces between their point of application and the bridge supports depend on the arrangement, orientation, and relative stiffness of

- Bridge members in the load path
- Bearing type (fixed or expansion)
- Substructure characteristics.

Example 4.8

The longitudinal design force for Cooper's E80 loading is required for each track of the open deck steel multibeam railway bridge shown in Figures E4.1

and E4.5. From Figure 4.12, it is determined that

- The longitudinal force due to train braking is $LF_B = 153.0$ kips per track on the entire bridge; because of relative span lengths and bearing arrangement, it may be equally distributed to each span as 76.5 kips.
- The longitudinal force due to train braking is $LF_B = 99.0$ kips per track on one span. However, this is an unlikely scenario considering the bridge length, train length, and distributed nature of train brake application.
- The longitudinal force due to locomotive traction is $LF_T = 237.2$ kips per track on the entire bridge; because of relative span lengths and bearing arrangement, it may be equally distributed to each span as 118.6 kips.
- The longitudinal force due to locomotive traction is $LF_T = 167.7$ kips per track on one span.

The longitudinal force due to locomotive traction is $LF_T = 167.7$ kips per track and may be used for superstructure design. The longitudinal forces are distributed through the superstructure to the bearings and substructures. Bearing component and substructure design will require consideration of these longitudinal forces. However, in this multibeam span, longitudinal forces of this magnitude will result in only small axial stresses in the longitudinal beams or girders, which may be disregarded in the design.

Example 4.9

The longitudinal design force for Cooper's E90 loading is required for the deck truss of the 1100 ft long single track 10 span steel bridge outlined in the data of Table E4.3. Each span has fixed and expansion bearings. All substructures have spans with adjacent fixed and expansion bearings.

- The longitudinal force due to train braking is $LF_B = (9/8)1365 = 1536$ kips on the entire bridge; it is distributed to the deck truss span as $(400/1100)(1536) = 558$ kips.
- The longitudinal force due to train braking is $LF_B = (9/8)525 = 591$ kips on the deck truss span. However, this is an unlikely scenario considering the bridge length, train length, and distributed nature of train brake application. Therefore, other portions of the bridge should be investigated for train braking. For example, the longitudinal force due to train braking, $LF_B = (9/8)(400/600)(765) = 574$ kips on the deck truss span when the train is on spans 7–10 only and $(9/8)(400/980)(1221) = 561$ kip when the train is on spans 1–7 only.

TABLE E4.3

Span	Type	Length (ft)
1	Through plate girder	100
2–6	Deck plate girder	80
7	Deck truss	400
8–10	Deck plate girder	100
Total		1100

- The longitudinal force due to locomotive traction is $LF_T = (9/8)829 = 933$ kips on the entire bridge. However, this is not likely (unless a string of powered accelerating/decelerating locomotives traverses the bridge) and other portions of the bridge should be investigated. For example, the longitudinal force due to locomotive traction, $LF_T = (9/8)(400/600)(612) = 459$ kips on the deck truss span when the train is on spans 7–10 only and $(9/8)(400/980)(783) = 359$ kips when the train is on spans 1–7 only.
- The longitudinal force due to locomotive traction is $LF_T = (9/8)500 = 563$ kips on the 400 ft deck truss span.

The longitudinal force due to train braking, $LF_T = 574$ kips, is likely to be used for design of the deck truss span.

As noted in Example 4.8, the distribution of longitudinal forces in the superstructure may be of little concern for some span types (e.g., multiple longitudinal beam and deck plate girder spans). However, for other types of superstructures, the longitudinal force path from rails to bearings is of considerable importance (e.g., floorbeams with direct fixation of track and span floor systems). The horizontal axial force resistance of deck plates from diaphragm behavior may preclude the need for bracing elements to carry longitudinal forces to the main girders or trusses. Nevertheless, in some open deck spans, specific consideration of the lateral bracing (traction bracing) requirements is necessary to adequately transfer longitudinal forces to the main girders or trusses for transfer to the substructures at the bearings. A typical instance where traction bracing may be required is within the panel adjacent to the fixed bearings in an open deck span with a stringer and floorbeam system supported each side of the track by long-span main girders or trusses. In order to preclude the torsional and/or lateral bending of floorbeams that might result from longitudinal forces transmitted by floor systems without connection to the lateral bracing (Figure 4.14a), traction bracing is used (Figure 4.14b). Traction bracing is provided through connection of the stringers to the lateral bracing and addition of a new transverse member (shown dashed in Figure 4.14b) between the stringers at the bracing connections. Provided the main girder or truss fixed bearings are adequate to transfer the longitudinal forces to the substructure, the traction bracing truss (Figures 4.14b and 4.15) will avoid lateral loading of floor beams (member 1–1 in Figure 4.15) since the stringers (members 2–3 in Figure 4.15) can carry no longitudinal force. Other traction bracing arrangements may be used in a similar manner at the fixed end of long single and multiple track spans to properly transmit longitudinal traction and braking forces to the bearings.

4.3.2.3 Centrifugal Forces

Centrifugal forces acting horizontally at the vehicle center of gravity (recommended as 8 ft above the top of the rails in AREMA, 2008) act on the moving live load as it traverses the curved track on a bridge, as shown in Figure 4.16. The centrifugal force corresponding to each axle load is

$$CF_A = \frac{m_A V^2}{R}, \tag{4.29}$$

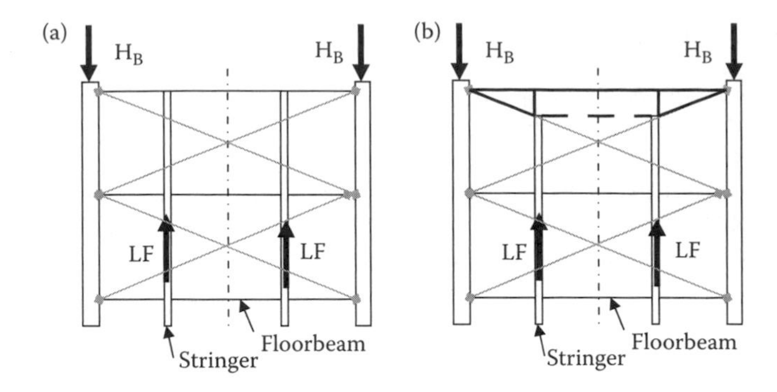

FIGURE 4.14 Plan of the floor system (a) without traction bracing and (b) with an additional member (dashed line) to create traction bracing.

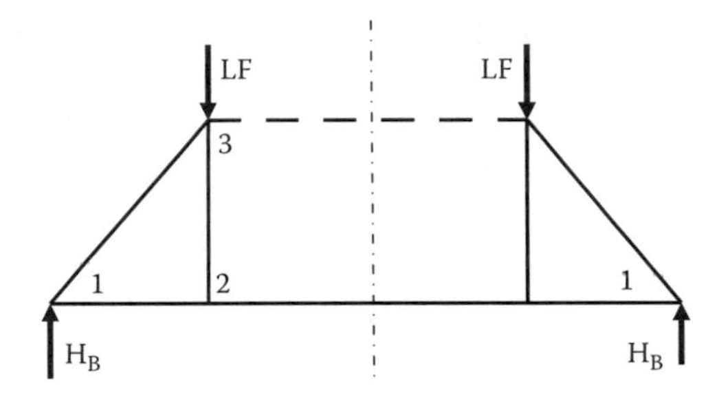

FIGURE 4.15 Traction frame truss for a single track span with two stringers per track.

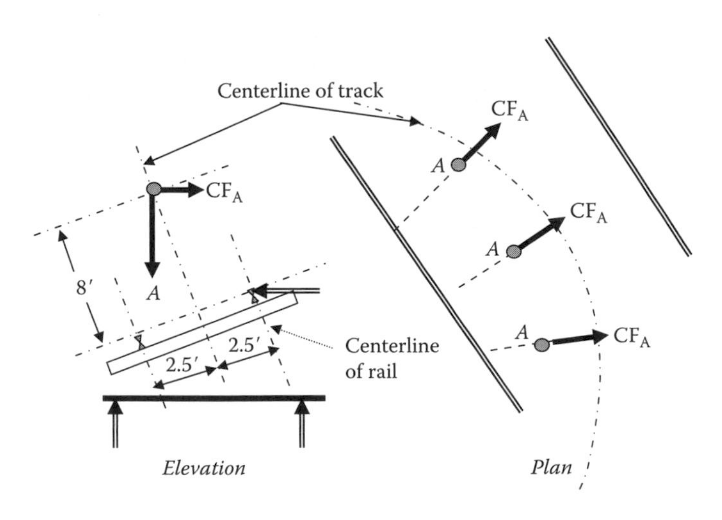

FIGURE 4.16 Centrifugal forces from a curved track.

where $m_A = A/g$; where A is the axle load, g is the acceleration due to gravity, V is the speed of the train, and R is the radius of the curve. This can be expressed as

$$CF_A = 0.0000117AV^2D, \tag{4.30}$$

where CF_A is the centrifugal force at each axle, kips; A is the axle load, kips; V is the speed of the train, mph; $R = 5730/D$, ft; D is the degree of the curve (see Chapter 3).

Due to the rail–wheel interface, the entire centrifugal force will be transmitted at the outer or high rail and, therefore, centrifugal effects are not considered for longitudinal members inside the curve. The calculation of centrifugal force on a curved railway bridge is shown in Example 4.10.

Example 4.10

A ballasted steel through plate girder railway bridge is to be designed with a 6° curvature track across its 70 ft span. The railroad has specified a 5 in superelevation based on a 40 mph operating speed. Both the tie depth and rail height of the track are 7 in. The girders are spaced at 16 ft and the plane at the top of the rails elevation is 2 ft-8 in above the deck surface at the track centerline. Determine the effects of the curvature on the design live load shear and bending moment for each girder.

Geometrical effects (see Example 3.1 in Chapter 3):

$$V_{out} = 0.903V_{LL+I},$$

$$V_{in} = 1.097V_{LL+I},$$

$$M_{out} = 0.916M_{LL+I},$$

$$M_{in} = 1.084M_{LL+I}.$$

Centrifugal effects:

$$CF_A/A = 0.0000117V^2D = 0.0000117(40)^2(6) = 0.112,$$

$(M_{CF}/M_{LL}) = (V_{CF}/V_{LL}) = \pm 0.112[(8+2.67)/(16/2)] = \pm 0.150$. (See Figure 4.16) (Note that the girder design forces due to centrifugal effects are independent of impact.)

Combined geometrical and centrifugal effects:
Since centrifugal forces are not affected by impact (rocking and dynamic vertical effects), the centrifugal effect on the girder design forces, which includes impact, must be reduced as follows:

$$\frac{M_{CF(out)}}{M_{LL+I}} = \frac{V_{CF(out)}}{V_{LL+I}} = \pm\frac{0.112}{(1+I)}\left(\frac{(8+2.67)}{(16/2)}\right) = \pm\frac{0.150}{(1+0.371)} = \pm 0.110.$$

Therefore, the following is determined for the shear, V_{LL+I}, and bending moment, M_{LL+I}, live load forces (combining impact, geometric and centrifugal force effects)

$$V_{\text{out}} = V_{\text{LL+I}}(0.903 + 0.110) = 1.013V_{\text{LL+I}},$$
$$V_{\text{in}} = 1.097V_{\text{LL+I}},$$
$$M_{\text{out}} = M_{\text{LL+I}}(0.916 + 0.110) = 1.026M_{\text{LL+I}},$$
$$M_{\text{in}} = 1.084M_{\text{LL+I}}.$$

4.3.2.4 Lateral Forces from Freight Equipment

In addition to the centrifugal lateral forces due to track curvature, lateral forces caused by track irregularities and the wheel–rail interface geometry must be considered in the design of steel railway bridges. The differential equation of horizontal (lateral) vibration, neglecting equivalent lateral motion viscous damping, is

$$EI_{\text{y}}\frac{\partial^4 \bar{z}(x,t)}{\partial x^4} + m\frac{\partial^2 \bar{z}(x,t)}{\partial t^2} = h_{\text{L}}(x,t), \tag{4.31}$$

where $\bar{z}(x,t)$ is the superstructure lateral deflection at distance x and time t, I_y is the lateral moment of inertia (with respect to the vertical axis), h_{L} is given by

$$h_{\text{L}}(x,t) = \sum_{i=1}^{N} \delta(x + s_i - Vt)H_{\text{L}_i}(t),$$

s_i is the distance from load $H_i(t)$ to the first load $H_1(t)$, and $H_{\text{L}i}(t)$ is the applied random horizontal lateral force (due to track irregularities and wheel–rail interface motion).

Assuming lateral, vertical, and torsional vibrations are uncoupled, the solution of Equation 4.31 for $\bar{z}(L/2,t)$ is (see Equation 4.13)

$$\bar{z}(L/2,t) = \frac{2H_{\text{L}}(t)}{mL[\omega_{\text{y}1}^2 - (\pi V/L)^2]}\left(\sin\frac{\pi Vt}{L} - \frac{\pi V}{\omega_{\text{y}1}L}\sin\omega_{\text{y}1}t\right), \tag{4.32}$$

where $\omega_{\text{y}1} = \pi^2/L^2\sqrt{EI_{\text{y}}/m}$.

Therefore, as even simplified dynamic methods are often inappropriate for routine or ordinary design, it is desirable to determine lateral forces from tests conducted on in-service bridges.

Recent tests concerning the dynamic lateral forces on in-service bridges (Otter et al., 2005) have confirmed that the AREMA (2008) design recommendation of a single moving lateral force of 25% of the heaviest axle of Cooper's E80 load is an appropriate representation of these effects.

The magnitude of lateral forces is of particular importance regarding the design of span lateral and cross bracing.* Therefore, in addition to the recommendation of

* Lateral loads from freight rail equipment are considered applied directly to bracing members (see Chapter 5) without producing lateral bending of supporting member flanges or chords.

a single moving lateral force of 25% of the heaviest axle of the Cooper's E80 load, a notional load of 200 lb/ft applied to the loaded chord or flange and 150 lb/ft on the unloaded chord or flange is recommended.

4.3.3 Distribution of Live Load

Unlike highway loads, which may move laterally across the bridge deck, railway live loads are generally fixed in lateral position. However, they are a longitudinal series of large magnitude concentrated wheel loads, and longitudinal and lateral distribution to the deck and supporting members must be considered.

4.3.3.1 Distribution of Live Load for Open Deck Steel Bridges

For open deck bridges, no longitudinal distribution is made and lateral distribution to supporting members is based on span cross-section geometry and type of lateral bracing system. Lateral bracing between longitudinal beams should be made with cross frames, or for spans with shallow beams, rolled beams, and/or close beam spacing, solid diaphragms.* The cross frames and diaphragms should not have a spacing exceeding 18 ft. In some cases, AREMA (2008) recommends that diaphragms and cross bracing be fastened to the beam or girder flanges. When the lateral bracing system meets these criteria and is properly designed for the lateral forces (see Chapters 5 through 7), all beams or girders supporting the track are considered as equally loaded.

4.3.3.2 Distribution of Live Load for Ballasted Deck Steel Bridges

For ballasted deck bridges, longitudinal and lateral distribution of live load to the deck is based on tests performed by the Association of American Railroads (AAR) (Sanders and Munse, 1969). Axle loads are distributed over a given width at a 2:1 (V/H ratio) distribution through ballast rock and the deck material, as shown in Figure 4.17.

The longitudinal deck distribution width, $(3' + d_b)$, should not exceed either 5 ft or the minimum axle spacing of the design load. The lateral deck distribution width, (Length of the tie $+ d_b$), should not exceed 14 ft or the distance between adjacent track centerlines or the width of the deck.

The longitudinal distribution of live load to members supporting the deck in the transverse direction (Figure 4.18) is given in terms of an effective beam spacing, which is dependent on deck material, beam span, and spacing, and for concrete decks, the stiffness of beams and deck, and the width of the deck.

$$P = \frac{1.15AD}{S}, \tag{4.33}$$

* Channels and coped flange wide flange shapes are often used for diaphragms between longitudinal beams. Plates are generally not used due to the absence of flanges and low bending strength.

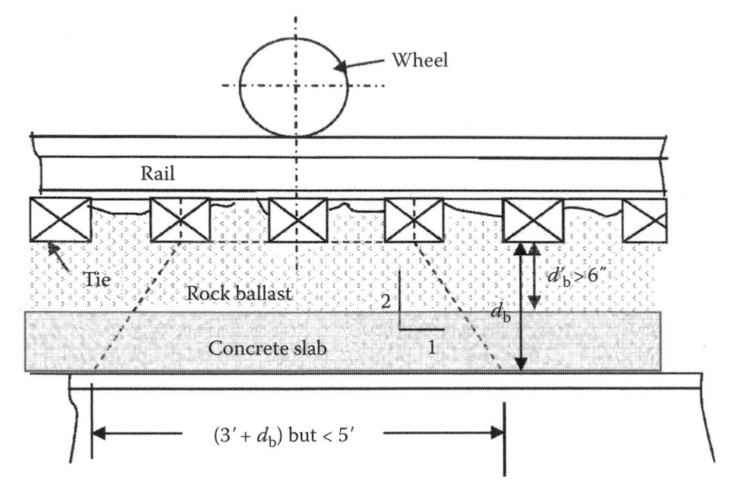

Longitudinal distribution to deck

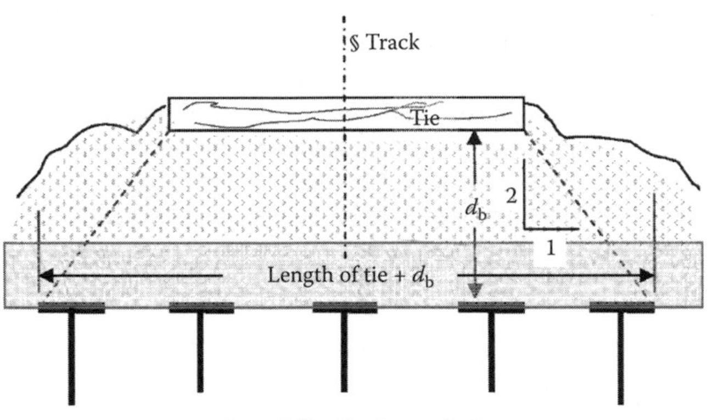

Lateral distribution to deck

FIGURE 4.17 Longitudinal and lateral distribution of live load to the deck on ballasted deck bridges.

where P is the portion of axle load on the transverse beam; A is the axle load; S is the axle spacing, ft; D is the effective beam spacing, ft, which

- for bending moment calculations with a concrete deck, is

$$D = d\left[\frac{1}{1 + (d/aH)}\right]\left(0.4 + \frac{1}{d} + \frac{\sqrt{H}}{12}\right),$$

  but $D \le d$ or S.

- for bending moment calculations with a steel plate deck, or for end shear in both concrete and steel decks, is

$$D = d,$$

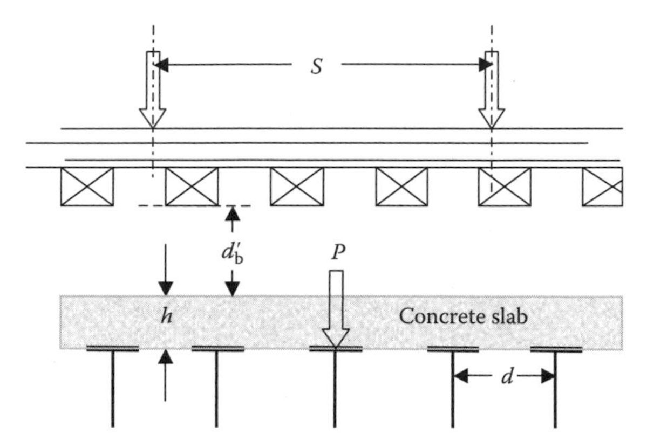

Longitudinal distribution to transverse members

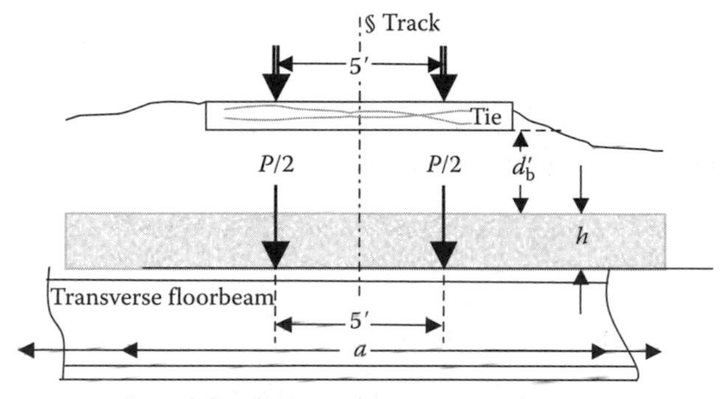

Lateral distribution to transverse members

FIGURE 4.18 Longitudinal and lateral distribution of live load on ballasted deck bridges with transverse floorbeams.

where d is the transverse beam spacing $\leq S$, ft (if $d > S$ then assume the with deck as simply supported between transverse beams). In the previous equation, a is the transverse beam span, ft, H is given by

$$H = \frac{nI_b}{ah^3} \text{ in./ft,}$$

n is the steel to concrete modular ratio, I_b is the transverse beam moment of inertia, in.^{4}, and h is the concrete slab thickness, in. No lateral distribution of load is made for transverse beams supporting ballasted decks (Figure 4.18).

Example 4.11

The longitudinal distribution of Cooper's E80 axle loads to 16 ft long transverse $W36 \times 150$ floorbeams spaced at 2.5 ft supporting a 7 in. thick reinforced

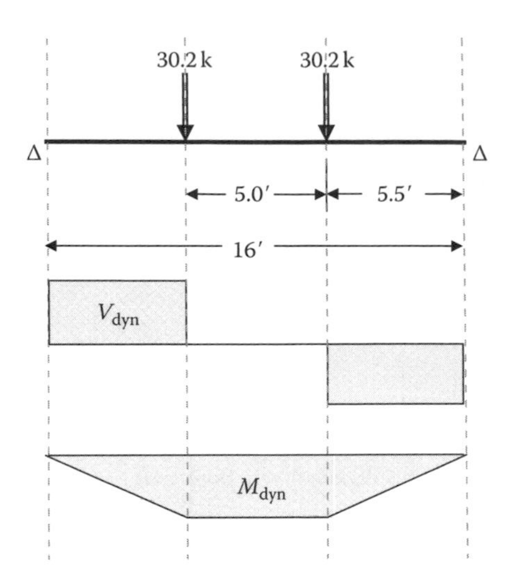

FIGURE E4.7

concrete deck slab is required (Figure E4.7).

$I_B = 9040 \text{ in.}^4$,

$n = 8$,

$H = (8)(9040)/[16(7^3)] = 13.18 \text{ in./ft}$,

$$D_M = \frac{2.50}{1 + [2.50/16(13.18)]}\left(0.4 + \frac{1}{2.50} + \frac{\sqrt{13.18}}{12}\right) = 2.72 \text{ ft, use } 2.50 \text{ ft},$$

$D_V = d = 2.50$ ft, and

$P = 1.15(80)(2.5)/5 = 46.0$ kips (for both shear and moment calculations).

Considering dynamic effects:

$\text{RE} = 0.20W(5)(100)/16W = 6.25\%$,

$I_V = 39.5\%$(Figure 4.9),

$P_{LL+I} = 0.90(1 + 0.063 + 0.395)46.0 = 60.3$ kips (30.2 kips for each rail),

$V_{LL+I} = 30.2$ kips, and

$M_{LL+I} = (5.5)30.2 = 166.1$ ft-kips.

4.4 OTHER STEEL RAILWAY BRIDGE DESIGN LOADS

In addition to dead load and live load effects, environmental forces (wind, thermal, and seismic events) and other miscellaneous forces related to serviceability and overall stability criteria must be considered in the design of steel railway bridges.

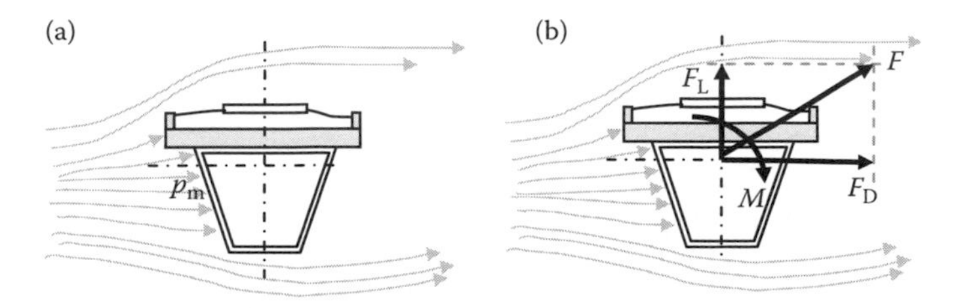

FIGURE 4.19 (a) Wind flow past a bluff body and (b) wind forces on a bluff body.

4.4.1 Wind Forces on Steel Railway Bridges

In contrast to long-span or flexible bridges (such as suspension or cable-stayed bridges), ordinary steel railway bridges (such as those composed of beam, girder, truss, and arch spans) need not consider aerodynamic effects* of the wind in design.† However, the aerostatic effects of the wind on the superstructure and moving train must be considered, particularly in regard to lateral bracing design.

A steady wind with uniform upstream velocity, V_u, flowing past a bluff body (such as the bridge cross section of Figure 4.19a) will create a maximum steady state local or dynamic pressure, p_m, in accordance with Bernoulli's fluid mechanics equation as

$$p_m = p_{amb} + \frac{1}{2}\rho V_u^2, \tag{4.34}$$

where p_{amb} is the ambient air pressure and is equal to 0 at atmospheric pressure; ρ is the air density; V_u is the upstream air speed.

However, the average dynamic pressure on the bridge span will be less than the maximum dynamic pressure given by Equation 4.34. Therefore, the dynamic pressure, p, at any point on the bluff body can be expressed as

$$p = C_p p_m = C_p\left(\tfrac{1}{2}\rho V_u^2\right), \tag{4.35}$$

where C_p is a dimensionless mean pressure coefficient that depends on the shape of the obstruction.

For example, if we assume a 100 mph wind speed (which may occur during gale and hurricane winds), Equation 4.34 yields a maximum dynamic pressure of 23.7 psf ($\rho = 0.0022$ slug/ft^3).

Design wind forces must be based on average dynamic wind pressures (i.e., reduced by the use of an appropriate pressure coefficient) calculated over an appropriate cross-sectional area. The design must also consider the effects of wind gusts.‡ It is beneficial, from a design perspective, to calculate design wind forces based on the maximum dynamic pressure, a characteristic area, and a dimensionless coefficient that includes

* The effects from dynamic behavior and buffeting.

† An equivalent static wind pressure is appropriate since the natural or fundamental frequency of the superstructure is substantially greater than the frequency of localized gust effects.

‡ Gust factors are generally between two and three for tall structures (Liu, 1991).

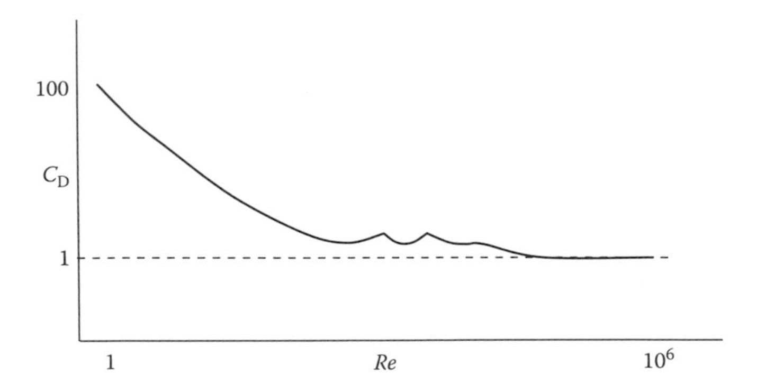

FIGURE 4.20a Typical relationship between the drag coefficient, C_D, and Reynold's number, *Re*.

the effects of bridge cross-sectional shape as well as the wind flow characteristics.* These coefficients are determined from tests and applied to the design process. If the dynamic pressure, p, is integrated over the surface of the bluff body, it will create a force, F, and a moment, M, as shown in Figure 4.19b. The force is resolved into horizontal (drag), F_D, and vertical (lift), F_L, forces. The equations for the forces and moments can then be expressed in a form similar to Equation 4.35 as

$$F_D = C_D\left(\frac{1}{2}\rho V_u^2\right)A_{RD}, \tag{4.36}$$

$$F_L = C_L\left(\frac{1}{2}\rho V_u^2\right)A_{RL}, \tag{4.37}$$

$$M = C_M\left(\frac{1}{2}\rho V_u^2\right)A_{RM}^2, \tag{4.38}$$

where C_D is the dimensionless drag coefficient that depends on span geometry and Reynolds number, *Re*. The Reynolds number is indicative of wind flow patterns related to inertial effects (*Re* large and C_D small) and viscous effects (*Re* small and C_D large). Figure 4.20a illustrates the typical relationship between C_D and *Re*. C_L is the dimensionless lift coefficient, C_M is the dimensionless moment coefficient,

$$Re = \frac{\rho V_u L_D}{\mu},$$

L_D is a characteristic length of the bridge or object for drag, A_{RD} is a characteristic area of the bridge or object for drag, A_{RL} is a characteristic area of the bridge or object for lift, A_{RM} is a characteristic area of the bridge or object for moment, and μ is the dynamic wind viscosity.

* Wind flow characteristics are described by the Reynolds number on a characteristic geometry, which is dependent on wind velocity and viscosity.

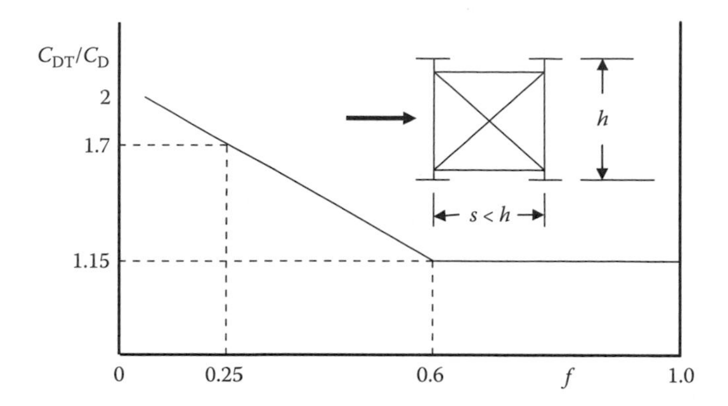

FIGURE 4.20b Typical relationship between the total drag coefficient on two girders or trusses, C_{DT}, and the drag coefficient for a single girder or truss, C_D.

The drag force or total wind thrust on a bluff body, such as a bridge cross section, created by the wind flow is of primary interest in the design of bridges. Therefore, drag coefficients are established by wind tunnel tests, which incorporate the effects of geometry and flow characteristics (as described by the Reynolds number), which may be used for design purposes. Drag coefficients for a solid element, such as a plate girder, are generally no greater than about 2.0 and drag coefficients for a truss are typically about 1.70 (Simiu and Scanlon, 1986). The difference is related primarily to the characteristic dimension of the effective area, generally taken as the area projected onto a plane normal to the wind flow. The solidity ratio, f, is defined as the ratio of the effective area to the gross area.

The effects of the usual pairing of girders, trusses, and arches in steel railway bridges must also be considered. Figure 4.20b illustrates the typical relationship between the drag coefficient relating to the total wind force on two girders or trusses, C_{DT}, to the drag coefficient for a single girder or truss, C_D, in terms of the solidity ratio, f, for spans with girder or truss spacing, s, no greater than the girder or truss height, h. Examples of the use of these drag coefficients are outlined in Examples 4.12 and 4.13.

Example 4.12

A 125 ft long ballasted deck steel deck plate girder span is shown in Figure E4.8. Determine the design wind force for a wind speed of 75 mph.

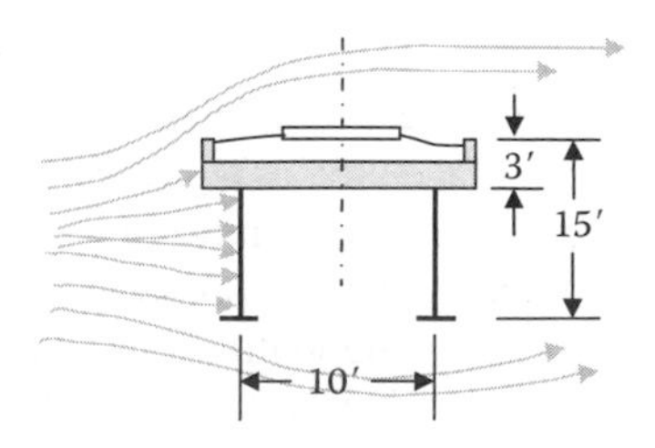

FIGURE E4.8

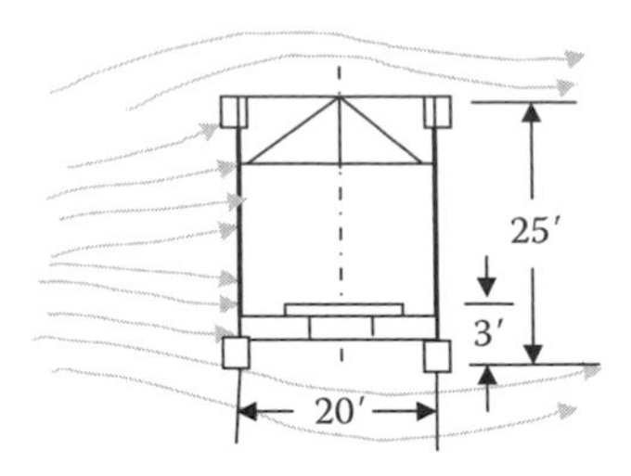

FIGURE E4.9

$r = 0.0022$ slug/ft^3
The solidity ratio $f = 1.00$ (plate girder)
$s/h = 0.67$
$C_{DT} = 1.15(C_D) = 1.15(2.0) = 2.3$ (Figure 4.20b)
$F_D = 2.3[1/2(0.0022)(110)^2]A_{RD} = [30.6(125)(15)/1000] = 57.4$ kips, not including the gust factor. If we assume a typical gust factor of 2.0, the design wind force $= 2.0(57.4) = 114.8$ kips.

Example 4.13

A 200 ft steel through truss railway span is shown in Figure E4.9. The solidity ratio, f, for this truss is 0.25. Determine the design wind force for a wind speed of 75 mph.

$s/h = 0.80$
$C_{DT} = 1.70(C_D) = 1.70(1.7) = 2.9$ (from Figure 4.20b)
$F_D = 2.9[1/2(0.0022)(110)^2]A_{RD} = [38.6(200)(25)(0.25)/1000] = 48.3$ kips, not including gust factor. If we assume a typical gust factor of 2.0, the design wind force $= 2.0(48.3) = 96.5$ kips.

The AREMA (2008) design recommendations for wind load on a loaded steel railway bridge superstructure assume that the maximum wind velocity at which trains can safely operate* will produce a wind pressure of 30 psf. The AREMA (2008) design recommendations for wind load on an unloaded steel railway bridge superstructure assume a maximum wind velocity corresponding to a typical hurricane event with a wind pressure of 50 psf (see Examples 4.14 and 4.15). In order to account for the effects of paired or multiple girders, these wind pressures are to be applied to a surface 50% greater than the projected surface area of a girder span. For truss spans the area is taken as the projected surface area of the windward truss plus the projected surface area of the leeward truss not shielded by the floor system.

The AREMA (2008) design recommendations also indicate that the load on the moving train is to be taken as 300 lb/ft at a distance 8 ft above the top of the rails.

* To avoid the overturning of empty cars.

Designers of railway bridges that carry high loads (e.g., double-stack rail cars) should review this recommendation.

Example 4.14

Determine the AREMA (2008) recommended design wind force for the unloaded girder span of Figure E4.8.

$$F_D = \frac{50(1.5)(15)(125)}{1000} = 140.6 \text{ kips.}$$

Example 4.15

Determine the AREMA-recommended design wind force for the unloaded truss span of Figure E4.9.

$$F_D = \frac{50(0.25)[(25-3)+25](200)}{1000} = 117.5 \text{ kips.}$$

4.4.2 Lateral Vibration Loads on Steel Railway Bridges

A dynamic analysis of lateral vibration similar to that performed in Section 4.3.2.1.2 for vertical vibration of the superstructure may be conducted. However, in order to simplify design procedures and ensure global rigidity of the superstructure, the AREMA (2008) recommendations include a notional vibration load to be resisted by the lateral bracing. Since the purpose of this design load is to ensure adequate lateral bracing (lateral stiffness to resist vibration from live load), it is not combined with other loads and forces (Waddell, 1916). Therefore, it is to be applied to the lateral bracing as an alternative to the wind load on a loaded railway bridge. This notional load is taken as 200 lb/ft on the loaded chord or flange of the superstructure (e.g., the top flange of a deck plate girder span) and 150 lb/ft on the unloaded chord or flange (e.g., the top chord of a through truss span).

Example 4.16

Determine the design wind load (including the notional vibration load) for the top and bottom lateral bracing of the 125 ft long ballasted deck steel deck plate girder railway span shown in Figure E4.8.

Unloaded span:

$W_T = (3+6)(50)(1.5) = 675$ lb/ft wind at the top lateral bracing,
$W_B = (6)(50)(1.5) = 450$ lb/ft wind at the bottom lateral bracing.

Loaded span:

$W_T = (3+6)(30)(1.5) + 300 = 705$ lb/ft wind at the top lateral bracing,
$W_B = (6)(30)(1.5) = 270$ lb/ft wind at the bottom lateral bracing,
$V_T = 200$ lb/ft vibration load at the top lateral bracing,

$V_B = 150$ lb/ft vibration load at the bottom lateral bracing.

Top lateral bracing design will be based on $W_T = 705$ lb/ft in addition to other lateral loads such as those due to live load (see Sections 4.3.2.3 and 4.3.2.4). Bottom lateral bracing design is based on $W_B = 450$ lb/ft.

4.4.3 Forces From the CWR on Steel Railway Bridges

Continuously welded rail is used in modern track construction because it diminishes dynamic effects (no impact forces due to joints in the rail), provides a smoother ride, and results in reduced rail maintenance and increased tie life. The rail may be fastened to the deck to provide lateral and longitudinal restraints.* The deck is also fastened to the superstructure to provide lateral and longitudinal restraints.†

The longitudinal forces generated due to restraint of thermal expansion and contraction of the rail may need to be considered in the design of some steel railway superstructures. The distribution of longitudinal forces through the superstructure may be of little concern for some superstructure types (e.g., multiple longitudinal beam spans and deck plate girder spans). However, for other types of superstructures and long spans, the longitudinal force path from rails to bearings may be of importance (e.g., floorbeams with direct fixation of track and some span floor systems). In addition, the CWR may experience internal stresses due to bridge span movements from thermal actions or live load bending. The magnitude of the CWR–bridge thermal interaction is governed by the following conditions:

- Movement of the bridge spans, in particular the maximum span length, which may freely expand in the bridge.
- The rail laying temperature (neutral temperature) and ambient temperature extremes at the bridge (the temperature ranges experienced by the rail and superstructure depend on neutral temperature, and maximum and minimum ambient temperatures).
- The type of bridge (open deck, ballasted‡) and bridge materials.
- The connection of rails-to-deck and deck-to-span interfaces.
- The cross-sectional area of the rail and the coefficient of thermal expansion of the bridge.
- The location of fixed and expansion bearings (spans with adjacent expansion bearings on the same pier create a long expansion length and generally provide the governing condition for design).

---

* Longitudinal restraint by elastic hold-down fasteners, friction, and/or rail anchors applied at the base-of-rail against the ties.

† Ballasted decks are generally rigidly connected to the superstructure. However, open deck spans may have various degrees of longitudinal restraint depending on deck-to-superstructure connection (see Chapter 3). Open decks are often fastened to the superstructure with bolts or "hook bolts" installed at regular intervals (e.g., every third tie).

‡ For ordinary ballast deck bridges, the differential thermal movements are generally accommodated by the ballast section.

The partial differential equation of horizontal motion from force equilibrium on a simply supported span bridge of constant mass and stiffness is

$$-\mathrm{EA}\frac{\partial^2 \overline{x}(x,t)}{\partial x^2} + m\frac{\partial^2 \overline{x}(x,t)}{\partial t^2} + c_x\frac{\partial \overline{x}(x,t)}{\partial t} = h(x,t), \tag{4.39}$$

where $\overline{x}(x,t)$ is the superstructure horizontal deflection at distance x and time t, EA is the axial stiffness of the span, c_x is the superstructure equivalent longitudinal viscous damping coefficient, and $h(x,t) = -k_\mathrm{d}\overline{x}(x,t)$ is the distributed longitudinal force due to thermal expansion transferred through an elastic rail-to-deck-to-superstructure system represented by an equivalent horizontal spring stiffness, k_d.* Longitudinal movement will typically occur primarily at the rail-to-deck or deck-to-superstructure interface depending on their respective degree of longitudinal restraint.† This model also oversimplifies the rail-to-deck-to-superstructure system with a single elastic horizontal stiffness. More sophisticated models may be developed‡ that use different elastic horizontal stiffnesses at the rail-to-deck and deck-to-superstructure interfaces.

Assuming negligible longitudinal viscous damping and neglecting superstructure longitudinal inertia effects (acceptable for ordinary steel railway superstructures), Equation 4.39 may be expressed as the differential equation (Fryba, 1996)

$$-\mathrm{EA}\frac{\mathrm{d}^2 \overline{x}(x)}{\mathrm{d}x^2} + k_\mathrm{d}\overline{x}(x) = 0, \tag{4.40}$$

which may be solved considering various failure criteria, such as:

- Safe rail gap (separation) on a bridge after fracture of the CWR. Rail fracture§ may occur due to cold weather contraction. The safe rail gap depends on individual railroad operating practice but is generally considered to be between 2 and 6 in.
- Safe stress in the CWR to preclude buckling.** Rail buckling, particularly at the typically weaker†† bridge approach track, may occur when rails on the bridge are highly longitudinally restrained such that large rail forces are created during hot weather rail expansion.

---

* A linear elastic spring is assumed for all levels of displacement in this model. Rail-to-deck and deck-to-superstructure interfaces may be more accurately modeled using bilinear springs, which following initial elastic behavior then acts perfectly plastic during steady state sliding friction displacement.

† For example, longitudinal movement may occur at the deck-to-superstructure interface for open deck beams and girders with smooth tops, and at the rail-to-deck interface for girders with substantial longitudinal resistance at the deck-to-superstructure interface (e.g., by rivet and bolt heads, restraint angles, and bars) positive deck connection or well-tensioned deck anchor bolts. Some modern elastic rail fasteners allow for longitudinal movement (no "hold-down" forces) at the rail-to-deck interface.

‡ Usually used in conjunction with computer-based FEA.

§ Modern North American heavy freight railroad CWR is considered to have a minimum fracture strength of about 300 kips.

** Modern North American heavy freight railroad CWR is considered to have a minimum safe buckling strength of about 150 kips.

†† Weaker lateral restraint behind abutment backwalls and approach track.

- Acceptable relative displacement between rail/deck and deck/span to preclude damage to deck and/or fasteners. Occurs due to excessive longitudinal movements (lower longitudinal stiffness) at either rail-to-deck or deck-to-superstructure.
- Avoidance of bearing component damage.

4.4.3.1 Safe Rail Separation Criteria

If the steel bridge is modeled as a series of spans with a distributed longitudinal force due to thermal expansion of rails transferred through an elastic deck system, the magnitude of the axial force in the CWR, $N(x)$, is

$$N(x) = \mathrm{EA_r}\left(\frac{\mathrm{d}\bar{x}(x)}{\mathrm{d}x} - \alpha\Delta t_\mathrm{c}\right), \tag{4.41}$$

where $\mathrm{EA_r}$ is the axial stiffness of the CWR; A_r is the cross-sectional area of the CWR (about 13 in.^{2} on typical heavy freight railroads); α is the coefficient of thermal expansion of the CWR and is $\sim 6.5 \times 10^{-6}/°\mathrm{F}$; Δt_c is the cold weather rail temperature change (with respect to neutral temperature).

Assuming zero displacement far from the rail break, $\bar{x}(\infty) = 0$, and zero force at the rail break, $N(0) = 0$, Equations 4.39 and 4.40 yield

$$\bar{x}(x) = \frac{-\alpha\Delta t_\mathrm{c}}{\lambda}\mathrm{e}^{-\lambda x}, \tag{4.42}$$

$$N(x) = -\mathrm{EA_r}\alpha\Delta t_\mathrm{c}(1 - \mathrm{e}^{-\lambda x}), \tag{4.43}$$

where $\lambda = \sqrt{k_1/\mathrm{EA_r}}$, k_1 is the longitudinal stiffness associated with a high strain rate event such as a rail breaking. It is generally about 1/2 of normal strain rate event (such as rail thermal expansion and contraction) stiffness.*

The separation of the CWR at fracture (assumed to occur over the expansion bearings) is

$$\Delta\bar{x}_\mathrm{s} = -\alpha\Delta t_\mathrm{c}\left(\frac{1}{\lambda_\mathrm{d}} + \frac{1}{\lambda_\mathrm{t}}\right), \tag{4.44}$$

where $\lambda_\mathrm{d} = \sqrt{k_\mathrm{d}/\mathrm{EA}}$,

$$\lambda_\mathrm{t} = \sqrt{\frac{k_\mathrm{t}}{\mathrm{EA}}},$$

k_d is the equivalent high strain rate event horizontal spring constant for the bridge deck, and k_t is the equivalent high strain rate event horizontal spring constant for the track approach. Figure 4.21 outlines the relationship of Equation 4.44, where $F_\mathrm{k} = 1 + \sqrt{(k_\mathrm{d}/k_\mathrm{t})}$.

* From Association of American Railrods, Transportation Technology Center, Inc. (AAR/TTCI) testing related to draft report of "Thermal Forces on Open Deck Steel Bridges," January, 2009.

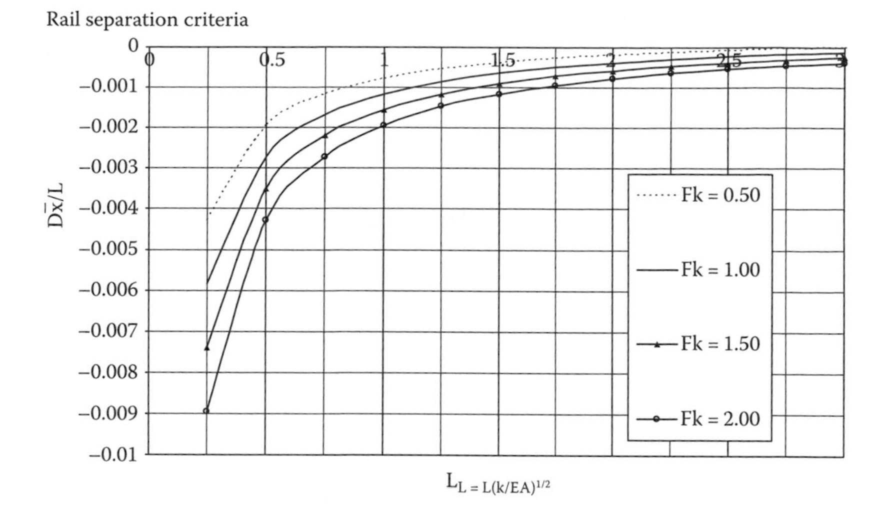

FIGURE 4.21 Typical relationships between rail separation, length of span, deck and track stiffness, and rail size for four stiffness ratios.

4.4.3.2 Safe Stress in the CWR to Preclude Buckling

Assuming a multiple span bridge with n spans of equal length, L, and alternating fixed and expansion bearings on substructures, Equation 4.39 with the boundary conditions (Figure 4.22),

- Zero displacement of the CWR away from the bridge [e.g., $\bar{x}_5(\infty) = 0$]
- Compatibility of displacements in the rail over bearings [e.g., $\bar{x}_2(L_2) = \bar{x}_3(0)$]
- Zero displacement at fixed bearings [e.g., $\bar{x}_7(0) = 0$]
- Rail force compatibility over bearings [e.g., $N_2(L_2) = N_3(0)$]
- Zero forces at expansion bearings [e.g., $N_6(L_6) = 0$]

may be solved to yield

$$\sigma_{\text{cwr}} = \frac{N_{n+1}(L)}{A_{\text{r}}} = -E\alpha\Delta t_{\text{h}}\left(1 + \frac{\alpha_0 \Delta T_{\text{h}}}{2\alpha\Delta t_{\text{h}}}(\lambda L - 1 + C_n)\right), \tag{4.45}$$

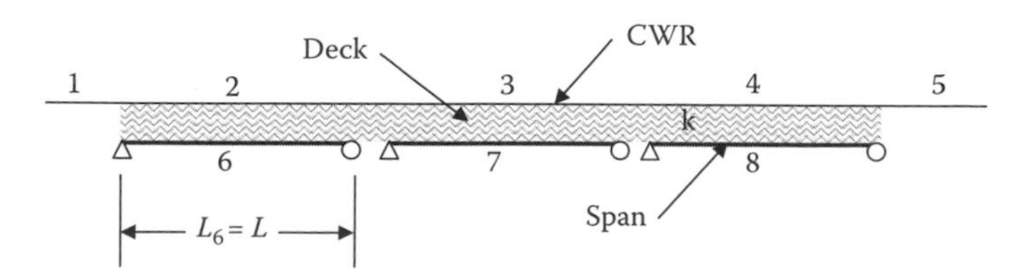

FIGURE 4.22 Three-span bridge model.

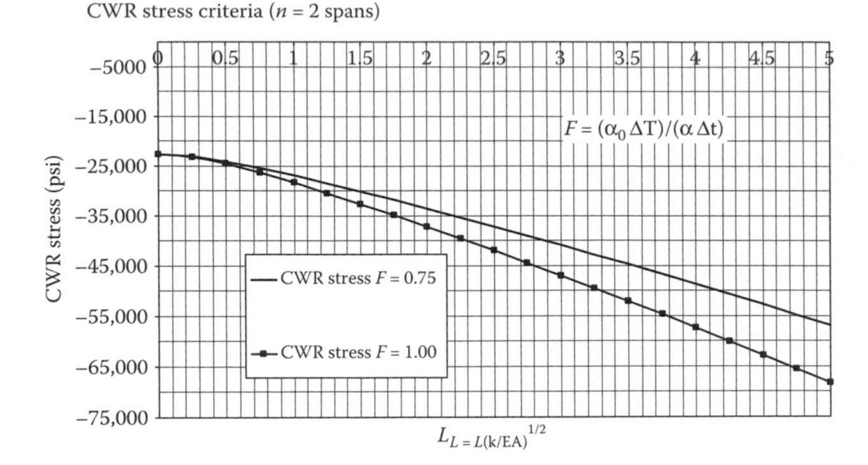

FIGURE 4.23 Typical relationships between stress in the CWR, length of span, deck and track stiffness, and rail size for two expansion/contraction ratios.

where $C_1 = e^{-\lambda L}$, $C_n = (\lambda L + C_{n-1})e^{-\lambda L}$ for $n \geq 2$, α_0 is the coefficient of thermal expansion of the bridge, Δt_h is the hot weather rail temperature change with respect to neutral temperature, ΔT_h is the hot weather bridge temperature change with respect to construction temperature, $\lambda = \sqrt{k_2/EA_r}$, and k_2 is the equivalent normal strain rate event horizontal spring constant for the rail-to-deck-to-superstructure system. Figure 4.23 outlines the relationship of Equation 4.45 for a two-span bridge with two expansion ratios.

The forces in the fixed bearings may also be determined from Equation 4.45 by considering

$$F_{abt} = N_3(0) - N_2(0), \tag{4.46}$$

$$F_{pier} = N_4(0) - N_3(0). \tag{4.47}$$

Figure 4.24 outlines the relationship of Equation 4.46 for a two-span bridge with two expansion ratios.

4.4.3.3 Acceptable Relative Displacement between Rail-to-Deck and Deck-to-Span

Assuming a multiple span bridge with n spans of equal length, L, and alternating fixed and expansion bearings on substructures, Equation 4.40 with the boundary conditions outlined in Section 4.4.3.2 may be solved to yield

$$\Delta\bar{x} = \bar{x}_{2(n+1)}(L) - \bar{x}_{n+1}(L) = \frac{\alpha_0 \Delta T}{2\lambda}(1 + \lambda L - C_n) \tag{4.48}$$

where ΔT is the bridge temperature change with respect to construction temperature, $\lambda = \sqrt{k/EA_r}$, and k_2 is the equivalent horizontal spring constant for the rail-to-deck-to-superstructure system.

Fixed bearing force at center pier (n = 2 spans)

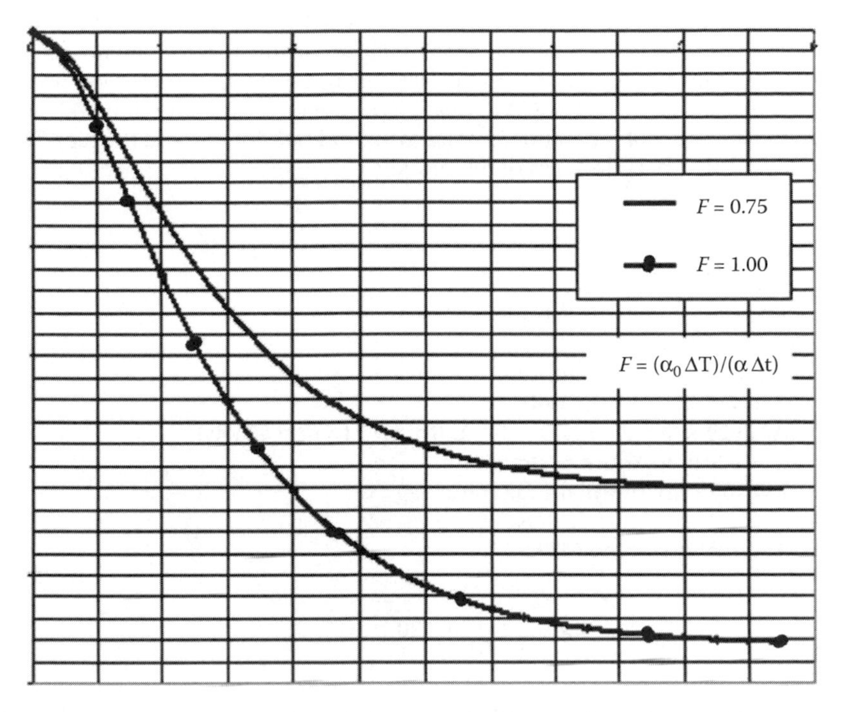

FIGURE 4.24 An example relationship between fixed bearing force, length of the span, deck and track stiffness, and rail size for two expansion/contraction ratios.

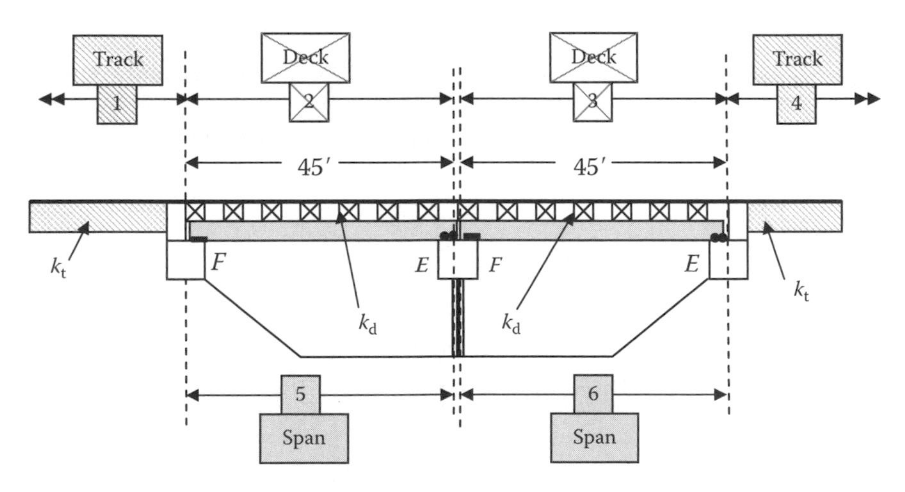

FIGURE E4.10

Example 4.17

The double track open deck steel multibeam railway bridge shown in Figure E4.1 comprises two 45 ft simple spans (Figure E4.10). The CWR with elastic rail fastenings is used on the friction-bolt fastened timber deck. Determine the maximum stress in the CWR, relative displacement between the rail and superstructure, rail separation, and longitudinal bearing force at the pier. The following are characteristics of the bridge:

$\Delta T_c = \Delta t_c = -100°F$
$\Delta t_h = 50°F$
$\Delta T_h = 40°F$
$\alpha_0 \Delta T_c = -5.00 \times 10^{-4}$(bridge)
$\alpha \Delta t_c = -6.50 \times 10^{-4}$(CWR)
$\alpha_0 \Delta T_h = 2.00 \times 10^{-4}$(bridge)
$\alpha \Delta t_h = 3.25 \times 10^{-4}$(CWR)
$EA_r = 29 \times 10^6(26) = 7.5 \times 10^8$ lb (for two typical CWRs)
$k_d = 400$ lb/in. (normal strain rate)
$k_t = 100$ lb/in. (normal strain rate)

Maximum stress in the CWR:

$$\lambda_d = \sqrt{\frac{k_d}{EA_r}} = 7.30 \times 10^{-4}\ \text{in.}^{-1},$$

$$\lambda_t = \sqrt{\frac{k_t}{EA_r}} = 3.65 \times 10^{-4}\ \text{in.}^{-1},$$

$$l_d L = 0.39.$$

Substitution into Equation 4.45 with $n = 2$ yields

$$\sigma_{cwr} = -9425\left[1 + 0.31\left(\lambda_d L - 1 + \left(\lambda_d L + e^{-\lambda_d L}\right)e^{-\lambda_d L}\right)\right],$$

$$\sigma_{cwr} = -9425\,[1 + 0.31\,(0.39 - 1 + (0.39 + 0.67)\,0.67)] = -9758 \text{ psi for two rails.}$$

Force in each rail = 9758(13)/1000 = 127 kips compression, OK.

Rail separation:

$k_d = 200$ lb/in. (rapid strain rate)
$k_t = 50$ lb/in. (rapid strain rate)

$$\lambda_d = \sqrt{\frac{k_d}{EA_r}} = 5.16 \times 10^{-4}\ \text{in.}^{-1},$$

$$\lambda_t = \sqrt{\frac{k_t}{EA_r}} = 2.58 \times 10^{-4}\ \text{in.}^{-1}.$$

Substitution into Equation 4.43 yields $\Delta\bar{x}_s = -6.50 \times 10^{-4}[(1/(5.16 \times 10^{-4})) + (1/(2.58 \times 10^{-4}))] = -3.78$ in., which may be excessive, requiring a longitudinally stiffer deck or track. For example, if a frozen ballast with $k_t = 100$ lb/in. (rapid strain rate) is considered, the rail gap is reduced to 3.0 in.

Relative displacement:
Substitution into Equation 4.48 with $n = 2$ yields

$$\Delta\bar{x} = 0.34\left(1 + \lambda L - (\lambda L + e^{-\lambda L})e^{-\lambda L}\right)$$

$\Delta\bar{x} = 0.34\,[1 + 0.39 - (0.39 + 0.67)\,0.67] = 0.23$ in., which is likely OK and will not cause deck fastener damage.

Fixed bearing force at the pier:
Substitution into Equation 4.47 yields

$$X_F = N_4(0) - N_3(0) = -EA_r\alpha\Delta t\left(\frac{\alpha_0\Delta T}{2\alpha\Delta t}(C_3 - C_2)\right)$$

$$= -487{,}500(0.385)(C_3 - C_2)$$

$l_d L = 0.39$
$C_1 = e^{-\lambda L} = 0.67$
$C_2 = (\lambda_d L + C_1)e^{-\lambda L} = 0.72$
$C_3 = (\lambda_d L + C_2)e^{-\lambda L} = 0.75$
$X_F = 187{,}688(0.75 - 0.72) = 5630$ lb for both bearings.

For bridges with short spans, the amount of thermal movement per span is small and generally easily accommodated by the normal tolerances of railroad track on bridges. It should be noted that the longitudinal stiffness values assumed for the rail-to-deck and deck-to-superstructure interfaces may not reflect values at a particular railroad location.

Example 4.18

An open deck steel deck truss bridge comprises a single 225 ft span. The CWR with elastic rail fastenings is used on the friction-bolt fastened timber deck. Determine the maximum stress in the CWR, relative displacement between the rail and superstructure, rail separation, and longitudinal bearing force at the abutment. The following are characteristics of the bridge:

$\Delta T_c = \Delta t_c = -70°F$
$\Delta t_h = 50°F$
$\Delta T_h = 40°F$
$\alpha_0\Delta T_c = -3.50 \times 10^{-4}$ (bridge)
$\alpha\Delta t_c = \alpha\Delta t_c = -4.55 \times 10^{-4}$ (CWR)
$\alpha_0\Delta T_h = 2.00 \times 10^{-4}$ (bridge)
$\alpha\Delta t_h = 3.25 \times 10^{-4}$ (CWR)
$EA_r = 29 \times 10^6(26) = 7.5 \times 10^8$ lb (for two typical CWRs)
$k_d = 400$ lb/in. (normal strain rate)
$k_t = 100$ lb/in. (normal strain rate)

Maximum stress in the CWR:

$$\lambda_d = \sqrt{\frac{k_d}{EA}} = 7.30 \times 10^{-4}\ \text{in.}^{-1}$$

$$\lambda_t = \sqrt{\frac{k_t}{EA}} = 3.65 \times 10^{-4}\ \text{in.}^{-1}$$

$$l_d L = 1.97.$$

Substitution into Equation 4.45 with $n = 1$ yields

$$\sigma_{cwr} = -9425\left(1 + 0.31(\lambda_d L - 1 + e^{-\lambda_d L})\right),$$

$$\sigma_{cwr} = -9425\,[1 + 0.31(1.97 - 1 + 0.14)] = -12{,}688\ \text{psi for both rails.}$$

Force in each rail = 12.688(13)/1000 = 165 kips compression; there is risk of rail buckling.

Rail separation:

$k_d = 200$ lb/in. (rapid strain rate)
$k_t = 50$ lb/in. (rapid strain rate)

$$\lambda_d = \sqrt{\frac{k_d}{EA_r}} = 5.16 \times 10^{-4}\ \text{in.}^{-1},$$

$$\lambda_t = \sqrt{\frac{k_t}{EA_r}} = 2.58 \times 10^{-4}\ \text{in.}^{1}.$$

Substitution into Equation 4.44 yields

$$\Delta\bar{x}_s = -4.55 \times 10^{-4}\left(\frac{1}{5.16 \times 10^{-4}} + \frac{1}{2.58 \times 10^{-4}}\right) = 2.65\ \text{in., OK.}$$

Relative displacement:
Substitution into Equation 4.48 with $n = 1$ yields

$$\Delta\bar{x} = 0.24(1 + \lambda L - e^{-\lambda L}),$$

$\Delta\bar{x} = 0.24\,(1 + 1.97 - 0.14) = 0.68$ in., which may be excessive requiring a longitudinally stiffer deck.

Fixed bearing force at the abutment:
Substitution into Equation 4.47 yields

$$X_F = N_3(0) - N_2(0) = -EA_r\alpha\Delta t\left(\frac{\alpha_0\Delta T}{2\alpha\Delta t}(C_2 - C_1)\right) = -131,250(C_2 - C_1),$$

$C_1 = e^{-\lambda L} = 0.14,$
$C_2 = (\lambda_d L + C_1)e^{-\lambda L} = 0.30,$
$X_F = 131{,}250(0.30 - 0.14) = 20{,}386$ lb for both bearings.

Example 4.19

In order to reduce the relative displacements at the rail-to-deck-to-superstructure system in Example 4.18, a fastening system on the bridge with greater horizontal elastic spring stiffness is proposed. Determine the maximum stress in the CWR, relative displacement between the rail and superstructure and rail separation.

$k_d = 850$ lb/in. (normal strain rate)
$k_t = 100$ lb/in. (normal strain rate)

Relative displacement:

$$\lambda_d = \sqrt{\frac{k_d}{EA}} = 1.06 \times 10^{-3}\ \text{in.}^{-1},$$

$$l_d L = 2.87.$$

Substitution into Equation 4.48 with $n = 1$ yields

$$\Delta\overline{x} = 0.16(1 + \lambda L - e^{-\lambda L}),$$

$\Delta\overline{x} = 0.16(1 + 2.87 - 0.056) = 0.63$ in.; relative displacement remains quite large.

Maximum stress in the CWR:
Substitution into Equation 4.45 with $n = 1$ yields

$$\sigma_{cwr} = -9425\left(1 + 0.31(\lambda_d L - 1 + e^{-\lambda_d L})\right)$$

$$\sigma_{cwr} = -9425\,[1 + 0.31(2.87 - 1 + 0.056)] = -15{,}052\ \text{psi for both rails.}$$

Force in each rail = 15,052(13)/1000 = 196 kips compression; the rail may buckle.

Rail separation:
$k_d = 425$ lb/in. (rapid strain rate)
$k_t = 50$ lb/in. (rapid strain rate)

$$\lambda_d = \sqrt{\frac{k_d}{EA_r}} = 7.53 \times 10^{-4}\ \text{in.}^{-1},$$

$$\lambda_t = \sqrt{\frac{k_t}{EA_r}} = 2.58 \times 10^{-4}\ \text{in.}^{-1}.$$

Substitution into Equation 4.44 yields

$$\Delta\overline{x}_s = -4.55 \times 10^{-4}\left(\frac{1}{7.53 \times 10^{-4}} + \frac{1}{2.58 \times 10^{-4}}\right) = 2.37\ \text{in., OK.}$$

Examples 4.18 and 4.19 illustrate that, for long open deck spans, there are conflicting design requirements that the rail-to-deck-to-superstructure connection be flexible enough to avoid excessive compressive stress in the CWR (that could precipitate buckling), and rigid enough to reduce rail separation and relative displacements at the rail-to-deck-to-superstructure interfaces.

Therefore, in order to allow for movement between the rail and superstructure while providing sufficient rail anchoring to preclude excessive relative displacements, the CWR may be anchored to only a portion of the span length. The portion of the span length to which the CWR is anchored should be adjacent to the fixed bearings to allow the necessary movement between the rail and superstructure (anchored CWR over the expansion bearing areas will resist the thermal movements of the span). The effect of this is illustrated in Example 4.20.

Example 4.20

In order to reduce the relative displacement between the rail and superstructure in Example 4.18, anchoring the CWR to only a portion of the rail is proposed. If only 1/3 of the span length (from fixed bearings) has the CWR anchored to the deck, determine the maximum stress in the CWR, relative displacement between the rail and superstructure and rail separation,

Maximum stress in the CWR:

$$\lambda_d = \sqrt{\frac{k_d}{EA}} = 7.30 \times 10^{-4}\ \text{in.}^{-1},$$

$$\lambda_t = \sqrt{\frac{k_t}{EA}}\ 3.65 \times 10^{-4}\ \text{in.}^{-1},$$

$$l_d L = (75)(12)(7.30 \times 10^{-4}) = 0.66.$$

into Equation 4.45 with $n = 1$ yields

$$\sigma_{cwr} = -9425\left(1 + 0.31(\lambda_d L - 1 + e^{-\lambda_d L})\right),$$

$$\sigma_{cwr} = -9425\,[1 + 0.31(0.66 - 1 + 0.52)] = -9937\ \text{psi for both rails.}$$

Force in each rail $= 9937(13)/1000 = 129$ kips compression, OK.

Relative displacement:

Substitution into Equation 4.48 with $n = 1$ yields

$$\Delta\bar{x} = 0.24\left(1 + \lambda L - e^{-\lambda L}\right),$$

$$\Delta\bar{x} = 0.24\,(1 + 0.66 - 0.52) = 0.27\ \text{in., OK.}$$

Rail separation:

Substitution into Equation 4.44 yields

$$\Delta\bar{x}_s = -4.55 \times 10^{-4}\left(\frac{1}{5.16 \times 10^{-4}} + \frac{1}{2.58 \times 10^{-4}}\right) = 2.65\ \text{in., OK.}$$

Fixed bearing force at the abutment:
Substitution into Equation 4.47 yields

$$X_F = N_3(0) - N_2(0) = -EA\alpha\Delta t\left(\frac{\alpha_0\Delta T}{2\alpha\Delta t}(C_2 - C_1)\right) = -131{,}250(C_2 - C_1),$$

$$C_1 = e^{-\lambda L} = 0.52,$$

$$C_2 = \left(\lambda_d L + C_1\right)e^{-\lambda L} = 0.61 = 0.61,$$

$$X_F = 131{,}250(0.61 - 0.52) = 11{,}585 \text{ lb for both bearings.}$$

4.4.3.4 Design for the CWR on Steel Railway Bridges

Based on similar considerations, AREMA (2008) and many railway companies establish standard practice for anchoring CWR to long open deck steel spans. In general, the recommended practice is to use longitudinal rail anchors on approaches, and near fixed ends of spans, allowing some movement near expansion ends of spans.*

4.4.4 Seismic Forces on Steel Railway Bridges

The level of seismic dynamic analysis required depends on the location and characteristics of the bridge.

An equivalent static analysis is often used in the analysis of ordinary steel railway bridges where the response to seismic forces is depicted primarily by the first or fundamental vibration mode. Steel railway bridges that may be analyzed by an equivalent static analysis are typically simply supported, not (or only slightly) skewed or curved, and have spans of almost equal length and supporting substructures of almost equal stiffness. Seismic forces in an equivalent static analysis are developed based on a period-dependent coefficient and the weight of the bridge. AREMA (2008) recommends the use of a seismic response coefficient and the uniform load method.†

The seismic forces on complex steel railway bridges are generally determined for use in a dynamic structural analysis.‡ These loads are typically represented by an elastic design seismic response spectrum. AREMA (2008) recommends the use of a normalized response spectrum based on the seismic response coefficient.

4.4.4.1 Equivalent Static Lateral Force

The equivalent static distributed lateral force, $p(x)$, applied to the steel superstructures of a railway bridge is

$$p(x) = C_n w(x) \tag{4.49}$$

where $C_n = (1.2ASD/T_n^{2/3}) \le 2.5AD$ is the seismic response coefficient for the nth mode of vibration and 5% damping ratio; $w(x)$ is the distributed weight of the

* Rail expansion joints are sometimes used on very long bridges or bridges with unusual bearing configuration (i.e., adjacent expansion bearings on long spans).
† For some bridges it may be appropriate to consider the multimode dynamic analysis method.
‡ AREMA Chapter 9 indicates that a modal analysis is appropriate for such railway bridges.

superstructure, A is the base acceleration ratio determined from appropriate geological sources* for the design return period,† S is the site coefficient between 1.0 and 2.0 depending on foundation soil conditions,‡ $D = [1.5/(0.4\xi + 1) + 0.5]$ is the damping adjustment factor to account for the actual superstructure percentage of critical damping, ξ,§ T_n is the natural period of the nth mode of vibration and is equal to $2\pi/\omega_n$, and ω_n is the natural frequency of the nth mode of vibration (see Tables 4.2 and 4.3 and Figure 4.8).

However, in some cases, the development of the equivalent static distributed lateral force based on the seismic response coefficient is inappropriate and consideration of loading based on site-specific information is required.**

The equivalent static lateral distributed force, $p(x)$, is calculated in two orthogonal directions (longitudinal and transverse for ordinary bridges). Following a linear elastic analysis†† in each direction, forces are distributed to superstructure members based on load path, support conditions, and stiffness. Since these member loads are orthogonal and uncorrelated, they must be combined‡‡ for design purposes. AREMA (2008) recommends the method often referred to as the 100%–30% rule (Equations 4.50a and b) to combine the seismic loads for member design.

$$\mathrm{EQ} = 1.00F_{\mathrm{T}} + 0.30F_{\mathrm{L}}, \tag{4.50a}$$

$$\mathrm{EQ} = 0.30F_{\mathrm{T}} + 1.00F_{\mathrm{L}}, \tag{4.50b}$$

where EQ is the combined seismic design force, F_{T} is the absolute value of the seismic force in the transverse direction, and F_{L} is the absolute value of the seismic force in the longitudinal direction.

4.4.4.2 Response Spectrum Analysis of Steel Railway Superstructures

The response spectrum used to represent the seismic loading of more complex steel superstructures is a plot of the peak value of the response as a function of the natural period of vibration of the superstructure. These are typically plotted for a particular damping ratio§§ and response (deformation, velocity, or acceleration). AREMA (2008) recommends the use of a normalized spectral response based on the seismic response

---

* For example, the U.S. Department of the Interior Geological Survey maps.

† The design return period depends on the earthquake event frequency and the limit state under consideration (serviceability, ultimate, or survivability).

‡ Rock, soil type, stratigraphy, depth, soil stiffness, and shear wave velocity are considered in the site coefficient.

§ Established from tests or other sources in the literature of structural dynamics. The percentage of critical damping for steel superstructures is often less than 5% and depends on materials, structural system/foundation, deck type and whether the structural response is linear elastic or post yield.

** For example, some bridges on soft-clays and silts where vibration modes greater than the fundamental mode have periods of less than 0.3 s and bridges near faults or in areas of high seismicity. In these cases, alternative equations, available in seismic design standards and guidelines, for C_n may apply.

†† Linear elastic analysis is used for the equivalent lateral force method at the serviceability limit state.

‡‡ These combined forces account for the directional uncertainty and simultaneous occurrence of the seismic design forces in members.

§§ Often established for a damping ratio (percentage of critical damping) of 5%.

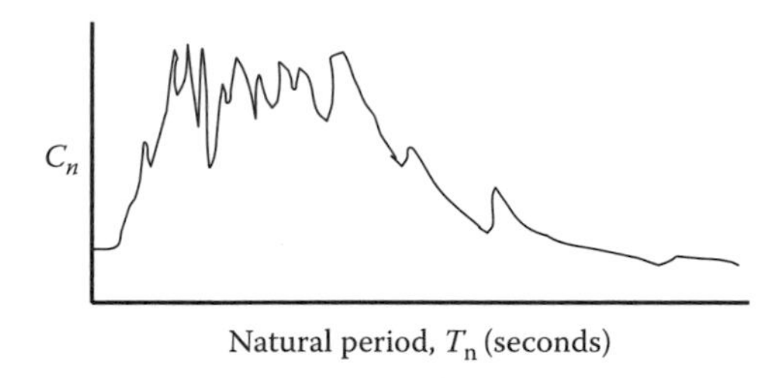

FIGURE 4.25 Typical actual response spectrum.

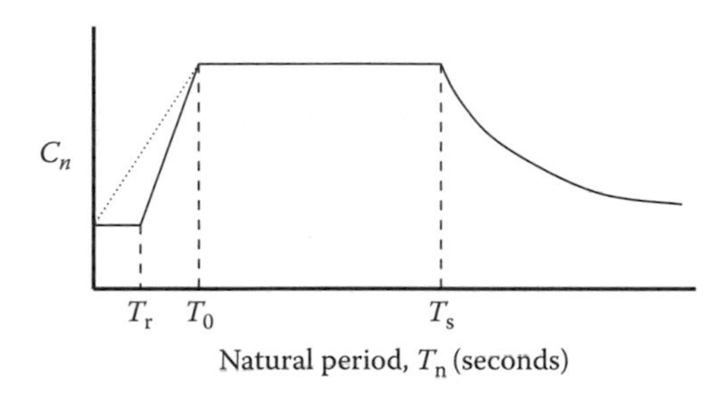

FIGURE 4.26 Typical design response spectrum.

coefficient. This is essentially a pseudo-acceleration* response spectrum normalized by the natural period of vibration, T_n. The actual pseudo-acceleration response spectrum for a given earthquake and the design pseudo-acceleration response spectrum will typically look like the plots of Figures 4.25 and 4.26, respectively.

AREMA (2008) recommends the following with respect to the normalized design response spectra: T_r is the maximum natural vibration period for essentially rigid response, $T_0 = 0.096S$, and $T_s = (0.48S)^{3/2}$.

However, dynamic analyses of railway bridges typically underestimate the actual natural vibration period and, therefore, the response of the bridge for low period structures. AREMA (2008) recommends a design response spectrum without reduced response (or C_n) below T_0 (Figure 4.27) unless the effects of foundation flexibility, foundation rotational movement, and lateral span flexibility were included in the dynamic analysis. In some cases, the development of the response spectra from the seismic response coefficient is inappropriate and consideration of loading based on site-specific response spectra is required.†

The response spectrum must be calculated in each orthogonal and uncorrelated direction (longitudinal and transverse) and, therefore, must be combined for design purposes. AREMA (2008) recommends either the square root sum of squares (SSRC)

* For steel bridge superstructures with low damping and short vibration periods the pseudo-acceleration response is a close approximation the actual acceleration response.

† For example, where $A \geq 0.2$ and $T_n \geq 0.7$ for bridges on very soft clays and silts, and for bridges on soft clays and silts where vibration modes greater than the fundamental mode have periods of less than 0.3s and bridges near faults or in areas of high seismicity. In these cases, alternative equations, available in seismic design standards and guidelines, for C_n may apply.

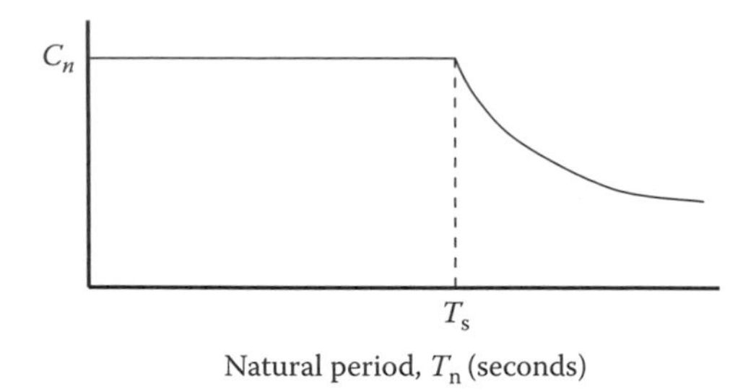

FIGURE 4.27 Typical AREMA design response spectrum used with simple dynamic analyses.

method (Equation 4.51) or the method often referred to as the 100%–30% rule (Equations 4.50a and b) to combine the seismic loads.

$$F = \sqrt{F_T^2 + F_L^2}. \tag{4.51}$$

Example 4.21

The normalized response spectrum is required for a 100 ft long steel girder span with the following properties:

Weight: 3000 lb/ft
$I_x = 100{,}000\ \text{in.}^4$
$I_y = 5000\ \text{in.}^4$
$\xi = 3\%$

The bridge is located where $A = 10\%$ and founded on material with $S = 1.0$ (rock)

$$T_s = [0.48(1.0)]^{3/2} = 0.33\ \text{s}$$

$$C_n = \frac{0.142}{T_n^{2/3}} \le 0.30.$$

The normalized response spectrum is shown in Figure E4.11.

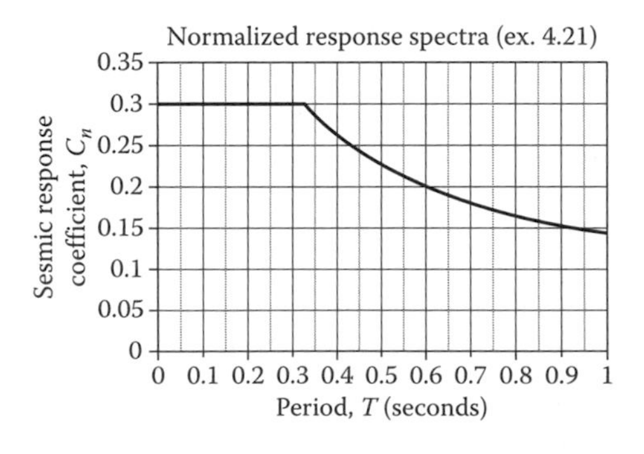

FIGURE E4.11

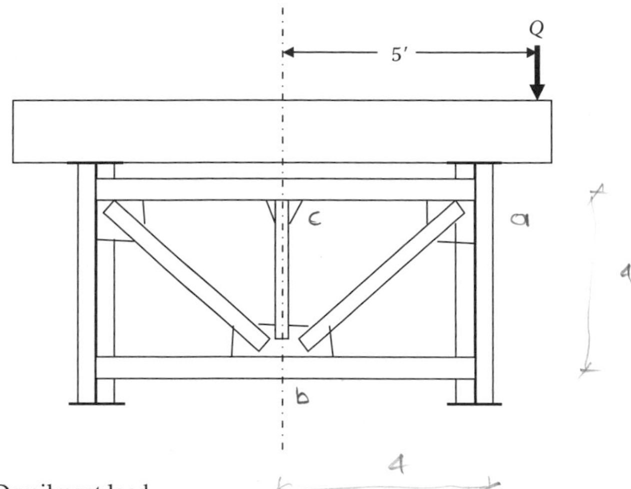

FIGURE 4.28 Derailment load.

4.4.5 Loads Relating to Overall Stability of the Superstructure

4.4.5.1 Derailment Load

Events such as train derailments on bridges are relatively infrequent. However, particularly on long bridges, train derailments can occur and create overall instability of individual spans. AREMA (2008) recommends an eccentric derailment load be used to ensure stability of spans. This derailment load, Q, is a single line of wheel loads, including impact, at a 5 ft eccentricity from the track centerline (Figure 4.28). It is used as a load case for the design of cross frames and diaphragms in beam and girder spans requiring lateral bracing.* AREMA (2008) recognizes that damage to some span elements may occur in these relatively extreme, but infrequent, events. A 50% increase in allowable stress is permissible when determining stresses in cross frames, diaphragms, anchor rods, or other members resisting overall instability of the span.

Example 4.22

Determine the Cooper's E80 derailment load forces in the brace frame modeled in Figure E4.12. The calculated impact factor for the span is 40%.

The derailment force applied to the cross frame at (a) is assumed to be transferred to the opposite girder through the cross frame members *ab* and *ac*. The force in the members is

$$P_{\text{ab}} = \frac{[80(1.40)/2(5)](8)[1 + ((5-4)/8)]}{\cos 45^\circ} = 142.6\,\text{kips compression.}$$

$$P_{\text{ac}} = 142.6 \sin 45^\circ = 100.8\,\text{kips tension.}$$

* The tendency for the span to "roll over" is prevented by lateral bracing between beams or girders.

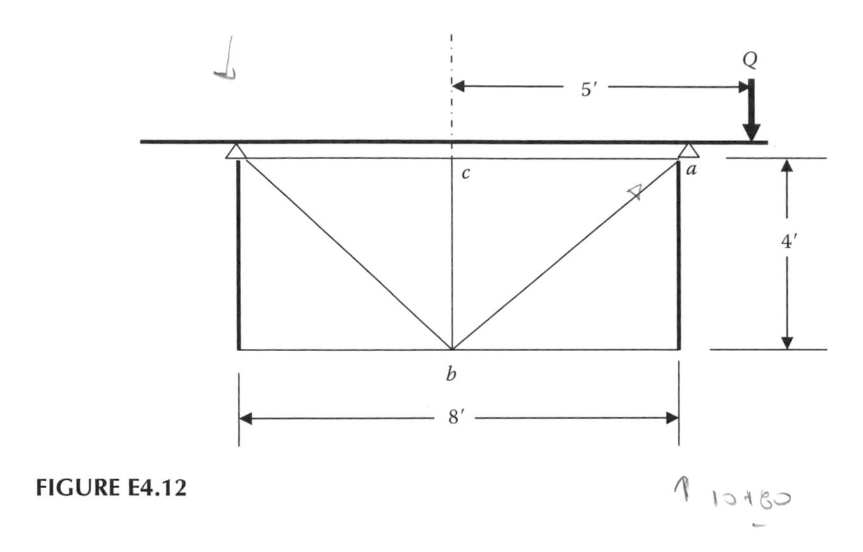

FIGURE E4.12

These forces are checked against the usual allowable stresses increased by 50%.

4.4.5.2 Other Loads for Overall Lateral Stability

The overall stability of the superstructure against wind, nosing, and centrifugal forces must also be ensured. The stability of spans and towers should be calculated using a live load, without impact, of 1200 lb/ft.* On multiple track bridges this live should be placed on the most leeward track on the bridge. A 50% increase in allowable stress is permissible when determining stresses anchor rods or other members resisting overall instability of the span.

4.4.6 Pedestrian Loads

Typical walkways for steel railway bridges consist of a steel grating or other system with nonslip surfaces. The walkway components are designed for a load of 85 psf and maximum deflection of 1/160 of the walkway span length. Guardrails for pedestrian walkways are typically designed for railing and postloads of 200 lb applied laterally or vertically at the location of maximum effect.

4.4.7 Load and Force Combinations for Design of Steel Railway Superstructures

AREMA does not provide explicit load combinations but incorporates combinations in various design recommendations (Sorgenfrei and Marianos, 2000). Table 4.5 outlines load combinations that apply to steel superstructure design found in various recommendations of AREMA.

* This represents a uniform load of empty rail cars.

TABLE 4.5
Load Combinations for Steel Railway Superstructure Design

Load Case	Load Combinations	Members	F_L
A1	DL + LL + I + CF	All members	1.00
A2	DL + LLT + I + CF	Truss web members	1.33
B1	DL + LL + I + W + LF + N + CWR	All members, except floorbeam hangers and high strength bolts	1.25
B1A	DL + LL + I + W + LF + N + CWR	Floorbeam hangers and high strength bolts	1.00
B2	DL + LLT + I + W + LF + N + CWR	Truss web members, except floorbeam hangers	1.66
C	(LL + I) range	All members	f_{fat}
D1	SL + N + CF	Members resisting overall instability	1.50
D2	Q	Members resisting overall instability	1.50
E1	DL + EQ	All members	1.50
E2	DL + LL + I + CF + EQ	Members in long bridges only	1.50
F	W or LV	Members loaded by wind only	1.00
G	DF	Cross frames, diaphragms, anchor rods	1.50
H1	DL	Members stressed during lifting or jacking	1.50
H2	DL	Members stressed during erection	1.25
H3	DL + W	Members stressed during erection	1.33

F_L = Allowable stress load factor (multiplier for basic allowable stresses), DL = Dead loads (self weight, superimposed dead loads, erection loads) (see Section 4.2), LL = Live loads (see Section 4.3.1), I = Impact (dynamic amplification) (see Section 4.3.2), CF = Centrifugal force (see Section 4.3.4), W = Wind forces (on live load and bridge) (see Section 4.4.1), LF = Longitudinal forces from equipment (braking and locomotive traction) (see Section 4.3.2.2), N = Lateral forces from equipment (nosing) (see Section 4.3.2.3), CWR = Forces from CWR (lateral and longitudinal) (see Section 4.4.3), EQ = Forces from earthquake (combined transverse and longitudinal) (see Section 4.4.4), DF = Lateral forces from out-of-plane bending and from load distribution effects (see Section 4.4.5.1), LV = "Notional" lateral vibration load (see Section 4.4.2), LLT = Live load that creates a total stress increase of 33% over the design stress (computed from load combination A1) in the most highly stressed chord member of the truss. This load ensures that web members attain their safe capacity at about the same increased live load as other truss members due to the observation that in steel railway trusses, the web members reach capacity prior to other members in the truss. This live load, LLT, is based on the requirements discussed in Chapter 5, Section 5.3.2.3.4, SL = Live load on leeward track of 1200 lb/ft without impact, I (see Section 4.4.5.2), Q = Derailment load, f_{fat} = Allowable stress based on member loaded length and fatigue detail category.

REFERENCES

American Railway Engineering Association (AREA), 1949, Test results on relation of impact to speed, *AREA Proceedings*, 50, 432–443.

American Railway Engineering Association (AREA), 1960, Summary of tests on steel girder spans, *AREA Proceedings*, 61, 151–178.

American Railway Engineering Association (AREA), 1966, Reduction of impact forces on ballasted deck bridges, *AREA Proceedings*, 67, 699.

American Railway Engineering and Maintenance-of-Way Association (AREMA), 2008, Steel structures, in *Manual for Railway Engineering*, Chapter 15, Lanham, MD.

American Society of Civil Engineers (ASCE), 2005, *Minimum Design Loads for Buildings and Other Structures*, ASCE/SEI 7-05, Reston, VA.

Beer, F.P. and Johnston, E.R., 1976, *Statics and Dynamics*, 3rd Edition, McGraw-Hill, New York.

Byers, W.G. 1970, Impact from railway loading on steel girder spans, *Journal of the Structural Division, ASCE*, 96(ST6), 1093–1103.

Carnahan, B., Luther, H.H., and Wilkes, J.O., 1969, *Applied Numerical Methods*, Wiley, New York.

Chopra, A.K., 2004, *Dynamics of Structures*, 2nd Edition, Prentice-Hall, New Jersey.

Clough, R.W. and Penzien, J., 1975, *Dynamics of Structures*, McGraw-Hill, New York.

Cooper, T., 1894, Train loadings for railroad bridges, *Transactions of the American Society of Civil Engineers*, 31, 174–184.

Dick, S.M., 2002, "Bending Moment Approximation Analysis for Use in Fatigue Life Evaluation of Steel Railway Girder Bridges," PhD Thesis, University of Kansas, Lawrence, KS.

Dick, S.M., 2006, *Estimation of Cycles for Railroad Girder Fatigue Life Assessment*, Bridge Structures, Taylor & Francis.

Foutch, D.A., Tobias, D., and Otter, D., 1996, *Analytical Investigation of the Longitudinal Loads in an Open-Deck Through-Plate-Girder Bridge*, Report R-894, Association of American Railroads.

Foutch, D.A., Tobias, D.H., Otter, D.E., LoPresti, J.A., and Uppal, A.S., 1997, *Experimental and Analytical Investigation of the Longitudinal Loads in an Open-Deck Plate-Girder Railway Bridge*, Report R-905, Association of American Railroads.

Fryba, L., 1972, *Vibration of Solids and Structures under Moving Loads*, Noordoff International, Groningen, Netherlands.

Fryba, L., 1996, *Dynamics of Railway Bridges*, Thomas Telford, London, UK.

Heywood, R., Roberts, W., and Boully, G., 2001, Dynamic loading of bridges, *Journal of the Transportation Research Board*, 1770, 58–66, National Academy Press, Washington, DC.

Inglis, C.E., 1928, *Impact in Railway-Bridges*, Minutes of Proceedings of the Institution of Civil Engineers, London, England.

Kreyszig, E., 1972, *Advanced Engineering Mathematics*, Wiley, New York.

Lin, J.H., 2006, Response of a bridge to a moving vehicle load, *Canadian Journal of Civil Engineering*, 33(1), 49–57.

Liu, H., 1991, *Wind Engineering—A Handbook for Structural Engineers*, Prentice-Hall, Englewood Cliffs, NJ.

LoPresti, J.A., Otter, D.A., Tobias, D.H., and Foutch, D.A., 1998, *Longitudinal Forces in an Open-Deck Steel Bridge*, Technology Digest 98-007, Association of American Railroads.

LoPresti, J.A. and Otter, D.A., 1998, *Longitudinal Forces in a Two-Span Open-Deck Steel Bridge at FAST*, Technology Digest 98-020, Association of American Railroads.

McLean, W.G. and Nelson, E.W., 1962, *Engineering Mechanics*, McGraw-Hill, New York.

Otter, D.E., LoPresti, J., Foutch, D.A., and Tobias, D.H., 1996, *Longitudinal Forces in an Open-Deck Steel Plate-Girder Bridge*, Technology Digest 96-024, Association of American Railroads.

Otter, D.E., LoPresti, J., Foutch, D.A., and Tobias, D.H., 1997, Longitudinal forces in an open-deck steel plate-girder bridge, *AREA Proceedings*, 98, 101–105.

Otter, D.E., Joy, R., and LoPresti, J.A., 1999, *Longitudinal Forces in a Single-Span, Ballasted-Deck, Plate-Girder Bridge*, Report R-935, Association of American Railroads.

Otter, D.E., Sweeney, R.A.P., and Dick, S.M., 2000, *Development of Design Guidelines for Longitudinal Forces in Bridges*, Technology Digest 00-018, Association of American Railroads.

Otter, D.E., Doe, B., and Belport, S., 2005, *Rail Car Lateral Forces for Bridge Design and Rating*, Technology Digest 05-002, Association of American Railroads.

Ruble, E.J., 1955, Impact in railroad bridges, *Proceedings of the American Society of Civil Engineers*, Seperate No. 736, 1–36.

Sanders, W.W. and Munse, W.H., 1969, Load distribution in steel railway bridges, *Journal of the Structural Division, ASCE*, 95(ST12), 2763–2781.

Simiu, E. and Scanlon, R.H., 1986, *Wind Effects on Structures*, Wiley, New York.

Sorgenfrei, D.F. and Marianos, W.N., 2000, Railroad bridges, in *Bridge Engineering Handbook*, Chen, W.F. and Duan, L. (Eds), CRC Press, Boca Raton, FL.

Sweeney, R.A.P., Oommen, G., and Le, H., 1997, Impact of site measurements on the evaluation of steel railway bridges, *International Association for Bridge and Structural Engineering Reports*, 76, 139–147.

Taly, N., 1998, *Design of Modern Highway Bridges*, McGraw-Hill, New York.

Tobias Otter, D.E. and LoPresti, J.A., 1998, Longitudinal Forces in Three Open-Deck Steel Bridges, *Proceedings of the 1998 AREMA Technical Conference*, Lanham, MD.

Tobias, D.H., Foutch, D.A., Lee, K., Otter, D.E., and LoPresti, J.A., 1999, *Experimental and Analytical Investigation of the Longitudinal Loads in a Multi-span Railway Bridge*, Report R-927, Association of American Railroads.

Tse, F.S., Morse, I.E., and Hinkle, R.T., 1978, *Mechanical Vibrations—Theory and Applications*, Allyn and Bacon, Boston, MA.

Uppal, A.S., Otter, D.E., and Joy, R.B., 2001, *Longitudinal Forces in Bridges due to Revenue Service*, Report R-950, Association of American Railroads.

Uppal, A.S., Otter, D.E., and Doe, B.E., 2003, *Impact Loads in Railroad Short Steel Bridge Spans*, Report R-964, Association of American Railroads.

Veletsos, A.S. and Huang, T., 1970, Analysis of dynamic response of highway bridges, *Journal of Engineering Mechanics, ASCE*, 96 (EM5), 593–620.

Waddell, J.A.L., 1916, *Bridge Engineering*, Vols. 1 & 2, Wiley, New York.

Yang, Y.B., Liao, S.S., and Lin, B.H., 1995, Impact formulas for vehicles moving over simple span continuous beams, *Journal of Structural Engineering, ASCE*, 121(11), 1644–1650.

5 Structural Analysis and Design of Steel Railway Bridges

5.1 INTRODUCTION

Elastic structural analysis procedures are used for steel railway bridge design based on the ASD methods of the AREMA (2008) *Manual for Railway Engineering*.* Strength (yield, ultimate, and stability), fatigue and fracture, and serviceability (deflection and vibration) criteria (or limit states) must be considered for safe and reliable steel railway bridge design.

Fatigue, or the failure of steel at nominal cyclical stresses lower than yield stress, is a phenomenon that occurs due to the cyclical nature of railway traffic and the presence of stress concentrations in the superstructure. Fracture behavior is primarily related to material (see Chapter 2) and fabricated detail characteristics. Therefore, it is not directly affected by design methodology.

Ordinary steel railway superstructure design is often governed by deflection (stiffness) and fatigue criteria. Since live load deflection and fatigue strength details are evaluated at service loads, ASD is generally appropriate for steel railway superstructures. Nevertheless, combined with the current state of knowledge concerning material behavior, a better understanding of the railway live load spectrum may precipitate a probabilistic reliability-based approach to future steel railway superstructure design. The design service life of railroad bridges is generally considered to be about 80 years.

* Recommended practices for the design of railroad bridges are developed and maintained by the American Railway Engineering and Maintenance-of-Way Association (AREMA, 2008). Chapter 15—Steel Structures provides detailed recommendations for the design of steel railway bridges for spans up to 400 ft in length, standard gage track (56.5″), and North American freight and passenger equipment at speeds up to 79 and 90 mph, respectively. The recommendations may be used for longer-span bridges with supplemental requirements. Many railroad companies establish steel railway bridge design criteria based on these recommended practices.

5.2 STRUCTURAL ANALYSIS OF STEEL RAILWAY SUPERSTRUCTURES

Railway live loads are a longitudinal series of moving concentrated axle or wheel loads (see Chapter 4) that are fixed with respect to lateral position.* The maximum elastic static normal stresses, shear stresses, and deformations in a steel superstructure member depend on the global longitudinal position of the railway live load. In addition, these maximum elastic static stresses are amplified due to dynamic effects† (Figure 5.1). The local longitudinal and lateral distribution of these moving loads to the deck and supporting members, as well as their dynamic effects, is considered in Chapter 4.

5.2.1 LIVE LOAD ANALYSIS OF STEEL RAILWAY SUPERSTRUCTURES

The static analysis of railway superstructures involves the determination of the maximum deformations and stresses caused by the moving loads. These maximum effects are influenced by the position of the moving load. Maximum effects are of primary interest but the designer must also carefully consider effects of the moving load at other locations on the span where changes of cross section, splices, fatigue effects,‡ and other considerations may require investigation.

Therefore, a structural analysis is required for multiple load positions to determine maximum effects. The necessary analytical effort may be reduced by careful consideration of the load configuration, the use of influence lines, and experience. Furthermore, if the concentrated load configuration remains constant (typical of Cooper's E and other design loads) the analyses may be carried out and prepared in tables, equations, and as equivalent uniform loads.

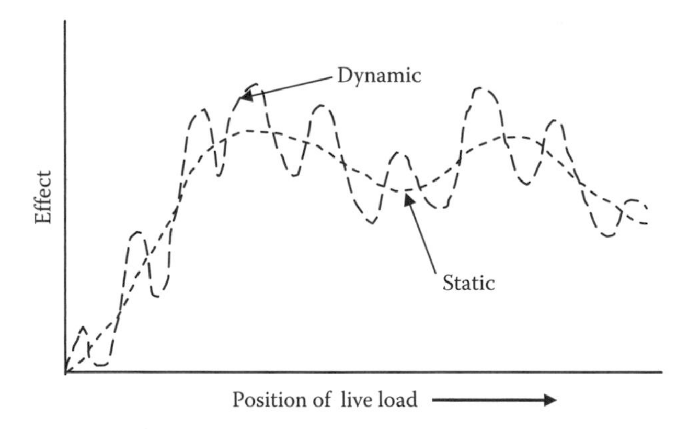

FIGURE 5.1 Static and dynamic effects on steel railway superstructures.

* By necessity, due to the steel wheel flange and rail head interface.

† The effects of inertial and damping forces are considered.

‡ For example, for some span lengths traversed by railway cars, stress ranges are greatest near the 1/4 point of the simple span length (Dick, 2002).

Modern structural engineering software has the ability to perform moving load analysis through stepping loads across the structure and performing the necessary calculations to provide elastic deformations and forces in the members. Many steel railway bridge spans are simply supported* and, therefore, statically determinate.† This enables the use of relatively simple computer programs and spreadsheets to determine the deformations and forces. For more complex superstructures (i.e., statically indeterminate superstructures‡), it may be necessary to utilize more sophisticated proprietary finite-element analysis software that enables moving load analysis. In any case, digital computing has made the analysis of structures for the effects of moving loads a routine component of the bridge design process. However, it is often necessary that individual members of a structure be investigated (e.g., during retrofit design or quality assurance design reviews) or relatively simple superstructures be designed. In these cases and in general, a rudimentary understanding of classical moving load analysis is beneficial to the design engineer.

For these reasons, the principles of moving load analysis for shear force and bending moment are developed in the chapter (these methods are also the basis of some software algorithms). The analyses are performed for beam and girder spans with loads applied directly to the longitudinal members or at discrete locations via transverse members (typically floorbeams in through girder and truss spans). The maximum shear force and bending moment in railway truss§ and arch** spans are also briefly outlined.

5.2.1.1 Maximum Shear Force and Bending Moment due to Moving Concentrated Loads on Simply Supported Spans

5.2.1.1.1 Criteria for Maximum Shear Force (with Loads Applied Directly to the Superstructure)

A series of concentrated loads applied directly to the steel beam or girder is typically assumed in the design of both open and ballasted deck spans.

The maximum shear force, V_C, at a location, C, in a simply supported span of length, L, traversed by a series of concentrated loads with resultant force at a distance, x_T, from one end of the span is (Figure 5.2)

$$V_C = P_T \frac{x_T}{L} - P_L, \tag{5.1}$$

* Some reasons for this are given in Chapter 3.

† The equations of static equilibrium suffice to determine forces in the structure.

‡ Typical of continuous and some movable steel superstructures.

§ The influence lines for simple span shear force and bending moment are useful for the construction of influence lines for axial force in truss web and chord members, respectively.

** Two-hinged arches (hinged at bases) are statically indeterminate and many steel railway arch superstructures are designed and constructed as three-hinged arches to create a statically determinate structure. For statically determinate arches, influence lines for axial forces in members may be constructed by superposition of horizontal and vertical effects. The influence lines for simple span bending moment are useful for the construction of the influence lines for the vertical components of axial force in arch members.

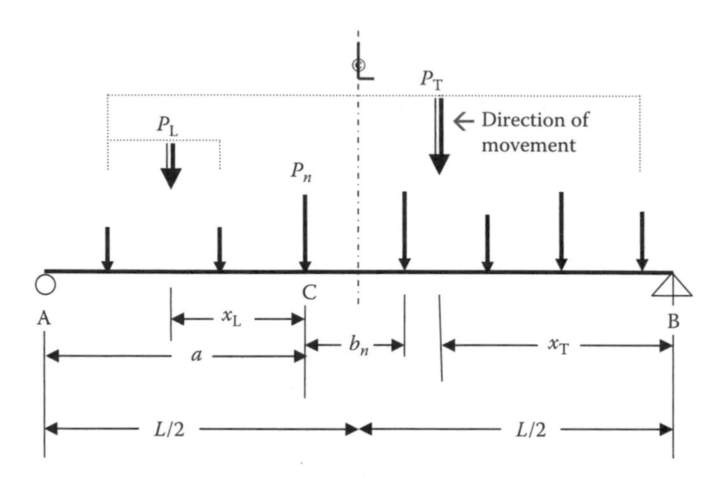

FIGURE 5.2 Concentrated moving loads applied directly to the superstructure.

where P_T is the total load on the span and P_L is the load to left of location C. Equation 5.1 indicates that V_C will be a maximum at a location where $P_T(x_T/L)$ is a maximum and P_L a minimum. If $P_L = 0$ the absolute maximum shear in the span occurs at the end of the span and is

$$V_A = P_T \frac{x_T}{L}. \tag{5.2}$$

For any span length, L, the maximum end shear, V_A, will be largest when the product $P_T x_T$ is greatest. Therefore, for a series of concentrated loads (such as the Cooper's E loading), the maximum end shear, V_A, must be determined with the heaviest loads included in P_T and these heavy loads should be close to the end of the beam (to maximize the distance, x_T).

This information assists in determination of the absolute maximum value of end shear force, which can be determined by a stepping the load configuration across the span (by each successive load spacing) until $P_T x_T$ causes a decrease in V_A. With the exception of end shear in spans between $L = 23$ and 27.3 ft, this occurs when the second axle* of the Cooper's design load configuration is placed at the end of the beam (location A in Figure 5.2). For spans between $L = 23$ and 27.3 ft, the maximum shear occurs with the fifth axle at the end of the span.

The maximum shear force at other locations, C, may be determined in a similar manner by considering a constant P_T moving from x_T to $x_T + b_n$ (where b_n is successive load spacing). In that case, the change in shear force, ΔV_C, at location C is

$$\Delta V_C = P_T \frac{b_n}{L} - P_L. \tag{5.3}$$

The relative changes in shear given by Equation 5.3 can be examined to determine the location of the concentrated loads for maximum shear at any location, C, in the span.

* The first driving wheel of the configuration.

5.2.1.1.2 Criteria for Maximum Shear Force (with Loads Applied to the Superstructure through Transverse Members)

A series of concentrated loads applied through longitudinal members (stringers) to transverse members (floorbeams) to the steel beam or girder is typically assumed in the design of open deck through spans. Ballasted deck superstructures that transfer load to the beams or girders by closely spaced transverse members without stringers may be treated as outlined in Section 5.2.1.1.1.

The maximum shear force, V_{BC}, in panel BC in a simply supported span of length, L, traversed by a series of concentrated loads (with resultant force at a distance, x_T) is (Figure 5.3)

$$V_{BC} = P_T \frac{x_T}{L} - \left(P_L + P_{BC}\left(\frac{c}{s_p}\right)\right). \tag{5.4}$$

The maximum shear force in panel BC may be determined by considering a constant P_T moving from x_T to $x_T + \Delta x_T$, where Δx_T is a small increment of movement of load assuming no change in concentrated forces on the span or within panel BC. In that case, the change in shear force, ΔV_{BC}, in panel BC, is

$$\Delta V_{BC} = \left(\frac{P_T}{L} - \frac{P_{BC}}{s_p}\right)\Delta x_T. \tag{5.5}$$

When $P_T/L = P_{BC}/s_p$, maximum shear in panel BC occurs as the change in shear changes sign (positive to negative). Therefore, the maximum shear occurs in panel BC when the average distributed load on the span, P_T/L, equals the average distributed load in panel BC, P_{BC}/s_p.

When $s_p = L/n$ (where n is the number of equal length panels)

$$\Delta V_{BC} = \left(\frac{P_T}{n} - P_{BC}\right)\frac{\Delta x_T}{s_p} \tag{5.6}$$

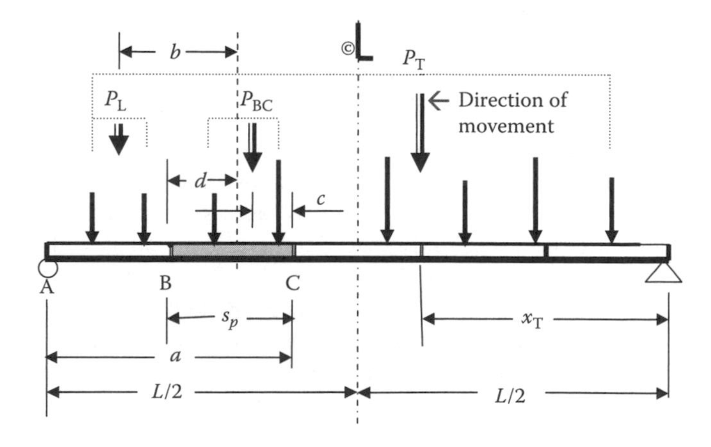

FIGURE 5.3 Concentrated moving loads applied to the superstructure at transverse members.

and $P_T/n - P_{BC} = 0$ for maximum shear in the panel. Therefore, the maximum shear in spans with equal length panels occurs in panel BC when the average panel load on the span, P_T/n, equals the load in panel BC, P_{BC}.

The relative changes in shear given by Equation 5.6 can be examined to determine the location of the concentrated loads for maximum shear in any panel on the span.

5.2.1.1.3 *Criteria for Maximum Bending Moment (with Loads Applied Directly to the Superstructure)*

A series of concentrated loads applied directly to the steel beam or girder is typically assumed in the design of both open and ballasted deck spans.

The maximum bending moment, M_C, at a location, C, in a simply supported span, of length, L, traversed by a series of concentrated loads with resultant at a distance, x_T, from one end of the span is (Figure 5.2)

$$M_C = \left(P_T \frac{x_T}{L}\right) a - (P_L)x_L, \tag{5.7}$$

where P_T is the total load on the span and P_L is the load to the left of location C.

The change in bending moment, ΔM_C, at location C as the constant force P_T moves from $x_T + \Delta x_T$ is

$$\Delta M_C = \left(P_T \frac{a}{L} - P_L\right) \Delta x_T, \tag{5.8}$$

where Δx_T is a small increment of movement of load assuming no change in concentrated forces on the span.

When $P_T/L = P_L/a$, maximum bending moment at location C occurs as the change in bending moment changes sign (positive to negative). Therefore, the maximum bending moment occurs at location C when the average distributed load on the span, P_T/L, equals the average distributed load to the left of location C, P_L/a.

The relative changes in bending moment given by Equation 5.8 can be examined to determine the location of the concentrated loads for maximum bending moment at any location, C, in the span.

5.2.1.1.4 *Criteria for Maximum Bending Moment (with Loads Applied to the Superstructure through Transverse Members)*

A series of concentrated loads applied through longitudinal members (stringers) to transverse members (floorbeams) to the steel beam or girder is typically assumed in the design of open deck through spans. Ballasted deck superstructures that transfer load to the beams or girders by closely spaced transverse members without stringers may be treated as outlined in Section 5.2.1.1.3.

The maximum bending moment, M_{BC}, in panel BC in a simply supported span of length, L, traversed by a series of concentrated loads (with resultant force at x_T) transferred to the span by stringers and transverse floorbeams is (Figure 5.3)

$$M_{BC} = \left(P_T \frac{x_T}{L}\right) a - (P_L)b - (P_{BC})\left(\frac{c}{s_p}\right) d. \tag{5.9}$$

The change in bending moment, ΔM_{BC}, in panel BC is

$$\Delta M_{BC} = \left(P_T \frac{a}{L} - \left(P_L + P_{BC} \frac{d}{s_p}\right)\right) \Delta x_T. \tag{5.10}$$

When $P_T(a/L) - \left(P_L + P_{BC}(d/s_p)\right) = 0$, maximum bending moment occurs in panel BC.

The relative changes in bending moment given by Equation 5.10 can be examined to determine the location of the concentrated loads for maximum bending moment at any panel in the span. The maximum bending moment will occur at the panel point nearest the center of the span.

5.2.1.1.5 *Maximum Bending Moment with Cooper's E80 Load*

The criteria for maximum shear force and bending moment in a simply supported span illustrate that loads must be stepped across the span and their effects investigated at the location of interest. In particular, the load position for maximum bending moment is of interest to engineers.

For live load configurations, such as Cooper's E80, that are expressed as a series of concentrated loads (with or without uniform load segments), the wheel load under which the maximum bending moment occurs may not be readily known by inspection, particularly on longer spans. In such cases, the development of a moment table or chart is of benefit for determining the maximum bending moments at any location along the span. For example, to determine the maximum moment at location C in Figure 5.4, the load configuration would have to be moved in many successive increments across the span. The construction of a moment table, for the particular live load configuration, makes such an iterative analysis unnecessary.

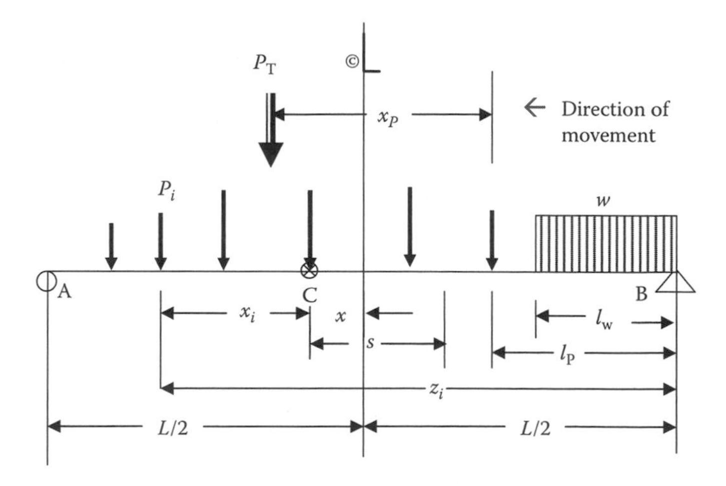

FIGURE 5.4 Bending moment at location C for a series of concentrated moving loads.

The bending moment, M_C, at any location, C, due to moving concentrated and uniform loads as shown in Figure 5.4 is

$$M_C = R_A\left(\frac{L}{2} - x\right) - \sum P_i x_i, \tag{5.11}$$

where $\Sigma P_i x_i$ is the sum of moments due to loads to the left of C. The left reaction, R_A, is

$$R_A = \frac{\sum P_i z_i + (w l_w^2/2)}{L}. \tag{5.12}$$

Substitution of Equation 5.12 into Equation 5.11 yields

$$M_C = \left(\frac{\sum P_i z_i + (w l_w^2/2)}{L}\right)\left(\frac{L}{2} - x\right) - \sum P_i x_i. \tag{5.13}$$

From Figure 5.4, the sum of the moments of concentrated loads about B is

$$\sum P_i z_i = P_T(x_P) + \left(\sum P_i\right) l_P. \tag{5.14}$$

Substitution of Equation 5.14 into Equations 5.12 and 5.13 yields

$$R_A = \frac{P_T(x_P) + (\sum P_i) l_P + (w l_w^2/2)}{L} \tag{5.15}$$

and

$$M_C = \left(\frac{P_T(x_P) + (\sum P_i) l_P + (w l_w^2/2)}{L}\right)\left(\frac{L}{2} - x\right) - \sum P_i x_i. \tag{5.16}$$

Equations 5.15 and 5.16 illustrate that to determine the end shear force and bending moment at any location in the simple span due to moving concentrated and uniform loads (such as the Cooper's E80 load), the following is required:

- The sum of the bending moments of all concentrated loads in front of, and about, the last concentrated load (at l_P from B in Figure 5.4) on the span, $P_T(x_P)$.
- The sum of all concentrated loads on the span, ΣP_i.
- The negative bending moment or the sum of the moments about C of all concentrated loads in front of C, $\Sigma P_i x_i$.

Since the Cooper's load pattern is constant, it is possible to develop charts and tables to readily determine the bending moment for various simple span lengths using Equation 5.16. Table 5.1 is developed for the wheel load (1/2 of axle load) of the Cooper's E80 live load. The legend to Table 5.1 outlines the methods used to

TABLE 5.1
Moment Table for Cooper's E80 Wheel Load

N	1	2	3	4	5	6	7	8	9	10	11	12	13	14	15	16	17	18
P	20	40	40	40	40	26	26	26	26	20	40	40	40	40	26	26	26	26
S_1	0	8	13	18	23	32	37	43	48	56	64	69	74	79	88	93	99	104
S_w	109	101	96	91	86	77	72	66	61	53	45	40	35	30	21	16	10	5
SP_1	20	60	100	140	180	206	232	258	284	304	344	384	424	464	490	516	542	568
SP_{18}	568	548	508	468	428	388	362	336	310	284	264	224	184	144	104	78	52	26
SM	0	320	840	1560	2480	3313	4273	5387	6640	7760	10,320	13,080	16,040	19,200	21,467	23,907	26,467	29,200
SM_{18}	32,728	30,548	26,508	22,668	19,028	15,588	13,586	11,714	9998	8412	7352	5552	3952	2552	1352	806	390	130
SM_{17}	29,888	27,808	23,698	20,328	16,888	13,648	11,776	10,034	8448	6992	6032	4432	3032	1832	832	416	130	
SM_{16}	27,178	25,198	21,558	18,118	14,878	11,838	10,096	8484	7028	5702	4842	3442	2242	1242	442	156		
SM_{15}	24,082	22,222	18,822	15,622	12,622	9822	8236	6780	5480	4310	3570	2410	1450	690	130			
SM_{14}	21,632	19,872	16,672	13,672	10,872	8272	6816	5490	4320	3280	2640	1680	920	360				
SM_{13}	17,456	15,876	13,036	10,396	7956	5716	4494	3402	2466	1660	1200	600	200					
SM_{12}	15,336	13,856	11,216	8776	6536	4496	3404	2442	1636	960	600	200						
SM_{11}	13,416	12,036	9596	7356	5316	3476	2514	1682	1006	460	200							
SM_{10}	11,696	10,416	8176	6136	4296	2656	1824	1122	576	160								
SM_9	9264	8144	6224	4504	2984	1664	1040	546	208									
SM_8	6992	6032	4432	3032	1832	832	416	130										
SM_7	5702	4842	3442	2242	1242	442	156											
SM_6	4310	3570	2410	1450	690	130												
SM_5	3280	2640	1680	920	360													
SM_4	1660	1200	600	200														
SM_3	960	600	200															
SM_2	460	200																
SM_1	160																	

N = Wheel number; P = Cooper's E80 wheel load, kips (1/2 axle load); S_1 = Distance, ft, from wheel N to wheel 1; S_w = Distance, ft, from wheel N to uniform train load of 4000 lb/ft; $S_1 + S_w = 109$ ft; SP_1 = Sum of wheel loads, kips, between and including wheel 1 to wheel N; SP_{18} = Sum of wheel loads, kips, between and including wheel N to wheel 18; SM = Sum of the moments, ft-kips, about wheel 1 of wheel loads between and including wheel 2 to wheel N; SM_{18} = Sum of the moments, ft-kips, about the beginning of the uniform load of wheel loads between and including wheel N to wheel 18; SM_k = Sum of the moments, ft-kips, about wheel load $(k + 1)$ of wheel loads between and including wheel N to wheel k.

determine the values shown. The use of Table 5.1 for determining maximum bending moments due to Cooper's E80 load is outlined in Examples 5.1 and 5.2.

Example 5.1

Determine the maximum bending moment per rail for Cooper's E80 load on a 60 ft long deck plate girder (DPG) span. The moment at the center, C, is assumed to be near the maximum bending moment in both location and magnitude.

A review of the Cooper's load configuration indicates that the maximum moment will likely occur under axles, $NP = 3, 4, 5, 12, 13,$ or 14 (Figure E5.1).

With $NP = 3$ (Cooper's load configuration wheel number 3)

From Table 5.1:

$x_1 = (S_1)_3 = 13$ ft

since $x1 \leq L/2 \leq 30$ ft; N1 = 1

support B is $(L/2 + x1) = 30 + 13 = 43$ ft from N1

since $(S_1)_8 = 43$ ft, $NL = 8$ and is over support B

$xL = 43–43 = 0$

NE is the last wheel on the span and is equal to $NL - 1 = 7$ when NL is over support B

$$R_B = \frac{\sum M_B}{L} = \frac{\sum M_{(NL-1),N1} + \sum P_{N1,NE}(xL)}{60} = \frac{\sum M_{7,1} + \sum P_{1,7}(xL)}{60}$$

$$= \frac{5702 + 232(0)}{60} = 95.03 \text{ kips},$$

$$M_C = R_B(L/2) - \sum M_{(NP-1),N1} = R_B(L/2) - \sum M_{2,1}$$

$$= 95.03(30) - 460 = 2391 \text{ ft-kips}.$$

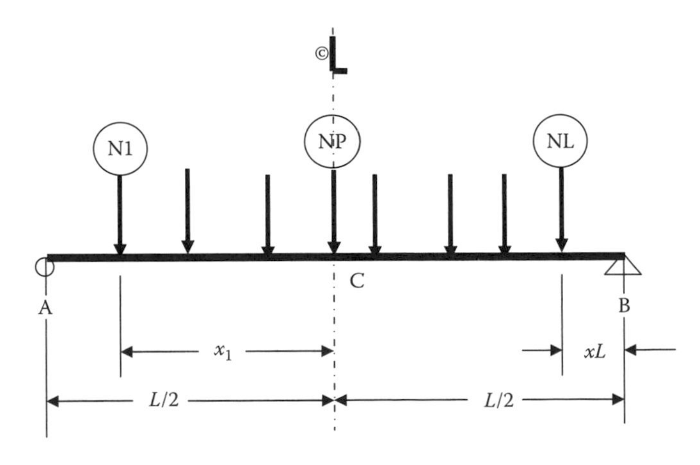

FIGURE E5.1

With $NP = 4$ (Cooper's load configuration wheel number 4)

From Table 5.1:

$x1 = (S_1)_4 = 18$ ft
since $x1 \leq L/2 \leq 30$ ft; N1 = 1
support B is $30 + 18 = 48$ ft from N1
since $(S_1)_9 = 48$ ft, $NL = 9$ and is over support B
$xL = 0$
NE is the last wheel on the span and is equal to $NL - 1 = 8$ when NL is over support B

$$R_B = \frac{\sum M_B}{L} = \frac{\sum M_{8,1} + \sum P_{1,8}(xL)}{60} = \frac{6992 + 258(0)}{60} = 116.5 \text{ kips},$$

$$M_C = R_B\left(\frac{L}{2}\right) - \sum M_{3,1} = 116.5(30) - 960 = 2536 \text{ ft-kips}.$$

With $NP = 5$ (Cooper's load configuration wheel number 5)

From Table 5.1:

$x1 = (S_1)_4 = 23$ ft
since $x1 \leq L/2 \leq 30$ ft; N1 = 1
support B is $30 + 23 = 53$ ft from N1
since $(S_1)_9 = 48$ ft, $NL = 9$ and is $xL = 53–48 = 5$ ft from support B
NE is the last wheel on the span and is equal to $NL = 9$ when NL is not over support B

$$R_B = \frac{\sum M_B}{L} = \frac{\sum M_{8,1} + \sum P_{1,9}(xL)}{60} = \frac{6992 + 284(5)}{60} = 140.2 \text{ kips},$$

$$M_C = R_B\left(\frac{L}{2}\right) - \sum M_{4,1} = 140.2(30) - 1660 = 2546 \text{ ft-kips}.$$

With $NP = 12$ (Cooper's load configuration wheel number 12)

From Table 5.1:

With $NP = 12$, the first wheel on the span = N1 = 8
$x1 = (S_1)_{12} - (S_1)_8 = 69 - 43 = 26$ ft
support B is $30 + 26 = 56$ ft from N1 = 8
With $NP = 12$, the last wheel on the span is $NL = 17$
$xL = (L/2) - [(S_1)_{17} - (S_1)_{12}] = 30 - (99 - 69) = 0$ ft and N17 is over support B
NE is the last wheel on the span and is equal to $NL - 1 = 16$ when NL is over support B

$$R_B = \frac{\sum M_B}{L} = \frac{\sum M_{16,8} + \sum P_{8,16}(xL)}{60} = \frac{8480 + (336 - 78)(0)}{60} = 141.3 \text{ kips},$$

$$M_C = R_B\left(\frac{L}{2}\right) - \sum M_{11,8} = 141.3(30) - 1682 = 2558 \text{ ft-kips}.$$

With $NP = 13$ (Cooper's load configuration wheel number 13)

From Table 5.1:

With $NP = 13$, the first wheel on the span $= N1 = 9$

$x1 = (S_1)_{13} - (S_1)_9 = 74 - 48 = 26$ ft

support B is $30 + 26 = 56$ ft from $N1 = 9$

With $NP = 13$, the last wheel on the span is $NL = 18$

$xL = (L/2) - [(S_1)_{18} - (S_1)_{13}] = 30 - (104 - 74) = 0$ ft and N18 is over support B

NE is the last wheel on the span and is equal to $NL - 1 = 17$ when NL is over support B

$$R_B = \frac{\sum M_B}{L} = \frac{\sum M_{17,9} + \sum P_{9,17}(xL)}{60} = \frac{8448 + (310 - 52)(0)}{60} = 140.8 \text{ kips},$$

$$M_C = R_B\left(\frac{L}{2}\right) - \sum M_{12,9} = 140.8(30) - 1636 = 2588 \text{ ft-kips}.$$

With $NP = 14$ (Cooper's load configuration wheel number 14)

From Table 5.1:

With $NP = 14$, the first wheel on the span $= N1 = 10$

$x1 = (S_1)_{14} - (S_1)_{10} = 79 - 56 = 23$ ft

support B is $30 + 23 = 53$ ft from $N1 = 10$

With $NP = 14$, the last wheel on the span, NL, is the beginning of the uniform load, w

$xL = (L/2) - [(S_1)_w - (S_1)_{14}] = 30 - (104 + 5 - 79) = 0$ ft from the beginning of the uniform load, w, to support B

NE is the last wheel on the span and is equal to the beginning of the uniform load, w

$$R_B = \frac{\sum M_B}{L} = \frac{\sum M_{18,10} + \sum P_{10,w}(xL)}{60} = \frac{8412 + (284)(0)}{60} = 140.2 \text{ kips},$$

$$M_C = R_B\left(\frac{L}{2}\right) - \sum M_{13,10} = 140.2(30) - 1660 = 2546 \text{ ft-kips}.$$

With $NP = 15$ (Cooper's load configuration wheel number 15)

From Table 5.1:

With $NP = 15$, the first wheel on the span $= N1 = 11$

$x1 = (S_1)_{15} - (S_1)_{11} = 88 - 64 = 24$ ft

support B is $30 + 24 = 54$ ft from $N1 = 11$

With $NP = 15$, the last wheel on the span, NL, is the end of 9 ft of the uniform load, w

$xL = (L/2) - [(S_1)_w - (S_1)_{15}] = 30 - (104 + 5 - 88) = 9$ ft from the beginning of the uniform load, w, to support B

NE is the last wheel on the span and is equal to 9 ft of the uniform load, w

$$R_B = \frac{\Sigma M_B}{L} = \frac{\Sigma M_{18,11} + \Sigma P_{11,w}(xL)}{60} = \frac{7352 + [(264)(9) + 4(9)(9/2)]}{60}$$

$$= 164.8 \text{ kips},$$

$$M_C = R_B\left(\frac{L}{2}\right) - \sum M_{14,11} = 164.8(30) - 2640 = 2305 \text{ ft-kips}.$$

The maximum bending moment is 2588 ft-kips. ($NP = 13$)

Example 5.2

Determine the bending moment per rail at location C under axles $NP = 3, 4$, and 13 of the Cooper's E80 load on a 60 ft long through plate girder span with a floor system comprising floorbeams and 20 ft long stringers (Figure E5.2).

With $NP = 3$ (Cooper's load configuration wheel number 3)

From Table 5.1:

$x1 = (S_1)_3 = 13$ ft
since $x1 \leq L/2 \leq 30$ ft; N1 = 1
support B is $(2L/3 + x1) = 40 + 13 = 53$ ft from N1
since $(S_1)_9 = 48$ ft, $NL = 9$, and $xL = 53 - 48 = 5$ ft from $NL = 9$ to support B
NE is the last wheel on the span and is equal to $NL = 9$

$$R_B = \frac{\sum M_B}{L} = \frac{\sum M_{(NL-1),N1} + \sum P_{N1,NE}(xL)}{60} = \frac{\sum M_{8,1} + \sum P_{1,9}(xL)}{60}$$

$$= \frac{6992 + 284(5)}{60} = 140.20 \text{ kips},$$

$$M_C = R_B\left(\frac{L}{3}\right) - \sum M_{(NP-1),N1} = R_B\left(\frac{L}{3}\right) - \sum M_{2,1}$$

$$= 140.2(20) - 460 = 2344 \text{ ft-kips}.$$

The superstructure with transverse floorbeams has a 100[1 − (2344/2391)] = 2.0% decrease in bending moment with $NP = 3$ at the location of maximum moment.

With $NP = 4$ (Cooper's load configuration wheel number 4)

From Table 5.1:

$x1 = (S_1)_4 = 18$ ft
since $x1 \leq L/2 \leq 30$ ft; N1 = 1
support B is $(2L/3 + x1) = 40 + 18 = 58$ ft from N1

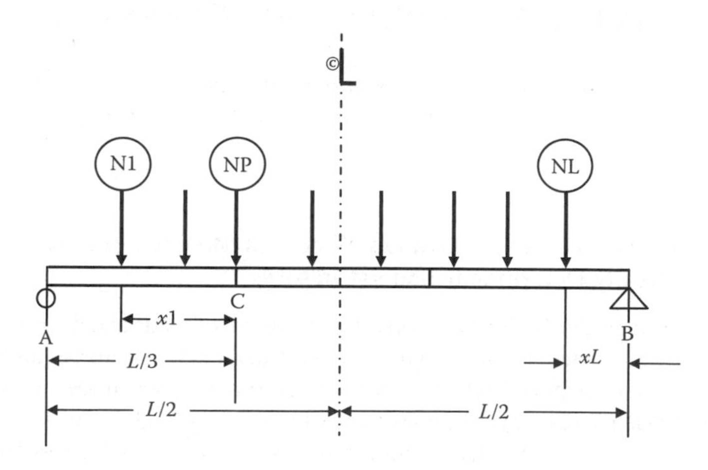

FIGURE E5.2

since $(S_1)_{10} = 56\,\text{ft}$, $NL = 10$, and $xL = 58 - 56 = 2\,\text{ft}$ from $NL = 10$ to support B

NE is the last wheel on the span and is equal to $NL = 10$

$$R_B = \frac{\sum M_B}{L} = \frac{\sum M_{(NL-1),N1} + \sum P_{N1,NE}(xL)}{60} = \frac{\sum M_{9,1} + \sum P_{1,10}(xL)}{60}$$

$$= \frac{9264 + 304(2)}{60} = 164.5\ \text{kips},$$

$$M_C = R_B\left(\frac{L}{3}\right) - \sum M_{(NP-1),N1} = R_B\left(\frac{L}{3}\right) - \sum M_{3,1}$$

$$= 164.5(20) - 960 = 2331\ \text{ft-kips}.$$

The superstructure loaded with transverse floorbeams has a $100[1 - (2331/2536)] = 8.1\%$ decrease in bending moment with $NP = 4$ at the location of maximum moment.

With $NP = 13$ (Cooper's load configuration wheel number 13)

From Table 5.1:

With $NP = 13$, the first wheel on the span $= \text{N1} = 10$

$x1 = (S_1)_{13} - (S_1)_{10} = 74 - 56 = 18\,\text{ft}$

support B is $40 + 18 = 58\,\text{ft}$ from $\text{N1} = 10$

With $NP = 13$, the last wheel on the span, NL, is the end of 5 ft of the uniform load, w

$xL = (2L/3) - [(S_1)_w - (S_1)_{13}] = 40 - (104 + 5 - 74) = 5\,\text{ft}$ from the beginning of the uniform load, w, to support B

NE is the last wheel on the span and is equal to 5 ft of the uniform load, w

$$R_B = \frac{\sum M_B}{L} = \frac{\sum M_{18,10} + \sum P_{10,w}(xL)}{60} = \frac{8412 + [(284)(5) + 4(5)(5/2)]}{60}$$

$$= 164.7\ \text{kips},$$

$$M_C = R_B\left(\frac{L}{3}\right) - \sum M_{12,10} = 164.7(20) - 960 = 2334\ \text{ft-kips}.$$

The superstructure with transverse floorbeams has a $100[1 - (2334/2588)] = 9.8\%$ decrease in bending moment with $NP = 13$ at the location of maximum moment.

5.2.1.2 Influence Lines for Maximum Effects of Moving Loads on Statically Determinate Superstructures

Influence lines facilitate both the appropriate placement of loads and determination of maximum effects in steel beam and girder superstructures (shear forces and bending moments), trusses (axial forces), and arches (axial forces, shear forces, and bending moments). Influence lines may be constructed for moving load analysis of statically determinate superstructures by moving a unit concentrated load across the superstructure and determining the value of an effect at each location. The construction of

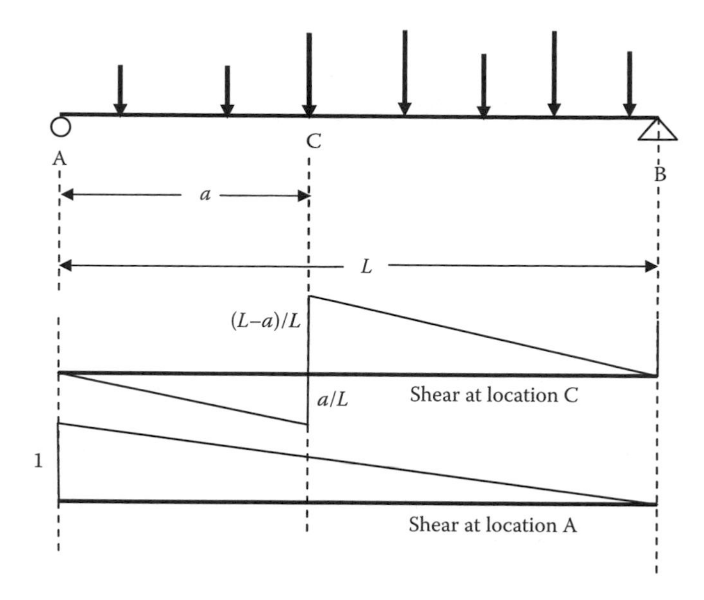

FIGURE 5.5 Influence lines for shear at locations C and A for concentrated moving loads applied directly to the superstructure.

influence lines may be simplified by determining the value of the effect at locations where changes in the influence line will occur (i.e., supports, panel points, hinges, etc.) and joining those locations with straight lines.*

5.2.1.2.1 Influence Lines for Maximum Shear Force and Bending Moment in Simply Supported Beam and Girder Spans

5.2.1.2.1.1 Maximum Shear Force (with Loads Applied Directly to the Superstructure) The influence lines for shear force at location C and at the end of the simple span (location A) are shown in Figure 5.5. They are developed by determining the shear force at location C and reaction at the end of the simple span (location A) with a unit load placed at locations A, B, and C.

5.2.1.2.1.2 Maximum Shear Force (with Loads Applied to the Superstructure through Transverse Members) The influence line for shear in panel BC of the simply supported span is shown in Figure 5.6. It is developed by determining the shear force at locations B and C with a unit load placed at locations A, B, C, and D; where, n is the number of panels, n_L is the number of panels left of panel BC, and n_R

* For axial force, shear force and bending moment in statically determinate structures, influence lines comprise straight-line segments. However, for deflections this is not the case.

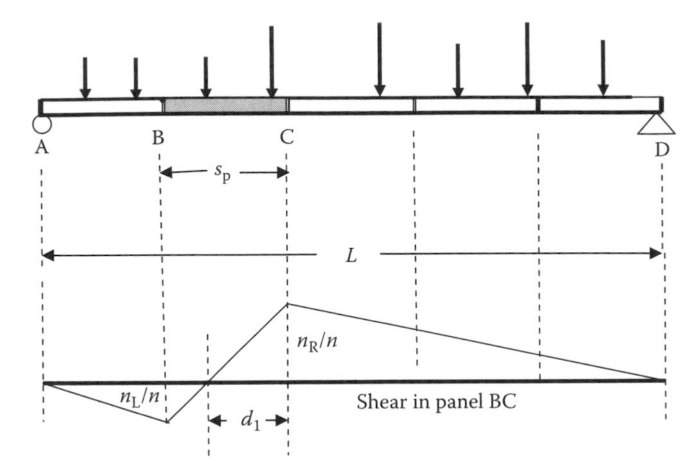

FIGURE 5.6 Influence line for shear in panel BC for concentrated moving loads applied to the superstructure at transverse members.

is the number of panels right of panel BC.

$$d_1 = \frac{(n_R/n_L)s_p}{(1 + (n_R/n_L))} = \frac{s_p n_R}{2((n_R/n_L) + 1)}, \tag{5.17}$$

$$L = n(s_p) = (n_L + n_R + 1)(s_p).$$

5.2.1.2.1.3 Maximum Bending Moment (with Loads Applied Directly to the Superstructure) The influence lines for bending at location C and at the center of the simple span are shown in Figure 5.7. They are developed by determining the bending moment at location C and at the center span with a unit load placed at locations A, B, and C.

5.2.1.2.1.4 Maximum Bending Moment (with Loads Applied to the Superstructure by Transverse Members) The influence lines for moment in panel BC (at distance d_2 from B) and at location C of the simple span are shown in Figure 5.8. They are developed by determining the bending moments at locations B and C with a unit load placed at locations A, B, C, and D. As shown in Figure 5.8, a reduction in bending moment occurs for superstructures loaded through transverse members.

5.2.1.2.1.5 Maximum Floorbeam Reactions for Loads on Simply Supported Stringers The influence line for floorbeam reaction at location C assuming simply supported stringer spans is shown in Figure 5.9. It is developed by determining the shear forces at locations B, C, and D with a unit load placed at locations B, C, and D. Since stringer spans are generally relatively short, the location of concentrated loads for maximum floorbeam reaction is usually quite obvious by inspection.

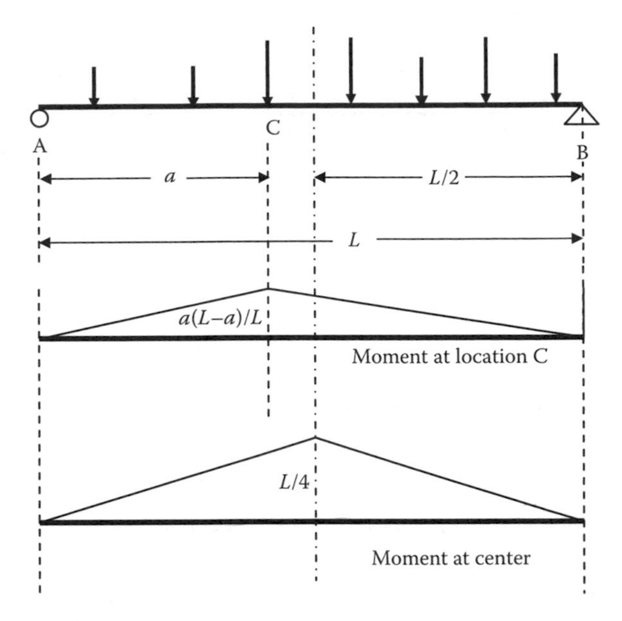

FIGURE 5.7 Influence lines for bending at location C and the center for concentrated moving loads applied directly to the superstructure.

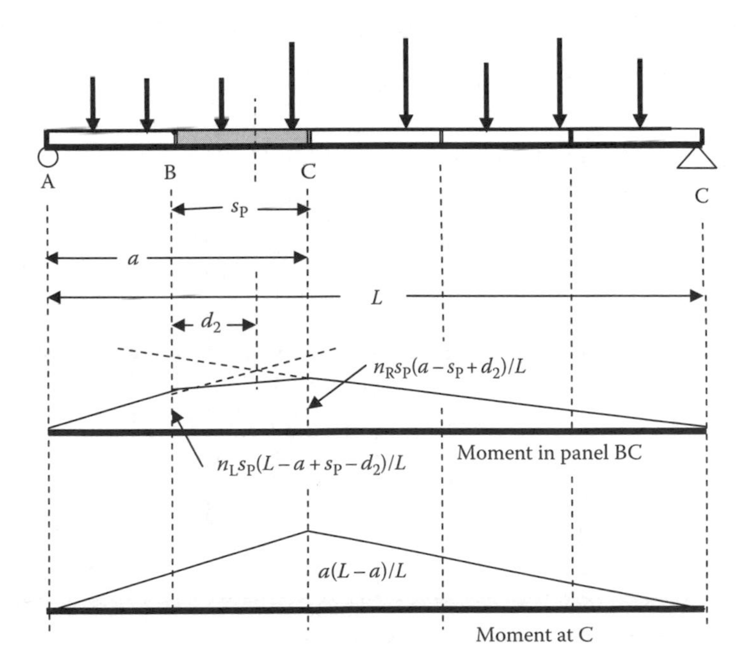

FIGURE 5.8 Influence lines for moment in panel BC and at location C for concentrated moving loads applied at transverse members.

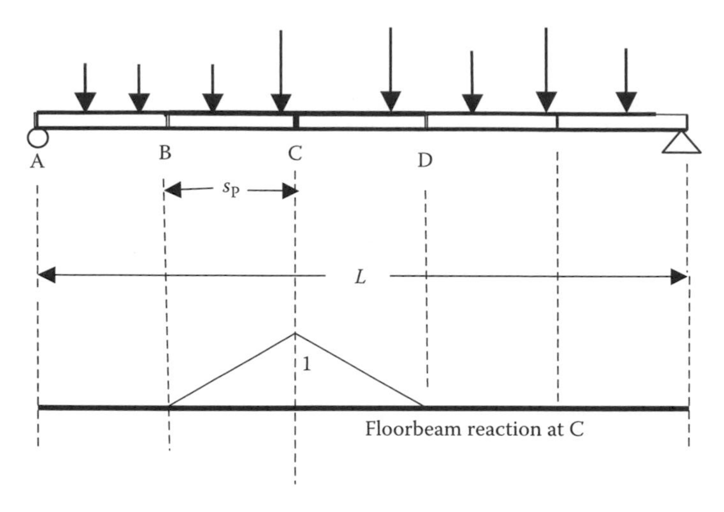

FIGURE 5.9 Influence line for floorbeam reaction at location C.

Example 5.3

Determine the maximum bending moment per rail under axle $NP = 5$ of the Cooper's E80 load on a 60 ft long DPG span. The moment at the center, C, is assumed to be near the maximum bending moment in both location and magnitude.

The influence line for the center span bending moments is shown in Figure E5.3.

The ordinates of the influence lines are

$$a = (7/30)15 = 3.50$$

$$b = (15/30)15 = 7.50$$

$$c = (20/30)15 = 10.00$$

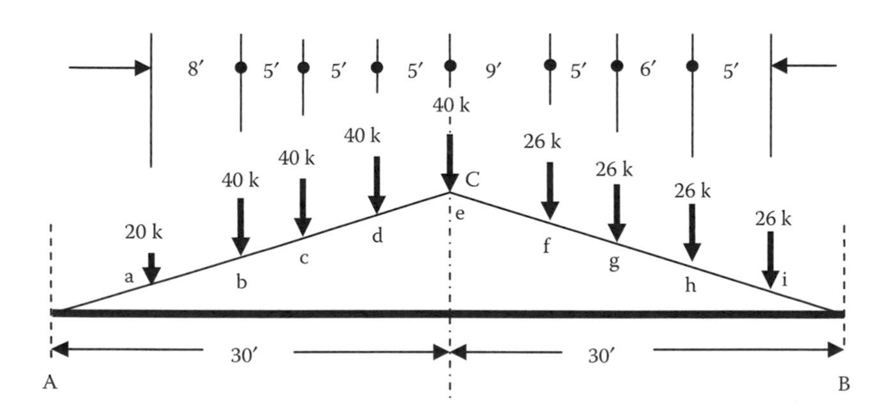

FIGURE E5.3

$$d = (25/30)15 = 12.50$$
$$e = 60/4 = 15.00$$
$$f = (21/30)15 = 10.50$$
$$g = (16/30)15 = 8.00$$
$$h = (10/30)15 = 5.00$$
$$i = (5/30)15 = 2.50$$
$$M_C = 20(3.50) + 40(7.50 + 10.00 + 12.50 + 15.00) + 26(10.50 + 8.00 + 5.00 + 2.50) = 2546 \text{ ft-kips.}$$

Example 5.4

Determine the bending moment per rail at location C under axle $NP = 3$ of the Cooper's E80 load on a 60 ft long through plate girder span with a floor system comprising floorbeams and 20 ft long stringers.

The influence line for bending moments at location C is shown in Figure E5.4.

The ordinates of the influence lines are

$$a = (7/20)13.33 = 4.67$$
$$b = (15/20)13.33 = 10.00$$

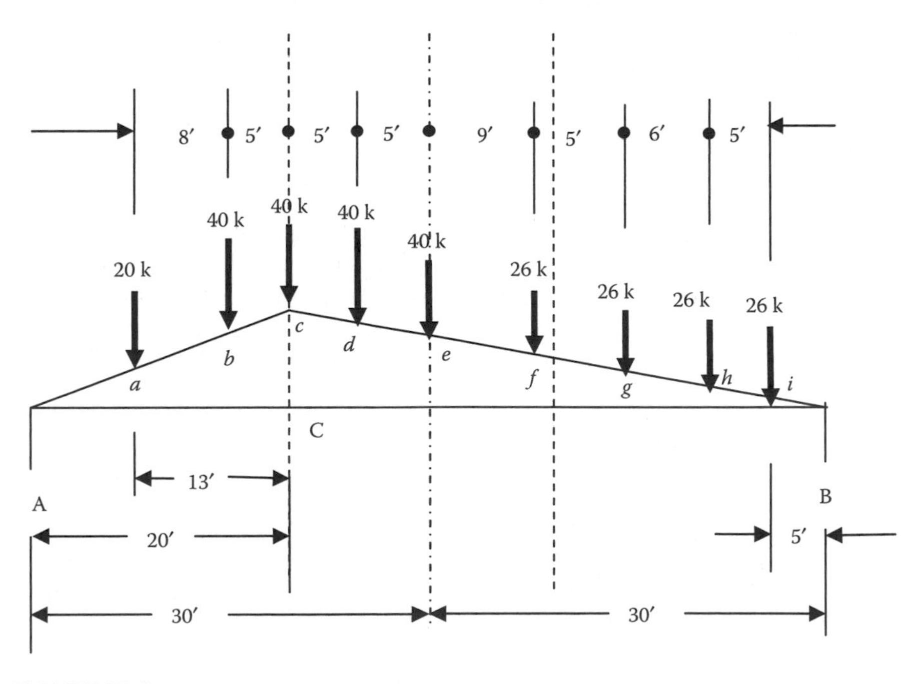

FIGURE E5.4

$$c = 40(20)/60 = 13.33$$

$$d = (35/40)13.33 = 11.67$$

$$e = (30/40)13.33 = 10.00$$

$$f = (21/40)13.33 = 7.00$$

$$g = (16/40)13.33 = 5.33$$

$$h = (10/40)13.33 = 3.33$$

$$i = (5/40)13.33 = 1.67$$

$$M_C = 20(4.67) + 40(10.00 + 13.33 + 11.67 + 10.00) + 26(7.00 + 5.33 + 3.33 + 1.67) = 2345 \text{ ft-kips.}$$

5.2.1.2.2 Influence Lines for Maximum Axial Forces in Statically Determinate Truss Spans

The influence lines developed in Section 5.2.1.2.1 for shear force and bending moment are useful in the construction of axial force influence lines for truss web and chord members, respectively. In addition, the consideration of the moving load effect at panel points simplifies the construction of axial force influence lines.

Influence lines for chord members may be constructed by considering free body diagrams and equilibrium of moments. Influence lines for web members may be constructed by considering free body diagrams and equilibrium of forces. The construction of influence lines for axial forces in members of simply supported truss spans is illustrated by examples of a Pratt truss and a Parker truss, respectively, in Examples 5.5 and 5.6.*

Example 5.5

Construct influence lines for the 156.38 ft eight-panel Pratt through truss in Figure E5.5. The influence lines are constructed by locating unit loads at appropriate locations and using the method of sections or the method of joints.

Determine influence lines for the reactions and members U1–U2, U3–L3, L1–L2, L3–L4, U1–L1, and U1–L2.

Section 1-1 may be isolated to determine the forces in members U1–U2 (Figure E5.5b), L1–L2 (Figure E5.5c), and U1–L2 (Figure E5.5d).

In Figure E5.5a: with unit load at L1 and taking moments about L2, the force in U1–U2 = [(1/8)(6)(19.55)]/27.25 = −0.54 (compression direction to balance reaction moment about L2).

In Figure E5.5b: with unit load at L2 and taking moments about L2, the force in U1–U2 = [(2/8)(6)(19.55)]/27.25 = −1.08 (compression direction to balance reaction moment about L2).

* These truss forms are often used for medium span steel railway bridges.

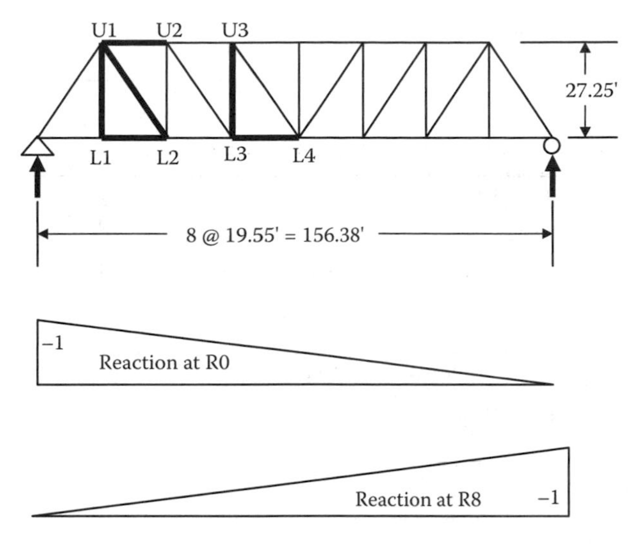

FIGURE E5.5a

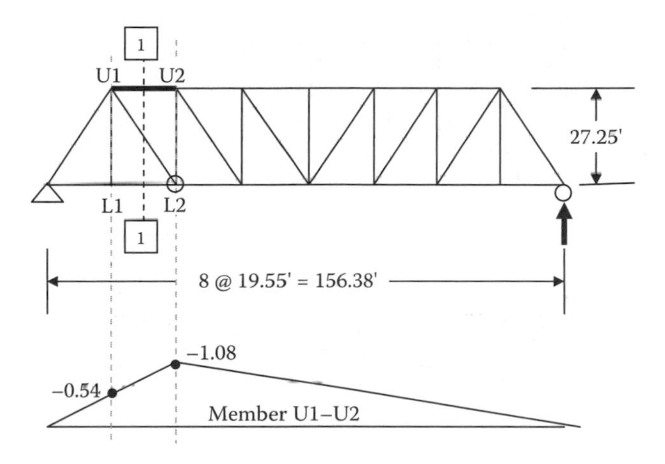

FIGURE E5.5b

In Figure E5.5c: with unit load at L1 and taking moments about U1, the force in L1–L2 = $[(1/8)(7)(19.55)]/27.25 = +0.63$ (tension direction to balance reaction moment about U1).

In Figure E5.5c: with unit load at L2 and taking moments about U1, the force in L1–L2 = $[(2/8)(7)(19.55) - (1)(19.55)]/27.25 = +0.54$ (tension direction to balance reaction moment about U1).

In Figure E5.5d: with unit load at L1 and summing horizontal forces in panel 1–2, the force in U1–L2 = $-(-0.54 + 0.63)(19.55^2 + 27.25^2)^{1/2}/19.55 = -0.15$.

In Figure E5.5d: with unit load at L2 and summing horizontal forces in panel 1–2, the force in U1–L2 = $-(-1.08 + 0.54)(19.55^2 + 27.25^2)^{1/2}/19.55 = +0.93$.

Section 2-2 may be isolated to determine the forces in member L3–L4 (Figure E5.5e).

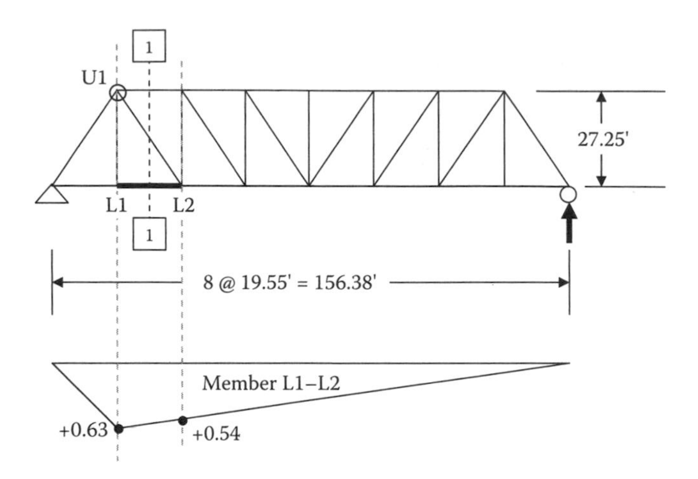

FIGURE E5.5c

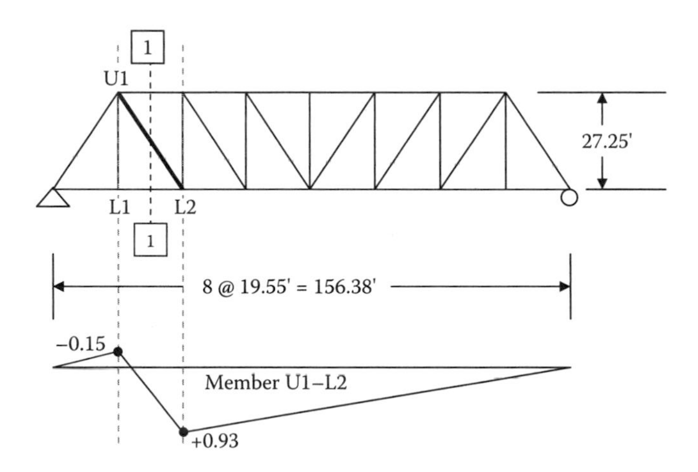

FIGURE E5.5d

In Figure E5.5e: with unit load at L3 and taking moments about U3, the force in L3–L4 = [(3/8)(5)(19.55)]/27.25 = +1.35 (tension direction to balance reaction moment about U1).

In Figure E5.5e: with unit load at L4 and taking moments about U3, the force in L3–L4 = [(4/8)(5)(19.55) − (1)(19.55)]/27.25 = +1.08 (tension direction to balance reaction moment about U1).

Section 3-3 may be isolated to determine the forces in member U3–L3 (Figure E5.5f).

In Figure E5.5f: with unit load at L3 and summing vertical forces in panel 3–4, the force in U3–L3 = +3/8 = +0.38.

In Figure E5.5f: with unit load at L4 and summing vertical forces in panel 3–4, the force in U3–L3 = +1/2 − 1 = −0.50.

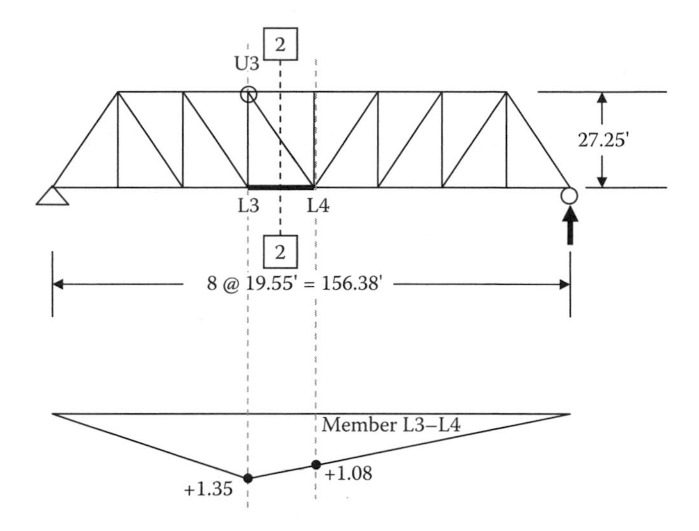

FIGURE E5.5e

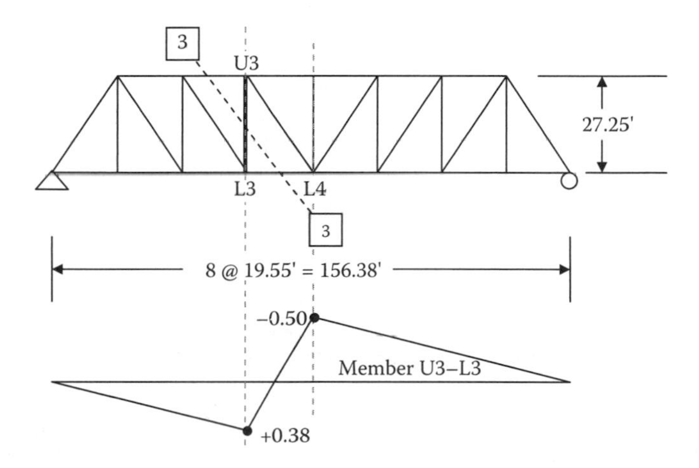

FIGURE E5.5f

The forces in member U1–L1 (Figure E5.5g) may be determined by the method of joints by locating unit loads at L0, L1, and L2.

The hanger U1–L1 is loaded only when moving loads are in adjacent panels of the hanger.*

Example 5.6

Construct influence lines for members U1–U2, U1–L2, and U2–L2 in the 240 ft six-panel curved-chord Parker through truss in Figure E5.6. The influence lines

* There are also increased impact effects for through truss hangers due to the short live load influence line (see Chapter 4).

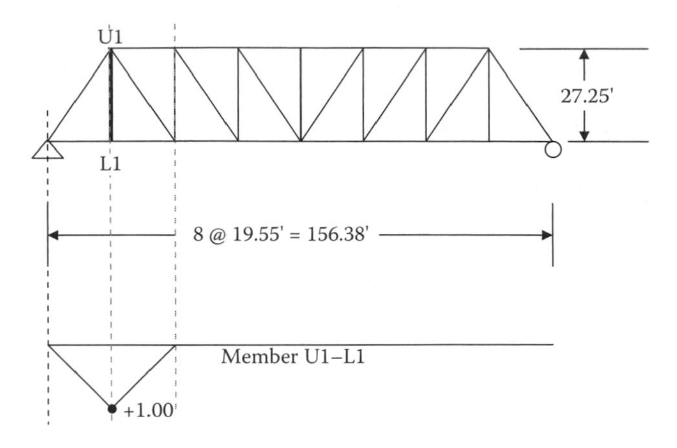

FIGURE E5.5g

are constructed by using the method of sections and locating unit loads at appropriate locations.

$$a_{\mathrm{p}} = \frac{40}{(36/28) - 1} - 40 = 100\,\mathrm{ft},$$

$$h_{\mathrm{p}} = (a_{\mathrm{p}} + 2(40))\left(\frac{28}{\sqrt{(a_{\mathrm{p}} + 40)^2 + 28^2}}\right) = 35.3\,\mathrm{ft},$$

$$\beta_{\mathrm{p}} = 45^\circ - \tan^{-1}\frac{28}{a_{\mathrm{p}} + 40} = 33.7^\circ,$$

$$L_{\mathrm{p}} = \sqrt{(a_{\mathrm{p}} + 40)^2 + 28^2}\cos(\beta_{\mathrm{p}}) = 118.8\,\mathrm{ft}.$$

Considering Section 1-1 in Figure E5.6: with unit load at L2 and taking moments about L2, the force in U1–U2 = $[-(4/6)(2)(40)]/35.3 = -1.51$ (compression direction to balance reaction moment about L2) (Figure E5.7).

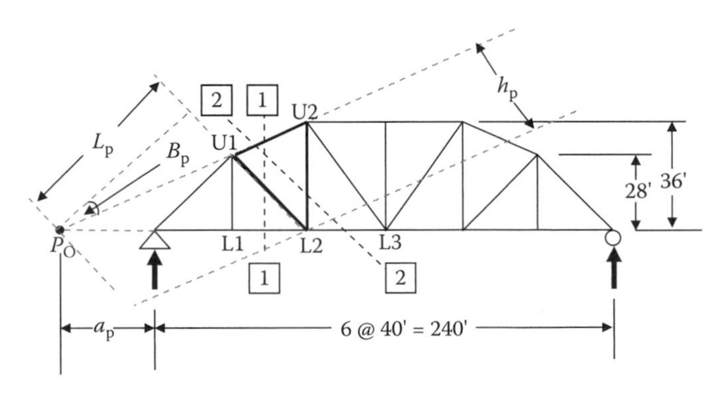

FIGURE E5.6

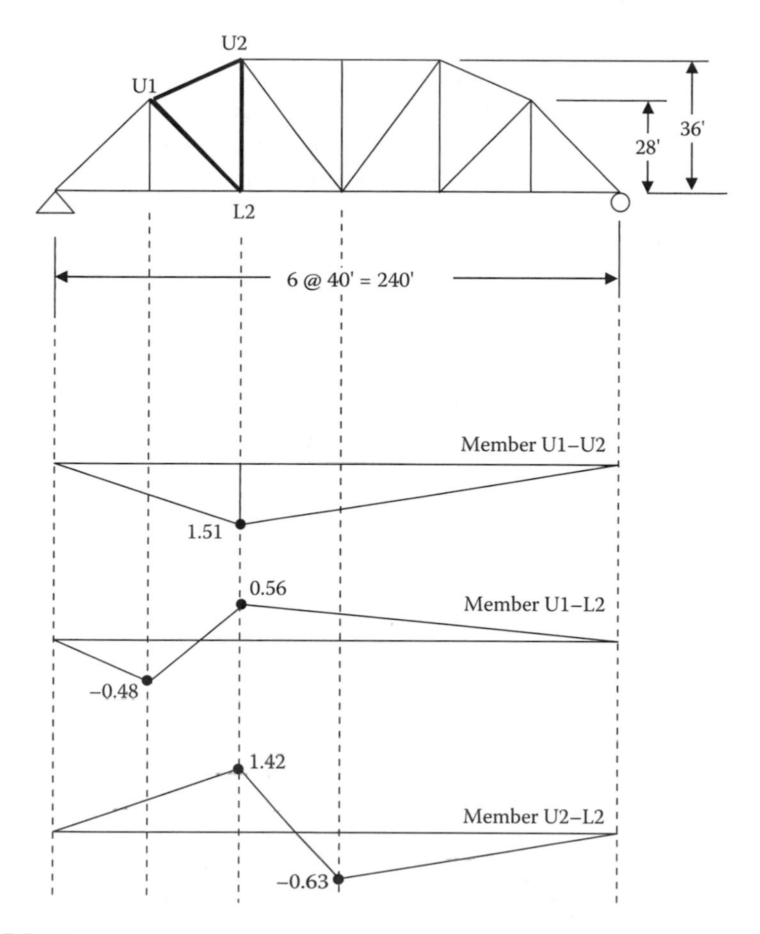

FIGURE E5.7

Considering Section 1-1 in Figure E5.6: with unit load at L1 and taking moments about P0, the force in U1–L2 = [−(1/6)(240 + 100)]/118.8 = −0.48.

Considering Section 1-1 in Figure E5.6: with unit load at L2 and taking moments about P0, the force in U1–L2 = (4/6)(100)/118.8 = 0.56 (Figure E5.7).

Considering Section 2-2 in Figure E5.6: with unit load at L2 and taking moments about P0, the force in U2–L2 = 2/6(100 + 240)/80 = 1.42.

Considering Section 2-2 in Figure E5.6: with unit load at L3 and taking moments about P0, the force in U2–L2 = −1/2(100)/80 = −0.63 (Figure E5.7).

The distance, h_p, in Figure E5.6 illustrates the effect of the "modified panel shear" created by the sloped chord, which participates in resisting the panel shear force.

Influence lines for other chord and web members of the truss may be constructed in a similar manner by applying unit loads at panel points and determining axial forces in members by the method of sections or the method of joints.

5.2.1.2.3 *Influence Lines for Maximum Effects in Statically Determinate Arch Spans*

Many steel railway arches are designed as three hinged to impose statically determinate conditions (Figure 5.10a). Statically determinate arches are often simpler to fabricate and erect, and are not subjected to temperature or support displacement induced stresses. The construction of influence lines for statically determinate arches can be made efficient by understanding the relationships between arch reactions, internal forces (shear, bending, and axial), and the influence lines obtained in Section 5.2.1.2.1 for shear and bending in simply supported spans.

5.2.1.2.3.1 Maximum Bending Moment, Shear Force, and Axial Force (with Moving Loads Applied Directly to the Arch) For the moving concentrated load, $P = 1$, a distance x_P from support A in Figure 5.10a,

$$R_A = \frac{L - x_P}{L}, \tag{5.18a}$$

$$R_B = \frac{x_P}{L}. \tag{5.18b}$$

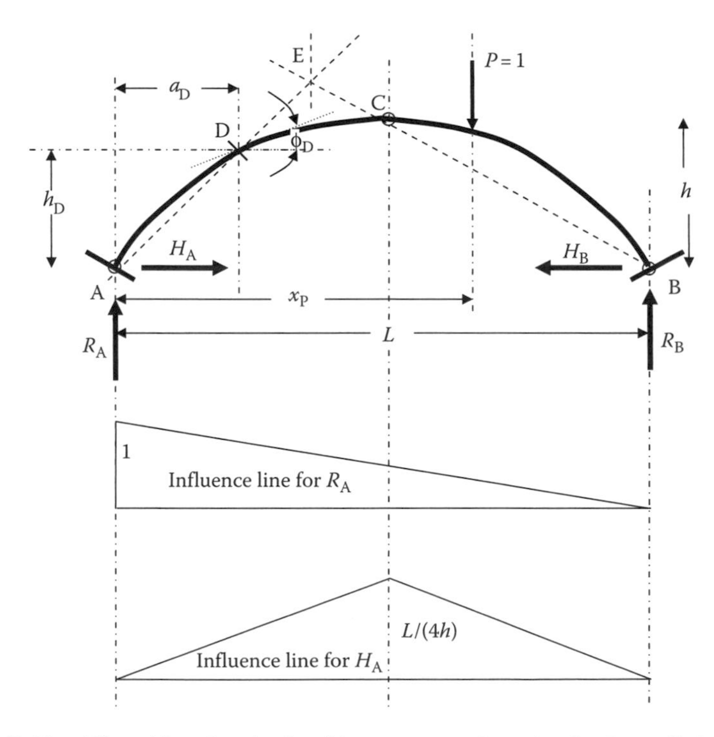

FIGURE 5.10a Three-hinged arch rib with concentrated moving loads applied directly to the rib.

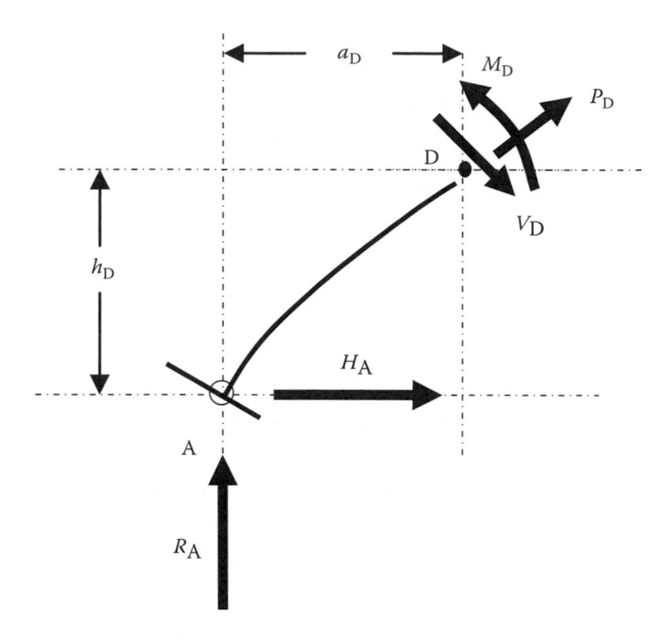

FIGURE 5.10b Free body diagram of arch rib from support A to point D.

Therefore, the influence line for the vertical components of the arch reactions, R_A and R_B, will be the same as those for a simply supported beam of length, L, as shown in Figure 5.10a.

If moments are taken about the arch crown pin (point C),*

$$H_A(h) = R_A\left(\frac{L}{2}\right). \tag{5.19}$$

Since $R_A(L/2)$ is the bending moment at point C in a simply supported span, the influence line for horizontal thrust reaction, H_A, is proportional (by the arch rise, h) to this simple span bending moment as shown in Figure 5.10a. Therefore, the criteria for the position of Cooper's load for maximum bending moment (see Section 5.2.1.1.3) can be used for the determination of maximum horizontal thrust.

The arch reactions may now be used to determine the internal shear force, bending moment, and axial force influence lines for the arch rib. From Figure 5.10b, the bending moment, M_D, at a location, D, is

$$M_D = R_A(a_D) - H_A(h_D). \tag{5.20}$$

Equation 5.20 indicates that the influence line for bending moment in the arch rib at location D can be obtained by subtracting the ordinates for the influence line for H_A (Figure 5.10a) multiplied by the distance h_D from the ordinates for simple beam

* It is the inclusion of the crown pin that enables this equilibrium equation to be written; thereby illustrating the benefits of statically determinate design and construction.

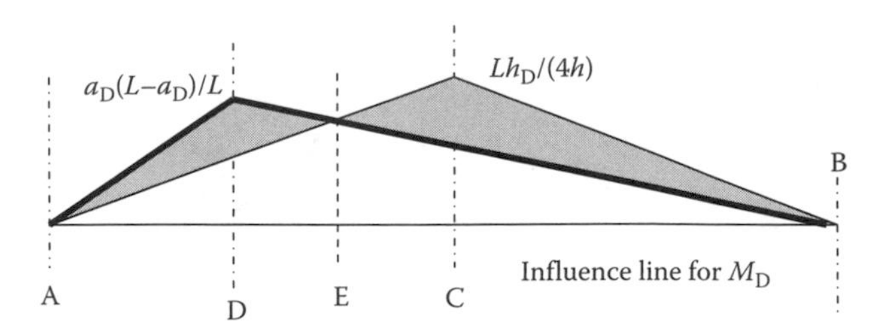

FIGURE 5.11 Influence line for bending moments at location D in three-hinged arch rib.

bending at location, D, described by $R_A(a_D)$. The construction of this influence line is shown in Figure 5.11. The ordinates (shaded areas) may be plotted on a horizontal line for ease of use in design.

From Figures 5.10a and b, the shear force, V_D, at a location, D, is

$$V_D = R_A \cos \phi_D - H_A \sin \phi_D. \tag{5.21}$$

Equation 5.21 indicates that the influence line for shear force in the arch rib at location D can be obtained by subtracting the ordinates for the influence line for H_A multiplied by $\sin \phi_D$ from the ordinates for simple beam shear at D multiplied by $\cos \phi_D$. The construction of this influence line is shown in Figure 5.12. Again, the ordinates (shaded areas) may be plotted on a horizontal line for ease of use in design. Location E is the position of the moving load that creates no shear force or bending moment in the arch at location D (Figure 5.10a).

From Figures 5.10a and b, the axial force, F_D, at a location, D, is

$$F_D = -R_A \sin \phi_D - H_A \cos \phi_D. \tag{5.22}$$

Equation 5.22 indicates that the influence line for axial force at location D in the arch rib can be obtained by adding the ordinates for the influence line for H_A multiplied by $\cos \phi_D$ to the ordinates for simple beam shear at D multiplied by $\sin \phi_D$. The construction of this influence line is shown in Figure 5.13. Again, the ordinates (shaded areas) may be plotted on a horizontal line for ease of use in design.

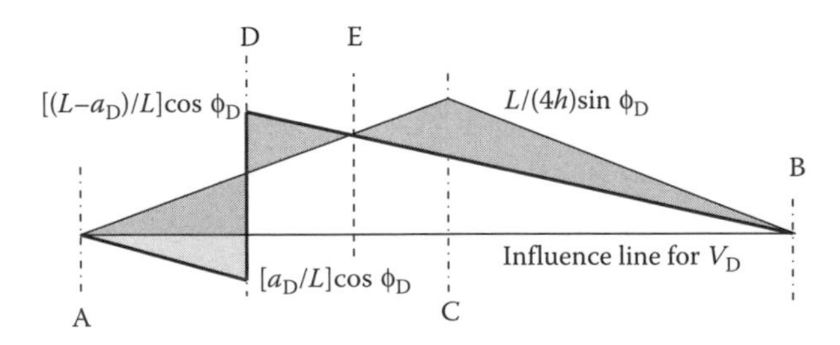

FIGURE 5.12 Influence line for shear forces at location D in three-hinged arch rib.

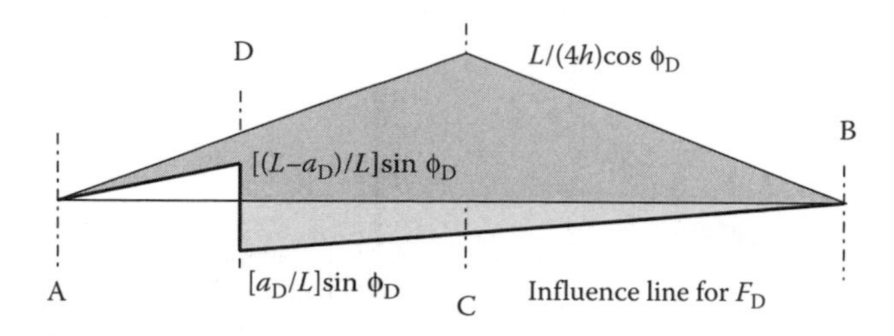

FIGURE 5.13 Influence line for axial force at location D in three-hinged arch rib.

5.2.1.2.3.2 Maximum Bending Moment, Shear Force, and Axial Force [with Loads Applied to the Arch by Transverse Members (Spandrel Columns or Walls)] Medium- and long-span steel railway bridges can be economically constructed of three-hinged arches with the arch rib loaded by spandrel members (Figure 5.14). Influence lines for arch spans with spandrel columns or vertical posts

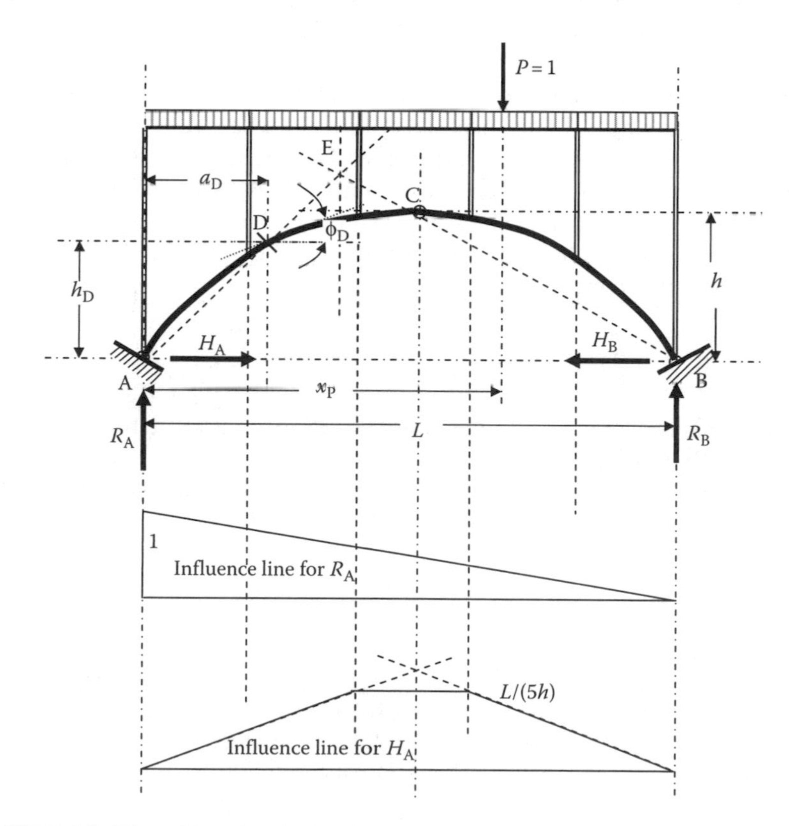

FIGURE 5.14 Three-hinged arch rib with concentrated moving loads applied to the rib at transverse members (e.g., spandrel columns).

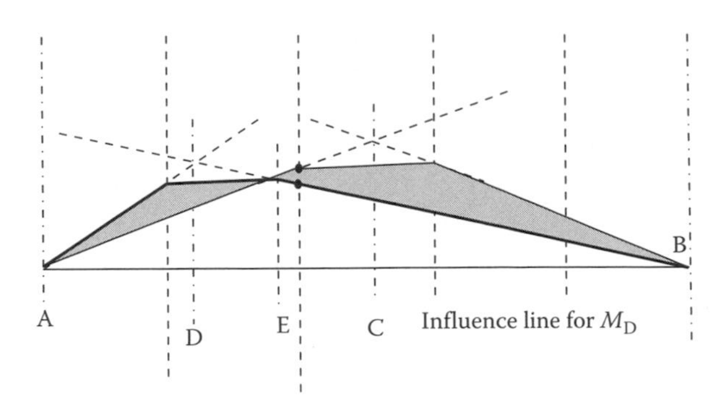

FIGURE 5.15 Influence line for bending moments at location D in three-hinged arch rib.

can be developed from influence lines for directly loaded arches in a manner analogous to simple spans with transverse members (floorbeams) (see Sections 5.2.1.1.2 and 5.2.1.1.4).

For example, with a pin at location C, the influence line for bending moment at D will be of the general form shown in Figure 5.15. The influence lines for other internal forces can be determined in a similar manner.

5.2.1.2.3.3 Maximum Axial Forces with Moving Loads on a Statically Determinate Trussed Arch Long-span steel railway bridges can be economically constructed of three-hinged arches with the arch rib replaced by a truss. The techniques used in Section 5.2.1.2.3.1 to determine maximum effects are useful for the construction of influence lines for trussed arches. The crown hinge is designed to achieve static determinacy with a bottom chord pin and top chord sliding arrangement* as shown in Example 5.7 and Figure E5.8.

Example 5.7

Determine the influence line for member U1–U2 in the 400 ft eight-panel deck trussed arch in Figure E5.8.

The force in the chord U1–U2 can be determined using Equation 5.20 by considering Section 1.1 and taking moments about L2.

$$F_{U1-U2} = \frac{M_D}{y_D} = \frac{R_A(a_D) - H_A(h_D)}{y_D} = \frac{R_A(a_D) - (L/4h)(h_D)}{y_D}.$$

The ordinate of the influence line at L2 provides $R_A(a_D) = (6/8)(100) = 75$. This component of the influence line is related to vertical reaction, R_A.

The ordinate of the influence line at L4 provides $(L/4h)(h_D) = (400/4(150))(45 + 55) = 66.67$. This component of the influence line is related to horizontal thrust reaction, H_A.

* Thereby, rending the force in one central top chord member as zero.

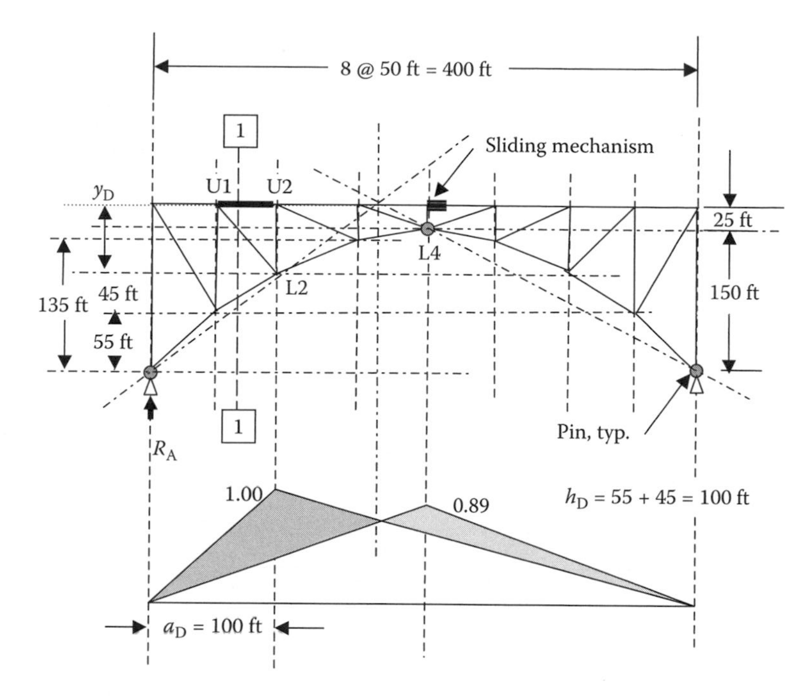

FIGURE E5.8

With $y_D = 175 - 45 - 55 = 75$ ft, the influence line for axial force in chord U1–U2 (shown by the shaded area in Figure E5.8) can be determined by the superposition of the influence lines for R_A and H_A.

5.2.1.2.4 *Influence Lines for Maximum Effects in Statically Determinate Cantilever Bridge Spans*

Long-span steel railway bridges may also be economically constructed as cantilever bridges (see Chapter 1). The economical relative lengths of the cantilever arm, L_c, anchor, L_a, and suspended, L_s, spans will vary with live to dead load bending moment ratio. For the relatively high live to dead load bending moment ratios of steel railway superstructures, typical L_a/L_c values of between 1 and 2 are used depending on the suspended span length, L_s. In steel railway superstructures, L_c/L_s values typically range from 0.4 to 2. The relative lengths of the cantilever arm, anchor, and suspended spans may also vary based on site conditions that dictate the location of piers at a crossing (see Chapter 3). Influence lines for cantilever superstructures may also be constructed by consideration of unit loads traversing the bridge. The ordinates of the influence lines are readily determined by calculation of the reaction, bending moment, and shear due to unit loads at locations where the influence lines change direction.

5.2.1.2.4.1 Cantilever Bridge Span Influence Lines [with Loads Applied Directly to the Superstructure] Influence lines for reactions at locations A and B, bending moment in the anchor span at location E and at location F in the cantilever

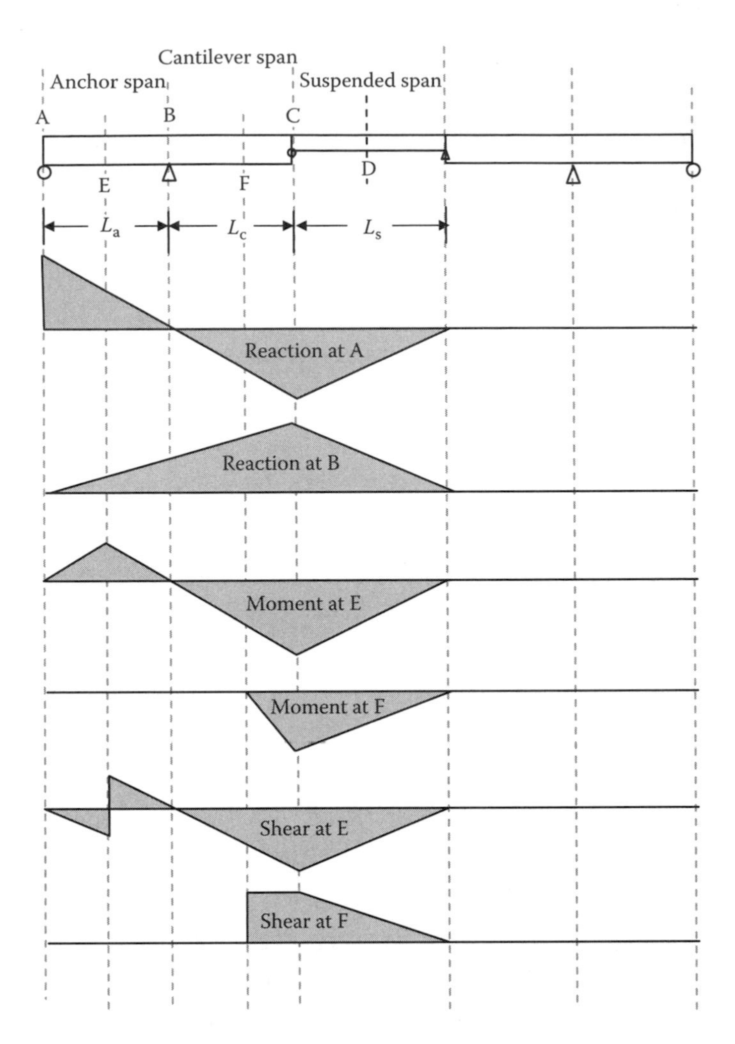

FIGURE 5.16 Influence lines for reactions, bending moments, and shear forces in anchor and cantilever spans with loads applied directly to the superstructure.

span may be constructed by considering effects of unit loads placed at locations A, B, and C as shown qualitatively* in Figure 5.16.

5.2.1.2.4.2 Cantilever Bridge Span Influence Lines [with Loads Applied to the Superstructure by Transverse Members (Floorbeams)] Influence lines for shear force and bending moment in the anchor span panel point A1–A2 and in the cantilever span panel point C2–C3 can be constructed by considering

* Qualitative influence lines are useful in both manual and electronic calculations of maximum or minimum effects to determine the approximate location of live load.

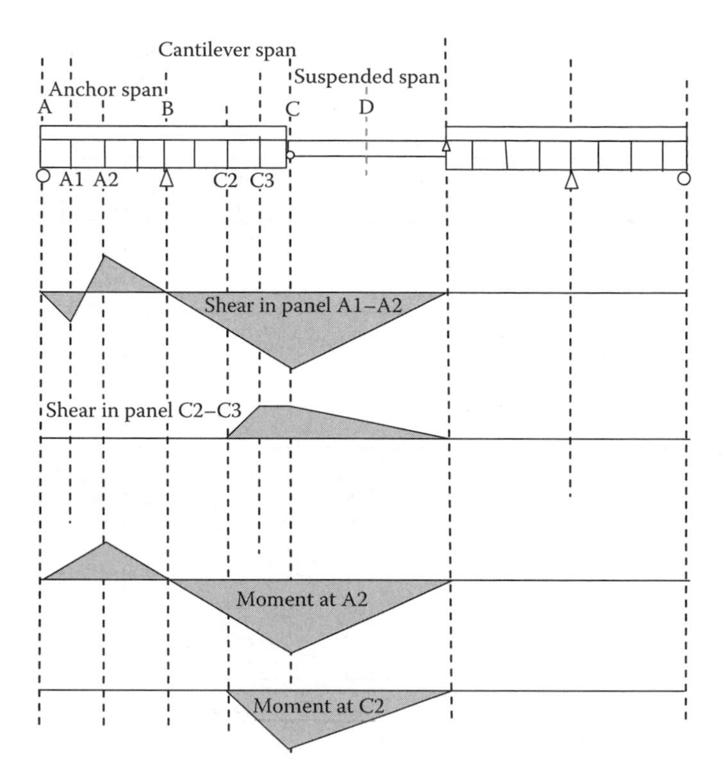

FIGURE 5.17 Influence lines for shear forces and bending moment in anchor and cantilever spans with panel points.

the effects of unit loads placed at locations A, A1, A2, B, C2, C3, and C (also shown qualitatively in Figure 5.17).

For long-span railway superstructures it is of further efficiency to utilize truss spans in cantilever bridges. The influence lines developed in Figure 5.17, in conjunction with those developed in Sections 5.2.1.2.1 and 5.2.1.2.2, are useful in the construction of axial force influence lines for cantilever bridge truss web and chord members. Also, as usual, the consideration of the moving load effect at panel points simplifies the construction of axial force influence lines. Influence lines constructed in this manner for axial forces in cantilever bridge truss members are shown in Example 5.8.

Example 5.8

Construct influence lines for members in panel point 2–3 in the anchor arm of the cantilever truss bridge in Figure E5.9.

By inspection and placement of unit loads at L0, L2, L3, and L6 and considering the hinge at the end of the cantilever and suspended spans, the influence lines for axial forces in L2–L3, U2–U3, and U2–L3 are shown in Figure E5.9. Influence lines for axial force in other members of the trusses may be constructed in a similar manner.

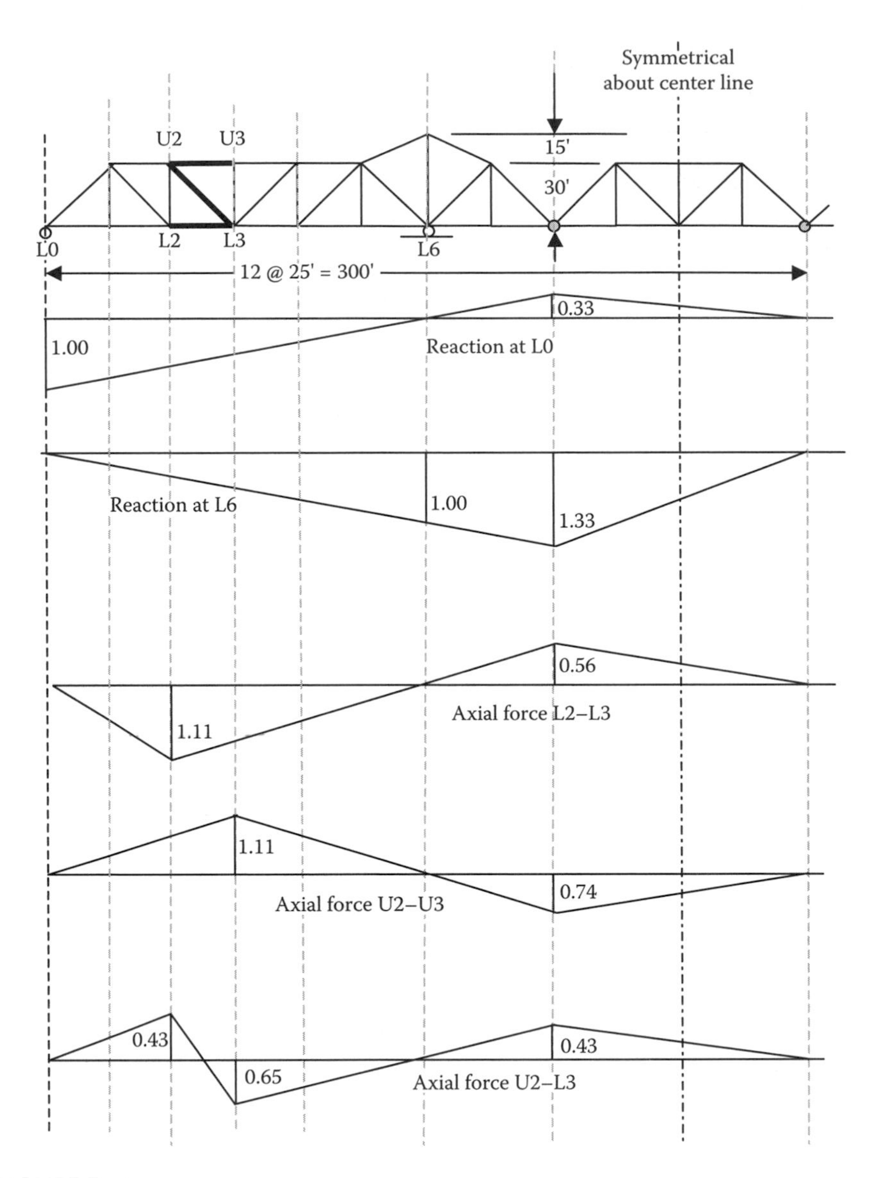

FIGURE E5.9

5.2.1.3 Equivalent Uniform Loads for Maximum Shear Force and Bending Moment in Simply Supported Spans

The methods outlined in Sections 5.2.1.1 and 5.2.1.2 require iteration that can be readily computerized. However, for concentrated design loads used on many bridge spans (e.g., Cooper's configuration) it is often beneficial to determine an equivalent uniform load, w_e, that represents the effects of the concentrated design loading.*

* Equivalent uniform loads are particularly useful for preliminary design.

5.2.1.3.1 *Maximum Shear Force in Simply Supported Spans [with Concentrated Moving Loads Applied Directly to the Superstructure (Figure 5.18)]*

Equating maximum shear force, V_C, from Equation 5.1 with the shear force, V_{Ce}, at location C from an equivalent uniform load, w_{ev}, yields

$$w_{ev} = \left(P_T\left(\frac{x_T}{L}\right) - P_L\right)\left(\frac{2L}{(L-a)^2}\right). \tag{5.23}$$

Equation 5.23 can be plotted for different P_T and P_L (which are dependent on load configuration and span length) at locations C on the span. Figure 5.19 shows the equivalent uniform load for shear force at the end, the 1/4 point and the center of the span for a Cooper's E80 series of concentrated moving wheel loads applied directly to the superstructure.

5.2.1.3.2 *Maximum Shear Force in Simply Supported Spans [with Concentrated Moving Loads Applied to the Superstructure by Transverse Members (Figure 5.20)]*

The location in the panel BC where the shear due to an equivalent uniform load, $V_{BCe} = 0$, is

$$d = \frac{(L-a)s_p}{L-s_p} = \frac{(L-a)}{n_p - 1}, \tag{5.24}$$

where $n_p = L/s_p$ is the number of equal length panels.

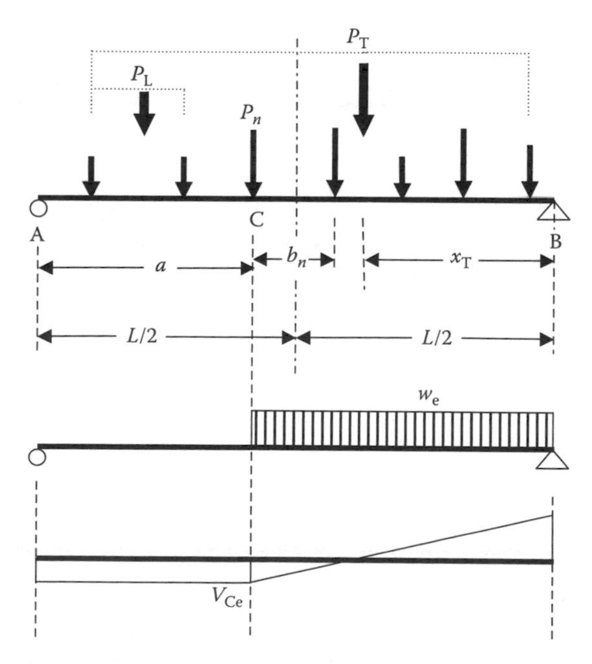

FIGURE 5.18 Equivalent uniform load for shear force for concentrated moving loads applied directly to the superstructure.

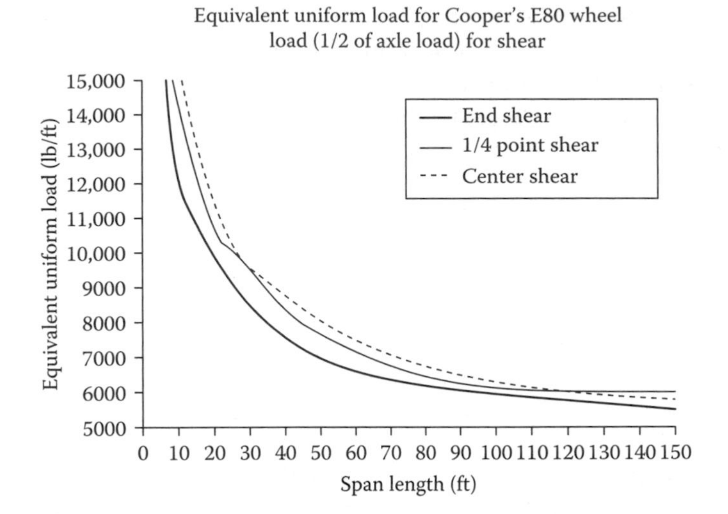

FIGURE 5.19 Equivalent uniform load for shear force for a Cooper's E80 series of concentrated moving wheel loads applied directly to the superstructure.

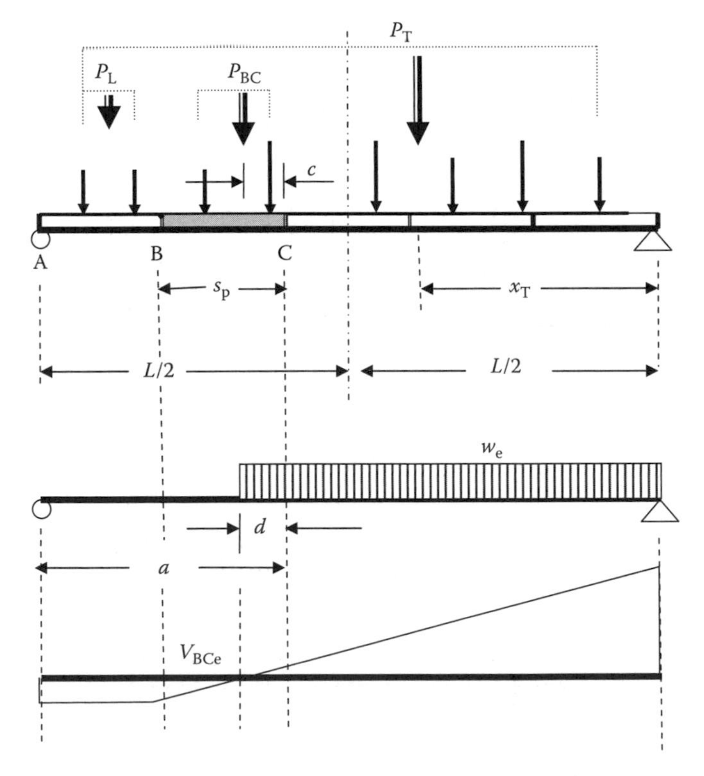

FIGURE 5.20 Equivalent uniform load for shear force for concentrated moving loads applied at transverse members to the superstructure.

Equating the maximum shear force, V_{BC}, from Equation 5.4 with the shear force, V_{BCe} in panel BC, from an equivalent uniform load, w_{ev}, yields

$$w_{ev} = \left(P_T\frac{x_T}{L} - \left(P_L + P_{BC}\left(\frac{c}{s_p}\right)\right)\right)\left(\frac{2L}{(L-a+d)^2}\right). \quad (5.25)$$

Equation 5.25 can be plotted for different P_T and P_L (which are dependent on load configuration and span length) and P_{BC} (which is dependent on load configuration and panel length) in different panels on the span (described by distances a and d). For a specific design load such as the Cooper's configuration, the value of $P_T(x_T/L) - (P_n + P_{BC}(c/s_p))$ can be calculated for various values of x_T and the equivalent uniform load for shear can be determined in various panels along the span. The equivalent uniform load for maximum shear in the panels will have the general form shown in Figure 5.19.

5.2.1.3.3 *Maximum Bending Moment in Simply Supported Spans [with Concentrated Moving Loads Applied Directly to the Superstructure (Figure 5.21)]*

Equating maximum bending moment, M_C, from Equation 5.7 with the bending moment, $M_{Ce} = [w_e a(L-a)]/2$, at location C from an equivalent uniform load,

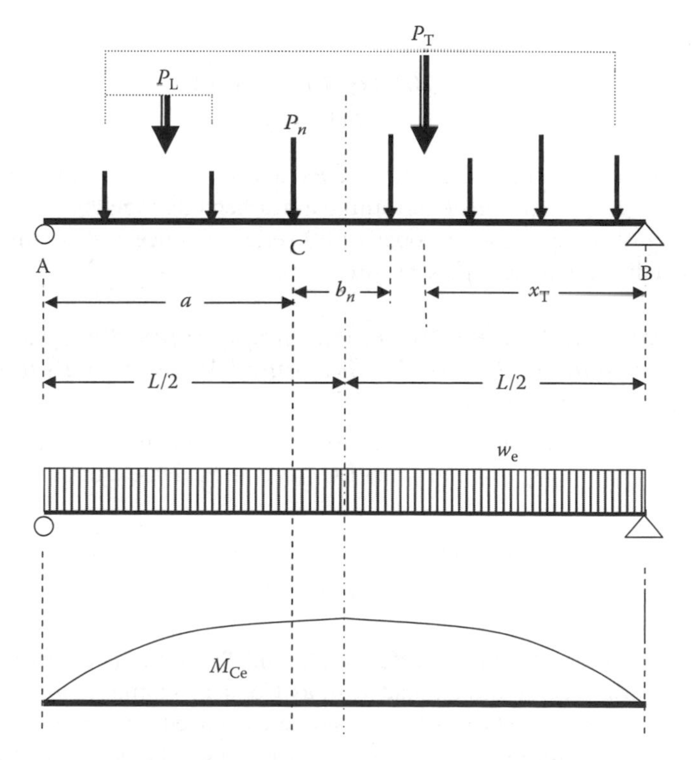

FIGURE 5.21 Equivalent uniform load for bending moment for concentrated moving loads applied directly to the superstructure.

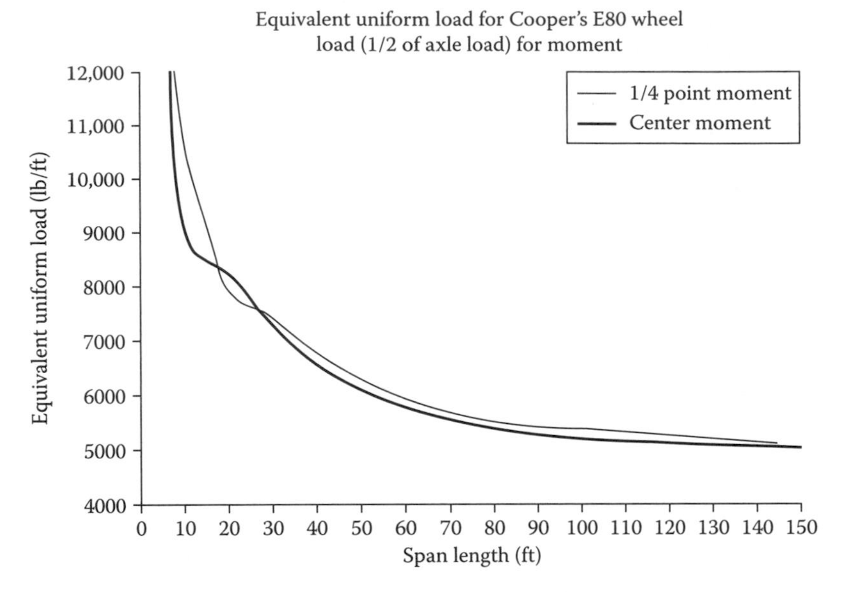

FIGURE 5.22 Equivalent uniform load for bending moment for a Cooper's E80 series of concentrated moving wheel loads applied directly to the superstructure.

w_{em}, yields

$$w_{\text{em}} = \frac{2(P_{\text{T}}(x_{\text{T}}/L)a - P_{\text{L}}x_{\text{L}})}{a(L-a)}. \tag{5.26}$$

Equation 5.26 can be plotted for different P_{T} and P_{L} at locations C on the span. Figure 5.22 shows the equivalent uniform load for bending moment at the 1/4 point and the center of the span for a Cooper's E80 series of concentrated moving wheel loads applied directly to the superstructure.

5.2.1.3.4 Maximum Bending Moment in Simply Supported Spans [with Concentrated Moving Loads Applied At Panel Points to the Superstructure (Figure 5.23)]

Equating the maximum bending moment, M_{BC}, from Equation 5.9 with the bending moment, $M_{\text{BCe}} = w_{\text{e}}s_{\text{p}}(s_{\text{p}} + a)$, in panel BC from an equivalent uniform load, w_{em}, yields

$$w_{\text{em}} = \frac{\left((P_{\text{T}}(x_{\text{T}}/L))\,a - (P_{\text{L}})b - (P_{\text{BC}})\,(c/s_{\text{p}})d\right)}{s_{\text{p}}(s_{\text{p}} + a)}. \tag{5.27}$$

Equation 5.27 can be plotted for different P_{T} and P_{L} and P_{BC} for various panels on the span. For a specific design load such as Cooper's configuration the value of $\left((P_{\text{T}}(x_{\text{T}}/L))\,a - (P_{\text{L}})b - (P_{\text{BC}})(c/s_{\text{p}})d\right)$ can be calculated for various values of x_{T} and the equivalent uniform load for bending moment can be determined in various panels along the span. The equivalent uniform load for maximum bending moments in the panels will have the general form shown in Figure 5.22.

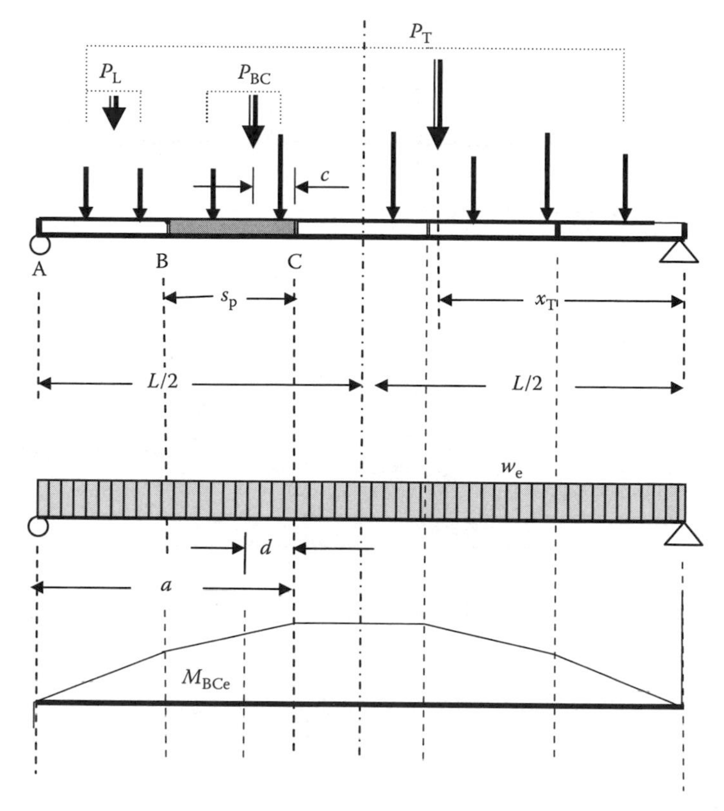

FIGURE 5.23 Equivalent uniform load for bending moment for concentrated moving loads applied at transverse members to the superstructure.

Example 5.9

Determine the maximum shear forces and bending moment per rail for Cooper's E80 load at the end, the 1/4 point and the center of a 60 ft long DPG span.

Using Figures 5.19 and 5.22 the maximum shear forces and bending moment per rail are given in Table E5.1.

TABLE E5.1

Location and Force	Equivalent Uniform Load (k/ft)	Loaded Length, l (ft)	Maximum Shear Force (kips)	Maximum Bending Moment (ft-kips)
End shear	6530 (Figure 5.19)	60	$wl/2 = 196$	—
1/4 point shear	7125 (Figure 5.19)	45	$wl^2/[(2)(60)] = 120$	—
Center shear	7425 (Figure 5.19)	30	$wl^2/[(2)(60)] = 55.7$	—
1/4 point moment	5955 (Figure 5.22)	60	—	$3wl^2/32 = 2010$
Center moment	5775 (Figure 5.22)	60	—	$wl^2/8 = 2599$

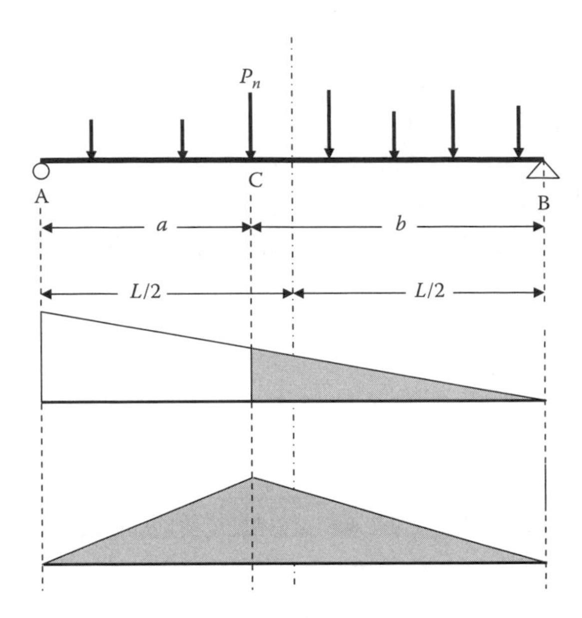

FIGURE 5.24 Determination of equivalent uniform loads for simple span shear and bending at location C.

Equivalent uniform loads for Cooper's and other locomotive and train live loads have been presented often in the early railway bridge design literature (Waddell, 1916; Ketchum, 1924).

5.2.1.3.5 Shear Force and Bending Moment at any Location in Simply Supported Spans [with Concentrated Moving Loads Applied Directly to the Superstructure (Figure 5.24)]

The use of uniform loads can be generalized for shear, bending, and floorbeam reaction at any location, C, on a simple span. The area under the shear influence line in Figure 5.24 is $b^2/2L$ and the area under the bending moment influence line in Figure 5.24 is $ab/2$. Therefore, the equivalent uniform load for Cooper's live load shear and bending moments, respectively, are

$$w_{\mathrm{ev}} = V_{\mathrm{LL}}\left(\frac{2L}{b^2}\right), \tag{5.28}$$

$$w_{\mathrm{em}} = M_{\mathrm{LL}}\left(\frac{2}{ab}\right). \tag{5.29}$$

The equivalent uniform load, w_{ev} or w_{em}, can be calculated for various span lengths, $L = a + b$, at location C (with $a < b$) and plotted to provide curves for use by design engineers. The curves will be of the general form shown in Figure 5.25. Curves such as these were prepared by the bridge engineer David B. Steinman* in 1915. The curves

* David B. Steinman also designed long-span suspension bridges and further developed J. Melan's "deflection theory" for suspension bridge design (Steinman, 1953; Petroski, 1995).

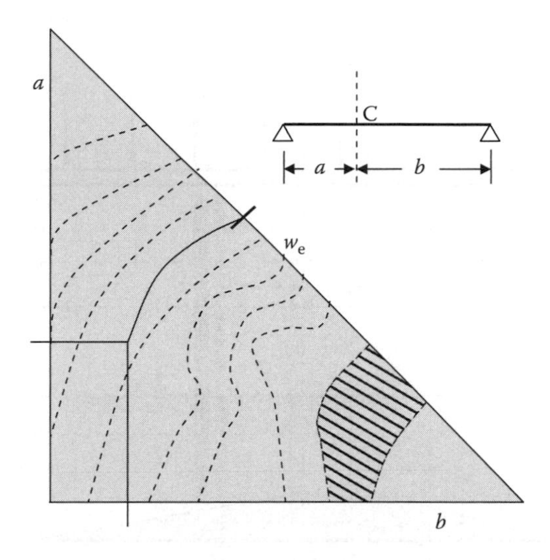

FIGURE 5.25 Schematic of generalized equivalent uniform loads for design live load shear, V_{LL}, bending moment, M_{LL}, and floorbeam reaction, R_{LL}.

(sometimes referred to as Steinman's charts) are available in many early bridge design handbooks, manuals, and texts, for example (Grinter, 1942).

5.2.1.3.6 Shear Force and Bending Moment at any Location in Simply Supported Spans [with Concentrated Moving Loads Applied at Panel Points in the Superstructure (Figure 5.26)]

The area under the shear influence line in Figure 5.26 is $[(a+b)(n_R/n)]/2$. Therefore, the equivalent uniform load for Cooper's live load shear is

$$w_{ev} = V_{LL}\left(\frac{2n}{(a+b)n_R}\right). \tag{5.30}$$

From Equation 5.17

$$a = \frac{s_p n_R}{2[(n_L/n_R)+1]} \tag{5.31}$$

and from Figure 5.26

$$b = n_R s_p, \tag{5.32}$$

where n is the number of panels ($n = 4$ in Figure 5.26), n_L is the number of panels left of the panel under consideration ($n_L = 1$ in Figure 5.26), n_R is the number of panels right of panel under consideration ($n_R = 2$ in Figure 5.26), and s_p is the uniform panel spacing ($s_p = L/n = L/4$ in Figure 5.26).

Substitution of Equations 5.31 and 5.32 into 5.30 yields

$$w_{ev} = V_{LL}\left(\frac{2s_p}{ab}\right). \tag{5.33}$$

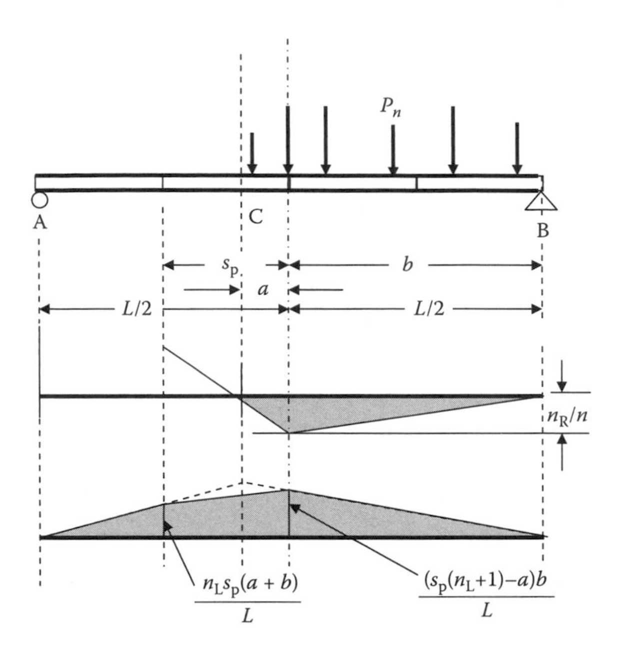

FIGURE 5.26 Determination of equivalent uniform loads for simple span shear and bending at location C.

The area under the bending moment influence line in Figure 5.26 is

$$\frac{m_1 s_p(n_L + 1) + m_2(s_p + b)}{2},$$

where

$$m_1 = (L - b - s_p)\left(\frac{a+b}{L}\right), \tag{5.34a}$$

$$m_2 = (L - b - a)\left(\frac{b}{L}\right). \tag{5.34b}$$

Therefore, the equivalent uniform load for Cooper's live load moment is

$$w_{em} = \frac{2M_{LL}}{m_1 s_p(n_L + 1) + m_2(s_p + b)} \tag{5.35}$$

and substitution of Equations 5.34a and 5.34b into Equation 5.35 yields

$$w_{em} = \frac{2LM_{LL}}{(a+b)[(L-b-s_p)(L-b)] + (L-b-a)[b(s_p+b)]}. \tag{5.36}$$

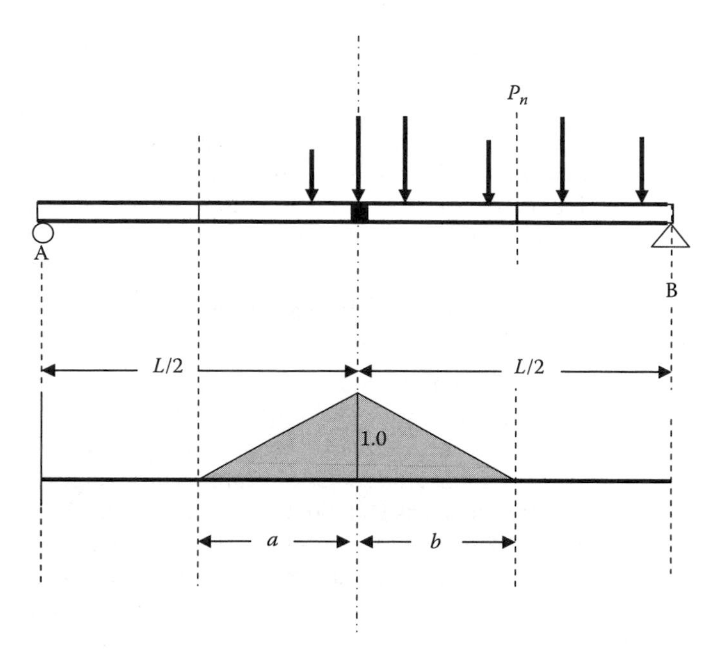

FIGURE 5.27 Determination of equivalent uniform loads for floorbeam reaction.

The equivalent uniform load, w_{ev}, for shear (Equation 5.33)* and, w_{em}, for bending moment (Equation 5.36)† can be calculated within various panels at location C ($a + b$ from the right side in Figure 5.26) using the same charts plotted for simple spans shown in Figure 5.25.

5.2.1.3.7 *Floorbeam Reaction at any Location in Simply Supported Spans [with Concentrated Moving Loads Applied at Panel Points (at Transverse Floorbeams) in the Superstructure (Figure 5.27)]*

The area under the shear influence line in Figure 5.27 is $(a + b)/2$. Therefore, the equivalent uniform load for Cooper's live load reaction is

$$w_{eR} = V_{LL}\left(\frac{2}{(a+b)}\right). \tag{5.37}$$

The equivalent uniform load, w_{eR}, can be calculated at various floorbeam locations with adjacent panel lengths a and b as shown in Figure 5.27, using the same charts plotted for simple spans shown in Figure 5.25.

The generalized equivalent uniform loads presented in Steinman's charts are useful in the design of usual steel girder and truss railway spans. However, despite the appeal of ease in design, the use of equivalent uniform live loads has never been prevalent in

---

* Equation 5.33 is of similar form as Equation 5.28.

† Equation 5.36 is of similar form as Equation 5.29 with L, b, and s_p being constant. If the constants are included together as K_1 and K_2, Equation 5.36 is $w_e = 2M_{LL}/\left[(a+b)K_1 + (L-b-a)K_2\right]$ and the similarity with Equation 5.29 is clear.

North America and most engineers develop shear forces and bending moments from an analysis of concentrated loads.

In order to encourage efficiency in the design process and avoid the use of charts and influence lines, digital computers, equations, and tables are useful. For usual bridge design projects (e.g., simply supported span bridges), equations and tables have been prepared for the Cooper's load configuration for the determination of maximum shearing forces, axial forces, and bending moments.

5.2.1.4 Maximum Shear Force and Bending Moment in Simply Supported Spans from Equations and Tables

Tabulated values for shear and bending moment at the end, the 1/4 point and the center of simple beam and girder spans from 5 to 400 ft are given in AREMA (2008). AREMA (2008) also provides equations for some span lengths for the shear and bending moment at the end, the 1/4 point and the center of simple beam and girder spans.

Tabulated values of shears in panels and moment at panel points for Pratt trusses with various panel lengths and number of panels have also been tabulated by railroad companies and are available in the railway bridge design literature, for example, Ketchum (1924). For typical railway truss span design these tables can save considerable computational effort.

For the design of complex steel bridges, such as continuous and cantilever steel spans, the use of influence lines and/or modern computer-based frame or finite-element analysis software may be required.

5.2.1.5 Modern Structural Analysis

Analytical methods for structures based on the theories and methods of applied elasticity and mechanics of materials are used for the determination of stress in usual steel railway superstructures. The modern digital computer and proliferation of computer software* based on these classical methods have been of great benefit to bridge engineers. When effectively utilized, such software not only enables the engineer to avoid many long and tedious calculations and attain greater speed and accuracy but also enhances the ability to perform multiple analyses for optimization purposes.

Complex, long span, and/or structures that require specialized analysis (e.g., dynamic structural analysis due to wind or seismic effects) generally require the use of finite-element method software. The finite-element method for structural analyses is most often based on displacement matrix methods. Computerized finite-element analysis is a powerful tool that enables a detailed stress analysis of structures. Engineers experienced in the concepts of matrix and finite-element analysis are generally required to review and assess the large quantity of data developed by these software programs. There are many proprietary specialized and general purpose finite-element method software programs available and many standard textbooks provide the theory and applications of the method, for example, Zienkiewicz (1977), Cook (1981), and Wilson (2004).

---

* For example, commercially available spreadsheets are relatively easy to program and are used extensively for structural analyses (Christy, 2006).

Modern trends in structural engineering software are toward integrated structural analysis, design, drafting, and fabrication. Some proprietary systems successfully integrate many of these functions, and it is likely that, as such integrated systems become even more "user-friendly" and reliable; they will be used even more frequently in structural engineering practice.

5.2.2 Lateral Load Analysis of Steel Railway Superstructures

The analysis of railway superstructures also involves the determination of the maximum deformations and stresses caused by lateral effects such as those due to moving loads (centrifugal and nosing), wind, and earthquakes.*

For usual steel railway bridge superstructures, lateral load effects may be determined by simplified analyses. This enables the use of hand calculations, relatively simple computer programs, and spreadsheets to determine the deformations and forces. For more complex superstructures more sophisticated computerized frame analysis software is often used.

5.2.2.1 Lateral Bracing Systems

Lateral forces on steel railway superstructures from wind, nosing, and centrifugal forces are generally transferred to the bearings and then substructures via bracing members in horizontal truss systems. Components of the horizontal bracing systems may also resist the buckling propensity of compression members such as girder top flanges or truss top chords in simply supported spans. Forces from horizontal truss systems that are not in the plane of the bearings are transferred to the substructures by end vertical (DPG spans and some deck truss spans) or portal (through truss and some deck truss spans) bracing systems. Knee braces are used to provide resistance to buckling of the compression flange and transfer wind forces from the top flanges to the bearings in through plate girder spans.

5.2.2.1.1 Horizontal Truss Bracing

Since, for usual steel railway bridges, the determination of lateral loads is approximate, it is reasonable to utilize simplifications regarding load distribution to the horizontal bracing systems. It is generally adequate to use a horizontal Pratt or Warren truss and apply lateral forces at the windward side of the lateral truss panel points. For bracing systems (horizontal trusses) with two diagonals in each panel (Pratt-type cross-bracing) the lateral shear can be assumed transferred equally between diagonals and the members are designed for both maximum tension and compression forces. When double bracing is not connected to the floor system or otherwise supported,† the diagonals can be assumed to act in tension only with transverse members (struts) in compression. For bracing systems with only a single diagonal in each panel (Warren-type bracing) the diagonals are also assumed to act as tension-only members.

---

* Wind and earthquake forces may also have longitudinal components.

† Therefore, relatively long and slender with a low critical buckling load and compressive force capacity.

The approximate determination of forces in lateral bracing systems is shown in Example 5.10.

5.2.2.1.1.1 Members in Top Lateral Systems In addition to lateral forces from wind, the top lateral system in through spans* requires bracing members that resist a transverse shear force of 2.5% of the total compressive axial force in the chord or flange at that panel point. The top lateral system in deck spans requires bracing to resist a transverse shear force of 2.5% of total compressive axial force in the chord or flange at that panel point in addition to other lateral forces from wind, nosing, and centrifugal forces. Deck span top lateral systems are often the heaviest bracing system required in steel railway superstructures. Concrete and steel plate decks may be used, which act as diaphragm, for resisting these lateral forces.

5.2.2.1.1.2 Members in Bottom Lateral Systems Lower lateral bracing is generally required when the span supports are at the bottom chord of a truss or bottom flange of a girder span. When the span supports are at the top chord of a truss† only struts at bottom panel points are strictly required. However, a nominal lateral bracing system is often employed for adequate overall lateral rigidity of the span.

The bottom lateral system in through spans may use the floorbeams as struts of the bracing system. The bracing is designed to resist lateral wind, nosing, and centrifugal forces. Concrete and steel decks may also be used to act as a diaphragm.

The bottom lateral bracing system in deck spans is lightly loaded by wind and, for short spans in particular, may not be required. However, in order to ensure overall rigidity of longer spans, a light bracing system (based on the maximum slenderness ratio for compression members) is often used. At a minimum, struts should be installed at each panel point in the chord or flange. AREMA (2008) recommends bottom lateral bracing for all deck spans greater than 50 ft long.

Example 5.10

The forces in the top and bottom lateral bracing system members of the through truss span in Figure E5.10 are required. The lateral wind and nosing forces, and compression forces required to be resisted for bracing of main compression members are as follows:

Wind load at the top chord = 350 lb/ft.
Wind load at top lateral bracing panels = 19.55(0.35) = 6.8 kips per panel.
Wind load at the bottom chord = 200 lb/ft.
Wind load on train = 300 lb/ft.
Wind load at the bottom lateral bracing panels = 19.55(0.5) = 9.8 kips per panel.

---

* Through spans such as plate girder and pony truss spans without room for top lateral bracing generally utilize knee brace frames to resist the transverse shear force of 2.5% of total compressive axial force in the chord or flange and the lateral forces from wind. An analysis of knee-braced through span transverse frames is outlined in Section 5.2.2.1.4.

† Such as a "fish-bellied" deck truss span.

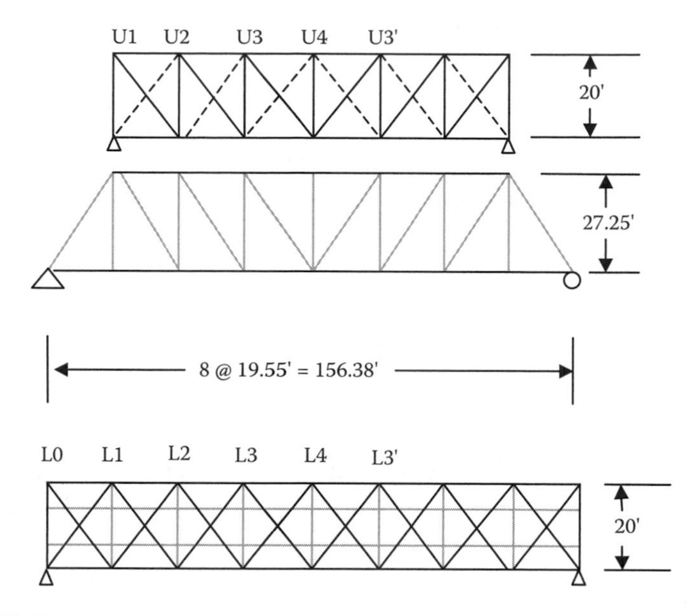

FIGURE E5.10

Cooper's E90 nosing load (lateral equipment load) at the bottom lateral bracing panels = 90/4 = 22.5 kips at any panel.

Bracing forces required to resist top chord buckling are shown in Table E5.2:

Top lateral bracing:

Due to their slenderness, top lateral bracing compressive members are assumed to be inactive and tension members only resist the panel forces. The top lateral bracing member forces are shown in Table E5.3.

Bottom lateral bracing:

Since the bracing members are connected to the floor system, both are assumed to equally participate in resisting panel shear forces. Therefore, each member is required to resist 50% of the panel shear force in both tension

TABLE E5.2

Panel Point	Total Axial Compression in Top Chord (kips)	Bracing Force (kips) (2.5% of Main Member Compressive Force)
U1	370	9.3
U2	640	16.0
U3	800	20.0
U4	850	21.3

TABLE E5.3

Panel	Shear (Wind) (kips)	Shear (Top Chord Compression) (kips)	Total Panel Shear (kips)	Force in Each Diagonal (kips)
U1–U2	$(6.8)(5+0.5+0.5)/2 = 20.4$	9.3	29.7	+41.6
U2–U3	$20.4 - 6.8 = 13.6$	16.0	29.6	+41.4
U3–U4	$13.6 - 6.8 = 6.8$	20.0	26.8	+37.5
U4–U3′	0	21.3	21.3	+29.8

TABLE E5.4

Panel	Shear (Wind) (kips)	Shear (Nosing) (kips)	Total Panel Shear (kips)	Force in Each Diagonal (kips)
L0–L1	$9.8(8)/2 = 39.2$	22.5	61.7	+/− 43.1
L1–L2	$39.2 - 9.8 = 29.4$	22.5	51.9	+/− 36.3
L2–L3	$29.4 - 9.8 = 19.6$	22.5	42.1	+/− 29.4
L3–L4	$19.6 - 9.8 = 9.8$	22.5	32.3	+/− 22.6
L4–L3′	0	22.5	22.5	+/− 15.7

and compression. The bottom lateral bracing member forces are shown in Table E5.4.

5.2.2.1.2 End Vertical and Portal Bracing

A supplemental structural system is required to transfer lateral forces from horizontal bracing members that are not located in the plane of the bearings to the horizontal lateral systems that are in the plane of the bearings. In through spans the lateral forces may be transferred through a system of knee braces or via sway and end portals. In deck spans a system of vertical cross frames or diaphragms may serve this purpose.* The vertical end bracing and portal bracing are required to carry the entire reaction from lateral loads to the substructures via the bearings.

Through truss end portal frames must be designed to transfer the total wind force reaction of the top lateral truss system, P_L, through flexure of end posts. The end posts are also required to resist additional axial forces from the portal action. In order to estimate the end portal effects on the end posts, it is assumed that horizontal reactions are equal at the bottom of the end posts and an inflection point is located midway between the bottom of the end post and the bottom of the portal bracing frame (often cross-braced or knee-braced in modern steel trusses) at a distance, $(h_e - h_p)/2$, as

* In deck truss spans supported at the top chord end portal bracing is used in the plane of the end diagonal member.

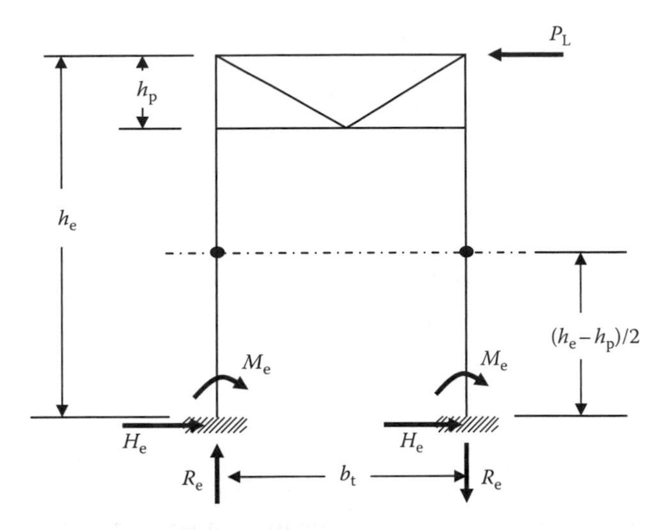

FIGURE 5.28 End post forces from through truss portal action.

shown in Figure 5.28. In this case, the vertical load, R_e, and horizontal shear, H_e, are

$$R_e = \frac{P_L((h_e + h_p)/2)}{b_t}, \tag{5.38}$$

$$H_e = \frac{P_L}{2}, \tag{5.39}$$

and the end post bending moment, M_e, due to the force, P_L, is estimated as

$$M_e = \frac{P_L((h_e - h_p)/2)}{2}. \tag{5.40}$$

The end portal bracing member forces can then be determined from free body equilibrium equations for one leg of the portal. The end portal bracing member forces for various portal configurations are shown in Examples 5.11 through 5.14. Example 5.15 outlines the analysis of a typical railway through truss span end portal.

Example 5.11

The axial forces in lattice portal frame members (Figure E5.11) are

$$P_A = \frac{\pm P_L((h_e + h_p)/2)}{2b_t}\left(\frac{\sqrt{h_p^2 + b_p^2}}{h_p}\right),$$

$$P_B = \frac{-P_L}{8h_p}\left(3\left(\frac{h_e + h_p}{2}\right) + 4h_p\right),$$

$$P_C = \frac{+3P_L}{8h_p}\left(\frac{h_e + h_p}{2}\right).$$

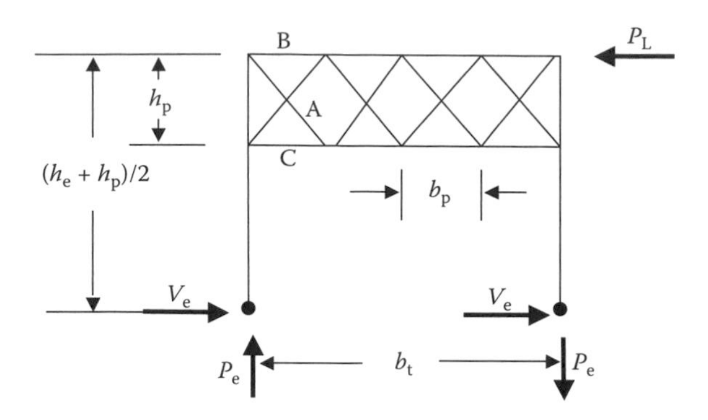

FIGURE E5.11 Lattice portal frame.

Example 5.12

The axial forces in cross-braced portal frame members (Figure E5.12) are

$$P_A = \frac{-P_L((h_e + 3h_p)/2)}{2h_p},$$

$$P_B = \frac{+P_L}{b_t}\left(\frac{h_e + h_p}{2}\right)\left(\frac{\sqrt{h_p^2 + b_t^2}}{h_p}\right),$$

$$P_C = \frac{-P_L}{2h_p}\left(\frac{h_e + h_p}{2}\right).$$

Example 5.13

The axial, shear, and bending forces in knee-braced portal frame members (Figure E5.13) are

$$P_A = -P_B = \frac{+P_L((h_e + h_p)/2)}{2h_p((b_t - d_t)/2)}\sqrt{h_p^2 + \left(\frac{b_t - d_t}{2}\right)^2},$$

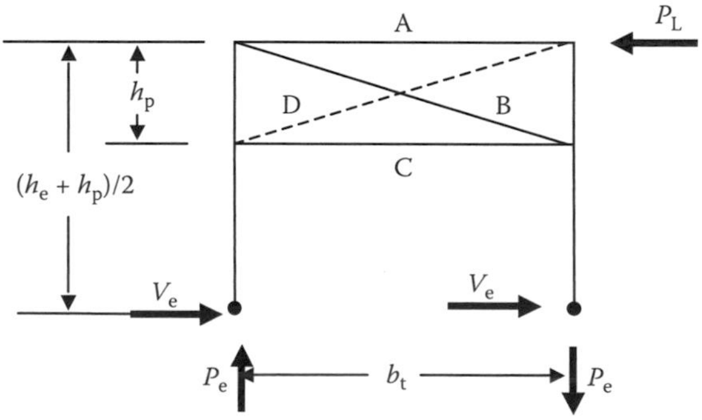

FIGURE E5.12 Cross-braced portal frame.

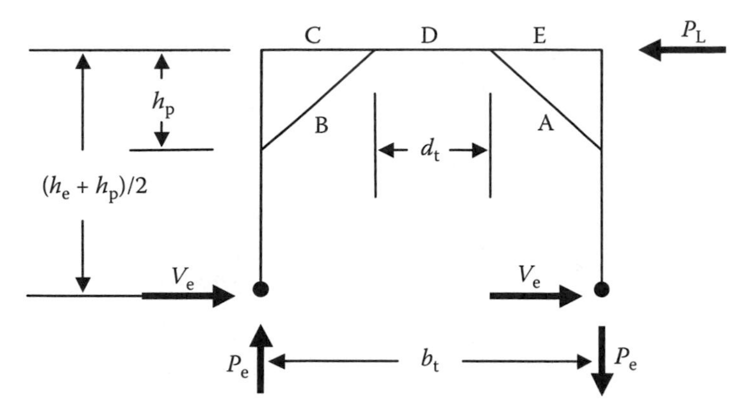

FIGURE E5.13 Knee-braced portal frame.

$$P_C = \frac{+P_L}{2}\left(\left(\frac{h_e + h_p}{2h_p}\right) - 1\right),$$

$$P_D = \frac{-P_L}{2},$$

$$P_E = \frac{-P_L}{2}\left(\left(\frac{h_e + h_p}{2h_p}\right) + 1\right),$$

$$V_C = V_E = \frac{-P_L\left((h_e + h_p)/2\right)}{b_t}\left(\left(\frac{b_t}{b_t - d_t}\right) - 1\right),$$

$$V_D = \frac{+P_L\left((h_e + h_p)/2\right)}{b_t},$$

$$M_E = \frac{P_L\left((h_e + h_p)/2\right)}{2b_t} d_t.$$

Example 5.14

The axial forces in triangular portal frame members (Figure E5.14) are

$$P_A = -P_B = \frac{+P_L\left((h_e + h_p)/2\right)}{b_t h_p}\sqrt{h_p^2 + \left(\frac{b_t}{2}\right)^2},$$

$$P_C = \frac{+P_L}{2}\left(\left(\frac{h_e + h_p}{2h_p}\right) - 1\right),$$

$$P_D = \frac{-P_L}{2}\left(\left(\frac{h_e + h_p}{2h_p}\right) + 1\right).$$

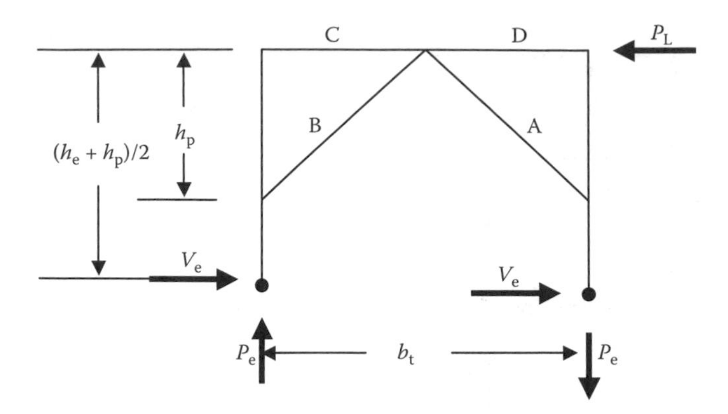

FIGURE E5.14 Triangular portal frame.

Example 5.15

The forces in the end portal bracing system (Figure E5.15) of the through truss span in Example 5.10 are required.

h_e = 33.5 ft (in-plane height of the portal)

h_p = 8.5 ft (in-plane height of the portal bracing system)

b_t = 20.0′ (spacing of trusses)

P_L = (6)19.55(0.35)/2 = 20.4 kips (top lateral truss wind force reactions transferred to the end portal frame)

R_e = 20.4[(42.0)/2]/20 = 21.4 kips (additional compression in the end post due to portal action)

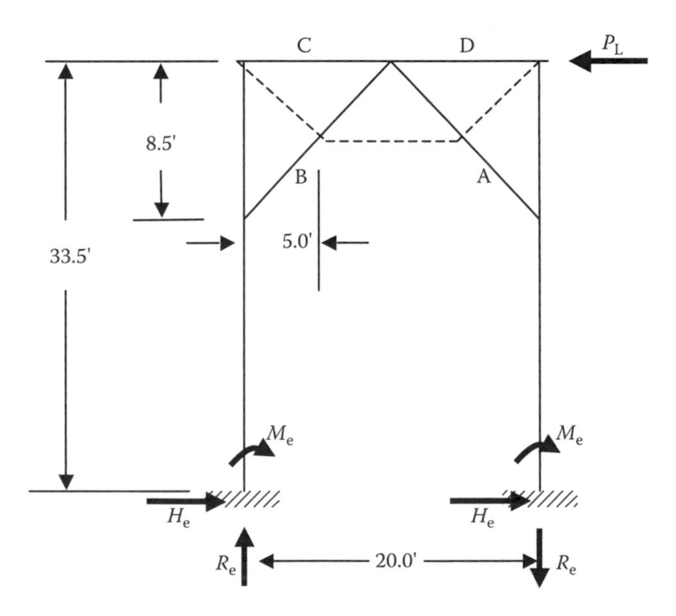

FIGURE E5.15 Braced triangular portal frame.

$H_e = 20.4/2 = 10.2$ kips (a small shear force that is generally neglected)
$M_e = 20.4[(25.0)/2]/2 = 127.5$ ft-kips (bending moment at the bottom of the portal frame end post).

The portal is of the triangular type (Example 5.14) and member forces are

$$P_A = -P_B = 20.4(1.62) = 33.0 \text{ kips}$$

$$P_C = 20.4(0.74) = 15.1 \text{ kips}$$

$$P_D = -20.4(1.74) = -35.5 \text{ kips}$$

The members shown as dotted lines may be designed for 2.5% of the compressive force in members A and B. However, as this force will be small (825 lb in this example), design based on compression member slenderness ratio criteria ($r_{min} \leq$ member length/120) will likely govern.

Other portal arrangement member forces may be determined in a similar approximate manner or by rigorous frame analysis. AREMA (2008) indicates that through truss spans should have portal bracing with knee braces (e.g., members A and B in Figure E5.15) as deep as clearances (see Chapter 3) will allow.

Cross frame members at the end of deck spans must transfer the reaction of the top lateral truss to the bearings and substructure. AREMA (2008) indicates that diaphragms may be used in lieu of cross frames for closely spaced shallow girders. Example 5.16 shows the analysis of a typical DPG vertical end brace frame.

Example 5.16

Determine the forces in the members of the end brace frame shown in Figure E5.16. The force P_L is given as 35.5 kips.

$P_L = 35.5$ kips (top lateral truss wind force and nosing reactions transferred to end portal frame).

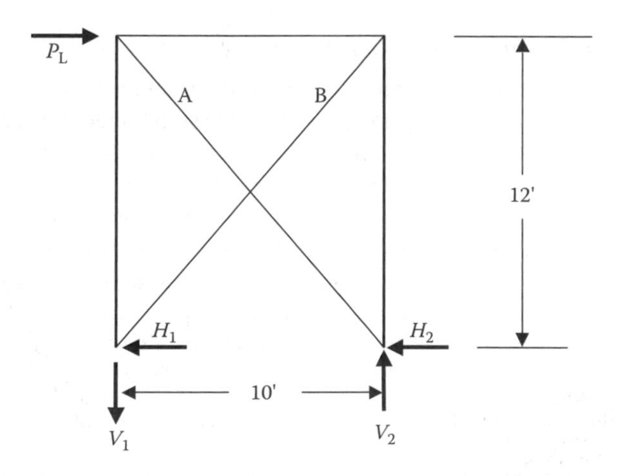

FIGURE E5.16 Deck span end frame.

Each brace is assumed to resist 1/2 of the horizontal shear. The force in each brace is estimated as

$$F_A = -F_B = -\frac{(35.5)}{2}\frac{\sqrt{12^2 + 10^2}}{10} = -27.7\ \text{kips}.$$

5.2.2.1.3 Intermediate Vertical and Sway Bracing

Intermediate vertical bracing in deck spans and sway bracing in through truss spans are required to provide compression chord or flange stability and adequate stiffness for serviceability conditions.

In through truss spans intermediate vertical sway bracing carries only small forces because of the negligible difference in relative lateral deformation of top and bottom lateral systems.* It is often estimated that 50% of panel load due to wind in addition to 2.5% of the total compressive axial force in the chord at the panel point is carried by the sway bracing. The analysis of forces may then proceed in a similar manner to that for the end portal frames of through truss spans (see Examples 5.11 through 5.14). Intermediate sway bracing is often designed as the knee-braced frame type (see Figure E5.13). Where estimated lateral forces are small it may be sufficient to proportion members based on maximum slenderness criteria for buckling. AREMA (2008) provides recommendations regarding types and geometry of through truss span sway bracing.

In deck spans, the intermediate cross frames or diaphragms provide for proper load distribution between main girders or trusses and, therefore, in addition to the lateral forces from equipment, wind and stability-related forces, must be designed to resist the forces induced by differential vertical deflections of trusses or girders.† AREMA (2008) indicates that for deck spans, diaphragms may be used in lieu of cross frames for closely spaced shallow girders. AREMA (2008) also provides the guidelines shown in Table 5.2 for recommended spacing of intermediate vertical brace frames.

5.2.2.1.4 Knee Bracing in Through Spans

The members (knee braces) that provide intermittent lateral bracing‡ to pony truss compression chords and through plate girder compression flanges must have adequate transverse elastic frame stiffness to ensure that the overall chord or flange has panel lengths with appropriate stiffness to attain the buckling load, P_c (Figure 5.29). Nodal points are created at each knee brace/frame (panel point) location if the transverse frame stiffness is very large. Conversely, if the transverse frame is too flexible, the entire compression chord or flange may buckle in a single half-wave. In structures

* Provided that there are no substantial live load eccentricities. Track eccentricity can create additional forces in the bracing members that may be determined by the simple tension member only assumption or by a more rigorous analysis.

† These forces can be particularly large in skewed spans (see Chapter 3) or spans with a substantial track eccentricity.

‡ Bracing of the compression chord or flange in the vertical direction is provided by truss and girder web members, respectively.

TABLE 5.2
Maximum Spacing of Intermediate Vertical Brace Frames in Deck Spans

Type of Bridge Deck	Maximum Vertical Brace Frame Spacing (ft)
Open deck construction (see Chapter 3)	18
Noncomposite steel–concrete ballasted decks (precast concrete, steel plate, solid timber) with top lateral bracing	18
Noncomposite steel–concrete ballasted decks (precast concrete, steel plate, solid timber) without top lateral bracing	12
Cast-in-place composite concrete decks	24

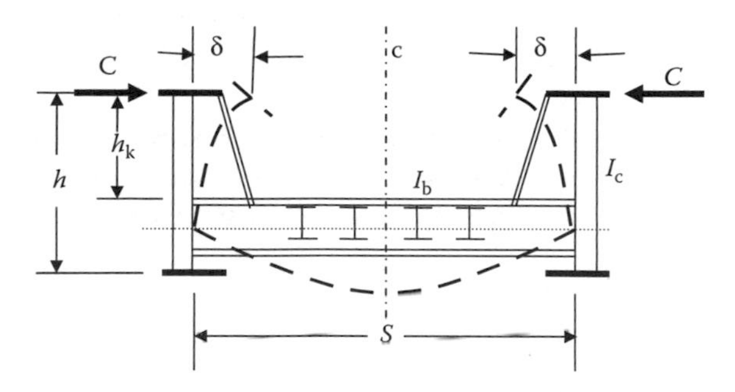

FIGURE 5.29 Through plate girder transverse frame behavior.

in-service, the buckled shape of the compression chord or flange is usually composed of many half-waves with length less than the distance between panel points (Bleich, 1952).

The lateral forces associated with resisting the compression chord or flange deformations can be estimated as the product of the buckling deformation and transverse frame elastic stiffness. The transverse frame elastic stiffness is expressed as an equivalent spring constant, C, developed by considering the stiffness contributions of the girders/knee brace, EI_c, and floorbeam/deck, EI_b, as (Galambos, 1988)

$$C = \frac{E}{h^2[(h/3I_c) + (S/2I_b)]}. \tag{5.41a}$$

If the floorbeam is very stiff in comparison to the vertical members of the transverse frame* ($I_b \gg I_c$),

$$C = \frac{3EI_c}{h^3}. \tag{5.41b}$$

* Which might be the case for some pony truss spans and girders without substantial knee braces.

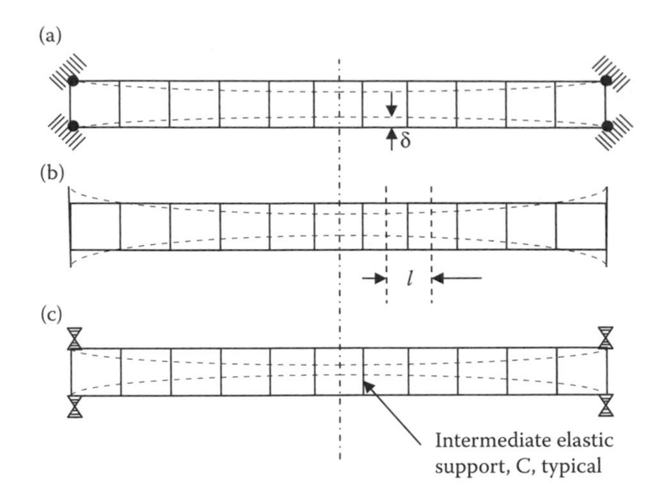

FIGURE 5.30 End restraint with intermediate transverse elastic frames: (a) pinned ends; (b) unrestrained ends; (c) elastic ends.

Assuming that the compression chord or flange is rigidly connected* at the ends and elastically supported at equally spaced transverse frames (girder/knee brace and floorbeam/deck),† the force (reaction), R_F, at the transverse frames is

$$R_F = C\delta. \tag{5.42}$$

Furthermore, assuming that the span buckles in a half-wave with continuously distributed elastic intermediate supports between ends of the span (Figure 5.30a), Engesser provided the solution for the required spring constant, C_{req}, as (Bleich, 1952)

$$C_{req} = \frac{F_{cr}^2 l}{4EI} = \frac{\pi^2 F_{cr}}{4k^2 l}, \tag{5.43}$$

where F_{cr} is the compression chord or flange critical buckling force (for entire chord/flange supported by transverse elastic frames or for length between the transverse frames with elastic end supports) and kl is the effective panel length.

However, because Equation 5.43 is only accurate when the half-length of the buckled chord or flange is greater than about $1.8l$, it is not applicable to short-span bridges or spans with only a few panel points. A considerably larger spring constant $(\pi^2 F_{cr}/k^2 l)$ is required if it is assumed that the ends of the girder or pony truss are laterally unsupported (Figure 5.30b) (Davison and Owens, 2003). This condition is unlikely and an analysis performed by Holt (1952, 1956) that provides for end supports modeled as cantilever springs is applicable to short spans (Figure 5.30c). The results of this analysis and associated design procedure are given in Galambos (1988). In such analyses, the compression area for through plate girders is generally taken as the area of the top flange and 1/3 of the web compression area.

* The assumption of pin-connected span ends will result in a nonconservative analysis for short spans.

† Also assuming that the chord or flange has a constant cross-sectional area and moment of inertia.

AREMA (2008) recommends that the lateral bracing of compression chords and flanges be designed for a transverse shear force, R_F, equal to 2.5% of the total axial force in both members in the panel. This "notional" force is recommended to ensure that intermediate knee brace/transverse frames are designed with adequate stiffness to prevent buckling failure. The bracing members may also have to be designed considering the shear force in the panel from lateral wind loads. The transverse shear force due to restraint of compression flange buckling, R_F, can then be determined for through plate girder (Figure 5.29) or pony spans, as

$$R_F = 0.025 A_f f_c, \tag{5.44}$$

where A_f is the area of the compression chord or flange and f_c is the compressive stress in the chord or flange.

Example 5.17

Determine the AREMA-recommended bracing design force for the knee braces (at a 3:1 slope) of the through plate girder span of Figure 5.29 with the following data:

$$h = 100 \text{ in.}$$

$$h_k = 75 \text{ in.}$$

$$S = 260 \text{ in.}$$

$$I_c = 150 \text{ in.}^3$$

$$I_b = 10{,}000 \text{ in.}^3$$

$$A_f = 45 \text{ in.}^2$$

$$A_w = 50 \text{ in.}^2 \text{ (the web plate area)}$$

$$f_c = 20 \text{ ksi}$$

$$L = 100 \text{ ft} = 1200 \text{ in.}$$

N_p is the number of panels and is equal to 10

$$l = 1200/10 = 120 \text{ in.}$$

$$R_F = 0.025 A_f f_c = 0.025(45)(20) = 22.5 \text{ kips}$$

Column load on the knee brace $= (22.5)(3) = 67.5$ kips
Bending moment at the floorbeam $= 22.5(75/12) = 140.6$ ft-kips.

Example 5.18

Determine the transverse stiffness of the through plate girder span compression flange bracing of Example 5.17 using the Engesser approach and the lateral deflection associated with the AREMA-recommended design force.

$$C = \frac{29{,}000}{100^2((100/3(150)) + (260/2(10{,}000)))} = 12.3\ \text{k/in.}$$

$$F_{cr} \sim 1.80(45 + (50/2/3))(20) \sim 1920\ \text{kips (using a safety factor of 1.80)}$$

$$C_{req} = \frac{\pi^2(1920)}{4(1200/10)} = 39.5\ \text{k/in.}$$

Since $C < C_{req}$, the transverse frames are not stiff enough to preclude excessive buckling deformations.

If, for example, the vertical member stiffness, I_c, was increased by the addition of a substantial knee brace so that $I_c = 750\ \text{in.}^4$

$$C = \frac{29{,}000}{100^2((100/3(750)) + (260/2(10{,}000)))} = 50.5\ \text{k/in.}$$

Using the shear force, $R_F = 22.5$ kips, from Example 5.17 (AREMA-recommended force)

$$\delta = \frac{R_F}{C} = \frac{22.5}{50.5} = 0.45\ \text{in.},$$

which appears reasonable for 100 ft long span lateral deflection (span/2666) and 6 ft-9 in frame wall (girder web/knee brace) cantilever tip deflection (height/160).

5.3 STRUCTURAL DESIGN OF STEEL RAILWAY SUPERSTRUCTURES

The structural design of members and connections in the superstructure may proceed once the bridge engineer has determined the loads (Chapter 4) on the superstructure and the internal member forces from structural analyses (Section 5.2) of appropriate load combinations.

Preliminary structural analyses use superstructure models developed through the planning and preliminary design process that are refined, as necessary, through the structural analysis process. The structural analysis may range from the routine analysis of statically determinate superstructures (reactions and internal forces determined from equilibrium) to continuous and more complex statically indeterminate structures (additional equations required such as compatible displacement equations, which require section properties and dimensions). Structural design (for strength and serviceability) involves material selection and determining the dimensions or section properties of members and connections in the superstructure. For statically indeterminate structures, an iterative analysis/design procedure is required. The potential failure modes of the steel superstructure require assessment in order to examine the strength and serviceability of the structure.

5.3.1 Steel Railway Superstructure Failure

Strength failure by fracture, yielding, or instability must be precluded. The von-Mises yield criterion (see Chapter 2) is appropriate for use in elastic strength design (Armenakas, 2006). Tension, compression, and shear yielding failure are based on the yield criterion of this failure theory. For structural design, allowable tension, compression, and shear stresses are based on the yield failure stress (typically the yield failure stress is divided by a safety factor to obtain the allowable stress). Tension members must also be designed considering ultimate stress fracture criteria. Compression members may become unstable prior to yielding and the effect is incorporated into elastic strength design procedures as effective reductions to the allowable compression stress (usually expressed as parabolic transition equations). The strength design of axial members, flexural members, and connections is discussed further in subsequent chapters.

Serviceability failure occurs as excessive elastic deformation or vibration, or fracture. Allowable service live load deflection criteria based on the length of the span are recommended by AREMA (2008), which will affect the stiffness design of the superstructure. Vibration effects on stresses are included in the empirically developed dynamic load increment (see Chapter 4) and vibration from wind is generally not a concern for the usually relatively stiff steel railway superstructures.* The deflection design of steel railway superstructures is discussed in greater detail later in this chapter.

Failure by fracture can be sudden or caused by accumulated damage from cyclical application of live loads over time. Sudden or brittle fracture is caused by preexisting flaws (cracks, notches, weld discontinuities, and areas where triaxial stresses are constrained) that create stress concentrations with high mean normal tensile stresses, which can cause failure before yielding.† Therefore, the failure may be a sudden fracture without evidence of yielding. Fracture susceptibility is generally more severe with dynamic loads, thick plates,‡ and low service temperatures. Design against brittle fracture is accomplished through the use of steel with adequate notch toughness§ for the design service temperature (which depends on geographical location**) and fabrication quality controls (see Chapters 2 and 3). AREMA (2008) recommends steel fracture toughness requirements†† for steel members considered as primary and for FCM. FCM are those members in tension where failure would result in failure of the entire structure (e.g., nonredundant structural members such as girders and trusses of many typical steel railway superstructures). The fracture toughness requirements for ordinary bridge design are based on relatively simple CVN testing‡‡in lieu

---

* Wind vibration is implicitly considered in the design of steel railway spans in accordance with AREMA (2008) by recommendation of a notional lateral load that provides for the design of sufficiently stiff lateral bracing systems.

† The von-Mises yield criterion is independent of mean normal (or hydrostatic) stresses.

‡ At a given temperature, thicker plates exhibit lower fracture toughness in elastic–plastic regions (crack tips) due to plane strain conditions.

§ Toughness can be interpreted as the energy required to cause fracture at a given temperature.

** Indicated as Zones 1, 2, and 3 in AREMA (2008).

†† Material with adequate toughness to initiate yielding prior to brittle fracture.

‡‡ In North America the CVN tests are generally carried out in accordance with ASTM A673.

of the more complex methods available by fracture mechanics testing and analysis (Anderson, 2005).

Failure by accumulated fatigue damage (initiation and propagation of small cracks) caused by repeated cycles of tensile stress is of primary concern in the design of steel railway superstructure members and connections. The fatigue life, or number of cycles to failure (generally taken as through-thickness fracture of a component), depends on the frequency and number of load cycles, load magnitude (in particular, stress range), member size, and member details. A fracture mechanics approach to fatigue design is not generally used for ordinary steel bridge design (Fisher, 1984; Kulak and Smith, 1995; Dexter, 2005). Therefore, the stress-life approach, recommended for the design of steel bridges by AREMA (2008), is outlined further in this chapter.

The serviceability criteria (or limit states) of deflection and fatigue often govern important aspects of the structural design of steel railway bridges.

5.3.2 Steel Railway Superstructure Design

5.3.2.1 Strength Design

The strength design of members and connections as recommended by AREMA (2008) is performed through elastic structural analyses and the ASD method. The ASD methodology divides the ultimate and yield stress of the steel by an FS to determine allowable stresses. Yield stress is associated with plastic deformation and ultimate stress with fracture. Internal stresses in members and connections must not be greater than the allowable yield or fracture criteria. As indicated in Chapter 2, the yield and ultimate stresses for tension, compression, and shear are all expressed in terms of the yield and ultimate tensile stresses.

The FS for tensile stresses recommended by AREMA (2008) ($9/5 = 1.80 \sim 1/0.55$) is greater than the typical allowable tensile stress FS ($5/3 = 1.67 = 1/0.60$) used in building or highway bridge ASD because of the high magnitude dynamic and cyclical live load regime of steel railway bridges. Further considerations relating to the larger FS for steel railway superstructures are the fracture (cold weather service), corrosion (industrial and wet environments), and damage susceptibility (railway and highway vehicle contact) of the superstructure due to location.

The FS for ASD design of compression members is generally taken as between 1.9 and 2.0 because of stability issues relating to unintended eccentricities and initial curvature of compression members. However, for short axial compression members that will yield prior to buckling, the FS corresponding to compressive yielding (related to tensile stresses, see Chapter 2) of $9/5 = 1.8$ could be used. A cubic polynomial equation (representing a quarter sine wave) is an appropriate transition function for a compression member FS (Salmon and Johnson, 1980) and can be applied to the AREMA (2008) recommended FS for axial compression stresses as shown in Figure 5.31, where K is the effective length factor (depends on compression member end condition) (see Chapter 6), L is the length of the member, r is the radius of gyration of the member, and C_{cr} is the limiting or critical value of (KL/r) at proportional limit ($0.50F_y$) to preclude instability in the elastic range (Euler buckling).

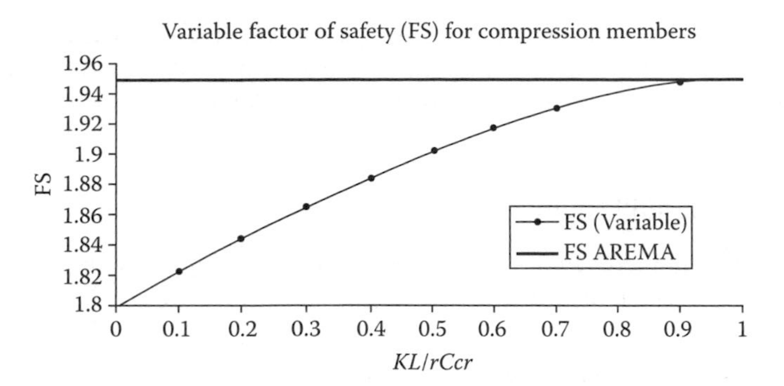

FIGURE 5.31 FS for compression members.

AREMA recommends an FS of 1.95 for axial compression members of all slenderness ratios. This may be appropriate unless the fabrication and erection can be carefully controlled to avoid eccentricities or other unintended secondary effects in axial compression members.

The allowable stress approach ensures all members behave elastically, which is appropriate for steel with its well-defined elastic behavior and tensile yield stress. Also, since stresses from various loads and load combinations are maintained within the elastic region of behavior, load superposition is possible.

However, the ASD FS does not consider the real uncertainties associated with loads or combinations of loads. AREMA (2008) recommends modification of the FS (modification of allowable stresses) for design load combinations based on the probability of the loads being applied concurrently to the member. Furthermore, the use of a single safety factor against yielding for many different loads within a load combination is a shortcoming of ASD. Methods that provide a probabilistic approach to the estimation of loads and member strength (partial safety factors) have, therefore, been adopted by many international building and bridge design guidelines, recommendations, codes, and specifications. However, the use of a single safety factor is not a significant shortcoming for ordinary steel railway superstructure design due to the relatively high live load to dead load ratio and the importance of deflection and fatigue criteria (both evaluated at service loads). In addition, ASD does not consider the localized yielding and load redistribution of steel structures at failure. Railroad operating requirements (see Chapter 3) make this a valid approach in regard to behavior at failure.

Beams, girders, trusses, arches, and frames are subjected to internal normal and shear stresses across cross sections caused by internal axial forces, shearing forces, torsional moments, and bending moments. Steel beam and girder design are based on shearing stresses and normal stresses caused by bending moments and shearing forces. Steel arch design is based on shearing stresses and normal stresses caused by bending moments, axial forces, and shearing forces. Steel truss design is concerned primarily with axial forces causing normal stresses, although eccentricities and secondary effects (e.g., due to deflections) might create additional normal (due to bending moments) and shearing stresses. Members and connections with internal stresses not greater than the allowable tension, compression, or shear allowable stresses recommended by AREMA (2008) are considered to be of safe and reliable design.

5.3.2.2 Serviceability Design

Serviceability criteria (or limit states) of deflection and fatigue are important aspects of the structural design of steel railway superstructures.

5.3.2.2.1 Deflection Criteria

Flexural deflections are calculated at the location of the maximum live load bending moment in a span. AREMA (2008) recommends that the maximum flexural deflection from live load including impact not exceed 1/640 of the span. Railroad companies and designers may further limit deflections based on span types (trusses, girders, and composite girder/beam spans*) and other operating practices.

The maximum flexural deflection in an ordinary simply supported span from live load including impact, $\Delta_{\mathrm{LL+I}}$, can be estimated considering an equivalent uniform load, $w_{\mathrm{e}\Delta}$, as

$$M_{\mathrm{LL+I}} = \frac{w_{\mathrm{e}\Delta}a(L-a)}{2} = \frac{w_{\mathrm{e}\Delta}L^2}{8} \quad \text{at} \quad a = L/2. \tag{5.45a}$$

Therefore,

$$w_{\mathrm{e}\Delta} = \frac{8M_{\mathrm{LL+I}}}{L^2}, \tag{5.45b}$$

where a is the distance to the location of interest (see Figure 5.21), $M_{\mathrm{LL+I}}$ is the maximum bending moment in the span due to live load including impact (at $a = L/2$), and L is the length of the simply supported span. Substitution of Equation 5.45b into the equation for the maximum deflection from a uniformly distributed load on a simple beam provides an estimate of the maximum flexural deflection due to live load including impact as

$$\Delta_{\mathrm{LL+I}} = \frac{5w_{\mathrm{e}\Delta}L^4}{384EI} = \frac{0.104M_{\mathrm{LL+I}}L^2}{EI}, \tag{5.46a}$$

where E is the modulus of elasticity and I is the gross moment of inertia (used for flexural member deflection calculations).

AREMA (2008) recommends, as do many other guidelines, codes, and specifications, that the maximum flexural deflection from live load including impact not exceed L/f_Δ of the span, where f_Δ is an integer established based on structural behavior and experience. Therefore, the minimum gross moment of inertia, I, of a simple

---

* To limit cracking and improve behavior of concrete decks.

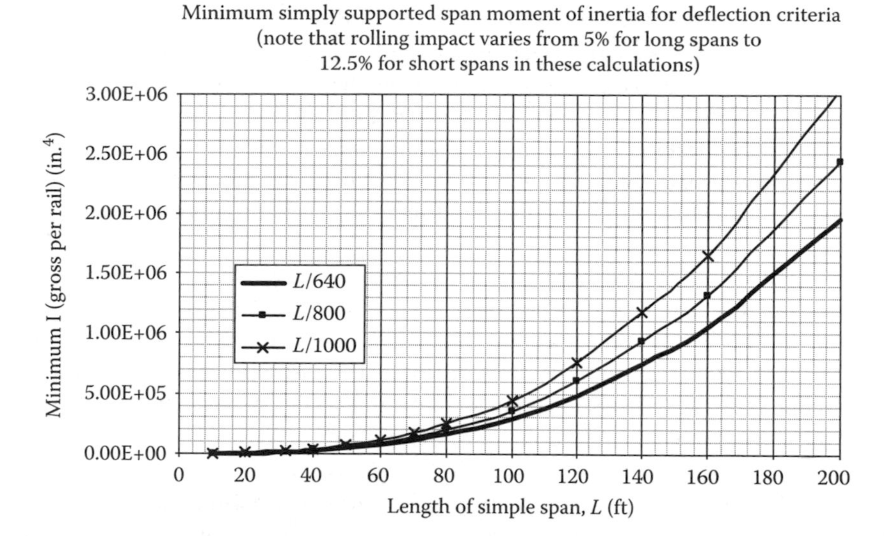

FIGURE 5.32 Stiffness design of simply supported spans for deflection criteria.

span to meet the deflection criteria is

$$I \geq \frac{M_{LL+I} L f_{\Delta}}{1934} \text{ in.}^4, \tag{5.46b}$$

where M_{LL+I} is the live load including impact bending moment for the span, L, ft-kips; L is the length of span, ft.

If the AREMA (2008)-recommended $f_{\Delta} = 640$ is used in Equation 5.46b, the minimum gross moment of inertia is $I \geq 0.33 M_{LL+I} L$, in.^{4}, for a simply supported beam or girder span. This relationship, and others using the deflection criteria $L/800$ and $L/1000$, are shown in Figure 5.32. It should be noted that the rolling impact used in Figure 5.32 varies from 5% for long spans to 12.5% for short spans and may require adjustment for particular span designs. Nevertheless, Figure 5.32 provides a ready estimate of minimum gross moment of inertia required to meet various deflection criteria for simply supported steel railway beam and girder spans.

The maximum deflection of a simply supported truss from live load including impact, Δ_{LL+I}, can be determined by calculating truss joint horizontal and vertical translations by the method of virtual work or through graphical means (Utku, 1976; Armenakas, 1988). However, today the use of matrix methods (stiffness or flexibility) and digital computer software enables routine calculation of truss deflections and member forces. The analysis of trusses for joint translation should use the gross area of truss members without perforated cover plates.* If truss members are designed with perforated cover plates the gross area should be reduced by the area determined by dividing the volume of a perforation by the spacing of perforations.

---

* Cover plates on truss members are usually used in compression chords and end posts.

5.3.2.2.2 Fatigue Design Criteria

The fracture* of steel by fatigue may be caused by modern high magnitude cyclical railway live loads. Fatigue cracks may initiate and then propagate at nominal tensile† cyclical stresses below the tensile yield stress at stress concentrations in the superstructure. The cyclical railway loading accumulates damage (which may be manifested as plastic deformation, crack initiation, and crack extension) at the stress concentrations that may precipitate fracture, leading to unserviceable deformations or failure at a certain number of cycles, N_f.

The high cycle‡ fatigue life, $N \leq N_f$, of a member or detail is determined by constant amplitude cyclical stress testing of specimens typical of steel bridge members and details. The testing of representative specimens makes the determination of stress concentration factors and consideration of residual stresses unnecessary for ordinary steel bridge design.§ Therefore, fatigue analysis and design may be performed at nominal stresses.

5.3.2.2.2.1 Railway Fatigue Loading The variable amplitude cyclical railway live load must be developed as an effective or equivalent constant amplitude cyclical design load because fatigue strength is established by constant amplitude cyclical stress testing of materials, components, members, and details. This effective cyclical fatigue design load must accumulate the same damage as the variable amplitude cyclical load over the total number of stress cycles to failure. The stress ratio, $R = S_{remin}/S_{remax}$, and stress range, $\Delta S_{re} = S_{remax} - S_{remin}$, may be used to describe constant amplitude loading (Figure 5.33). Constant amplitude loads are used for fatigue testing. They are often performed with $R = 0$ (cyclical tension with $S_{min} = 0$) or $R = -1$ (fully reversed with $S_{max} = -S_{min}$). The mean stress** is $S_{remean} = (S_{remax} + S_{remin})/2$.

The variable amplitude cyclical load corresponding to the AREMA (2008) design load mid-span bending moment is shown in Figure 5.34. The uniformly distributed load (8000 lb/ft for Cooper's E80 design live load) creates no change in stress and is shown truncated in Figure 5.34. Actual variable amplitude freight rail traffic loads measured on in-service bridges are more complex, often making the assessment (life cycle analysis) of existing bridges difficult from a fatigue perspective.†† The number of stress range cycles and their magnitudes can be determined directly from load

---

* The fracture limit state can be defined in various ways, such as crack propagation to some critical length or number of cycles to appearance of a visible crack (generally considered to be on the order of 1–5 mm in the stress-life approach to fatigue). It is also defined as when initiated fatigue cracks propagate through the thickness of the component, member, or detail.

† At members and details with a net applied tensile stress, since there is no fatigue cracking in purely compression regions that never experience tension stress.

‡ High cycle or long life fatigue analysis is appropriate for steel railway bridge design.

§ However, the use of nominal stresses without stress concentration factors should be carefully reviewed in areas of high stress gradients.

** Since constant amplitude cyclical fatigue testing of members and details includes the effects of stress concentrations and residual stresses (present from rolling, forming, fabricating, and welding operations), the fatigue life is not influenced by mean stress effects and the range of stress is important.

†† Also, lack of records regarding historical rail traffic may hinder assessment.

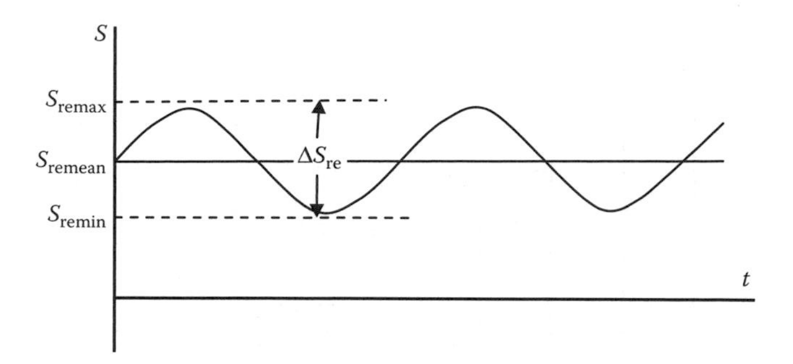

FIGURE 5.33 Constant amplitude cyclical loading.

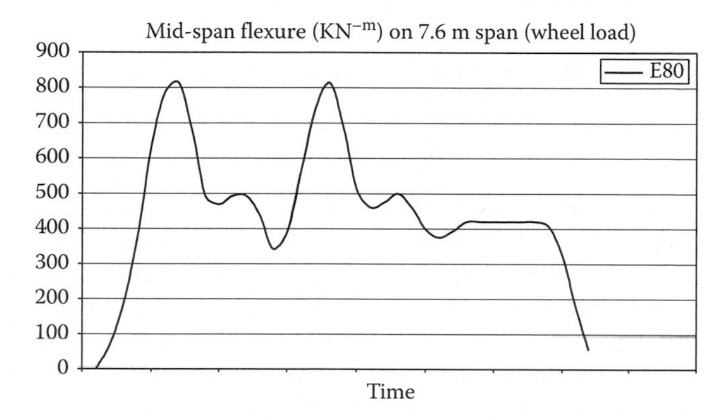

FIGURE 5.34 Mid-span bending moment for Cooper's E80 load traversing a 25 ft simply supported bridge span.

traces in elastic structures. Areas near the 1/4 span length* and locations of change in section may also be important for the determination of the maximum number of stress cycles and their magnitude.

The effective stress cycles must accumulate the same damage as the variable amplitude stress cycles over the total number of stress cycles to failure. Therefore, a damage accumulation rule is required. There are many damage accumulation rules, but it is usual to apply the Palmgren-Miner (Miner, 1945) linear damage accumulation rule because, even though the sequence and interaction of load cycle effects are not accounted for, the linear damage rule provides good agreement with test results (Stephens et al., 2001). The rule is also independent of the stress magnitude. Also, where residual stresses are high† (typical of modern steel railway bridge fabrications), mean stress effects are negligible‡ and the stress range magnitude is of

* Because of the relationship between span length and car length during the cycling of spans, locations around the 1/4 point may govern the maximum number of stress range cycles and magnitude (Dick, 2002).

† Typically at yield stress level.

‡ Dead load is unimportant since mean stress effects are negligible.

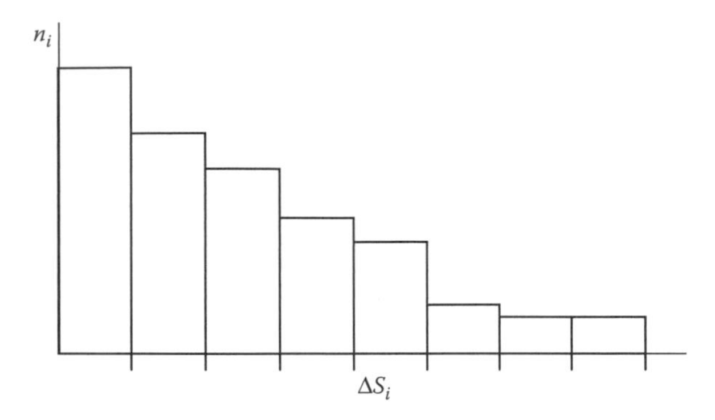

FIGURE 5.35 Frequency distribution histogram of stress ranges.

principal importance. The Palmgren-Miner linear damage accumulation rule is

$$\sum \frac{n_i}{N_i} = 1.0, \tag{5.47}$$

where n_i is the number of cycles at stress range level, ΔS_i, developed from an appropriate cycle counting method. There are also many techniques for counting the cycles of variable amplitude load or stress traces. However, for many structures, the rainflow method appears to provide the best results (Dowling, 1999). Instead of counting only the tensile portion of stress range cycles, AREMA (2008) recommends counting all live load stress range cycles as a complete tensile cycle (even those with a compressive component due to stress reversal). This is appropriate because near flaws and details the member will be subjected to a fully effective tensile stress range cycle due to superposition of tensile residual stress. This is analogous to raising the mean stress such that the entire stress range cycle is in tension. A frequency distribution histogram for the numbers of stress range cycles, n_i, can be developed from rainflow cycle counting, as shown schematically in Figure 5.35.

N_i is the number of cycles to failure at stress range level, ΔS_i. A log–log straight line relationship exists between N_i and ΔS_i (Basquin, 1910). This relationship is also observed in constant amplitude fatigue testing of members and details (Kulak and Smith, 1995).

Crack growth behavior, as defined by the Paris-Erdogan power law,* can be used to establish a relationship between stress range and number of cycles to failure. The crack growth rate is

$$\frac{\mathrm{d}a}{\mathrm{d}N} = C\Delta K^m, \tag{5.48}$$

where a is the crack length, N is the total number of constant amplitude stress range cycles causing the same fatigue damage as the total number of variable amplitude

* This is a log–log linear relationship. At crack growth rates below log–log linear behavior, a threshold exists below which cracks will not propagate. At crack growth rates above log–log linear behavior, fracture occurs at critical stress intensity equal to the fracture toughness of steel (Barsom and Rolfe, 1987; Anderson, 2005).

stress range cycles, N_v, m is a material constant established from regression analysis of test data as $m = 3$ for structural steel, C is a material constant established from regression analysis of test data, $\Delta K = C_\mathrm{K} \Delta S_\mathrm{re} \sqrt{\pi a}$ is the change in stress intensity factor for effective stress range, ΔS_re, ΔS_re is the effective constant amplitude stress range distribution that causes the same amount of fatigue damage as the variable amplitude stress range distribution, and C_K is the constant depending on shape and size of crack, edge conditions, stress concentration, and residual stresses (Pilkey, 1997).

Integration of Equation 5.48 yields

$$N = \frac{1}{C} \int_{a_\mathrm{i}}^{a_\mathrm{f}} \frac{da}{\Delta K^m}, \tag{5.49}$$

where a_i is the initial crack length and a_f is the final crack length.

Substitution of $\Delta K = C_\mathrm{K} \Delta S_\mathrm{re} \sqrt{\pi a}$ into Equation 5.49 yields

$$N = \frac{\left(\sqrt{\pi} \Delta S_\mathrm{re}\right)^{-m}}{C} \int_{a_\mathrm{i}}^{a_\mathrm{f}} \frac{da}{\left(C_\mathrm{K} \sqrt{a}\right)^m}. \tag{5.50}$$

Since C, C_K, and $a = a_\mathrm{i}$ ($a_\mathrm{f} \gg a_\mathrm{i}$, therefore neglect terms with a_f because of $-m$ power) are constant (Kulak and Smith, 1995),

$$N = A(\Delta S_\mathrm{re}^{-m}), \tag{5.51}$$

where A is a constant depending on detail and established from regression analysis of test data (see Section 5.3.2.2.2.2).

Equation 5.51 illustrates that the number of cycles to failure, N, for steel bridge members or details is very sensitive to the effective stress range, ΔS_re. Equation 5.51 also provides the number of cycles at failure at stress range level, ΔS_i, as

$$N_i = A(\Delta S_i^{-m}). \tag{5.52}$$

Substitution of Equation 5.52 into Equation 5.47 yields

$$\sum \frac{n_i}{N_i} = \sum \frac{n_i}{A(\Delta S_i^{-m})} = \sum \frac{\lambda_i N}{A(\Delta S_i^{-m})} = \sum \frac{\lambda_i N}{A(\Delta S_i^{-m})} = 1, \tag{5.53}$$

where

$$\lambda_i = \frac{n_i}{\sum n_i} = \frac{n_i}{N}$$

and substitution of Equation 5.51 into Equation 5.53 yields

$$\sum \frac{\lambda_i \Delta S_\mathrm{re}^{-m}}{\Delta S_i^{-m}} = 1 \tag{5.54}$$

or

$$\Delta S_{\mathrm{re}} = \left(\sum \lambda_i \Delta S_i^m\right)^{1/m}. \tag{5.55}$$

Equation 5.55 with $m = 3$ is the root mean cube (RMC) probability density function describing the effective constant amplitude stress range distribution, ΔS_{re}, that causes the same amount of fatigue damage as the design variable amplitude stress range distribution (e.g., the mid-span bending moment for Cooper's E80 load shown in Figure 5.34). No FS is applied since the Palmgren-Miner linear damage accumulation rule is considered relatively accurate for service level highway and railway live loads (Fisher, 1984). An example of determining the effective constant amplitude stress range from a variable amplitude load is shown in Example 5.19.

Example 5.19

Determine the effective constant amplitude stress range, ΔS_{re}, for the variable amplitude stress spectrum shown in Figure E5.17 (the variable amplitude Cooper's E80 design loading on a 25 ft long span shown in Figure 5.34).

The peak stresses corresponding to the variable amplitude loads are shown in Table E5.5 and Figure E5.18. Rainflow cycle counting is performed as indicated in Figure E5.18 and Table E5.6.

Four complete stress range cycles (eight half-cycles) are present ($N = 4$). Calculation of the effective constant amplitude stress range, ΔS_{re}, is shown in Table E5.7.

$$\Delta S_{\mathrm{re}} = \sum g_i (\Delta S)^3 = (509.7)^{1/3} = 8.0\,\mathrm{ksi}.$$

Equation 5.55 indicates that the railway fatigue design load must be expressed in terms of the number of cycles and magnitude of load. The fatigue design load recommended by AREMA (2008) is based on analyses of continuous unit freight

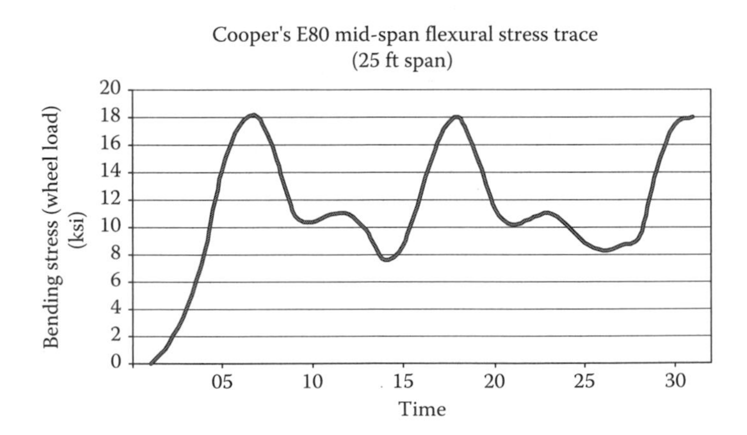

FIGURE E5.17

TABLE E5.5

Stress Peak[a]	Stress (ksi)
A	18.0
B	10.9
C	10.2
D	7.6
E	18.0
F	10.1
G	11.0
H	8.3

[a] See Figure E5.18.

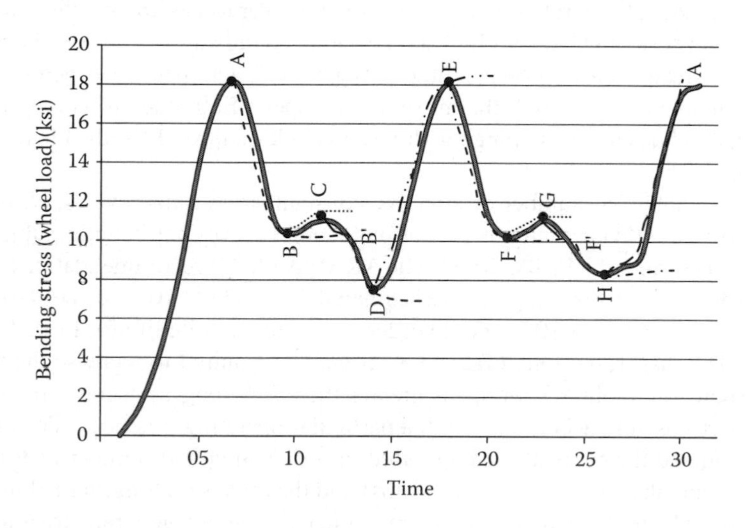

FIGURE E5.18

TABLE E5.6

Half-Cycle	Rainflow Stop Criteria	Stress Range (ksi)
ABB′D	Next peak E equal to peak A	18.0 − 7.6 = 10.4
BC	Next valley at D greater than valley at B	10.9 − 10.2 = 0.70
CB′	Encounter previous rainflow ABB′D	10.9 − 10.2 = 0.70
DE	End of stress history	18.0 − 7.6 = 10.4
EFF′H	Next peak A equal to peak E	18.0 − 8.3 = 9.7
FG	Next valley at H greater than valley at F	11.0 − 10.1 = 0.90
GF′	Encounter previous rainflow EFF′H	11.0 − 10.1 = 0.90
HA	End of stress history	18.0 − 8.3 = 9.7

TABLE E5.7

i	Stress Range, ΔS_i	Cycles, n_i	$g_i = n_i/N$	$g_i(\Delta S)^3$
1	0.70	1	0.25	0.086
2	0.90	1	0.25	0.182
3	9.7	1	0.25	228.2
4	10.4	1	0.25	281.2
S	—	4	1.00	509.7

trains typical of grain, coal, and other bulk commodity traffic on North American and other heavy haul railways.

It is based on locomotives and equipment with maximum axle loads of 80,000 lb. In addition, for loaded lengths or spans greater than 100 ft, it is based on a maximum equivalent uniform load of 6000 lb/ft (see Section 5.2.1.3). These loads are characteristic of modern train traffic that typically create shear forces and bending moments equivalent to between Cooper's E50 and E80 load on railway spans (see Chapter 4).* Therefore, since cycles corresponding to typical characteristic load geometry and loaded length are considered, the maximum Cooper's E80 load stress may be used for the fatigue design stress range with design cycles adjusted for the characteristic load magnitude.

The recommended number of effective constant stress range cycles, N, is based on an analysis of loaded lengths for various member types and lengths subjected to a 110 car unit train (AREMA, 2008). The AREMA (2008) recommendations assume variable amplitude stress range cycles estimated from 1.75×10^6 trains (60 trains per day over a design life of 80 years), in order to provide infinite life for loaded lengths or spans less than 100 ft long (Table 5.3). It may be required to increase the number cycles shown in Table 5.3 for spans greater than 75 ft long to account for specific load patterns used in accordance with a particular operating practice.† The analyses also considered the cyclical loading based on orientation and number of tracks for transverse members (generally floorbeams) and the effects on longitudinal members by transverse loads applied directly at, or within, panel points (generally truss hangers, subdiagonals, and web members). For spans or loaded lengths greater than 300 ft, a more detailed analysis by influence lines (see Section 5.2.1.2) or using structural analysis computer software may be required. The adjusted equivalent number of constant amplitude design stress range cycles, N, considering an E60 characteristic load magnitude, is

$$N = N_{\mathrm{v}}\left(\frac{\Delta S_{\mathrm{E60}}}{\Delta S_{\mathrm{E80}}}\right)^m = N_{\mathrm{v}}\left(\frac{60}{80}\right)^3 = 0.42N_{\mathrm{v}}, \tag{5.56}$$

* For longer spans (greater than about 50 or 75 ft depending on car lengths), modern unit freight train traffic typically creates forces equivalent to about Cooper's E50–E60. For shorter spans, modern unit freight train traffic can generate forces equivalent to about Cooper's E60–E80.

† For example, it is theoretically possible to generate 55 cycles on spans almost 100 ft long with a repeating load pattern of two loaded and two unloaded rail cars (AREMA, 2008).

TABLE 5.3
Variable Amplitude Stress Range Cycles per Train

Span Length, L (ft)	Variable Amplitude Stress Range Cycles per Train	Total Variable Amplitude Stress Range Cycles, N_v
$L > 100$	3	5.3×10^6
$100 >= L > 75$	6	10.5×10^6
$75 >= L > 50$	55	96.3×10^6
$50 >= L$	110	192.5×10^6

Source: AREMA, 2008, Manual for Railway Engineering, Chapter 15. With permission.

where N_v is the total number of variable amplitude load cycles, ΔS_{E60} is the stress range from Cooper's E60 load (characteristic of modern freight train loads), and ΔS_{E80} is the stress range from Cooper's E80 design load. Table 5.3 (based on 1.75×10^6 trains over the bridge design life) and Equation 5.56 provide the adjusted number of equivalent constant amplitude stress range cycles over the superstructure design life as shown in Table 5.4.

AREMA (2008) recommends reductions in the fatigue design live load as a means of considering the lower number of cycles on lightly traveled railway lines. On railway lines with less than 5 MGT (million gross tons) per mile per year, AREMA (2008) recommends a fatigue design load based on Cooper's E40. On railway lines with 5–15 MGT per mile per year, stress ranges from a Cooper's E65 design load are recommended. Therefore, for all traffic levels, the number of effective or equivalent constant amplitude live load stress ranges in Table 5.4 can be used to develop appropriate allowable fatigue stress ranges for the design of steel railway superstructure members and details.

5.3.2.2.2.2 Fatigue Strength of Steel Railway Bridges AREMA (2008) recommends a stress-life approach for fatigue strength evaluation of members and details. This is appropriate for high cycle stress range magnitudes that are generally low enough to preclude the need of considering yield effect (predominantly elastic strains

TABLE 5.4
Constant Amplitude Stress Range Cycles

Span Length, L (ft)	Equivalent Constant Amplitude Stress Range Cycles Over Member or Detail Life, N
$L > 100$	2.2×10^6
$100 >= L > 75$	4.4×10^6
$75 >= L > 50$	40.7×10^6
$50 >= L$	81.3×10^6

and little plastic deformation). Linear elastic fracture mechanics (LEFM) methods are also applicable to steel railway bridges but are not often used in ordinary steel bridge design because of lack of information regarding initial crack shapes and sizes with which to conduct crack growth rate analyses.

Fatigue damage accumulation occurs at stress concentrations in tension zones* making location and detail characteristics of prime importance. These characteristics are compiled within various fatigue detail categories† in AREMA (2008) based on the number of cycles to "failure,"‡ N, at various constant amplitude stress range tests, ΔS_{re}. Since railway live load is applied as a high cycle (long life) load, testing must also be conducted at high cycle constant amplitude stress ranges. The allowable fatigue stress for design of a particular detail is based on a probabilistic analysis (without FS) of high cycle test data and, therefore, it is appropriate to perform fatigue design at service load levels. Also, since stress concentration effects are accounted for within the various fatigue detail categories, a nominal applied stress approach for fatigue design is recommended in AREMA (2008).

The allowable fatigue stress range for design, ΔS_{rall}, from Equation 5.51, depends on the number of equivalent constant amplitude stress range cycles over the member or detail life, N, as

$$\Delta S_{rall} = \left(\frac{A}{N}\right)^{1/m} \tag{5.57}$$

or

$$\text{Log}(N) = \text{Log}(A) - m\,\text{Log}(\Delta S_{rall}), \tag{5.58}$$

which is plotted in Figure 5.36 for $m = 3$ and various values of constant A (as shown in Table 5.5 for fatigue detail categories A, B, B′, C, D, E, and E′). The constant A is established from regression analysis of test results such that Equation 5.58 describes S–N behavior for details with 95% confidence limits for 97.5% survival (2.5% probability of failure). Testing has also indicated that there is a constant amplitude fatigue limit (CAFL) stress range, ΔS_{CAFL}, below which no fatigue damage accumulates.§ The CAFL is also shown in Table 5.5 and by the horizontal lines in Figure 5.36.

2.0×10^6 cycles is considered an infinite life condition in terms of fatigue testing (Taly, 1998). In Table 5.4, the number of applied equivalent constant amplitude stress range cycles over the member or detail life, N, clearly exceeds 2.0×10^6 cycles for loaded lengths less than 100 ft. Therefore, the allowable fatigue stress range for loaded lengths or spans less than 100 ft is limited to the to CAFL stress range (Table 5.5), which provides for infinite life.

---

* Therefore, the presence of residual tensile stresses from rolling or welding processes may be important.

† These are designated as A, B, B′, C, D, E, and E′ details according to the number of constant amplitude stress cycles to "failure" at a given stress range.

‡ "Failure" in terms of fatigue design does not mean failure as defined by the strength limit state. Fatigue "failure" is a criterion based on data at some standard deviation (generally, 2 or 2.5) from the mean of test data for the member or detail (AREMA, 2008 uses a standard deviation of 2.5).

§ Even a small number of cycles exceeding the CAFL may effectively render it as nonexistent. Therefore, fatigue design ensures that all design live load stress ranges are below the CAFL.

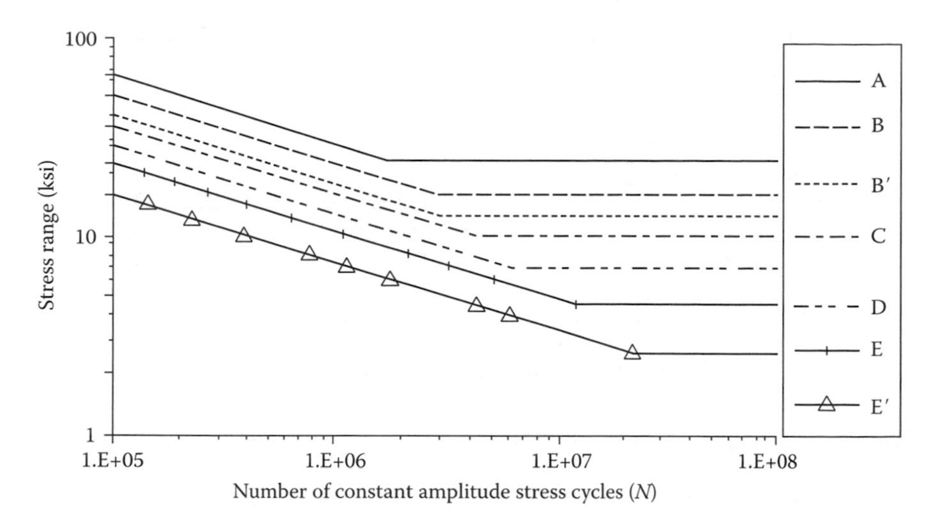

FIGURE 5.36 Constant amplitude *S–N* curves for fatigue detail categories.

A review of Tables 5.4 and 5.5 indicates that limiting the allowable fatigue stress range to the CAFL stress range for Category D, E, and E′ details appears to be conservative for spans between 75 and 100 ft long ($100' >= L > 75'$) (see Figure 5.37). However, because it is relatively easy to obtain much more than six stress range cycles for some load conditions on spans between 75 and 100 ft long, the apparent conservatism may not be realized and is acceptable for routine bridge design.

Table 5.6 shows the maximum number of variable amplitude stress cycles per train at the CAFL for various fatigue detail categories. Table 5.6 indicates that limiting the allowable fatigue stress range to CAFL is conservative for some fatigue category details (conservative for cycles shown underlined) in spans between 75 and 100 ft long

TABLE 5.5
Number of Constant Amplitude Stress Range Cycles of CAFL for Fatigue Detail Categories

Fatigue Detail Category	*A* (ksi^3)	CAFL (ksi) (Allowable Fatigue Stress Range for $L =< 100'$)	*N*, Constant Amplitude Cycles to Failure at CAFL
A	2.5×10^{10}	24	1.8×10^6
B	1.2×10^{10}	16	2.9×10^6
B′	6.1×10^9	12	3.5×10^6
C	4.4×10^9	10	4.4×10^6
D	2.2×10^9	7	6.4×10^6
E	1.1×10^9	4.5	1.2×10^7
E′	3.9×10^8	2.6	2.2×10^7

Source: AREMA, 2008, *Manual for Railway Engineering*, Chapter 15. With permission.

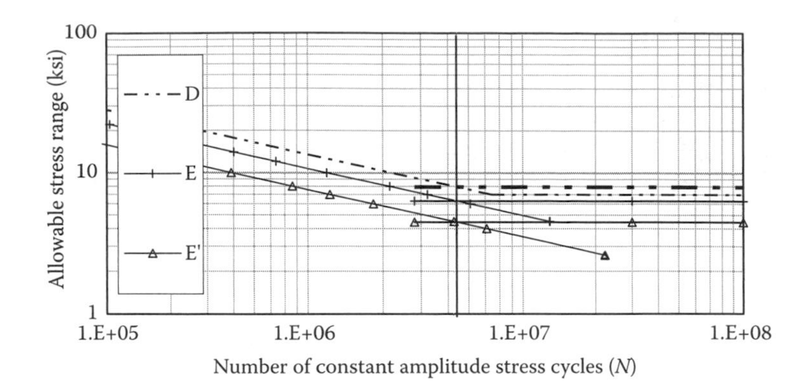

FIGURE 5.37 Allowable fatigue stress range at 4.4 × 10^6 cycles for category D, E, and E′ details.

TABLE 5.6
Maximum Number of Variable Amplitude cycles per Train at the CAFL for Various Fatique Detail Categories

	Maximum Number of Cycles per Train at CAFL (10^6)		
Fatigue Detail Category	**E60 Characteristic Loading**	**E55 Characteristic Loading**	**E50 Characteristic Loading**
A	3	3	4
B	4	5	7
B′	5	6	8
C	6	8	11
D	9	11	15
E	16	21	29
E′	30	39	53

with an applied number of effective constant amplitude cycles of 4.4 × 10^6 (based on six variable amplitude stress range cycles per train). The analysis is performed for various typical characteristic loads. It is not unreasonable, for example, that trains with a characteristic load of about E55 may cause 39 cycles of live load with some frequency on spans between 75 and 100 ft long. Therefore, the AREMA (2008) recommendation to limit the allowable fatigue stress range to the appropriate fatigue category detail CAFL is appropriate for loaded lengths or spans less than 100 ft long.

Table 5.4 indicates that for loaded lengths or spans greater than 100 ft in length the allowable fatigue stress range may be based on the detail strength at 2.0 × 10^6 constant amplitude stress range cycles as

$$\Delta S_{\text{rall}} = \left(\frac{A}{2 \times 10^6}\right)^{1/3}. \tag{5.59}$$

TABLE 5.7
Allowable Fatigue Stress Range at 2,000,000 Cycles (Used for Loaded Lengths or Spans Greater Than 100 ft Long)

Fatigue Detail Category	Allowable Fatigue Stress Range at 2,000,000 Cycles
A	23 (AREMA uses 24, which is CAFL)
B	18
B′	14.5
C	13
D	10
E	8
E′	5.8

Using Table 5.5 and Equation 5.59, the allowable fatigue stress range, S_{rall}, for loaded lengths or spans greater than 100 ft long is shown in Table 5.7.

AREMA (2008) also recommends fatigue detail Category F for shear stress on the throat of fillet welds. The allowable fatigue stress range is 9 ksi for loaded lengths or spans greater than 100 ft and 8 ksi for loaded lengths or spans less than 100 ft.* It may be acceptable to consider this as a fatigue detail Category E detail, provided that adequate weld throat is provided by recommended minimum sizes or strength requirements, because cracking will occur in the base metal at the weld toe (Dexter, 2005).

Mechanical fasteners designed in accordance with AREMA (2008) (see Chapter 9) will generally not experience fatigue failure prior to the connection or member base metal. Therefore, AREMA (2008) contains no recommendations concerning allowable fatigue shear stress ranges for fasteners.

AREMA (2008) also recommends that fatigue detail Category E and E′ details should not be used in FCM. Caution regarding the use of Category D details is also expressed. It is generally good practice that designers avoid any poor fatigue details in all main carrying members and, particularly, in nonredundant or FCM members.

Example 5.20 outlines the selection of a fatigue detail category that governs the design of a member.

Example 5.20

The limiting fatigue detail category for the 25 ft long stringer of Example 5.19 is required. For Cooper's E80 live load $\Delta S_{\mathrm{re}} = 8.0$ ksi. However, the AREMA (2008) recommended that fatigue design load is adjusted based on actual typical freight traffic cycles to enable the use of the maximum stress (stress range with zero minimum stress). The maximum stress is 18.0 ksi.

* These allowable fatigue shear stress ranges for fillet welds are determined from an S–N curve line with a slope $m > 3$.

For a member with a 25 ft long loaded length, the allowable fatigue stress range is the CAFL stress range (greater than 2×10^6 cycles). Therefore, the 25 ft long stringers should not contain any details less than Category A (as shown in Table 5.5 and Figure 5.36).

5.3.2.3 Other Design Criteria for Steel Railway Bridges

There are other specific design criteria relating to both the strength and serviceability design of steel railway bridges that require consideration by the design engineer.

5.3.2.3.1 Secondary Stresses in Truss Members

Truss members may be designed as axial members (see Chapter 6) provided that secondary forces from truss distortion,* force eccentricity, end conditions (unsymmetrical connections), or other effects do not create excessive bending stresses in the members. AREMA (2008) recommends that truss members be designed as axial members provided that secondary forces do not create stresses in excess of 4 ksi in tension members and 3 ksi in compression members. For secondary stresses in excess of 4 or 3 ksi for tension or compression members, respectively, the excess is superimposed on main primary stresses and the member designed as a combined axial and flexural member (see Chapter 8). Combined axial and flexural stress† can occur in many superstructure members. For steel railway superstructures, the members most likely to be subjected to combined stresses are

- Truss members with secondary stresses in excess of 4 or 3 ksi for tension or compression members, respectively.
- End posts in trusses, which are subject to bending and axial forces due to portal bracing effects superimposed on the axial compression as a member in the main truss (see Section 5.2.2.1.2).
- Truss hangers‡ are stressed primarily in axial tension but the effects of out-of-plane bending must also be investigated in their design. Hanger allowable fatigue stress range is of critical importance due to the cyclical tensile live load regime and relatively short influence line.
- Girders and trusses with external steel prestressing cables (Dunker et al., 1985; Troitsky, 1990).

5.3.2.3.2 Minimum Thickness of Material

Material thickness is related to strength and serviceability. AREMA (2008) recommends that steel members should not have any components less than 0.335 in thick (with the exception of fillers), but some design engineers specify a greater minimum material thickness (often 3/8 or 1/2 in.). Gusset plates used to connect chord and web

---

* Truss distortion effects on a member are generally negligible for relatively slender members where the width of the member parallel to the plane of distortion is less than 10% of the member length.

† The classic example is the so-called "beam-column."

‡ Truss vertical members without diagonals at the bottom chord panel point.

members in trusses should be proportioned for the force transmitted, but should not be less than 0.50 in thick (see Chapter 9).

Where components are subject to corrosive conditions, they should be made thicker than otherwise required (as determined by judgment of the design engineer) or protected against corrosion by painting or metallic coating (usually zinc or aluminum). Atmospheric corrosion resistant (weathering) steel (see Chapter 2) does not provide protection against corrosion by standing water, and/or often wet or corrosive environments. Therefore, the design engineer should also carefully consider drainage holes and deck drainage in the design of a bridge.

5.3.2.3.3 Camber

Camber is a serviceability-related criterion. AREMA (2008) recommends that plate girder spans in excess of 90 ft long be cambered for dead load deflection. Trusses are recommended for greater camber based on dead load deflection plus the deflection from a uniform live load of 3000 lb per track foot at each panel point.*

5.3.2.3.4 Web Members in Trusses

AREMA (2008) recommends that truss web members and their connections be designed for the live load that increases the total stress by 33% over the design stress in the most highly stressed chord member of the truss. This live load ensures that web members attain their safe capacity at about the same increased live load as other truss members due to the observation that, in steel railway trusses, the web members reach capacity prior to other members in the truss (Hardesty, 1935). This recommendation is reflected by design load cases A2 and B2 in Table 4.5 of Chapter 4. An example of the calculation is shown in Example 6.4 of Chapter 6. In that example, the increase in gross area, A_g, and effective net area, A_e, are less than 1.5% based on this recommended design load case. In parametric studies of some recent railway bridge designs the effect was similar (Conway, 2003). In any case, when tensile stress ranges are present, fatigue criteria often governs truss web member design (also see Example 6.4).

REFERENCES

American Railway Engineering and Maintenance-of-Way Association (AREMA), 2008, Steel structures, in *Manual for Railway Engineering*, Chapter 15, Lanham, MD.

Anderson T.L., 2005, *Fracture Mechanics*, 3rd Edition, Taylor & Francis, Boca Raton, FL.

Armenakas, A.E., 1988, *Classical Structural Analysis*, McGraw-Hill, New York.

Barsom, J.M. and Rolfe, S.T., 1987, *Fatigue and Fracture Control in Steel Structures*, 2nd Edition, Prentice-Hall, Englewood Cliffs, NJ.

Basquin, O.H., 1910, Exponential law of endurance tests, *Proceedings,* American Society for Testing Materials, Vol. 10, Part 2, ASTM, West Conshohocken, PA.

Bleich, F., 1952, *Buckling Strength of Metal Structures*, 4th Edition, McGraw-Hill, New York.

Christy, C.T., 2006, *Engineering with the Spreadsheet*, ASCE Press, Reston, VA.

* This can be accomplished during fabrication by vertically offsetting truss joints by changing the length of the truss members.

Clough, R.W. and Penzien, J., 1975, *Dynamics of Structures*, McGraw-Hill, New York.

Conway, W.B., 2003, *Article 1.3.16 of Chapter 15*, Communication with AREMA Committee 15.

Cook, R.D., 1981, *Concepts and Applications of Finite Element Analysis*, 2nd Edition, Wiley, New York.

Davison, B. and Owens, G.W., 2003, *Steel Designer's Manual*, The Steel Construction Institute, Blackwell Publishing, Oxford, UK.

Dexter, R.J., 2005, Fatigue and fracture, in *Structural Engineering Handbook*, Chapter 34, W.F. Chen and E.M. Lui, Eds, Taylor & Francis, Boca Raton, FL.

Dick, S.M., 2002, *Bending Moment Approximation Analysis for Use in Fatigue Life Evaluation of Steel Railway Girder Bridges*, PhD Thesis, University of Kansas, Lawrence, KS.

Dunker, K.F., Klaiber, F.W., and Sanders, W.W., 1985, *Design Manual for Strengthening Single-Span Composite Bridges by Post-tensioning*, Iowa State University Engineering Research Report, Ames, IA.

Dowling, N.E., 1999, *Mechanical Behavior of Materials*, 2nd Edition, Prentice Hall, Upper Saddle River, NJ.

Fisher, J.W., 1984, *Fatigue and Fracture in Steel Bridges*, Wiley, New York.

Galambos, T.V. (Ed.), 1988, Centrally loaded columns, in *Guide to Stability Design Criteria for Metal Structures*, Chapter 3, McGraw-Hill, New York.

Grinter, L.E., 1942, *Theory of Modern Steel Structures*, Vol. 1, Macmillan, New York.

Hardesty, S., 1935, Live loads and unit stresses, *AREA Proceedings*, 36, 770–773.

Holt, E.C., 1952, *Buckling of a Pony Truss Bridge*, Column Research Council, Report No. 2.

Holt, E.C., 1956, *The Analysis and Design of Single Span Pony Truss Bridges*, Column Research Council, Report No. 3.

Ketchum, M.S., 1924, *Structural Engineer's Handbook*, Part 1, McGraw-Hill, New York.

Kulak, G.L. and Smith, I.F.C., 1995, *Analysis and Design of Fabricated Steel Structures for Fatigue*, University of Alberta Structural Engineering Report No. 190, Edmonton, AB.

Miner, M.A., 1945, Cumulative damage in fatigue, *Journal of Applied Mechanics*, (67), New York.

Norris, C.H., Wilbur, J.B., and Utku, S., 1976, *Elementary Structural Analysis*, 3rd Edition, McGraw-Hill, New York.

Petroski, H., 1995, *Engineers of Dreams*, Random House, New York.

Pilkey, W.D., 1987, *Stress Concentration Factors*, 2nd Edition, Wiley, New York.

Steinman, D.B., 1953, *Suspension Bridges*, Wiley, New York.

Stephens, R.I., Fatemi, A., Stephens, R.R., and Fuchs, H.O., 2001, *Metal Fatigue in Engineering*, 2nd Edition, Wiley, New York.

Troitsky, M.S., 1990, *Prestressed Steel Bridges Theory and Design*, Van Nostrand Reinhold, New York.

Waddell, J.A.L., 1916, *Bridge Engineering*, Vol. 1, Wiley, New York.

Wilson, E.L., 2004, *Static and Dynamic Analysis of Structures*, Computers and Structures, Berkeley, CA.

Zienkiewicz, O.C., 1983, *The Finite Element Method*, 3rd Edition, McGraw-Hill, New York.

6 Design of Axial Force Steel Members

6.1 INTRODUCTION

Members designed to carry primarily axial forces are found in steel railway bridges as truss members [chords, vertical members (hangers and posts), and diagonal members (web members and end posts)], span and tower bracing members, steel tower columns, spandrel columns in arches, knee braces, and struts. These members may be in axial tension, compression, or both (due to stress reversal from moving train and wind loads) and must be designed considering yield strength and serviceability criteria. Members in axial tension must also be designed for the fatigue and fracture limit states, and members in compression must also resist instability. Furthermore, some axial tension and compression members are subjected to additional stresses due to flexure* and must be designed for these combined stresses (see Chapter 8).

6.2 AXIAL TENSION MEMBERS

Axial tension main members in steel railway superstructures are often fracture critical and nonredundant. Therefore, the strength (yielding and ultimate) and fatigue limit states require careful consideration during design. Brittle fracture is considered by appropriate material selection, detailing, and fabrication quality assurance (see Chapters 2, 3, and 5).

6.2.1 STRENGTH OF AXIAL TENSION MEMBERS

The strength of a tension member is contingent upon yielding of the gross section, A_g, occurring prior to failure (defined at ultimate strength) of the effective net section, A_e, or

$$F_y A_g \leq \phi F_u A_e, \tag{6.1}$$

* Typically from bending forces due to end conditions (frame action, connection fixity, and eccentricity), presence of transverse loads, and/or load eccentricities.

where F_y is the tensile yield stress of steel, F_u is the ultimate tensile stress of steel, and $\phi = 0.85$ is the connection strength capacity reduction factor (Salmon and Johnson, 1980).

AREMA (2008) uses a safety factor of 9/5 which, when substituted into Equation 6.1, results in

$$0.56F_yA_g \leq 0.47F_uA_e. \tag{6.2}$$

The net area, A_n, is determined from the gross area, A_g, with connection holes removed, and may require further reduction to an effective net area, A_e, to account for the effects of stress concentrations and eccentricities at connections. Therefore, the allowable strength of the tension member, T_{all}, is

$$T_{all} = 0.55F_yA_g \tag{6.3a}$$

or

$$T_{all} = 0.47F_uA_e. \tag{6.3b}$$

The design of the tension member should be established based on the lesser T_{all} given by Equation 6.3a or 6.3b.

6.2.1.1 Net Area, A_n, of Tension Members

The net area, A_n, is determined at the cross section of the member with the greatest area removed for perforations or other openings in the member.* The gross area, A_g, across a bolted tension member connection is reduced by the holes. The net area at the connection is at the potential tensile failure line, w_{nc}, of the least length. The length of potential failure lines at connections is (Cochrane, 1922)

$$w_{nc} = w_g - \sum_{i=1}^{N_b} d_b + \sum_{j=1}^{N_{b-1}} \frac{s_j^2}{4g_j}, \tag{6.4}$$

where w_g is the gross length across the connection (gross width of the axial member), N_b is the number of bolt holes in the failure line, d_b is the effective diameter of bolt holes = bolt diameter + 1/8 in., s is the hole stagger or pitch (hole spacing in a direction parallel to load), and g is the hole gage (hole spacing in a direction perpendicular to load).

The net area is

$$A_n = w_{nc}(t_m) = A_g - \left(\sum_{i=1}^{N_b} d_b - \sum_{j=1}^{N_{b-1}} \frac{s_j^2}{4g_j} \right)(t_m), \tag{6.5}$$

where t_m is the thickness of the member.

* Perforations and openings are stress raisers, and require consideration in fatigue design.

The calculation of net area is shown in Example 6.1.

Example 6.1

Member U1–L1 is connected with gusset plates to the bottom chord of the truss in Figure E6.1 by 7/8 in. diameter ASTM A325 bolts, as shown in Figure E6.2. Determine the net area of the member if it is comprised of two laced C 12 × 30 channels.

For a C 12 × 30 channel:

$$A = 8.82\ \text{in.}^2$$

$$t_w = 0.51\ \text{in.}$$

$$b_f = 3.17\ \text{in.}$$

$$t_f = 0.50\ \text{in.}$$

Path A–B–D–E: $A_n = 8.82 - 2(1)(0.51) = 7.80\ \text{in.}^2$

Path A–B–C–D–E: $A_n = 8.82 - \{3(1) - 2[3^2/4(2.5)]\}(0.51) = 8.21\ \text{in.}^2$

Path A–B–C–E1: $A_n = 8.82 - \{2(1) - 1[3^2/4(2.5)]\}(0.51) = 8.26\ \text{in.}^2$

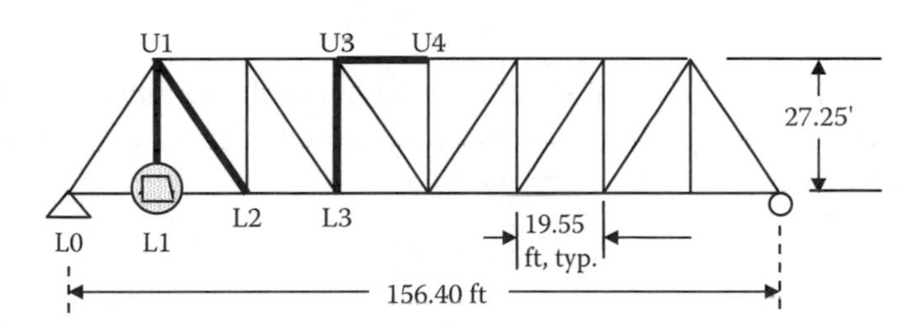

FIGURE E6.1

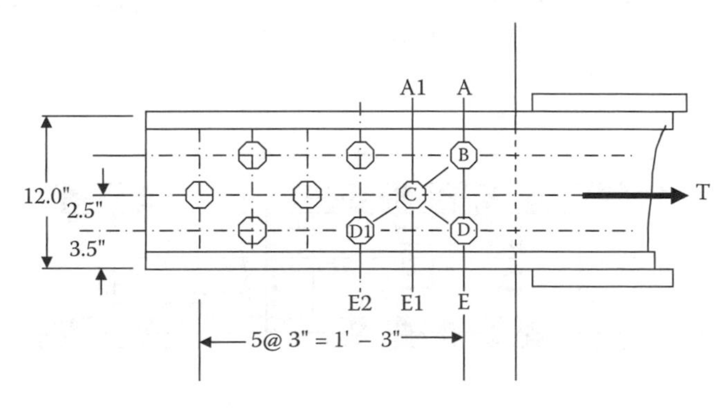

FIGURE E6.2

Path A–B–C–D1–E2: $A_n = 8.82 - \{3(1) - 2[3^2/4(2.5)]\}(0.51) = 8.21$ in.^{2}

Path A1–C–E1: $A_n = 8.82 - 1(1)(0.51) = 8.31$ in.^{2}

Path A1–C–D1–E2: $A_n = 8.82 - \{2(1) - 1[3^2/4(2.5)]\}(0.51) = 8.26$ in.^{2}

Therefore, for the member U1–L1: $A_n = 2[7.80] = 15.60$ in.^{2} (88.4% of A_g).

6.2.1.2 Effective Net Area, A_e, of Tension Members

Shear lag occurs at connections when the tension load is not transmitted by all of the member elements to the connection. Therefore, at tension member connections with elements in different planes (splices, flanges of channels with web only connected, angles with only one leg connected, webs of I-sections with only the flanges connected), an effective area is determined to reflect that the tensile force is not uniformly distributed across the net area at the connection (Figure 6.1). Shear lag effects are related to the length of the connection and the efficacy of the tension member with respect to the transfer of forces on the shear plane between the member and the connection plate (Munse and Chesson, 1963). The connection efficiency, U_c, is described by the ratio of the eccentricity, e_x, (the distance between the center of gravity of connected member elements and the shear plane) and connection length, L_c, as

$$U_c = \left(1 - \frac{e_x}{L_c}\right). \tag{6.6}$$

Therefore, when a joint is arranged such that not all of the member elements in the connection are fastened with bolts or with a combination of good quality longitudinal and transverse welds, the effective net area, A_e, is

$$A_e = U_c A_n, \tag{6.7}$$

where A_n is the net area (see Section 6.2.1.1) (gross area for welded connections) and U_c is the connection efficiency or shear lag reduction factor ≤ 0.90.

AREMA (2008) recommends shear lag reduction factors, U_c, between 0.75 and 1.00, depending on weld length for connections between members and plates that use only longitudinal welds.

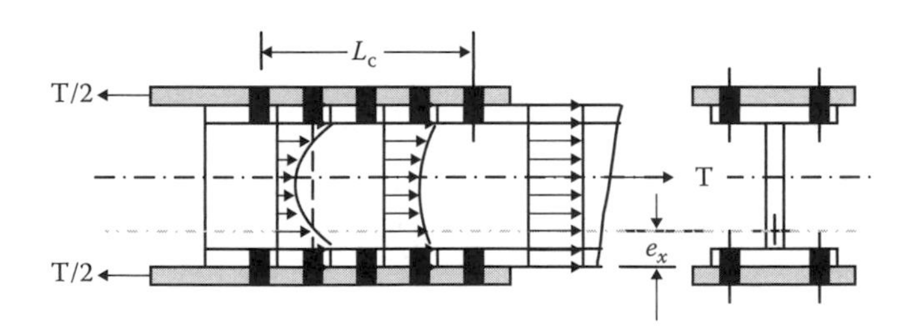

FIGURE 6.1 Shear lag at tension connection.

Angle members connected by only one leg are particularly susceptible to shear lag. AREMA (2008) recommends shear lag reduction factors, U_c, between 0.60 and 0.80, depending on the number of bolts per fastener line in the connection.*

For members that are continuous through a joint (such as truss chord members continuous across several panels) or connections where load transfer between the chord segments is efficient, AREMA (2008) indicates that shear lag may not be of concern and U_c may be effectively taken as 1.00. However, in these circumstances, engineering judgment may indicate that consideration of $A_e = 0.90 A_n$ is appropriate for design (Bowles, 1980).

The calculation of effective net area is shown in Example 6.2.

Example 6.2

Member U1–L1 is connected to the bottom chord of the truss in Figure E6.1 with 7/8 in. diameter ASTM A325 bolts in gusset plates, as shown in Figures E6.3a and b. Determine the effective net area for strength design of the member if it is comprised of a W 12 × 79 rolled section.

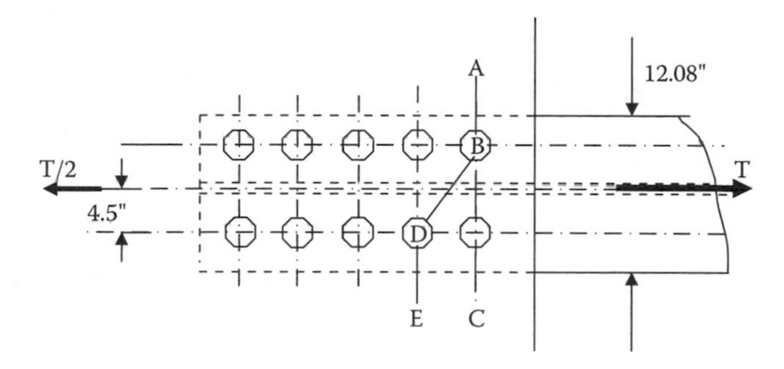

FIGURE E6.3a

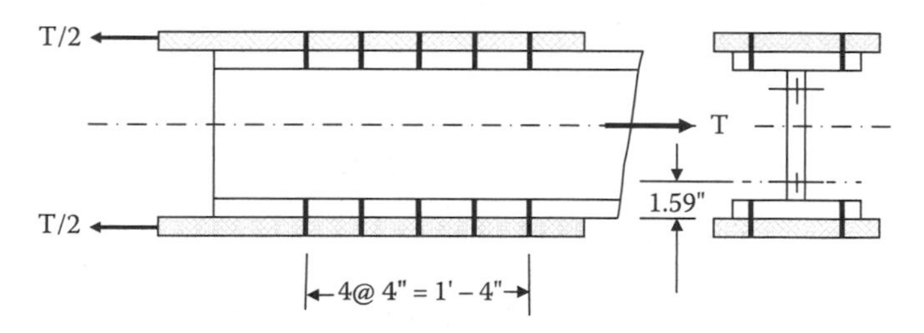

FIGURE E6.3b

* The larger value of 0.80 is used for angles when there are four or more bolts per fastener line in the connection (i.e., a relatively long connection).

For a $W12 \times 79$ section:

$$A = 23.2\ \text{in.}^2$$
$$t_w = 0.47\ \text{in.}$$
$$b_f = 12.08\ \text{in.}$$
$$t_f = 0.735\ \text{in.}$$
$$h_w = 10.61\ \text{in.}$$
$$d = 12.38\ \text{in.}$$

With elastic properties:

$$I_x = 662\ \text{in.}^4$$
$$S_x = 107\ \text{in.}^3$$
$$r_x = 5.34\ \text{in.}$$
$$I_y = 216\ \text{in.}^4$$
$$S_y = 35.8\ \text{in.}^3$$
$$r_y = 3.05\ \text{in.} = r_{min}$$

Net area:

$$\text{Path A–B–C: } A_n = 23.2 - 4(1)(0.735) = 20.26\ \text{in.}^2$$

$$\text{Path A–B–D–E: } A_n = 23.2 - 2\{2(1) - [4^2/4(9)]\}(0.735) = 20.91\ \text{in.}^2$$

Effective net area:

$$e_x \sim \{(0.735)(12.08)(0.735/2) + (0.47)(12.38/2 - 0.735)[(12.38/2 - 0.735)/2 + 0.735]\}/\{(0.735)(12.08) + (0.47)(12.38/2 - 0.735)\} = 1.59\ \text{in.}$$

$$L_c = 16\ \text{in.}$$

$$U_c = 1 - (1.59/16) = 0.90 \leq 0.90,\ \text{OK}$$

$$A_e = 0.90(20.26) = 18.23\ \text{in.}^2$$

$$L/r_{min} = (27.25)(12)/3.05 = 107.$$

6.2.2 Fatigue Strength of Axial Tension Members

The fatigue strength of an axial tension member is

$$T_{fat} = S_{rfat} A_{efat} \tag{6.8}$$

where A_{efat} is the effective gross or net area of only the member elements that are directly connected (e.g., the flange elements in Figure 6.1). This reduction accounts for shear lag effects for fatigue design, which occur at stress levels below fracture.*

* Shear lag for strength design is evaluated at stress levels near fracture.

For slip-resistant (or friction-type) bolted connections,* the effective net area, A_{efat}, is taken as the gross area of only the member elements that are directly connected. S_{rfat} is the allowable fatigue stress range depending on the number of design cycles, and connection and fabrication details of the tension member. For design, the number of cycles is generally assumed at greater than 2,000,000 for single-track bridges with relatively short influence lines (for spans $\leq$100 ft as recommended by AREMA, 2008). Since fatigue design is based on nominal stresses, S_{rfat} is recommended for various fatigue detail categories (A, B, B′, C, D, E, E′, and F) depending on connection or detail geometry (see Chapter 5). Table 6.1 indicates the allowable fatigue stress ranges used for the design of tension members at connections for greater than 2,000,000 stress cycles.† The allowable fatigue stress ranges for detail categories consider stress concentrations related to member discontinuities (such as change in section) (Figure 6.2a) or apertures in the member (such as bearing connection holes, access, or drainage openings) (Figure 6.2b). The magnitude of the stress concentrations may be determined by elasticity theory and fracture mechanics methods or by use of published stress concentration factors,‡ K_t, (Pilkey, 1997; Anderson, 2005; Armenakas, 2006).

Member transition fillets of usual dimensions that might be considered for steel bridge tension members may result in stress concentration factors in the order of $K_t = 1.5 - 2.0$. Figure 6.2b illustrates that, for the simple case of uniaxial tension in a flat plate with a single circular hole, stress concentration factors may reach $K_t = 3.0$. Table 6.2 indicates AREMA (2008) fatigue design detail categories and corresponding stress concentration factors (Sweeney, 2006).

However, since the allowable fatigue stress ranges recommended by AREMA (2008) are based on nominal stress test results, it is usually not necessary to explicitly consider stress concentration factors in the design of axial tension members unless the design details are particularly severe. Therefore, the designer must carefully consider the use of transitions, apertures, or other discontinuities in the detailing of axial tension

TABLE 6.1
Allowable Fatigue Stress Range for Number of Design Stress Range Cycles >2,000,000

Member or Connection Condition	S_{rfat} (ksi)
Plain member	24
Bolted slip resistant connection	16
Partial penetration groove and fillet welded connection	2.6–10
Full penetration weld connection	2.6–16

* Recommended for main member connections and all connections subject to stress reversal and/or cyclical loading.

† For welded connections the allowable fatigue stress range depends on type of stress, direction of stress, direction of weld, weld continuity, and transition details. AREMA (2008) makes recommendations considering these factors.

‡ Often obtained by photo-elastic testing.

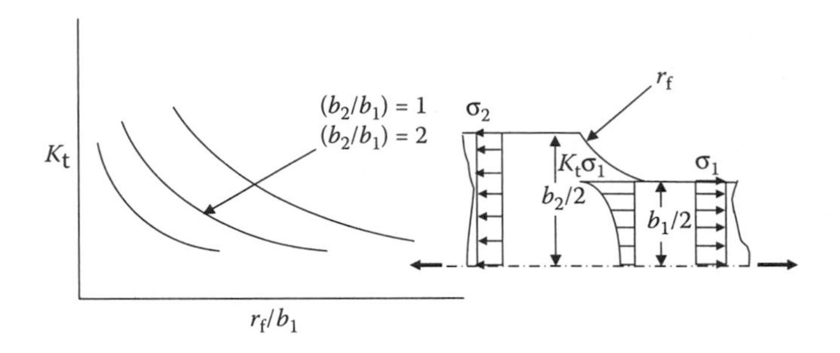

FIGURE 6.2a Stress concentration factors for a flat bar with transition fillets in axial tension.

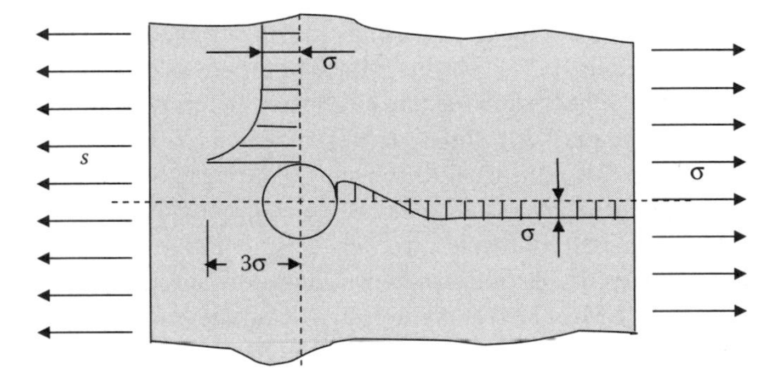

FIGURE 6.2b Stress concentration factors for a flat bar with round holes in axial tension.

members subjected to fluctuating stresses and ensure that an allowable fatigue stress range based on the appropriate detail category is used.

6.2.3 Serviceability of Axial Tension Members

In order to preclude excessive deflection (e.g., sag of long members under self-weight) or vibration (e.g., from wind loads on bracing members), the flexibility of axial tension members must be limited. The maximum slenderness between points of support, L_u,

TABLE 6.2
Stress Concentration Factors for AREMA (2008) Fatigue Detail Categories

Fatigue Detail Category	Stress Concentration Factor, K_t
A	1.00
B	1.15
C	2.35
C′	2.75
D	3.6
E	4.8

is recommended by AREMA (2008) as

$$\frac{L_u}{r_{min}} \leq 200, \tag{6.9}$$

where $r_{min} = \sqrt{I_{min}/A_g}$, and I_{min} is the minimum moment of inertia of the tension member in bending about an axis between supports, L_u.

Example 6.3

Determine the design criteria for load case A1 (Table 4.5) for member U1–L1 in Figure E6.1 for Cooper's E80 load. Use Grade 50 ($F_y = 50$ ksi) steel with ultimate stress, F_u, of 70 ksi. Assume a connection as shown in Example 6.2.

The forces in member U1–L1 are

Dead load force = DL = +10.82 kips

Maximum live load force = LL1 = +128.8 kips (see Chapter 5)

Minimum live load force = LL2 = 0

Maximum Live Load Impact = 39.28% (L = 19.55′ see Chapter 4)

Mean Live Load Impact = 0.40(39.28)% = 15.71% (see Chapter 4)

Range of live load force = LL_{range} = LL1 + LL2 = +128.80 + (0.00) = 128.80 kips

Load combinations:

$P_{rangeLL+I} = 1.157(128.8) = 149.0$ kips

$P_{max} = +10.82 + 1.393(128.80) = +190.2$ kips

Strength considerations:

$$A_g \geq \frac{190.2}{0.55(50)} \geq 6.92 \text{ in.}^2 \text{ (tensile yielding)}$$

$$A_n \geq \frac{190.2}{0.47(70)} \geq 5.78 \text{ in.}^2 \text{ (fracture)}$$

due to strength-related shear lag effects at the connection

$$A_e \geq \frac{5.78}{U_c} \geq 6.42 \text{ in.}^2 \text{ (fracture).}$$

Fatigue considerations (including fatigue shear lag):

$A_{efat} \geq (149.0/16) \geq 9.31$ in.^{2} for the gross area of connected elements of the member (Table 6.1) (AREMA (2008) recommends slip resistant connections for main members such as U1–L1).

$A_e \geq (149.0/24) \geq 6.21$ in.^{2} for the net area of the member away from the connection (Table 6.1)

Stiffness considerations:

$$\frac{L}{r_{min}} \leq 200$$

$$r_{min} \geq \frac{(27.25)(12)}{200} \geq 1.64 \text{ in.}$$

Select a member with the following section properties:

Minimum Gross Area of the member $= A_g \geq 6.92\,\text{in.}^2$ (tensile yielding)

Minimum Gross Area for portions of the member connected $= A'_g \geq 9.31\,\text{in.}^2$ (fatigue)

Minimum Effective Net Area of the member at the connection $= A_e \geq 6.42\,\text{in.}^2$ (fracture)

Minimum Net Area of the member away from the connection $= A_n \geq 6.21\,\text{in.}^2$ (fatigue)

Minimum Radius of Gyration of the member $= r_{min} \geq 1.64$ in.

Example 6.4

Determine the design criteria for load cases A1 and A2 (Table 4.5) for member U1–L2 in Figure E6.1 for Cooper's E80 load. Use Grade 50 ($F_y = 50$ ksi) steel with ultimate stress, F_u, of 70 ksi.

Forces in member U1–L2 are

Dead load force = DL = +54.56 kips

Maximum live load force = LL1= +332.4 kips

Minimum live load force = LL2 = −7.4 kips

Maximum live load impact = 20.75% ($L = 156.4'$ see Chapter 4)

Mean live load impact = 0.65(20.758)% = 13.49% (see Chapter 4)

Range of live load force $= LL_{range} = +332.4 - (-7.4) = 339.8$ kips [AREMA (2008) recommends that all stress ranges be considered as tensile stress ranges, due to the potential for preexisting mean tensile stresses—see Chapters 5 and 9].

Load combinations:

$P_{rangeLL+I} = 1.135(339.8) = 385.7$ kips

$P_{max} = +54.56 + 1.208(+332.4) = 456.1$ kips.

AREMA (2008) recommends that web members and their connections be designed for 133% of the allowable stress using the live load LLT that will increase the total maximum chord stress by 33% (see Chapters 4 and 5).

The maximum chord forces are

Maximum dead load force in chord = 98.10 kips.

Total maximum chord force (for Cooper's E80 live load) $= P_{chordmax} = 98.10 + 1.208(557.2) = 771.2$ kips (same impact factor for chord and diagonal members in 156.4′ through truss) (see Chapter 5).

Therefore,

$$\text{LLT} = \left(\frac{1.33(771.2)}{1.208(557.2)} - 98.1\right)(\text{E80}) = 1.39(\text{E80}) = \text{E111}.$$

$P^1_{max} = +54.56 + 1.208(+332.4)(122/80) = 666.4$ kips.

Strength considerations:

$$A_g \geq \frac{456.1}{0.55(50)} \geq 16.6 \text{ in.}^2 \text{ (tensile yielding)}$$

$$A_e \geq \frac{456.1}{0.47(70)} \geq 13.9 \text{ in.}^2 \text{ (fracture)}$$

due to strength-related shear lag effects at the connection (assuming $U_c = 0.90$)

$$A_e \geq \frac{13.9}{U_c} \geq 15.4 \text{ in.}^2 \text{ (fracture)}$$

from the requirement related to live load LLT

$$A_g \geq \frac{611.7}{1.33(0.55)(50)} \geq 16.7 \text{ in.}^2 \text{ (tensile yielding)}$$

$$A_e \geq \frac{611.7}{1.33(0.47)(70)} \geq 14.0 \text{ in.}^2 \text{ (fracture)}$$

due to strength-related shear lag effects of LLT at the connection

$$A_e \geq \frac{14.0}{U_c} = 15.6 \text{ in.}^2 \text{ (fracture).}$$

Fatigue considerations:

$A_{efat} \geq (385.7/16) \geq 24.1$ in.^{2} for the gross area of connected elements of the member

$A_e \geq (385.7/24) \geq 16.1$ in.^{2} for the net area of the member away from the connection.

Stiffness considerations:

$$\frac{L}{r_{min}} \leq 200$$

$$r_{min} \geq \frac{\left(\sqrt{(27.25)^2 + (19.55)^2}\right)(12)}{200} \geq 2.01 \text{ in.}$$

Select a member with the following section properties:

Minimum gross area of the member $= A_g \geq 18.2$ in.^{2} (tensile yielding)

Minimum gross area for portions of the member connected $= A'_g \geq 24.1$ in.^{2} (fatigue)

Minimum net effective area of the member at the connection $= A_e \geq 15.6$ in.^{2} (fracture)

Minimum net area of the member away from the connection $= A_e \geq 16.1$ in.^{2} (fatigue)

Minimum radius of gyration $= r_{min} \geq 2.01$ in.

6.2.4 Design of Axial Tension Members for Steel Railway Bridges

Tension members in steel railway bridges may be comprised of eyebars, cables, structural shapes, and built-up sections. Eyebars are not often used in modern bridge superstructure fabrication, and suspension or cable-stayed bridges are not common for freight railway structures due to flexibility concerns (see Chapter 1). Structural shapes such as W, WT, C, and angles are frequently used for steel railway bridge tension members.

It is often necessary to fabricate railway bridge tension members of several structural shapes due to the large magnitude railroad loads and tension members that undergo stress reversals. Bending effects (see Chapter 8) and connection geometry constraints may also dictate the use of built-up tension members. The components must be adequately fastened together to ensure integral behavior of the tension member. In cases where a box-type member is undesirable, such as where the ingress of water is difficult to preclude, open tension members are used. Built-up open tension members are often fabricated with lacing bars and stay (tie or batten) plates or perforated cover plates.* Shear deformation in tension members, which is primarily due to self-weight and wind loads, is relatively small and AREMA (2008) recognizes this by providing nominal recommendations for lacing bars and stay plates.

Lacing bar width should be a minimum of three times the fastener diameter to provide adequate edge distance and the thickness for single flat bar lacing bars should be at least 1/40 of the length† for main structural members and 1/50 of the length for bracing members. Stay plates should be used at the ends of built-up tension members and at intermediate locations where lacing bar continuity is interrupted due to the connection of other members.‡ The length of the stay plates at the ends of laced bar built-up tension members must be at least 85% of the distance between connection lines across the member. The length of the stay plates at intermediate locations of laced bar built-up tension members must be at least 50% of the distance between connection lines across the member.

The thickness of perforated cover plates should be at least 1/50 of the length between closet adjacent fastener lines. Perforated cover plate thickness is based on transverse shear, V, at centerline of the cover plate. The maximum transverse shear stress, τ_V, at the center of the cover plate is

$$\tau_V = \frac{3V}{2bt_{pc}}, \tag{6.10}$$

where b is the width of the perforated cover plate and t_{pc} is the thickness of the perforated cover plate. Therefore, the longitudinal shear force, V', over the distance between centers of perforations or apertures, l_p, is

$$V' = \tau_V(l_p t_{pc}) = \frac{3Vl_p}{2b} \tag{6.11}$$

---

* Perforated cover plates are most commonly used for modern built-up truss members.

† Lacing bar length is the distance between fastener centers.

‡ For example, stay plates are used each side of members that interrupt lacing bars.

and the shear stress over the net area of the plate between the centers of perforations, τ_{pc}, is

$$\tau_{pc} = \frac{V'}{(l_p - c)t_{pc}} = \frac{3Vl_p}{2bt_{pc}(l_p - c)}, \tag{6.12}$$

where c is the length of the perforation.

Rearrangement of Equation 6.12 yields

$$t_{pc} = \frac{3Vl_p}{2\tau_{all}b(l_p - c)}, \tag{6.13}$$

where τ_{all} is the allowable shear stress ($0.35F_y$ recommended by AREMA, 2008).

The transverse shear force, V, is generally small in tension members, and the perforated cover plate thickness is primarily dependent on requirements for axial tension and recommended minimum material thickness (see Chapter 5).

The net section through the perforation of the plate is included in the member net area.

Example 6.5

Use the AREMA (2008) recommendations to select bolted lacing bars and stay plates for tension member U1–L1 of Example 6.1. The member cross section is comprised of two C 12° × 30 channels 12 in. apart (Figure E6.4).

Lacing bars:

Minimum width for single flat lacing bars = 3.5 in.

Use approximately 60° angle between the lacing bar and the longitudinal axis of the hanger.

Minimum thickness of the bar = (1/40)(15/ sin 60°) = 0.43 in.

For 7/8 in diameter bolts, the minimum bar width = 3(7/8) = 2.63 in.

Stay plates:

Use at ends and intermediate locations where lacing is not present.

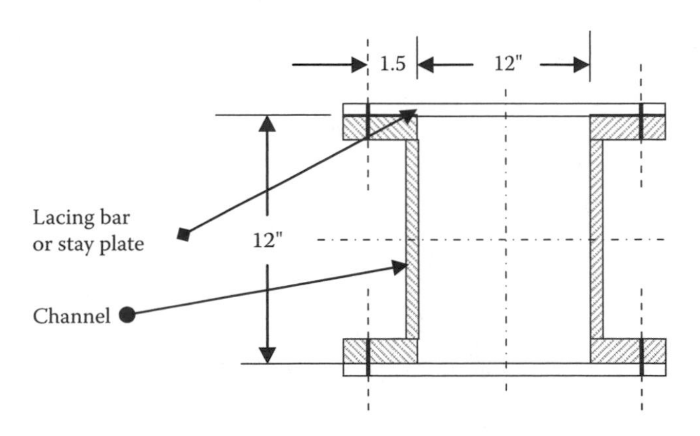

FIGURE E6.4

Minimum end plates length = (2/3)(1.25)(15) = 12.5 in.
Minimum intermediate plates length = (3/4)(12.5) = 9.38 in.
Minimum thickness of stay plate = (1/50)(15) = 0.30 in.
For 7/8 in. diameter Bolts, maximum fastener spacing = 4(7/8) = 3.5 in.
Minimum stay plate length = 10 in. (minimum of three bolts per side of the stay plate).

The bridge designer will use these minimum requirements to select design dimensions of lacing bars and stay plates.

6.3 AXIAL COMPRESSION MEMBERS

Axial compression main members in steel railway superstructures are often nonredundant. Therefore, the strength (yielding and stability) limit state requires careful consideration during design.

6.3.1 Strength of Axial Compression Members

The strength of a steel compression member is contingent upon its susceptibility to instability or buckling. For very short members, failure is governed by yield stress, F_y.* However, for members with greater slenderness, inelastic or elastic instability, depending on the degree of slenderness, will govern the failure at a critical buckling force, P_{cr}.

6.3.1.1 Elastic Compression Members

Long and slender members will buckle at loads with compressive stresses below the proportional limit, F_p (Figure 6.3). The magnitude of this elastic critical buckling force depends on the stiffness, length, and end conditions of the compression member; as well as imperfections in loading and geometry.

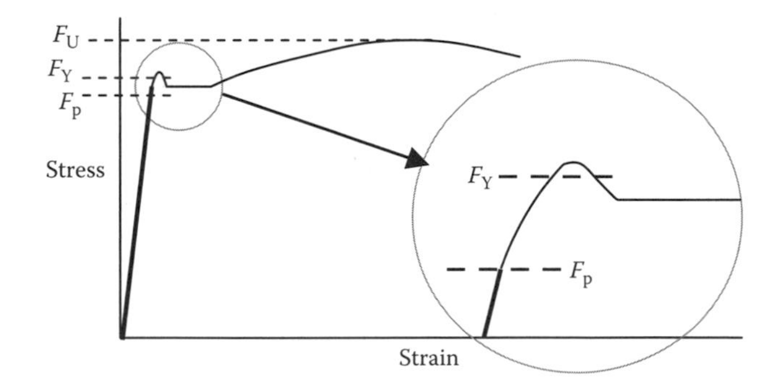

FIGURE 6.3 Typical stress–strain curve for structural steel.

* Compressive yield stress is almost equal to tensile yield stress (see Chapter 2).

6.3.1.1.1 *Elastic Buckling with Load, P, Applied along the Centroidal Axis of the Member*

Assuming that

- The member has no geometric imperfections (perfectly straight)
- Plane sections remain plane after deformation
- Flexural deflection is considered only (shear deflection is neglected)
- Hooke's law is applied
- Member deflections are small,

the differential equation of the deflection curve is

$$\frac{\mathrm{d}^2y(x)}{\mathrm{d}x^2} + k^2y(x) = U, \tag{6.14}$$

where $y(x)$ is the lateral deflection of compression member, U depends on the effects on load, P, of the compression member end conditions, and

$$k^2 = \frac{P}{EI},$$

where P is the load applied at the end and along the centroidal axis of the compression member. The solution of Equation 6.14 for the elastic critical buckling force, P_{cr}, is readily accomplished by consideration of the appropriate boundary conditions (Wang et al., 2005). The elastic critical buckling force for various member end conditions is shown in Table 6.3. Figures 6.4 and 6.5 illustrate the various compression member end conditions in Table 6.3. The critical buckling force can be expressed as

$$P_{\mathrm{cr}} = \frac{\pi^2 EI}{(KL)^2}. \tag{6.15}$$

TABLE 6.3
Elastic Critical Buckling Force for Concentrically Loaded Members with Various End Conditions

End Condition	U	P_{cr}
Both ends pinned (Figure 6.4a)	0	π^2EI/L^2
Both ends fixed (Figure 6.4b)	$-(V/EI)x + M/EI$	$4\pi^2EI/L^2$
One end fixed and other end free (Figure 6.4c)	$k^2\Delta$	$\pi^2EI/4L^2$
One end hinged and other end fixed (Figure 6.4d)	$(M/EIL)x$	$2.046(\pi^2EI/L^2)$
One end guided and other end fixed (Figure 6.5a)	$P\Delta/2EI$	π^2EI/L^2
One end hinged and other end guided (Figure 6.5b)	0	$\pi^2EI/4L^2$

L = Length of member between end supports, V = Shear force in member, M = Bending moment at end of member, Δ = Lateral deflection at free or guided end of member with other end fixed.

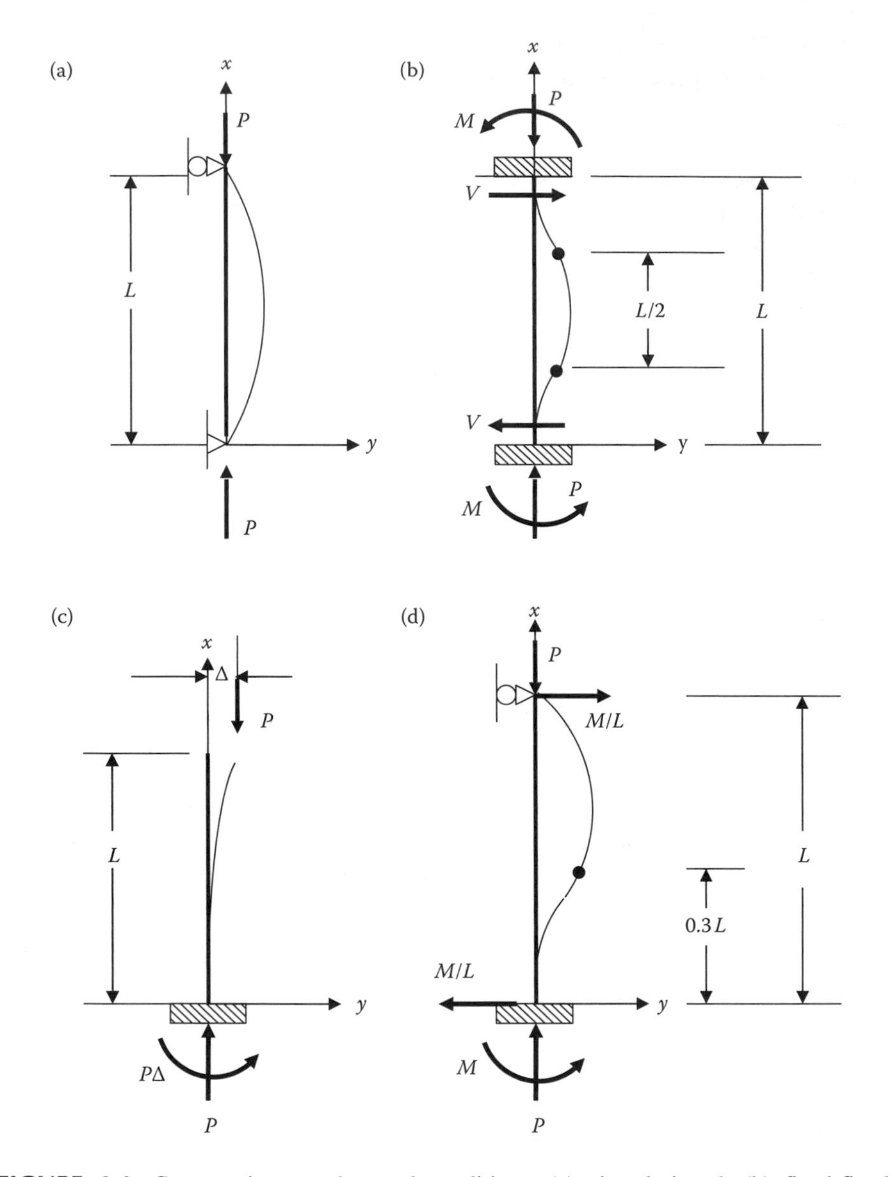

FIGURE 6.4 Compression member end conditions: (a) pinned-pinned; (b) fixed-fixed; (c) fixed-free; (d) fixed-hinged.

Considering $I = A_g r^2$, the critical buckling stress may be obtained from Equation 6.15 as

$$F_{cr} = \frac{P_{cr}}{A_g} = \frac{\pi^2 E}{(KL/r)^2}. \tag{6.16}$$

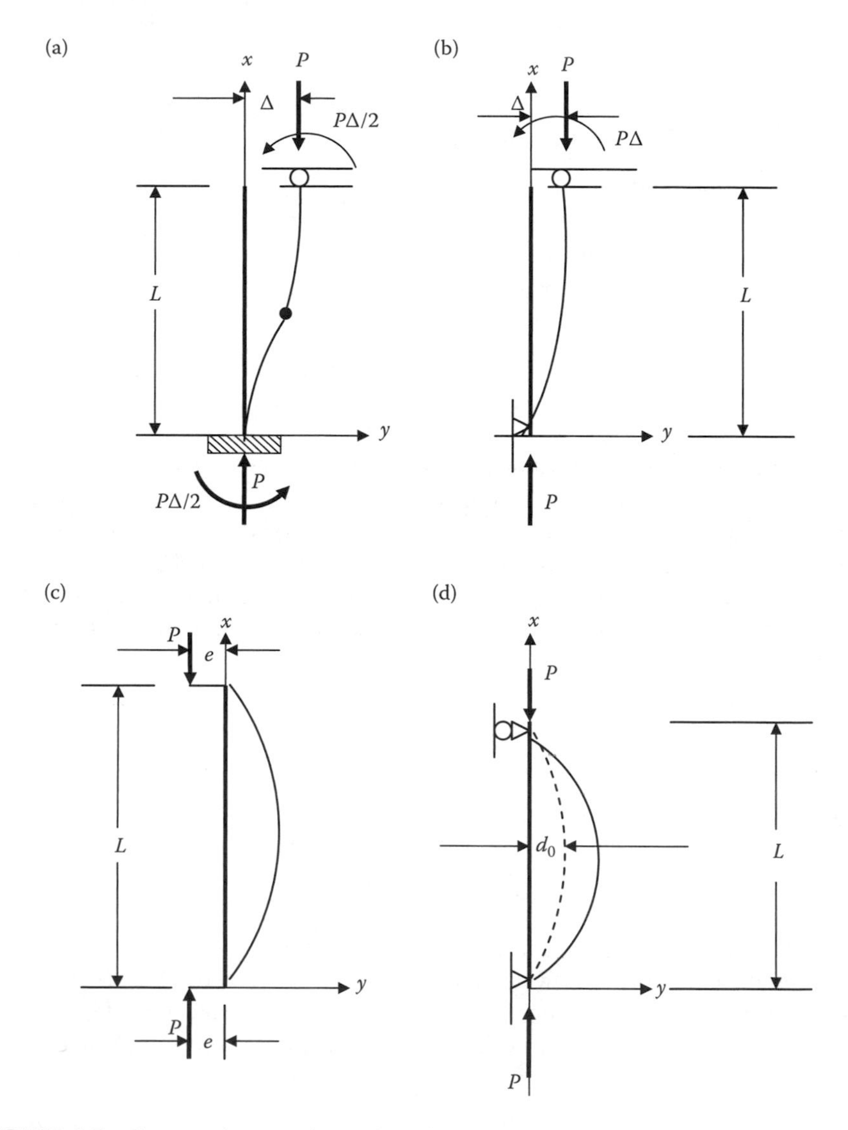

FIGURE 6.5 Compression member end and initial conditions: (a) fixed-guided; (b) hinged-guided; (c) eccentric load; (d) geometric imperfection.

The rearrangement of Equation 6.15 yields

$$K = \text{effective length factor} = \frac{\pi}{L}\sqrt{\frac{EI}{P_{\text{cr}}}}. \tag{6.17}$$

Values of K from Equation 6.17 for the various end conditions and corresponding critical buckling force, P_{cr}, in Table 6.3 are shown in Table 6.4. However, due to the ideal conditions of the mathematical model, which may not be representative of the actual end conditions, the effective length factors in Table 6.4 are generally increased

TABLE 6.4
Effective Length Factors for Various Compression Member End Conditions

End Condition	K
Both ends pinned (Figure 6.4a)	1.00
Both ends fixed (Figure 6.4b)	0.50
One end fixed and other end free (Figure 6.4c)	2.00
One end hinged and other end fixed (Figure 6.4d)	0.70
One end guided and other end fixed (Figure 6.5a)	1.00
One end hinged and other end guided (Figure 6.5b)	2.00

for the ASD of compression members. For railway trusses with moving live loads, the forces in members framing into the end of a member under consideration will be less than maximum when the member under consideration is subject to maximum force. Therefore, the ideally pinned end condition is not established because this force arrangement imposes rotational restraints at the end of the member under consideration. For members with equal rotational restraint at each end, an approximate effective length factor has been developed as (Newmark, 1949)

$$K = \frac{\overline{C}+2}{\overline{C}+4}, \tag{6.18}$$

where

$$\overline{C} = \frac{\pi^2 EI}{LR_\text{k}},$$

where R_k is the equivalent rotational spring constant. For members and end conditions typically used in steel railway trusses, Equation 6.18 provides $K = 0.75 - 0.90$. Furthermore, theoretical solutions for truss members indicate that, for constant cross-section chord members, the effective length factor, K, can be estimated as

$$K = \sqrt{1 - \frac{5}{4n}}, \tag{6.19}$$

where n is the number of truss panels (typically, $K = 0.85 - 0.95$).

The same studies also indicated that, for web members typically used in steel railway trusses, the effective length factor, K, is generally between 0.70 and 0.90 (Bleich, 1952).

AREMA (2008) recommends two effective length factors, K, to represent actual steel railway bridge compression member end conditions. For true pin-end connections, $K = 0.875$ is recommended. For all other end conditions (with bolted or welded end connections), AREMA (2008) recommends $K = 0.75$ for design purposes.

A safety factor must be applied to Equations 6.15 and 6.16 to account for small load eccentricities and geometric imperfections. AREMA (2008) uses a factor of safety,

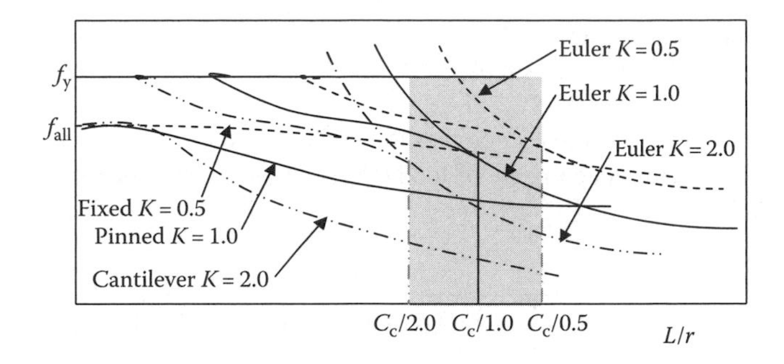

FIGURE 6.6 Effect of end restraint on allowable stresses and slenderness values at elastic (Euler) buckling.

FS, of 1.95 to arrive at the allowable compressive strength, C_{all}, of

$$C_{all} = \frac{P_{cr}}{1.95} = \frac{0.514\pi^2 EI}{(KL)^2} \tag{6.20}$$

or

$$F_{all} = \frac{C_{all}}{A_g} = \frac{F_{cr}}{1.95} = \frac{0.514\pi^2 E}{(KL/r)^2}. \tag{6.21}$$

Elastic buckling, described by Equation 6.21, will occur at values of $KL/r \geq C_c$. C_c is defined by the intersection of the Euler buckling curve (Figure 6.6) with a transition curve from compressive yielding, f_y, as shown by the vertical lines in Figure 6.6 for members with $K = 0.5, 1.0$, and 2.0. The transition curve represents the effects of eccentricities, initial imperfections, and residual stresses introduced during fabrication and erection of steel railway bridge compression members.* For elastic buckling (at large KL/r), the degree of member end restraint, expressed in terms of the effective length factor, K, greatly affects the allowable compressive stress, f_{all}, as shown within the shaded area in Figure 6.6.

If they exist, explicit consideration of relatively large load eccentricities and/or geometric imperfections must be made for long and slender compression members.

6.3.1.1.2 *Elastic Buckling with Load Applied Eccentric to the Centroidal Axis of the Member*

Assuming that

- The member has no geometric imperfections (perfectly straight)
- Plane sections remain plane after deformation
- Flexural deflection is considered only (shear deflection is neglected)

---

* The transition curve describes the inelastic buckling of members with KL/r less than that for elastic buckling but greater than the maximum KL/r value for compressive yielding.

- Hooke's law is applied
- Member deflections are small,

the solution of the differential equation of the deflection curve (Equation 6.14), where $U = -k^2 e$ and e is the eccentricity of load (Figure 6.5c), is the secant formula (Chen and Lui, 1987)

$$P_{cr} = \frac{P_y}{1 + \left[(ec/r^2)\sec(\pi/2)\sqrt{P_{cr}L^2/\pi^2 EI}\right]}, \quad (6.22)$$

where $P_y = A_g F_y$, c is the distance from the neutral axis to the extreme fiber of the member cross section, and r is the radius of gyration of the member cross section.

The secant formula was considered appropriate for inelastic buckling of members from initial curvature and load eccentricity. However, it does not include the consideration of residual stresses, which are of considerable importance in modern steel structures. Therefore, Equation 6.22 is no longer used to determine the critical buckling force of compression members. The equation was used in the AREMA recommended practice prior to 1969, but was discontinued as a basis for compression member design because of the difficulty associated with its use and indication that Euler-type formulas are appropriate for eccentrically loaded compression members (AREMA, 2008).

6.3.1.1.3 *Elastic Buckling of Members with Geometric Imperfections (Initial Out-of-Straightness)*

Assuming that

- The member is concentrically loaded
- Plane sections remain plane after deformation
- Flexural deflection is considered only (shear deflection is neglected)
- Hooke's law is applied
- Member deflections are small,

the solution of the differential equation of the deflection curve (Equation 6.14), where $U = -k^2\delta_0 \sin(\pi x/L)$ (Figure 6.5d), is the Perry–Robertson formula (Chen and Lui, 1987)

$$P_{cr} = \frac{P_y}{1 + (\delta_0 c/r^2)(1/(1 - (P_{cr}L^2/\pi^2 EI)))}, \quad (6.23)$$

where δ_0 is the out-of-straightness at the middle of the member (Figure 6.5d). For low values of L/r (generally less than about 60), out-of-straightness geometric imperfections are usually not an important design consideration.

6.3.1.2 Inelastic Compression Members

Steel railway bridge members of the usual length and slenderness will buckle at loads above the proportional limit, F_p, (Figure 6.3) when some cross-section fibers have already been yielded before the initiation of instability. Therefore, the effective modulus of elasticity is less than the initial value. This nonlinear behavior occurs primarily as a result of residual stresses* but may also be a result of initial curvature and force eccentricity.

These material and/or geometric imperfections (or nonlinearities) are considered by replacing the elastic modulus, E, with an effective modulus, E_{eff}. Therefore, inelastic critical buckling force solutions are analogous to those shown in Table 6.3 with elastic modulus, E, replaced with the effective modulus, E_{eff}, so that

$$P_{\text{cr}} = \frac{\pi^2 E_{\text{eff}} I}{(KL)^2}. \tag{6.24}$$

Engesser (see Chapter 1) proposed both the tangent modulus, E_t (Equation 6.25), and the reduced modulus, E_r (Equation 6.26), for the effective modulus. The tangent modulus is

$$E_t = \frac{d\sigma}{d\varepsilon} = E\left(\frac{F_y - \sigma}{F_y - c\sigma}\right). \tag{6.25}$$

The reduced modulus, for symmetric I-sections (and neglecting web area), is (Timoshenko and Gere, 1961)

$$E_r = \frac{2EE_t}{E - E_t}, \tag{6.26}$$

where $d\sigma$ is the change in stress, $d\varepsilon$ is the change in strain, σ is the applied stress = P/A, $c = 0.96 - 0.99$ for structural steel. The reduced modulus is less than the tangent modulus, E_t, as shown in Figure 6.7.

An inelastic compression member theory was also proposed by Shanley (Tall, 1974). The theory indicates that actual inelastic compression member behavior lies between that of the tangent and the reduced modulus load curves. However, because test results are closer to the tangent modulus curve values (Chen and Lui, 1987), the

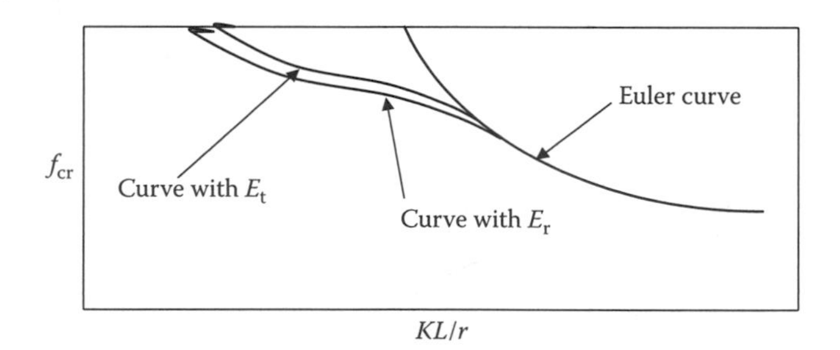

FIGURE 6.7 Typical compression member curves.

* Since residual stresses are most affected by size, the use of high-strength steel can make their effect relatively smaller. Annealing to reduce residual stresses (heat treatment) may increase the strength of a compression member.

tangent modulus is often used in the development of modern inelastic compression curves and equations.

The tangent modulus theory (Equation 6.25) includes material imperfection considerations but it does not explicitly consider the effects of geometric imperfections (member out-of-straightness) and residual stresses in compression members.

Geometric imperfections (unintentional member out-of-straightness and eccentricity) have a detrimental effect on the inelastic critical buckling force of compression members of relatively large slenderness. The American Institute of Steel Construction (AISC, 1980) ASD provisions recognize this by increasing the factor of safety, FS, to 115% of 5/3 for compression members with an effective slenderness ratio at the value for Euler elastic buckling, $C_c = KL/r$. This variable FS is

$$\mathrm{FS} = \frac{5}{3} + \frac{3}{8}\frac{(KL/r)}{C_c} - \frac{1}{8}\left(\frac{(KL/r)}{C_c}\right)^3 \quad \text{for } (KL/r) \le C_c \tag{6.27}$$

Geometric imperfections are also implicitly recognized in the AREMA (2008) recommendations through the use of a higher, although constant, FS for axial compression (FS = 1.95) than what is used for axial tension (FS = 1.82). A similar cubic polynomial as Equation 6.27 was used in Chapter 5 to investigate a variable FS using the AREMA (2008) criteria. However, due to the potential for geometric imperfections to create greater instability for members loaded with relatively large-magnitude live loads, the higher factor of safety is likely to be appropriate even for less slender compression members in railway bridges.

The rolling of structural steel plates and shapes, and fabrication bending, cutting, and/or welding procedures may create residual stresses that affect the inelastic critical buckling stress in a compression member. The pattern of compressive and tensile residual stresses is very dependent on member cross section and dimensions. The presence of varying residual stresses will affect the material compressive stress–strain curve (Figure 6.8) and establish a different effective modulus of elasticity in each direction across a compression member cross section. If the tangent modulus is taken as the effective modulus of elasticity, it will differ depending on the direction

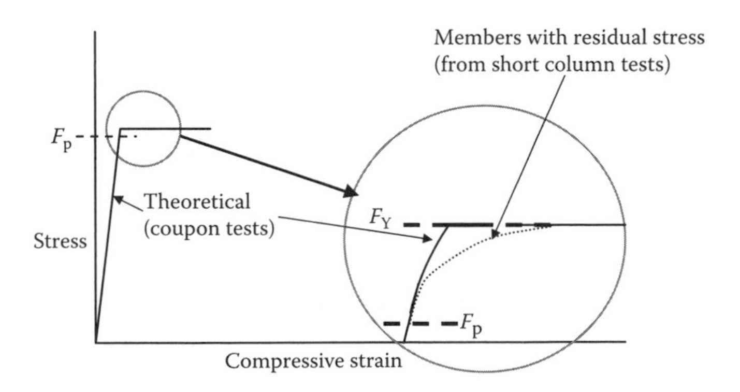

FIGURE 6.8 Typical compressive stress–strain curve for structural steel.

of buckling and will underestimate compression member strength. Therefore, buckling direction (weak or strong axis) must be considered independently to determine compression member strength when allowing for residual stresses.

The Column Research Council (CRC) conducted tests and analytical studies of weak and strong axis inelastic buckling with linear and parabolic residual stress distributions across the compression member cross section. These studies revealed that, within the inelastic range, the compression member curves (see Figure 6.9a) were parabolic. The residual stresses used in the CRC studies were about $0.3F_y$. However, the value of $0.5F_y$ is used in order to conservatively represent the residual stresses and provide a smooth transition to the Euler elastic buckling curve at $KL/r = C_c$ (elastic behavior and buckling below the proportional limit of $0.5F_y$). Therefore, for $KL/r < C_c$, the Johnson parabola (Equation 6.28) may be used to represent inelastic behavior (Tall, 1974). The Johnson parabola is

$$F_{cr} = F_y - B(KL/r)^2. \tag{6.28}$$

The value of the constant B with $F_p = 0.5F_y$ and $F_r = F_y - F_p$ (Figure 6.8) is (Bleich, 1952)

$$B = \frac{F_r}{F_y \pi^2 E}(F_y - F_r) = \frac{1}{4\pi^2 E} \tag{6.29}$$

and the inelastic critical buckling stress is

$$F_{cr} = F_y\left(1 - \frac{F_y}{4\pi^2 E}\left(\frac{KL}{r}\right)^2\right) \tag{6.30}$$

or

$$\frac{F_{cr}}{F_y} = 1 - 0.25\lambda_c^2, \tag{6.31}$$

where $\lambda_c = \left(\frac{KL}{r}\right)\sqrt{\frac{F_y}{\pi^2 E}}$.

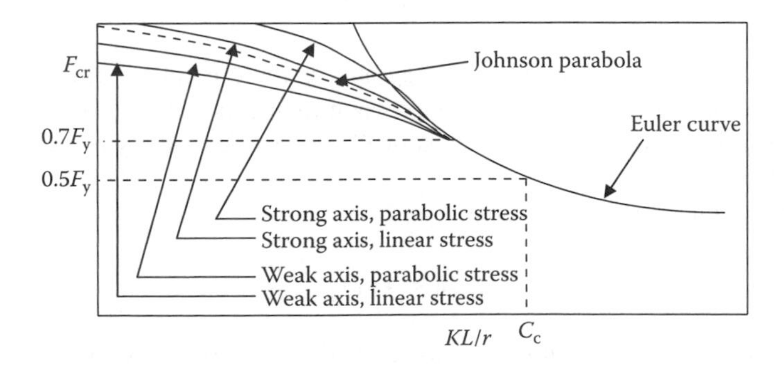

FIGURE 6.9a Weak and strong axis compression member curves using linear and parabolic residual stress distributions.

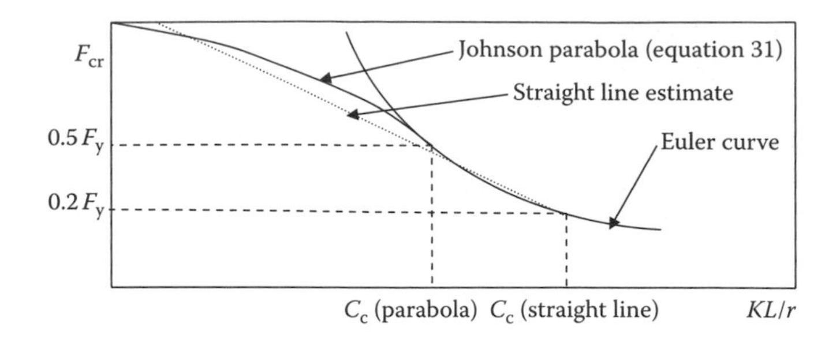

FIGURE 6.9b Parabolic and linear compression member curves.

The uncertainties involved in the determination of K will result in overestimates for F_{cr}, particularly when KL/r is between 40 and 100 (AREMA, 2008). To mitigate this, AREMA (2008) adopts a conservative straight line approximation of

$$F_{cr} = F_y - B(KL/r). \tag{6.32}$$

The value of the constant, B (the slope of the line), can be established through the development of compression member curves using variable and constant safety factors with Equation 6.30. In this manner, using an appropriate safety factor, AREMA (2008) recommends an allowable compression stress (units are lb and in.) of

$$F_{all} = \frac{F_{cr}}{FS} = 0.60F_y - \left(17{,}500\frac{F_y}{E}\right)^{3/2}\left(\frac{KL}{r}\right). \tag{6.33}$$

This curve is made to intersect the Euler elastic curve at $0.20F_y$ (Figure 6.9b) in order to conservatively represent the effects of eccentricities, initial imperfections, and residual stresses introduced during the fabrication and erection of steel railway bridge compression members. Inserting $F_{cr} = 0.20F_y$ into Equation 6.33 yields

$$\frac{KL}{r} = 5.034\sqrt{\frac{E}{F_y}}. \tag{6.34}$$

Equation 6.34 is the limiting slenderness ratio or the critical buckling coefficient, C_c, to preclude elastic buckling of compression members with the AREMA (2008) straight line approximation. The values of C_c for various steel yield strengths are given in Table 6.5.

The allowable compressive force, C_{all}, is

$$C_{all} = F_{all}A_g. \tag{6.35}$$

The CRC curve (Equation 6.30) with the variable factor of safety used by AISC ASD (Equation 6.27) and the AREMA (2008) curve (Equation 6.33) are shown in Figure 6.10 for steel with $F_y = 50$ ksi.

TABLE 6.5
Critical Buckling Coefficients

Steel Yield Stress (F_y) (ksi)	Critical Buckling Coefficient (C_c)
36	143
44	129
50	121

6.3.1.3 Yielding of Compression Members

When $F_{all} = 0.55F_y$, the slenderness ratio, KL/r, from Equation 6.33 is

$$\frac{KL}{r} \leq 0.629\sqrt{\frac{E}{F_y}}. \tag{6.36}$$

Compression members are very short and without potential for instability for slenderness below that given by Equation 6.36. The value of Equation 6.36 for various steel yield strengths is given in Table 6.6. Therefore, the allowable compressive force, C_{all}, based on yielding is

$$C_{all} = 0.55F_yA_g. \tag{6.37}$$

6.3.1.4 Compression Member Design in Steel Railway Superstructures

Equations 6.35 (used with either 6.21 or 6.33) and 6.37 encompass the AREMA (2008) recommendations for allowable compressive stress considering elastic stability, inelastic stability, and compressive yielding, respectively. Equation 6.34 provides

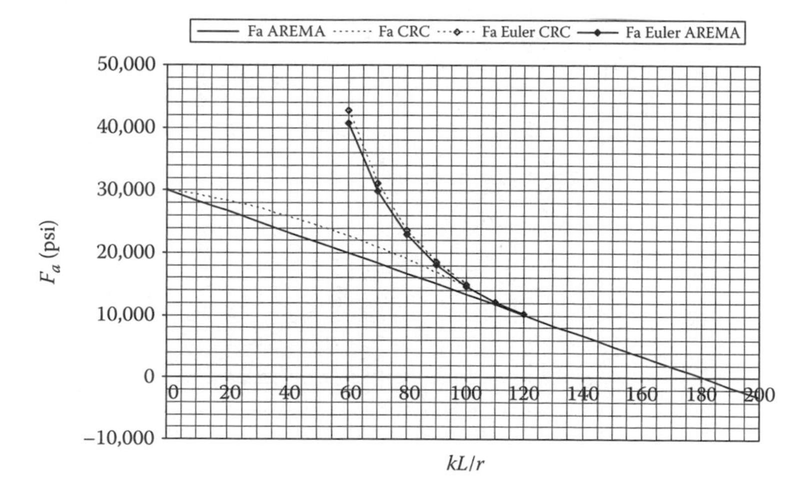

FIGURE 6.10 CRC parabolic and AREMA linear compression member curves.

TABLE 6.6
Short Compression Member Buckling Coefficient

Steel Yield Stress (F_y) (ksi)	KL/r From Equation (6.36)
36	18
44	16
50	15

the value of KL/r that delineates elastic and inelastic stability and Equation 6.36 provides the value of KL/r that delineates inelastic and yielding behavior. The AREMA (2008) strength criterion for axial compression members with steel yield strength of 50 ksi is shown in Figure 6.11 with some other compression member design criteria.

6.3.2 Serviceability of Axial Compression Members

Limiting the compression member slenderness ratio based on effective length, KL, to that of Equation 6.34 precludes the possibility of elastic buckling in order to avoid sudden stability failures in steel railway superstructures. However, slenderness ratio, L/r, must also be limited to values that will preclude excessive vibration or deflection, which is of particular concern for compression member stability (e.g., to avoid excessive secondary flexural compressive stress due to member curvature).

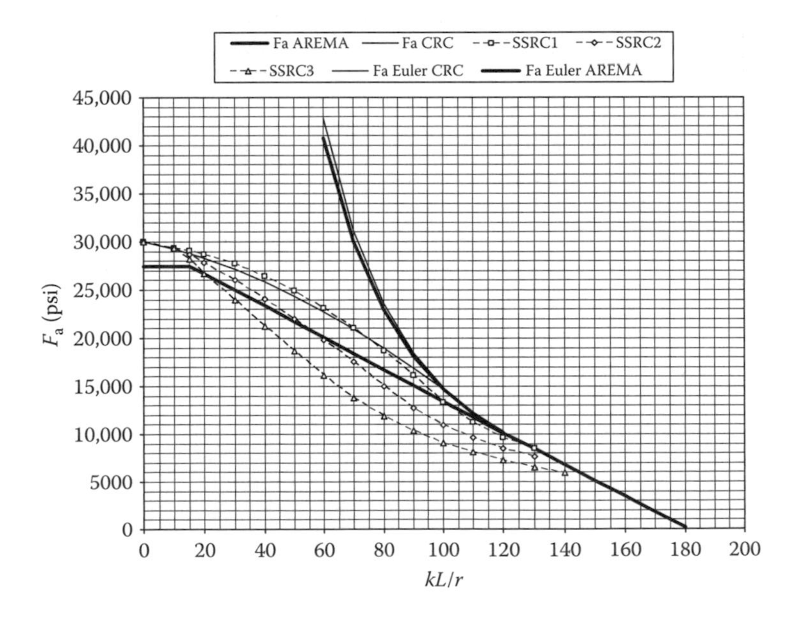

FIGURE 6.11 Compression member strength curves (F_y = 50 ksi).

AREMA (2008) recommends that

$$\frac{L}{r_{min}} \le 100 \text{ for main compression members,} \tag{6.38}$$

$$\frac{L}{r_{min}} \le 120 \text{ for wind and sway bracing compression members.} \tag{6.39}$$

Example 6.6

Determine the design criteria for load case A1 (Table 4.5) for member U3–U4 in Figure E6.1 for Cooper's E80 load. Use Grade 50 ($F_y = 50$ ksi) steel with ultimate stress, F_u, of 70 ksi (see Chapter 2).

Maximum forces in member U3–U4 (see Chapter 5) due to

Dead load force = DL = −98.10 kips

Maximum live load force with load moving across truss = $LL_{max} = 0.00$ kips

Minimum live load force with load moving across truss = $LL_{min} = -557.2$ kips

Maximum live load impact = 20.75% ($L = 156.38'$ Figure 4.5)

$P_{max} = -98.10 + 1.21(-557.2) = -771.2$ kips.

Stiffness considerations:

$$\frac{L}{r_{min}} \le 100$$

$$r_{min} \ge \frac{(19.55)(12)}{100} \ge 2.35 \text{ in. (main member of truss)}$$

$$\frac{KL}{r_{min}} \le 75 \text{ (bolted end connection).}$$

Strength considerations:

$F_{cr(min)} = 0.60(50{,}000) - (30.17)^{3/2}(75) = 30{,}000 - 12{,}430 = 17{,}570$ psi at $KL/r = 75$ (Figure 6.11) and

$$A_g = \frac{771{,}200}{17{,}570} = 43.9 \text{ in.}^2$$

(stability) (for maximum allowable slenderness).

6.3.3 Axial Compression Members in Steel Railway Bridges

It is often necessary to fabricate railway bridge compression members of several structural shapes due to large magnitude railroad loads and potential for instability. Bending effects and connection geometry constraints may also dictate the use of built-up compression members. The components must be adequately fastened to ensure integral behavior of the compression member. In cases where a box-type member is undesirable, such as where the ingress of water is difficult to preclude, open compression members are used.* Built-up open compression members are often fabricated with lacing bars and stay (tie or batten) plates or perforated cover plates.

6.3.3.1 Buckling Strength of Built-up Compression Members

Only bending deformations were considered in the development of Equation 6.24 for buckling strength, P_{cr}. The effect of shear forces was neglected. For solid section compression members, this is appropriate. However, for built-up compression members, the shear force may create deformations of the open section that reduce the overall stiffness and, thereby, reduce the buckling strength. Therefore, the curvature of the compression member due to shear must be included in Equation 6.14 to determine the critical buckling load, $\overline{P}_{cr}$, for built-up compression members. The shear force is (Figure 6.12)

$$V(x) = P \sin \varphi(x) = P\frac{\mathrm{d}y(x)}{\mathrm{d}x} \tag{6.40}$$

with curvature, γ_v, given as

$$\gamma_v = \frac{\beta}{A_g G_{eff}}\frac{\mathrm{d}V(x)}{\mathrm{d}x} = \frac{\beta P}{A_g G_{eff}}\frac{\mathrm{d}^2 y(x)}{\mathrm{d}x^2}, \tag{6.41}$$

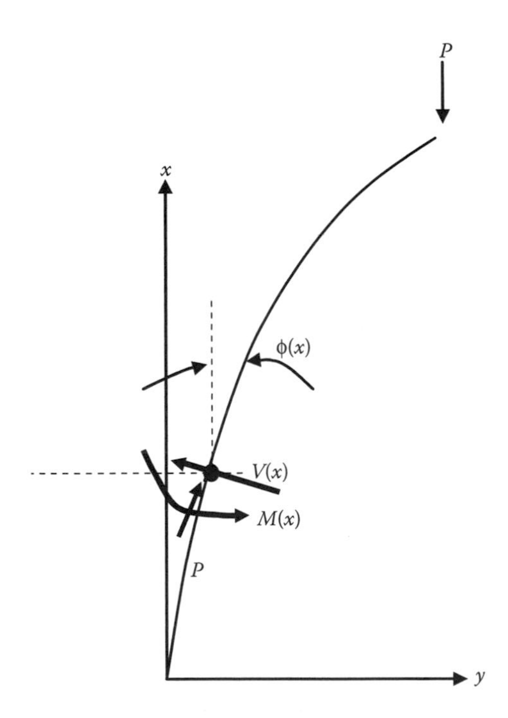

FIGURE 6.12 Bending and shear forces at compression member cross section.

* This is generally the case for railway compression members exposed to rain, ice, and snow.

where β is a numerical factor to correct for nonuniform stress distribution across the cross section of the compression member, A_g is the gross cross-sectional area of the compression member, G_{eff} is the effective shear modulus equal to $E_{eff}/(2(1+\upsilon))$, and υ is the Poisson's ratio $= 0.3$ for steel.

The inclusion of Equation 6.41 into Equation 6.14 with $E = E_{eff}$ yields the differential equation

$$\frac{d^2y(x)}{dx^2} + \frac{k^2y(x)}{(1+(\beta P/A_gG_{eff}))} = \frac{U}{(1+(\beta P/A_gG_{eff}))} \tag{6.42}$$

with solution analogous to Equation 6.15 of

$$\overline{P}_{cr} = \frac{\pi^2 E_{eff} I}{(\alpha KL)^2} = \frac{P_{cr}}{\alpha^2}, \tag{6.43}$$

where P_{cr} is the critical buckling load for compression member with gross cross-sectional area, A_g, moment of inertia, I, and length, L (see Equation 6.15), and

$$\alpha = \sqrt{\left(1 + \frac{\beta P_{cr}}{A_gG_{eff}}\right)}. \tag{6.44}$$

Equation 6.43 may be written as

$$\overline{P}_{cr} = \frac{P_{cr}}{(1+(\beta P_{cr}/A_gG_{eff}))}. \tag{6.45}$$

Equation 6.45 illustrates that the critical buckling load, $\overline{P}_{cr}$, for built-up compression members can be readily determined based on the critical buckling load, P_{cr}, for closed members of the same cross-sectional area, A_g.

The majority of steel railway superstructure compression members are slender and connected with modern fasteners and are assumed to be pin connected at each end ($K = 0.75$). Therefore, the critical buckling strength of built-up compression members of various configurations (using lacing and batten bars, and perforated cover plates) with pinned ends will be considered further.

Equation 6.44 may be written as

$$\alpha = \sqrt{\left(1 + \frac{\beta P_{cr}}{A_gG_{eff}}\right)} = \sqrt{1+\Omega P_{cr}}, \tag{6.46}$$

where

$$\Omega = \frac{\beta}{A_gG_{eff}} = \frac{2\beta(1+\nu)}{A_gE_{eff}} = \frac{1}{P_\Omega}. \tag{6.47}$$

The value of Ω is determined through investigation of the deformations of the lacing bars, batten plates, and/or perforated cover plates caused by lateral displacements from shear force, V. The results of such investigations for various built-up compression members are presented in the next sections (see e.g., Timoshenko and Gere, 1961).

6.3.3.1.1 *Critical Buckling Strength of Laced Bar Built-up Compression Members without Batten Plates (Pinned at Each End) (Figure 6.13a and b)*

Shear is resisted by pin-connected truss behavior of lacing bars (for double lacing, consider tension resistance only).

$$\Omega = \frac{1}{A_{\mathrm{lb}} E_{\mathrm{eff}} \sin\phi \cos^2\phi}, \tag{6.48a}$$

$$\alpha = \sqrt{1 + \Omega P_{\mathrm{cr}}} = \sqrt{1 + \left(\frac{13.2}{(L/r)^2}\right)\left(\frac{A_{\mathrm{g}}}{A_{\mathrm{plb}}}\right)\frac{1}{\sin\phi\cos^2\phi}}, \tag{6.48b}$$

where A_{plb} is the cross-sectional area of diagonal lacing bars in each panel of the member.

For single lacing, $A_{\mathrm{plb}} = A_{\mathrm{lb}} = t_{\mathrm{lb}} w_{\mathrm{lb}}$.

For double lacing, $A_{\mathrm{plb}} = 2A_{\mathrm{lb}} = 2t_{\mathrm{lb}} w_{\mathrm{lb}}$.

Here t_{lb} is the thickness of the lacing bar, w_{lb} is the width of the lacing bar, r is the radius of gyration of the compression member $= \sqrt{I/A_{\mathrm{g}}}$, and Φ is the angle of the lacing bar from the line perpendicular to the member axis (should be about 30° for single lacing and 45° for double lacing).

6.3.3.1.2 *Critical Buckling Strength of Built-up Compression Members with Batten Plates Only (Pinned at Each End) (Figure 6.13c)*

Shear is resisted by flexure of the batten plates and the main member elements.

$$\Omega = \frac{ab}{12E_{\mathrm{eff}} I_{\mathrm{bb}}} + \frac{a^2}{24E_{\mathrm{eff}} I} \tag{6.49a}$$

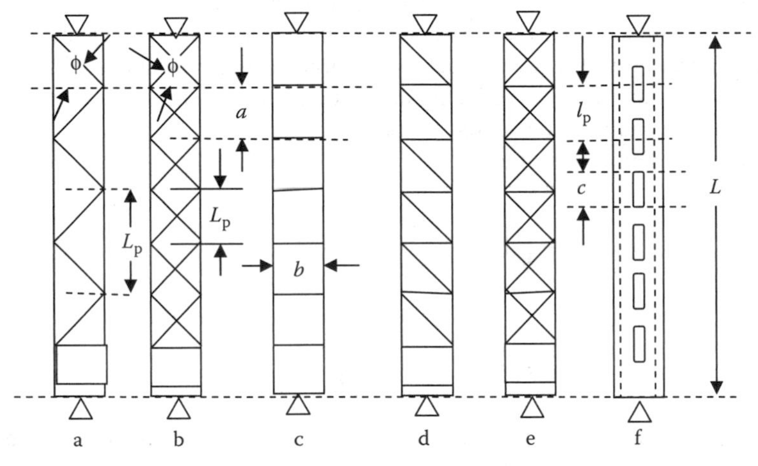

FIGURE 6.13 Various built-up compression members comprised of lacing bars with and without batten plates, and perforated cover plates (note that, for clarity, stay plates required at the ends of laced bar members are only shown at the bottom of compression members a, b, d, and e).

$$\alpha = \sqrt{1+\Omega P_{cr}} = \sqrt{1+\left(\frac{1.10}{(L/r)^2}\right)\left(\left(\frac{A_g}{A_{bb}}\right)\frac{ab}{r_{bb}^2}+\frac{a^2}{2r^2}\right)}, \tag{6.49b}$$

where a is the distance between the centroids of batten plates, b is the distance between the centroids of the main compression elements of the member (effective batten plate length), I_{bb} is the moment of inertia of the batten plate $= t_{bb}(w_{bb})^3/12$, t_{bb} is the batten plate thickness, w_{bb} is the batten plate width, A_{bb} is the batten plate cross-sectional area $= t_{bb}w_{bb}$, and $r_{bb} = \sqrt{I_{bb}/A_{bb}}$.

If the shear rigidity of the batten plates is small, reduction of the built-up compression member critical buckling force will result. Inclusion of the batten plate shearing strain into Equation 6.49a yields

$$\Omega = \frac{ab}{12E_{eff}I_{bb}} + \frac{a^2}{24E_{eff}I} + \frac{\beta a}{A_{bb}G_{eff}b} \tag{6.49c}$$

$$\alpha = \sqrt{1+\Omega P_{cr}} = \sqrt{1+\left(\frac{1.10}{(L/r)^2}\right)\left(\left(\frac{A_g}{A_{bb}}\right)\left(\frac{ab}{r_{bb}^2}+\frac{31.2a}{b}\right)+\frac{a^2}{2r^2}\right)}. \tag{6.49d}$$

6.3.3.1.3 *Critical Buckling Strength of Laced Bar Built-up Compression Members with Batten Plates (Pinned at Each End)** *(Figure 6.13d and e)*

Shear is resisted by pin-connected truss behavior of lacing bars (for double lacing, consider tension resistance only).

$$\Omega = \frac{1}{A_{lb}E_{eff}\sin\phi\cos^2\phi} + \frac{b}{A_{bb}E_{eff}a}, \tag{6.50a}$$

$$\alpha = \sqrt{1+\Omega P_{cr}}$$

$$= \sqrt{1+\left(\frac{13.2}{(L/r)^2}\right)\left(\left(\frac{A}{A_{lb}}\right)\left(\frac{1}{\sin\phi\cos^2\phi}\right)+\left(\frac{A_g}{A_{bb}}\right)\left(\frac{1}{\tan\phi}\right)\right)}. \tag{6.50b}$$

6.3.3.1.4 *Critical Buckling Strength of Built-up Compression Members with Perforated Cover Plates (Pinned at Each End) (Figure 6.13f)*

Most built-up compression members in modern steel railway superstructures are comprised of main elements connected by perforated cover plates. Shear is resisted by flexure of the main member elements because the perforated cover plates act as rigid

---

* This equation was developed and then later used by Engesser in connection with investigations into the collapse of the Quebec Bridge (see Chapter 1).

batten plates between the perforations.

$$\Omega = \frac{9c^3}{32 l_p E_{eff} I}, \tag{6.51a}$$

$$\alpha = \sqrt{1 + \Omega P_{cr}} = \sqrt{1 + \left(\frac{3.71}{(L/r)^2}\right)\left(\frac{c^3}{l_p r^2}\right)}, \tag{6.51b}$$

where l_p is the distance between the centers of perforations and c is the length of the perforation. The perforation length, c, can be expressed in terms of the distance between the centers of perforations, l_p, so that

$$\alpha = \sqrt{1 + \left(\frac{3.71\gamma^3}{(L/r)^2}\right)\left(\frac{l_p}{r}\right)^2}, \tag{6.51c}$$

where $\gamma = c/l_p$.

6.3.3.1.5 Design of Built-up Compression Members

The allowable compressive stress is

$$F_{all} = \frac{\overline{P_{cr}}}{FS}. \tag{6.52}$$

The overall critical buckling strength of the built-up compression member, $\overline{P_{cr}}$, (Equation 6.43) is contingent upon the main element web plates, cover plates, lacing bars, and/or batten plates being of adequate strength and stability as individual components (local buckling).

In order to ensure that the webs of main elements of built-up compression members do not buckle prior to $\overline{P_{cr}}$, a minimum web plate thickness, t_w, is recommended by AREMA (2008) as

$$t_w \geq \frac{0.90b\sqrt{F_y/E}}{\sqrt{F_{all}/f}} \text{ (in.)}, \tag{6.53}$$

where $F_{all}/f \leq 4, f$ is the calculated compressive stress in the member.

Also, in order to ensure that cover plates for main elements of built-up compression members do not buckle prior to $\overline{P_{cr}}$, a minimum cover plate thickness, t_{cp}, is recommended by AREMA (2008) as

$$t_{cp} \geq \frac{0.72b\sqrt{F_y/E}}{\sqrt{F_{all}/f}} \text{ (in.)}. \tag{6.54}$$

The thickness of lacing bars, stay plates, batten plates, and perforated cover plates must also be considered to complete the design of built-up compression members.

6.3.3.1.6 *Design of Lacing Bars, Stay Plates, Batten Plates, and Perforated Plates*

AREMA (2008) recommends that lacing bars, batten plates, and perforated plates be designed for a total shear force normal to the member in the plane of the lacing bars, batten plates, or cover plates comprised of self-weight, wind, and 2.5% of the axial compressive force in the member. The shear forces related to self-weight and wind are generally small and may be neglected in many cases. The shear force normal to the member in the plane of the lacing bars, batten plates, or cover plates related to the axial compressive force can be estimated by considering the conditions of Figure 6.5c. The solution of the differential equation of the deflection curve (Equation 6.14), where $U = Pe$, is (Bowles, 1980)

$$y(x) = e\left(\tan\frac{k^2L}{2}\sin k^2y(x) + \cos k^2y(x) - 1\right). \tag{6.55}$$

Differentiating Equation 6.55 yields

$$\frac{\mathrm{d}y(x)}{\mathrm{d}x} = k^2e\tan\frac{k^2L}{2} \tag{6.56}$$

and substitution of Equation 6.56 into Equation 6.40 yields

$$V = Pk^2e\tan\frac{k^2L}{2}. \tag{6.57}$$

AREMA (2008) recommends

$$k^2e\tan\frac{k^2L}{2} = 0.025 \tag{6.58}$$

so that

$$V = 0.025P, \tag{6.59}$$

where

$$V \geq \frac{A_\mathrm{r}F_\mathrm{y}}{150} \geq \frac{PF_\mathrm{y}}{150F_\mathrm{all}} \tag{6.60}$$

and $A_\mathrm{r} = P/F_\mathrm{all}$.

Equation 6.60 was developed considering force eccentricity, initial curvature, and flexure of the compression member (Hardesty, 1935). Figure 6.14 illustrates that the AREMA (2008) recommendation for minimum shear force (Equation 6.60) ensures that shear forces with relatively greater proportion to the axial compressive force are used for the design of weaker compression members (more slender members that approach the Euler buckling behavior) when $F_\mathrm{all}/F_\mathrm{y} < 0.26$.

The shear force, V, forms the basis of the lacing bar, batten plate, and perforated cover plate design for built-up compression members.

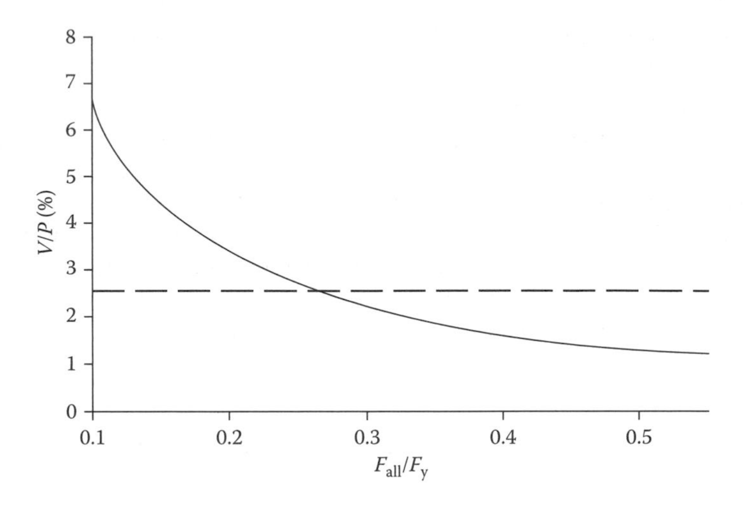

FIGURE 6.14 AREMA minimum shear force for built-up compression member design.

6.3.3.1.6.1 Lacing Bars for Compression Members The spacing of lacing bars along the main member must be designed to preclude buckling of portions of the main member elements between the lacing bar connections. AREMA (2008) limits the slenderness ratio, L_p/r_p, of elements between lacing bar connections to 2/3 of the member slenderness ratio, L/r. This is appropriate in order to consider not only the local buckling effects over the length, L_p, but the interaction between global and local buckling (or compound buckling) (Duan et al., 2002). AREMA (2008) recommends that the lacing bar spacing be such that

$$\frac{L_p}{r_p} \leq 40 \leq \frac{2L}{3r}, \tag{6.61}$$

where L_p is the length of main member element between lacing bar connections (see Figure 6.13a and b) and is equal to $2a$ (for single lacing), and a (for double lacing), and r_p is the minimum radius of gyration of the main member element.

The lacing bars on each side of the main member must be designed to resist the shear force $V/2$ in the plane of the lacing bars. Therefore, the force in each lacing bar is

$$P_{lb} = \frac{V}{2\cos\phi}. \tag{6.62}$$

The critical buckling stress and minimum cross-sectional area of the bar, A_{lb}, can be determined, using Equation 6.33 with $K = 1.0$, as

$$F_{all} = 0.60F_y - \left(17{,}500\frac{F_y}{E}\right)^{3/2}\left(\frac{L_{lb}}{r_{lb}}\right), \tag{6.63}$$

where $L_{lb} = \sqrt{a^2 + b^2}$ for single lacing systems (Figure 6.13a and d) (length of the lacing bar between connections at the main member), $L_{lb} = 0.70\sqrt{a^2 + b^2}$ for double

TABLE 6.7
Minimum Lacing Bar Thickness

Member	t_{lbmin} Single Lacing	t_{lbmin} Double Lacing
Main	$L_{\text{lb}}/40$	$L_{\text{lb}}/60$
Bracing	$L_{\text{lb}}/50$	$L_{\text{lb}}/75$

lacing systems (Figure 6.13b and e) (equivalent length of the lacing bar between connections at the main member), and $r_{\text{lb}} = \sqrt{I_{\text{lb}}/A_{\text{lb}}} = 0.29t_{\text{lb}}$ for flat lacing bars. Double lacing systems are used to reduce lacing bar slenderness and thickness. In particular, AREMA (2008) recommends that double lacing connected at the center be used when $b > 15$ in. and the lacing bar width is less than 3.5 in.

The minimum thickness of lacing bars based on the slenderness recommendations of AREMA (2008) is indicated in Table 6.7. In order to ensure adequate edge distance for connections and accommodate the AREMA (2008) recommendation that the lacing bar connection bolt diameter does not exceed 1/3 of the lacing bar width, $w_{\text{lb}}/3$, the minimum lacing bar width should be $2\frac{5}{8}$ in. for 7/8 in. diameter bolts.

6.3.3.1.6.2 Stay Plates for Built-up Compression Members Stay plates must be used at the ends of laced bar built-up compression members and at locations where the lacing bars are interrupted (e.g., at a connection with another member).

The length of the stay plates at the ends of laced bar built-up compression members must be at least 25% greater than the distance between connection lines across the member (distance b in Figure 6.13). The length of the stay plates at intermediate locations of laced bar built-up compression members must be at least 75% of the distance between connection lines across the member. The minimum thickness of the stay plate, t_{sp}, as recommended by AREMA (2008) is shown in Table 6.8.

The center-to-center spacing of bolts must not exceed four bolt diameters and not less than three bolts should be used to connect stay plates to main elements of the built-up compression member. Welded stay plates shall utilize a minimum 5/16 in. continuous fillet weld along the stay plate longitudinal edges.

Example 6.7 outlines the design of a laced built-up compression member.

TABLE 6.8
Minimum Stay Plate Thickness

Member	Minimum Stay Plate Thickness (t_{sp})
Main	$b/50$
Bracing	$b/60$

6.3.3.1.6.3 Batten Plates for Compression Members Built-up compression members using only batten plates (Figure 6.13c) are not generally used* in steel railway superstructure compression elements due to strength and stability requirements. When batten plates are used in conjunction with laced bar compression members (Figure 6.13d and e), they may be designed for minimum slenderness considerations since the lacing bars are assumed to resist the applied shear forces.

However, if used without lacing bars, the spacing of batten plates, a, along the main member must also be designed to preclude the buckling of portions of main member elements between the batten plate connections. Based on the AREMA (2008) recommendation for lacing bar spacing (Equation 6.62), it appears reasonable that the batten plate spacing be such that

$$\frac{a_{\mathrm{bp}}}{r_{\mathrm{p}}} \leq 40 \leq \frac{2L}{3r}, \tag{6.64}$$

where $a_{\mathrm{bp}} = a - w_{\mathrm{bp}} + 2e_{\mathrm{bp}}$, where e_{bp} is the edge distance to first fastener in the batten plate and w_{bp} is the width of the batten plate.

Furthermore, when batten plates are used without lacing bars, the batten plates and main member elements† on each side of the main member must be designed to resist bending created by the shear force $V/2$ in each panel between batten plates.

The force creating bending in each batten plate, P_{bp}, is

$$P_{\mathrm{bp}} = \frac{1}{2}\frac{V}{2}\left(\frac{a}{b}\right) = \frac{Va}{4b}. \tag{6.65}$$

Therefore, the bending moment, M_{bp}, in each batten plate is

$$M_{\mathrm{bp}} = P_{\mathrm{bp}}\left(\frac{b}{2}\right) = \frac{Va}{8} \tag{6.66}$$

such that, based on an allowable stress of $0.55F_{\mathrm{y}}$, the minimum thickness of batten plate, t_{bp}, is

$$t_{\mathrm{bp}} \geq \frac{1.36Va}{F_{\mathrm{y}}(w_{\mathrm{bp}})^2} \geq \frac{Pa}{29.3F_{\mathrm{y}}(w_{\mathrm{bp}})^2}, \tag{6.67}$$

where P is the compressive force $= fA_{\mathrm{g}}$, and f is the calculated compressive stress in the member.

The bending moment in the main member element, M_{p}, is

$$M_{\mathrm{p}} = \frac{1}{2}\frac{V}{2}\left(\frac{a}{2}\right) = \frac{Va}{8}. \tag{6.68}$$

* For this reason AREMA (2008) does not contain any specific recommendations relating to the use of batten plates.

† This criterion for main member elements will usually govern the design of built-up compression members using only batten plates.

The bending stress in each main member element from shear forces applied on each side of the member is then calculated to ensure that it does not exceed $0.55F_y$.*

Also, when batten plates are used without lacing bars, the batten plate spacing, a, is critical in regard to the overall stability of the built-up compression member, as indicated by Equations 6.49b and 6.49d.

The minimum thickness recommended in Table 6.8 for stay plates for main and bracing members also appears appropriate for batten plates. Also, if used, the minimum width and connection geometry of batten plates should consider the criteria outlined for stay plates.

Example 6.7 outlines the design of a batten plate built-up compression member.

6.3.3.1.6.4 Perforated Cover Plates for Compression Members In order to avoid the fabrication cost of laced bar built-up compression members and the bending inefficiencies of using batten plates, perforated cover plates are often used for built-up compression members in modern railway steel superstructures.

AREMA (2008) recommends, to ensure stability of the main member elements at perforations, that the effective slenderness ratio, C_{pc}, about the member axis (Figure 6.15) not exceed 20 or 33% of the member slenderness ratio about an axis perpendicular to the plane of the perforation. The effective slenderness ratio is

$$C_{pc} = \frac{c}{r_{pc}} \leq 20 \leq \frac{L}{3r}, \tag{6.69}$$

where c is the length of the perforation, and

$$r_{pc} = \sqrt{\frac{I_{pc}}{A_{pc}}},$$

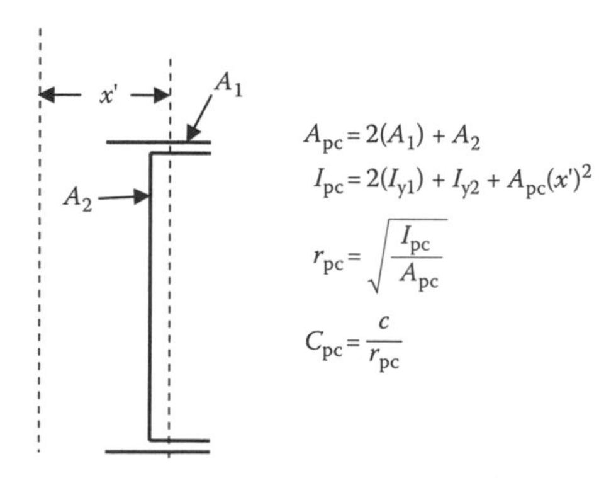

FIGURE 6.15 Element of compression member at a perforation.

* The main member elements between batten plates can be considered as flexural compression members with $L/r = 0$ (Tall, 1974).

where I_{pc} is the moment of inertia of half the member (one "flange") about the member axis at the center of the perforation, and A_{pc} is the area of half the member (one "flange").

AREMA also presents other recommendations related to perforated cover plates for built-up compression member design as

$$c \leq 2w_{perf}, \tag{6.70}$$

$$c \leq l_p - b', \tag{6.71}$$

where w_{perf} is the width of the perforation, l_p is the distance between the centers of perforations, and b' is the width of the perforated plate between the inside lines of fasteners.

The thickness of perforated cover plates should be governed by the largest of the following expressions relating to local plate stability criteria:

$$t_{pc} \geq b'/50' \tag{6.72}$$

$$t_{pc} \geq 1.17(b' - w_{perf})\sqrt{F_y/E}, \tag{6.73}$$

$$t_{pc} \geq \frac{0.90b'\sqrt{F_y/E}}{\sqrt{F_{all}/f}} \text{(in.)}, \tag{6.74}$$

where $F_{all}/f \leq 4$.

Perforated cover plate thickness is also based on transverse shear, V, at the centerline of the cover plate. The minimum perforated cover plate thickness (Equation 6.13) is

$$t_{pc} \geq \frac{3Vl_p}{2\tau_{all}b(l_p - c)}, \tag{6.75}$$

where τ_{all} is the allowable shear stress ($0.35F_y$ recommended by AREMA, 2008). The gross section through the perforation (with only the perforated area removed) of the plate is included in the member gross area for compression design.

Example 6.7 outlines the design of a perforated cover plate built-up compression member.

Example 6.7

Design a 19.55 ft long compression member to resist a 550 kip load as a solid, built-up laced bar, built-up batten plate, and built-up perforated cover plate member.

Design of the solid section
Compression member: W14 × 90 rolled section

$$A = 26.5 \text{ in.}^2$$

$$I_x = 999 \text{ in.}^4$$

$$S_x = 143\text{ in.}^3$$

$$r_x = 6.14\text{ in.}$$

$$I_y = 362\text{ in.}^4$$

$$S_y = 49.9\text{ in.}^3$$

$$r_y = 3.70\text{ in.}$$

$$\frac{L}{r_{\min}} = \frac{(19.55)(12)}{3.70} = 63.4 \leq 100\text{ Ok}$$

$$\frac{KL}{r_{\min}} = \frac{0.75(19.55)(12)}{3.70} = 47.6 \leq 5.034\sqrt{\frac{E}{F_y}} \leq 121$$

$$\frac{KL}{r_{\min}} = 47.6 \geq 0.629\sqrt{\frac{E}{F_y}} \geq 15$$

$$F_{\text{all}} = 0.60F_y - \left(17{,}500\frac{F_y}{E}\right)^{3/2}\left(\frac{KL}{r_{\min}}\right) = 30 - 0.166\left(\frac{KL}{r_{\min}}\right)$$

$$= 30 - 7.9 = 22.1\text{ ksi}$$

$$P_{\text{all}} = F_{\text{all}}(A) = 22.1(26.5) = 586 \geq 550\text{ ksi OK.}$$

Design of the laced section

Compression member: $2 - C15 \times 50$ channels laced $6\frac{1}{2}$ in. back to back with 4 in. $\times$ (1/2) in. lacing bars as shown in Figure E6.5.

For $C15 \times 50$ channel:

$$A = 14.7\text{ in.}^2$$

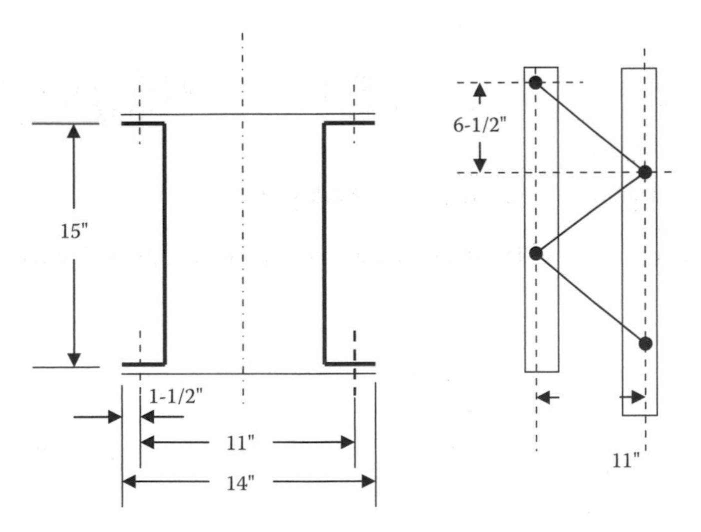

FIGURE E6.5

$$I_x = 404\ \text{in.}^4$$
$$S_x = 53.8\ \text{in.}^3$$
$$r_x = 5.24\ \text{in.}$$
$$I_y = 11.0\ \text{in.}^4$$
$$S_y = 3.78\ \text{in.}^3$$
$$r_y = 0.867\ \text{in.}$$
$$\bar{x} = 0.80\ \text{in.}$$

For laced member:

$$A = 2(14.7) = 29.4\ \text{in.}^2$$
$$I_x = 2(404) = 808\ \text{in.}^4$$
$$I_y = 2(11.0) + 2(14.7)(3.25 + 0.80)^2 = 504\ \text{in.}^4$$
$$r_y = \sqrt{\frac{504}{29.4}} = 4.14\ \text{in.}$$
$$\frac{L}{r_{\min}} = \frac{(19.55)(12)}{4.14} = 56.7 \le 100\ \text{OK}$$
$$\frac{KL}{r_{\min}} = \frac{0.75(19.55)(12)}{4.14} = 42.5 \le 5.034\sqrt{\frac{E}{F_y}} \le 121$$
$$\frac{KL}{r_{\min}} = 42.5 \ge 0.629\sqrt{\frac{E}{F_y}} \ge 15.$$

Therefore,

$$F_{\text{all}} = 0.60F_y - \left(17{,}500\frac{F_y}{E}\right)^{3/2}\left(\frac{KL}{r_{\min}}\right) = 30 - 0.166\left(\frac{KL}{r_{\min}}\right) = 30 - 7.1$$
$$= 22.9\ \text{ksi}$$
$$P_{\text{all}} = F_{\text{all}}(A) = 22.9(29.4) = 675 \ge 550\ \text{kips OK.}$$

If the effects of shear deformation are included (Equations 6.43 and 6.48b),

$$\alpha = \sqrt{1 + \left(\frac{13.2}{(L/r)^2}\right)\left(\frac{A}{A_{\text{plb}}\sin\phi\cos^2\phi}\right)}$$
$$= \sqrt{1 + \left(\frac{13.2}{(56.7)^2}\right)\left(\frac{29.4}{2(4)(0.5)\sin 30.6^\circ \cos^2 30.6^\circ}\right)} = 1.04$$
$$\overline{P}_{\text{all}} = \frac{P_{\text{all}}}{\alpha^2} = \frac{675}{1.08} = 625 \ge 550\ \text{kips OK.}$$

Check design of 4 in. × (1/2) in. lacing bars at 30.6° to horizontal:

$$t_{lb} \geq \frac{\sqrt{6.5^2 + 11^2}}{40} = \frac{12.8}{40} = 0.32 \text{ in. OK}$$

$$V = 0.025P = 0.025(550) = 13.8 \geq \frac{PF_y}{150F_{all}} \geq \frac{550(50)}{150(22.9)} \geq 8.0 \text{ kips OK}$$

$$P_{lb} = \frac{V}{2\cos 30.6°} = \frac{13.8}{2\cos 30.6°} = 8.0 \text{ kips}$$

$$F_{all} = 30 - 0.166\left(\frac{(K)L_{lb}}{r_{lb}}\right) = 30 - 0.166\left(\frac{(1.0)12.8}{0.29t_{lb}}\right) = 30 - 0.166\left(\frac{(1.0)12.8}{0.29(0.5)}\right)$$

$$= 30 - 14.7 = 15.3 \text{ ksi}$$

$$P_{all} = (15.3)(4)(0.5) = 30.6 \text{ kips} > 8.0 \text{ kips OK.}$$

Check design of the main member with lacing bars at 30.6° to horizontal:

$L_p/r_p = 2(6.5)/0.867 = 15 \leq 40$ OK (length of the main member element (channel) between lacing bar connections)

$$\frac{L_p}{r_p} = 15 \leq \frac{2}{3}(56.7) \leq 37.8 \text{ OK.}$$

Design of the battened section

Compression member: 2 – C15 × 50 channels battened $6\frac{1}{2}$ in. back to back with 6 in. × (1/2) in. batten plates as shown in Figure E6.6

C15 × 50 channel and battened member section properties and allowable compression force (675 kips) are same as that for the laced section design.

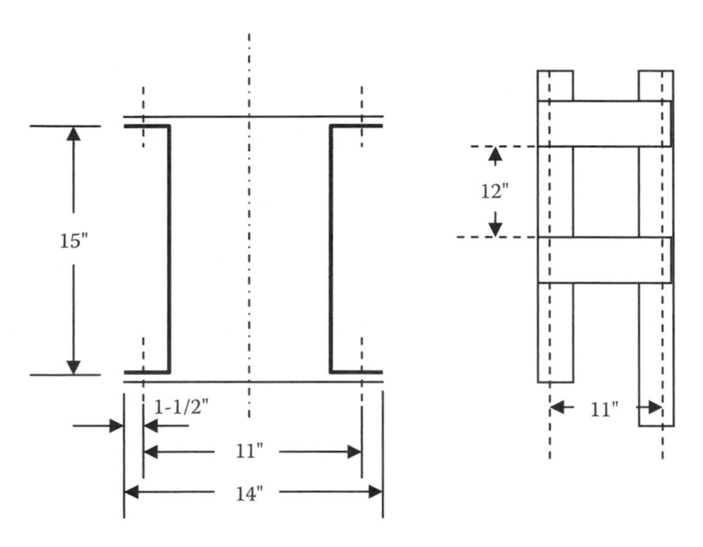

FIGURE E6.6

If the effects of shear deformation are included (Equations 6.43 and 6.49d),

$$\alpha = \sqrt{1+\left(\frac{1.10}{(L/r)^2}\right)\left(\left(\frac{A}{A_{\mathrm{bb}}}\right)\left(\frac{ab}{r_{\mathrm{bb}}^2}+\frac{31.2a}{b}\right)+\frac{a^2}{2r^2}\right)}$$

$$= \sqrt{1+\left(\frac{1.10}{(56.7)^2}\right)\left(\left(\frac{29.4}{2(6)(0.5)}\right)\left(\frac{(18)(11)}{(6^2/12)}+\frac{31.2(18)}{11}\right)+\frac{(18)^2}{2(4.14)^2}\right)} = 1.10$$

$$\overline{P}_{\mathrm{all}} = \frac{P_{\mathrm{all}}}{\alpha^2} = \frac{675}{1.20} = 562 \geq 550 \text{ kips OK.}$$

Check design of 6 in. × (1/2) in. batten plates at 18 in. center-to-center spacing:

$$t_{\mathrm{bp}} \geq \frac{b}{40} = \frac{11}{50} = 0.22 \text{ in. OK}$$

$$t_{\mathrm{bp}} \geq \frac{Pa}{29.3F_{\mathrm{y}}(w_{\mathrm{bp}})^2} \geq \frac{550(18)}{29.3(50)(6)^2} = 0.20 \text{ in. OK.}$$

Check design of the main member with 6 in batten plates at 18 in. center-to-center spacing: $L_{\mathrm{p}}/r_{\mathrm{p}} = (12+3)/0.867 = 17.3 \leq 40$ OK (length of the main member element (channel) between closest connections of adjacent batten plates)

$$\frac{L_{\mathrm{p}}}{r_{\mathrm{p}}} = \frac{(12+3)}{0.867} = 17.3 \leq \frac{2}{3}(56.7) \leq 37.8 \text{ OK}$$

$$M_{\mathrm{p}} = \frac{Va}{8} = \frac{13.8(18)}{8} = 31.1 \text{ in.-kips}$$

$$f_{\mathrm{p}} = \frac{31.1}{3.78} = 8.2 \text{ ksi}$$

$$F_{\mathrm{all}} = 30 - 0.166\left(\frac{KL}{r_{\mathrm{min}}}\right) = 30 - 0.166\left(\frac{(1.0)18}{0.867}\right)$$

$$= 30 - 3.5 = 26.5 \geq 8.2 \text{ ksi OK.}$$

Design of the perforated cover plated section (4 in. × 8 in. perforations at 18 in. center to center)

Compression member: 2 − C12 × 30 channels 6 in. back to back connected with 1/2 in. thick perforated cover plates as shown in Figure E6.7.

For C12 × 3 channel:

$$A = 8.82 \text{ in.}^2$$

$$I_x = 162 \text{ in.}^4$$

$$S_x = 27.0 \text{ in.}^3$$

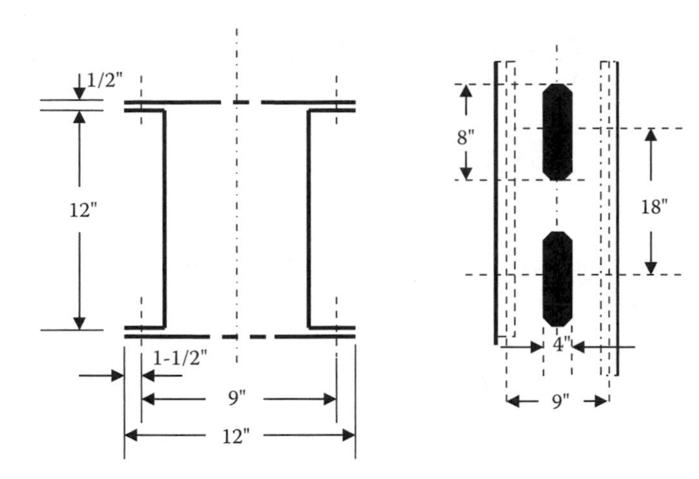

FIGURE E6.7

$$r_x = 4.29\text{ in.}$$

$$I_y = 5.14\text{ in.}^4$$

$$S_y = 2.06\text{ in.}^3$$

$$r_y = 0.763\text{ in.}$$

$$\bar{x} = 0.67\text{ in.}$$

For cover plated member:

$$A = 2(8.82) + 2(12)(0.5) - 2(4)(1/2) = 25.6\text{ in.}^2$$

$$I_x = 2(162) + 2(8)(0.5)(6.25)^2 + 2(8)(0.5)^3/12 = 637\text{ in.}^4$$

$$r_x = \sqrt{\frac{637}{25.6}} = 5.00\text{ in.}$$

$$I_y = 2(5.14) + 2(8.82)(3 + 0.67)^2 + 2(2)(4)^3(0.5)/12 + 2(2)(4)(0.5)(2 + 2)^2$$

$$= 387\text{ in.}^4$$

$$r_y = \sqrt{\frac{387}{25.6}} = 3.89\text{ in.}$$

$$\frac{L}{r_{\min}} = \frac{(19.55)(12)}{3.89} = 60.4 \le 100\text{ OK}$$

$$\frac{KL}{r_{\min}} = \frac{0.75(19.55)(12)}{3.89} = 45.3 \le 5.034\sqrt{\frac{E}{F_y}} \le 121$$

$$\frac{KL}{r_{\min}} = 45.3 \ge 0.629\sqrt{\frac{E}{F_y}} \ge 15$$

$$F_{all} = 0.60F_y - \left(17{,}500\frac{F_y}{E}\right)^{3/2}\left(\frac{KL}{r_{min}}\right) = 30 - 0.166\left(\frac{KL}{r_{min}}\right) = 30 - 7.5$$
$$= 22.5 \text{ ksi}$$
$$P_{all} = F_{all}(A) = 22.5(25.6) = 576 \geq 550 \text{ kips OK.}$$

If the effects of shear deformation are included (Equations 6.43 and 6.51b),

$$\alpha = \sqrt{1 + \left(\frac{3.71}{(L/r)^2}\right)\left(\frac{c^3}{I_p r^2}\right)} = \sqrt{1 + \left(\frac{3.71}{(60.4)^2}\right)\left(\frac{8^3}{18(3.89)^2}\right)} = 1.00$$
$$\overline{P}_{all} = \frac{P_{all}}{\alpha^2} = \frac{576}{1.00} = 576 \geq 550 \text{ kips OK.}$$

Check design of (1/2) in. cover plates with 4 in. × 8 in. perforations at 18 in. center-to-center spacing as shown in Figure E6.7.

The properties of half the member at the center of the perforation about its own axis are (see Figure E6.8):

$$z' = \frac{2(4)(0.5)(4) + 8.82(3.67)}{2(4)(0.5) + 8.82} = 3.77 \text{ in.}$$
$$A_{pc} = 2(4)(0.5) + 8.82 = 12.82 \text{ in.}^2$$
$$I_{pc} = 5.14 + 8.82(3.77 - 3.67)^2 + 2(0.5)(4)^3/12 + 2(4)(0.5)(4 - 3.77)^2$$
$$= 10.77 \text{ in.}^4$$
$$r_{pc} = \sqrt{\frac{10.77}{12.82}} = 0.92 \text{ in.}$$
$$C_{pc} = \frac{8}{0.92} = 8.7 \leq 20 \text{ OK}$$
$$C_{pc} = 8.7 \leq \frac{L}{3r} \leq \frac{1}{3}(60.4) \leq 20.1 \text{ OK}$$
$$c = 8 \leq 2w_{perf} \leq 2(4) \leq 8 \text{ in. OK}$$
$$c = 8 \leq l_p - b' \leq 18 - 9 \leq 9 \text{ in. OK}$$ (b' is the width between the inside lines of fasteners)
$$t_{pc} \geq \frac{b}{50} \geq \frac{9}{50} \geq 0.18 \text{ in. OK}$$
$$t_{pc} \geq 1.17(b - w_{perf})\sqrt{\frac{F_y}{E}} \geq 1.17(9 - 4)\sqrt{\frac{F_y}{E}} \geq 0.24 \text{ in. OK}$$

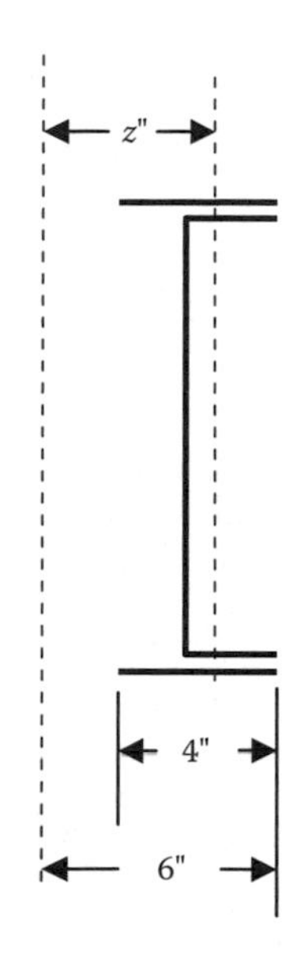

FIGURE E6.8

$$t_{pc} \geq \frac{0.90b\sqrt{F_y/E}}{\sqrt{F_{all}/f}} \geq \frac{0.90(9)\sqrt{F_y/E}}{\sqrt{22.0/(600/29.6)}} \geq 0.32\,\text{in. OK}$$

$$\sqrt{\frac{F_{all}}{f}} = 1.04 \leq 2\ \text{OK}$$

$$t_{pc} \geq \frac{3Vl_p}{2\tau_{all}b(l_p - c)} \geq \frac{3(15)(18)}{2(0.35)(50)(9)(18-8)} = 0.26\,\text{in. OK.}$$

A summary of the compression members is shown in Table E6.1.

The benefits of the perforated plate contribution to the compression member area and moment of inertia, and the relative weakness of battened only compression members is apparent in Table E6.1.

TABLE E6.1

Compression Member	Allowable Force (kips)	Gross Area of Member (Through Perforation for Perforated Plates) (in.^{2})
Solid	586	26.5 (wide flange only)
Laced	625	29.4 (channels only)
Battened	562	29.4 (channels only)
Perforated plated	576	25.6 (channels 17.64)

REFERENCES

American Institute of Steel Construction (AISC), 1980, *Manual of Steel Construction*, 8th Edition, AISC, Chicago, IL.

American Railway Engineering and Maintenance-of-Way Association (AREMA), 2008, Steel structures, in *Manual for Railway Engineering*, Chapter 15, Lanham, MD.

Anderson, T.L., 2005, *Fracture Mechanics*, Taylor & Francis, Boca Raton, FL.

Armenakas, A.E., 2006, *Advanced Mechanics of Materials and Applied Elasticity*, Taylor & Francis, Boca Raton, FL.

Bowles, J.E., 1980, *Structural Steel Design*, McGraw-Hill, New York.

Chen, W.F. and Lui, E.M., 1987, *Structural Stability*, Elsevier, New York.

Cochrane, V.H., 1922, *Rules for Rivet Hole Deductions in Tension Members*, November, Engineering News Record, New York.

Duan, L., Reno, M., and Uang, C.-M., 2002, Effect of compound buckling on compression strength of built-up members, *AISC Journal of Engineering*, First quarter, Chicago, IL.

Hardesty, S., 1935, Live loads and unit stresses, *AREA Proceedings*, 36, 770–773.

Munse, W.H. and Chesson, E., 1963, Riveted and bolted joints: Net section design, *ASCE Journal of the Structural Division*, 83(ST1), 107–126.

Newmark, N.M., 1949, A simple approximate formula for effective end-fixity of columns, *Journal of Aeronautical Sciences*, 16(2), 116.

Pilkey, W.D., 1997, *Peterson's Stress Concentration Factors*, Wiley, New York.

Salmon, C.G. and Johnson, J.E., 1980, *Steel Structures Design and Behavior*, Harper & Row, New York.

Sweeney, R.A.P., 2006, What's important in railroad bridge fatigue life evaluation, *Bridge Structures*, 2(4), Taylor & Francis, Oxford, UK.

Tall, L. (Ed), 1974, *Structural Steel Design*, 2nd Edition, Wiley, New York.

Timoshenko, S.P. and Gere, J.M., 1961, *Theory of Elastic Stability*, McGraw-Hill, New York.

Wang, C.M., Wang, C.Y., and Reddy, J.N., 2005, *Exact Solutions for Buckling of Structural Members*, CRC Press, Boca Raton, FL.

7 Design of Flexural Steel Members

7.1 INTRODUCTION

Members designed to carry primarily bending forces or flexure are typically found in steel railway bridges as girders, beams, floorbeams, and stringers. These beam and girder members experience normal tensile, normal compressive and shear stresses and are designed considering strength and serviceability criteria.

Flexural members must be designed in accordance with the AREMA (2008) allowable stress method to resist normal tensile stresses based on yield strength at the net section. Flexural members subjected to cyclical or fluctuating normal tensile stress ranges must be designed with due consideration of stress concentration effects and metal fatigue. The members must also be designed for strength to resist shear and compressive normal stresses with due consideration of stability.

Flexural member serviceability design for stiffness is achieved by respecting live load deflection limits for simple spans used in freight rail operations. Lateral rigidity to resist vibration and other effects (lateral loads from track misalignments, rail wear, track fastener deterioration) is generally provided by designing for the lateral forces recommended by AREMA (2008).

Steel beams and girders for railway spans can be of noncomposite material (steel, timber, or independent concrete deck) or composite materials (integral concrete deck) design. Noncomposite beams and girders are used for the design of both open and ballasted deck spans. Composite design provides a ballasted deck span. The relative merits of each system are discussed in Chapter 3.

7.2 STRENGTH DESIGN OF NONCOMPOSITE FLEXURAL MEMBERS

Steel girders, beams, floorbeams, and stringers are designed as noncomposite flexural members. They must be designed for internal normal flexural and shear stresses created by combinations of external actions (see Table 4.5).

7.2.1 Bending of Laterally Supported Beams and Girders

Elastic strains in beams and girders with at least one axis of symmetry that undergo bending with small deformations, and where plane sections through the beam longitudinal axis remain plane, will have a linear distribution. Furthermore, it is assumed that Poisson's effect and shear deformations can be neglected due to practical beam member geometry (Wang et al., 2000). Therefore, for an elastic design where stress is proportional to strain (Hooke's Law), the distribution of stress is shown in Figure 7.1. It should be noted that no instability or stress concentration effects are considered.

Equilibrium of moments (ignoring signs because tension and compression are easily located by inspection) results in

$$M = 0 = \int \left(\frac{\bar{y}}{c}\sigma_c\right) \mathrm{d}A(\bar{y}) = \frac{\sigma_c}{c}\int \bar{y}^2\,\mathrm{d}A = \frac{\sigma_c}{c}I_x. \tag{7.1}$$

Therefore, the maximum bending stress is

$$\sigma_{\max} = \frac{Mc}{I_x} = \frac{M}{S_x}, \tag{7.2}$$

where

$$S_x = \frac{I_x}{c}.$$

The AREMA (2008) allowable stress design uses a factor of safety against the tensile yield stress of 1.82. The required section modulus of the beam or girder is then

$$S_x \geq \frac{M}{0.55F_y}, \tag{7.3}$$

where M is the externally applied bending moment; $\bar{y}$ is the distance from the neutral axis to the area under consideration; c is the distance from the neutral axis to the

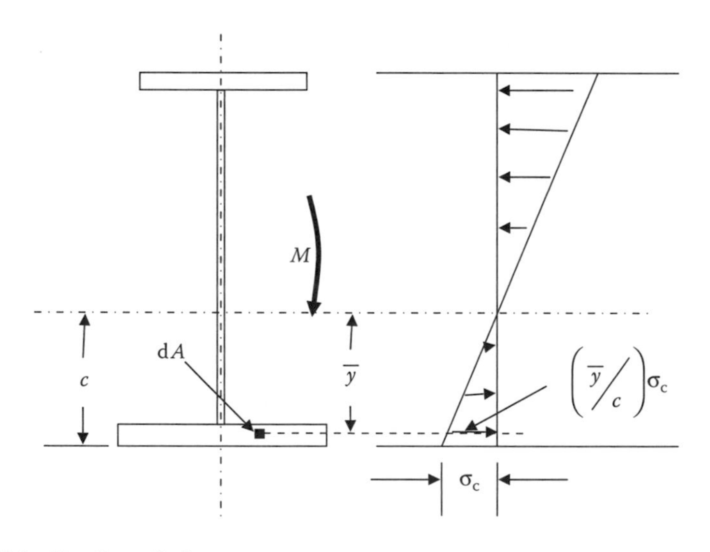

FIGURE 7.1 Bending of a beam.

extreme fiber; $\mathrm{d}A$ is the infinitesimal area under consideration; σ_c is the stress at the extreme fiber of the beam; σ_{max} is the maximum stress (at the top or bottom extreme fiber); F_y is the specified steel yield stress; $I_x = \int y^2 \, \mathrm{d}A$ is the vertical bending moment of inertia about the beam or girder neutral axis; and $S_x = I_x/c$ is the vertical bending section modulus about the beam or girder neutral axis.

Equation 7.3 enables the determination of the section modulus based on allowable tensile stress. However, the equation does not address instability in the compression region. If compression region stability (usually by lateral support of compression flange) is sustained, Equation 7.3 may be used to determine both the required net and gross section properties of the beam. Lateral support of the compression flange may be provided by a connected steel or concrete deck, and/or either diaphragms, cross bracing frames, or struts at appropriate intervals. However, if the compression flange is laterally unsupported, instability must be considered as it may effectively reduce the allowable compressive stress.

7.2.2 Bending of Laterally Unsupported Beams and Girders

If the compression flange of a beam or girder is not supported at adequately close intervals, it is susceptible to lateral–torsional instability prior to yielding and may not be able to fully participate in resisting bending moment applied to the beam or girder.

In addition to the vertical translation or deflection, y, simply supported doubly symmetric elastic beams subjected to uniform bending will buckle with lateral translation, w, and torsional translation or twist, ϕ, as shown in Figure 7.2. It is assumed that $I_x \gg I_y$ so that vertical deformation effects may be neglected with respect to the lateral deformation. It is also assumed that vertical deformation has no effect on torsional twist and the effect of prebuckling on in-plane deflections may be ignored because $EI_x \gg EI_y \gg GJ \gg EI_w/L^2$. The equilibrium equation of out-of-plane bending (in terms of flexural resistance) is

$$M\frac{\mathrm{d}^2\phi(x)}{\mathrm{d}x^2} = -EI_y\frac{\mathrm{d}^4 w(x)}{\mathrm{d}x^4} \tag{7.4}$$

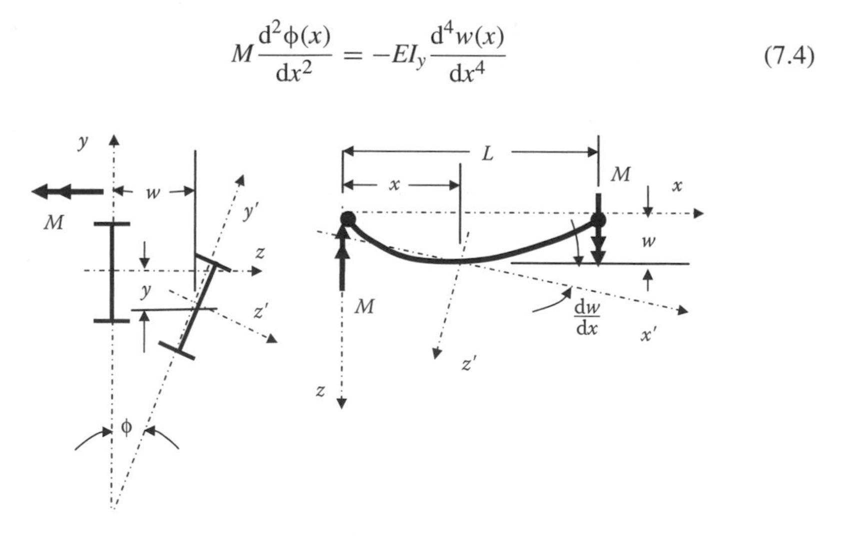

FIGURE 7.2 Bending of a beam in buckled position.

and the equation of equilibrium for torsion (in terms of warping and twisting resistance) is

$$M\frac{d^2w(x)}{dx^2} = -EI_w\frac{d^4\phi(x)}{dx^4} + GJ\frac{d^2\phi(x)}{dx^2} \tag{7.5}$$

with boundary conditions

$$w(0) = w(L) = \frac{d^2w(0)}{dx^2} = \frac{d^2w(L)}{dx^2} = \phi(0) = \phi(L) = \frac{d^2\phi(0)}{dx^2} = \frac{d^2\phi(L)}{dx^2} = 0, \tag{7.6}$$

where L is the length of the beam or girder between lateral supports ($L = 0$ when members are continuously laterally supported at compression flanges).

Equations 7.4 and 7.5 are satisfied when (Trahair, 1993)

$$\frac{w(L/2)}{w(x)} = \frac{\phi(L/2)}{\phi(x)} = \frac{1}{\sin(\pi x/L)} \tag{7.7}$$

and

$$M_{cr} = \pi\sqrt{\frac{\pi^2E^2I_wI_y}{L^4} + \frac{EI_yGJ}{L^2}}, \tag{7.8}$$

where $\phi(x)$ is the angle of twist about the shear center axis; E is the tensile modulus of elasticity (~29,000 ksi for steel) (see Chapter 2); G is the shear modulus of elasticity $= E/2(1 + \upsilon)$, where υ is Poisson's ratio (0.3 for steel); and J is the torsional constant (depends on element dimensions). Equations for some common cross sections are given in Table 7.1. $w(x)$ is the lateral deflection along the x-axis; $I_y = \int x^2\, dA$ is the lateral bending moment of inertia about the beam or girder vertical axis of symmetry; and $I_w = \int \omega^2\, dA = C_w$ is the torsional moment of inertia or warping constant (ω is defined in terms of the position of the shear center and thickness of the member). Warping constant values are available in many references (Roark and Young, 1982; Seaburg and Carter, 1997). Equations for some common cross sections are given in Table 7.1. M_{cr} is the critical lateral–torsional buckling moment.

For an I section with equal flanges (see Table 7.1)

$$I_y = Ar_y^2 \approx 2\frac{t_fb^3}{12}, \tag{7.9}$$

$$I_w = C_w = \frac{h^2t_fb^3}{24} = \frac{h^2I_y}{4}, \tag{7.10}$$

$$J = 0.3At_f^2, \tag{7.11}$$

$$S_x = \frac{2Ar_x^2}{d}, \tag{7.12}$$

$$r_x \approx 0.4d \text{ (radius of gyration in the vertical direction)}, \tag{7.13}$$

TABLE 7.1
Torsional Warping Constants for Common Cross Sections

Cross Section Denotes Position of Shear Center	Warping Constant, C_w Torsion Stiffness Constant, J
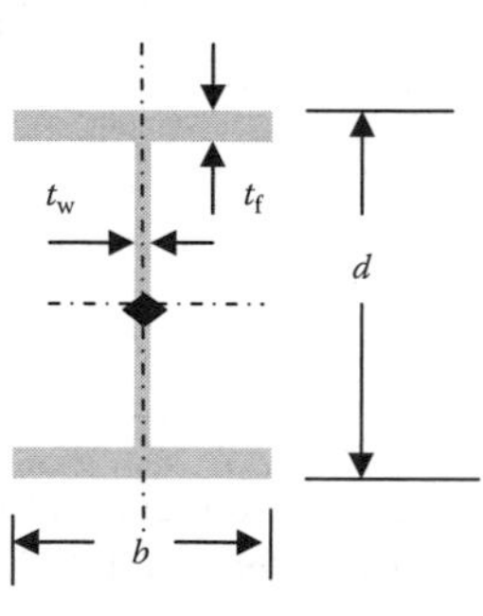	$h = d - t_f$ $C_w = \frac{h^2 t_f b^3}{24}$ $J = 1/3(2bt_f^3 + ht_w^3)$
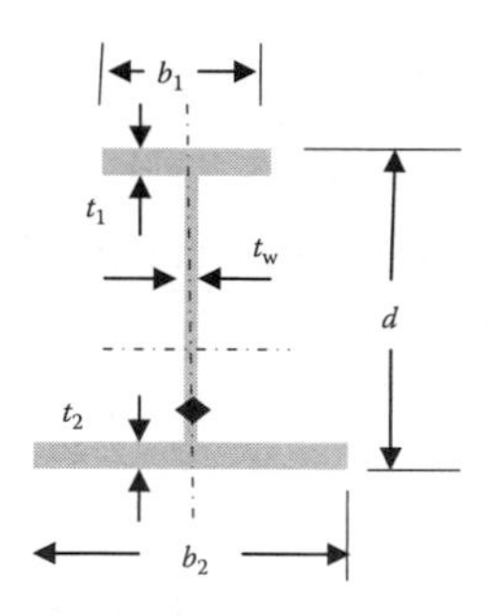	$h = d - (t_1/2) - (t_2/2)$ $C_w = \frac{h^2 t_1 t_2 b_1^3 b_2^3}{12(t_1 b_1^3 + t_2 b_2^3)}$ $J = 1/3\left(t_1^3 b_1 + t_2^3 b_2 + t_w^3 h\right)$
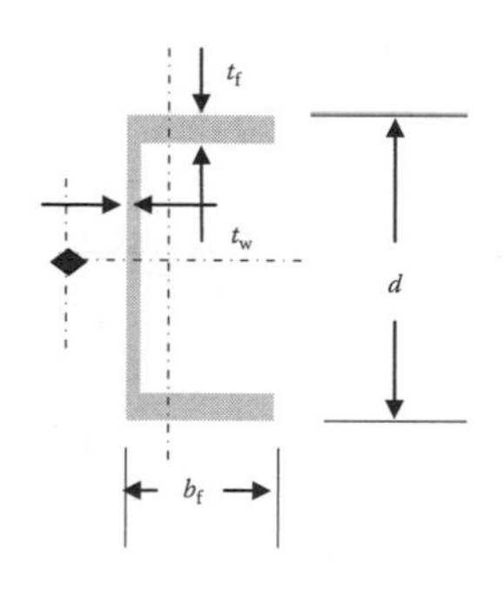	$h = d - t_f$ $b = b_f - t_w/2$ $C_w = \frac{t_f b^3 h^2}{12}\left(\frac{3bt_f + 2ht_w}{6bt_f + ht_w}\right)$ $J = 1/3(2bt_f^3 + ht_w^3)$
b_2, t_2, t_1, b_1	$b = b_1 - t_2/2$ $h = b_2 - t_1/2$ $C_w = \frac{1}{36}(b^3 t_1^3 + h^3 t_2^3)$ (for small t_1 and t_2 $C_w \sim 0$) $J = 1/3(bt_1^3 + ht_2^3)$

$$r_y \approx 0.2b \text{ (radius of gyration in the lateral direction on the compression side of the neutral axis)}, \tag{7.14}$$

$$h \approx d, \tag{7.15}$$

$$G = \frac{E}{2(1+\upsilon)} = 0.38E. \tag{7.16}$$

Substitution of Equations 7.9 through 7.16 into Equation 7.8 yields

$$f_{cr} = \sqrt{\left(\frac{15.4E}{(L/r_y)^2}\right)^2 + \left(\frac{0.67E}{Ld/bt_f}\right)^2}, \tag{7.17}$$

where $f_{cr} = M_{cr}/S_x$ is the critical lateral–torsional buckling stress.

The first term in Equation 7.17 represents the warping torsion effects and the second term describes the pure torsion effects. For torsionally strong sections (shallow sections with thick flanges) the warping effects are negligible and

$$f'_{cr} = \left(\frac{0.67E}{Ld/bt_f}\right) = \left(\frac{0.21\pi E}{Ld/bt_f}\right) = \left(\frac{0.24\pi E}{Ld(\sqrt{1+\upsilon})/bt_f}\right). \tag{7.18}$$

For torsionally weak sections (deep sections with thin flanges and web, typical of railway plate girders) the pure torsion effects are negligible and

$$f''_{cr} = \left(\frac{15.4E}{(L/r_y)^2}\right) = \left(\frac{1.56\pi^2 E}{(L/r_y)^2}\right), \tag{7.19}$$

which is analogous to determining the elastic (Euler) column strength of the flanges.

Using a factor of safety of $9/5 = 1.80$, Equations 7.18 and 7.19 for torsionally strong and weak sections, respectively, are

$$F'_{cr} = \frac{f'_{cr}}{1.80} = \left(\frac{0.13\pi E}{Ld(\sqrt{1+\upsilon})/bt_f}\right) \tag{7.20}$$

and

$$F''_{cr} = \frac{f''_{cr}}{1.80} = \left(\frac{0.87\pi^2 E}{(L/r_y)^2}\right). \tag{7.21}$$

However, due to residual stress, unintended load eccentricity, and fabrication imperfections, axial compression member strength is based on an inelastic buckling strength parabola when $F''_{cr} \geq F_y/2$ and an Euler (elastic) buckling curve for $F''_{cr} \leq F_y/2$ (Figure 7.3, also see Chapter 6 on axial compression member behavior).

When $F''_{cr} = (0.55F_y)/2$, Equation 7.21 (elastic buckling curve) is equal and tangent to the inelastic buckling strength parabola (transition curve). From Equation 7.21 the slenderness, L/r_y, at $(0.55F_y)/2$ is

$$\frac{L}{r_y} = 5.55\sqrt{\frac{E}{F_y}}. \tag{7.22}$$

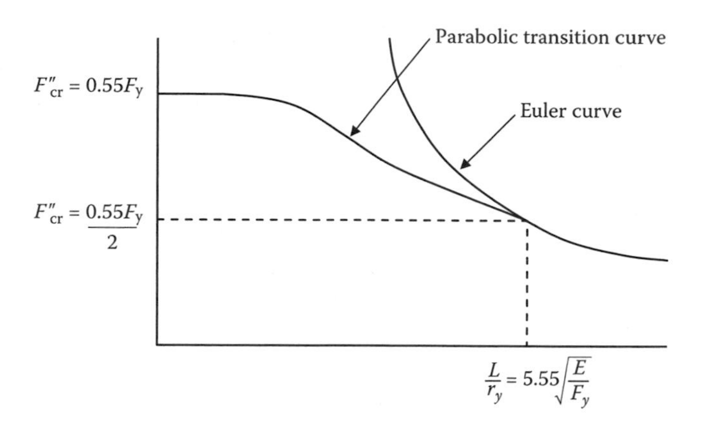

FIGURE 7.3 Lateral–torsional buckling curve for flexural compression.

The parabolic transition equation is

$$F''_{cr} = A - B\left(\frac{L}{r_y}\right)^2, \tag{7.23}$$

which becomes

$$F''_{cr} = 0.55F_y - \frac{0.55F_y^2}{6.3\pi^2 E}\left(\frac{L}{r_y}\right)^2, \tag{7.24}$$

with

$$A = F''_{cr} = 0.55F_y \quad (\text{when } L/r_y = 0),$$

$$B = \frac{0.55F_y^2}{6.3\pi^2 E} \quad (\text{when } F''_{cr} = 0.55\ F_y/2 \text{ and } L/r_y \text{ is given by Equation 7.22}).$$

AREMA (2008) recommends a conservative approach using Equations 7.20 and 7.24 independently and adopting the larger of the two buckling stresses, F'_{cr} or F''_{cr}, for the design of flexural members. AREMA (2008) also restricts beam and girder slenderness to $L/r_y \leq 5.55\sqrt{E/F_y}$ in order to preclude Euler buckling.

It should be noted that Equations 7.20 and 7.24 are developed based on the assumption of a uniform moment (no shear forces). Moment gradients relating to concentrated or moving load effects on simply supported beams and girders can be considered through the use of modification factors (Salmon and Johnson, 1980). Modification factors, C_b, based on loading and support conditions are available in the literature on structural stability. The equations for pure and warping effects are then

$$F'_{cr} = \left(\frac{0.13\pi E C_b}{Ld(\sqrt{1+\upsilon})/bt}\right), \tag{7.25}$$

$$F''_{cr} = 0.55F_y - \frac{0.55F_y^2}{6.3\pi^2 E C_b}\left(\frac{L}{r_y}\right)^2. \tag{7.26}$$

AREMA (2008) conservatively neglects this effect ($C_b = 1$) and uses Equations 7.20 and 7.24 as the basis for steel beam and girder flexural design because the actual moment gradient along the unbraced length of a beam or girder is difficult to assess for moving train live loads.

7.2.3 Shearing of Beams and Girders

Shear stresses will exist due to the change in bending stresses at adjacent sections (Figure 7.4). Equilibrium of moments and neglecting infinitesimals of higher order leads to

$$\mathrm{d}M - V\,\mathrm{d}x = 0, \tag{7.27}$$

$$V = \frac{\mathrm{d}M}{\mathrm{d}x} = \text{shear force}. \tag{7.28}$$

Referring to Figure 7.5

$$F = -\frac{M}{I_x}\int_{\substack{\text{shaded}\\ \text{area}}} y\,\mathrm{d}A = -\frac{MQ}{I_x}. \tag{7.29}$$

The change in force, F, acting normal to the shaded area in Figure 7.5 is the shear flow, q, or

$$q = \frac{\mathrm{d}F}{\mathrm{d}x} = -\frac{(\mathrm{d}M/\mathrm{d}x)Q}{I_x} = -\frac{VQ}{I_x} \tag{7.30}$$

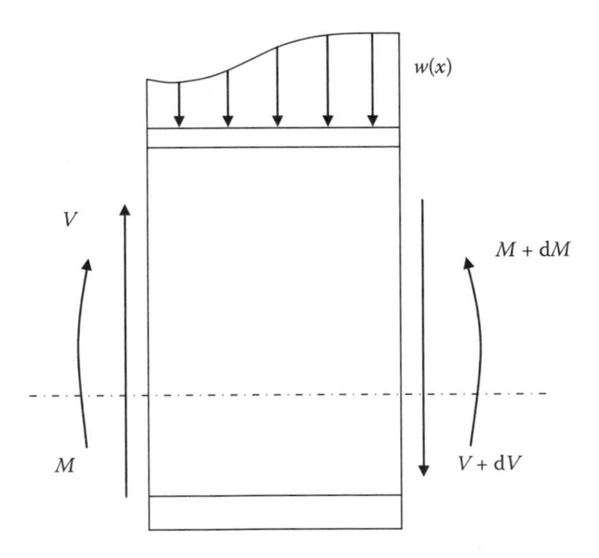

FIGURE 7.4 Shearing of a beam.

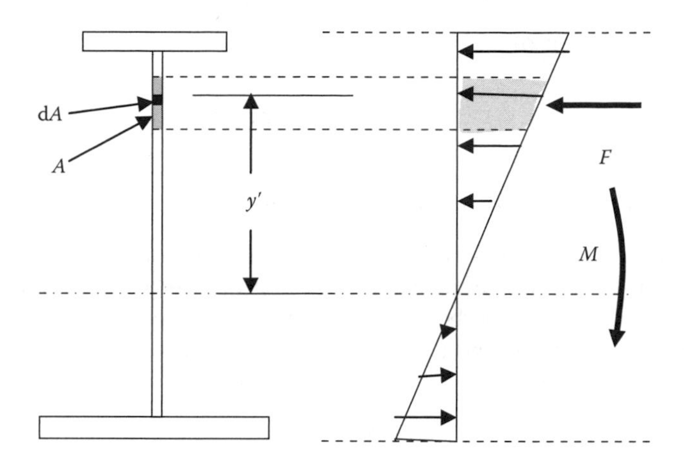

FIGURE 7.5 Shear flow in a beam.

and the shear stress is

$$\tau = \frac{\mathrm{d}F/\mathrm{d}x}{t} = \frac{VQ}{I_x t}, \tag{7.31}$$

where t is the thickness of steel at the area, A, under consideration; $Q = Ay'$ is the statical moment of area, A, about the neutral axis.

Equation 7.31 results in a shear stress profile through the girder as shown in Figure 7.6. AREMA (2008) recommends the determination of the average shear stress computed based on the area of the beam or girder web ($t_{avg} = V/A_w$, where A_w is the area of the web) to simplify steel beam and girder I-cross section design. As shown in Figure 7.6, this is an accurate (although slightly nonconservative) approach to shear design.

The AREMA (2008) allowable stress design uses an allowable shear stress based on tensile yield stress ($\tau_y = F_y/\sqrt{3}$, see Chapter 2). The required gross web area of

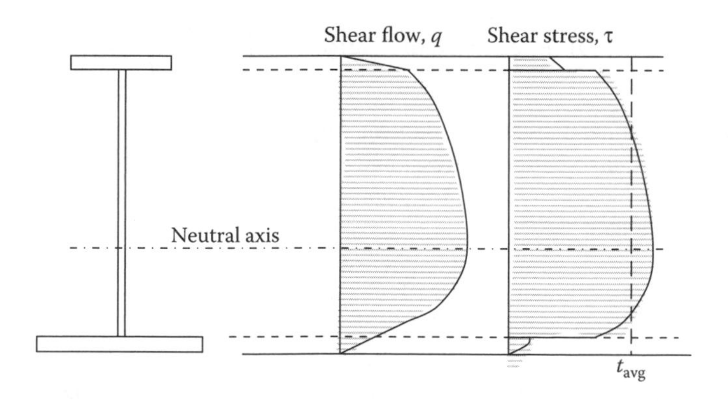

FIGURE 7.6 Shear stresses in an I-beam.

the I beam or girder is then

$$A_{\mathrm{w}} \geq \frac{V}{0.55F_{\mathrm{y}}/\sqrt{3}} \approx \frac{V}{0.35F_{\mathrm{y}}}. \tag{7.32}$$

7.2.4 Biaxial Bending of Beams and Girders

Biaxial bending is not generally a concern for ordinary steel railway longitudinal beams and girders. However, biaxial bending of stringers and floorbeams or unsymmetrical bending of floorbeams* may warrant consideration (see Chapter 8).

Stresses in perpendicular principal directions may be superimposed at critical symmetric sections. This is shown in Equations 7.33 and 7.34.

$$1.0 \leq \pm\frac{M_x}{F_{\mathrm{b}x}S_x} \pm \frac{M_y}{F_{\mathrm{b}y}S_y}, \tag{7.33}$$

$$F_v \leq \pm\frac{V_xQ_x}{I_xt_x} \pm \frac{V_yQ_y}{I_yt_y}, \tag{7.34}$$

where $F_{\mathrm{b}x}$, $F_{\mathrm{b}y}$ are the allowable bending stresses in the x and y directions, respectively; F_v is the allowable shear stress.

7.2.5 Preliminary Design of Beams and Girders

For planning purposes, preliminary proportioning of plate girders may be necessary before detailed design. Preliminary dimensions may be needed for estimating the weight and cost and for assessing the site geometry constraints. Preliminary proportions may also be required in order to estimate splice requirements and for fabrication, shipping, and erection methodologies.

There are many techniques used by experienced bridge engineers to assess the preliminary dimensions of beams and girders. One method is by simplification of the bending resistance into that resisted by equal flanges and webs separately as (Figure 7.7)

$$M = M_{\mathrm{f}} + M_{\mathrm{w}} = bt_{\mathrm{f}}(d - t_{\mathrm{f}})(f_{\mathrm{bf}}) + \left(\frac{t_{\mathrm{w}}(h)^2}{6}\right)(f_{\mathrm{bw}}), \tag{7.35}$$

where M_{f} is the moment carried by flanges, M_{w} is the moment carried by the web, f_{bf} is the allowable flange bending stress, and f_{bw} is the allowable web bending stress. Assuming that $f_{\mathrm{bf}} \sim f_{\mathrm{bw}} \sim f_{\mathrm{b}}$ and, for usual railway beams and girders, $h \sim (d - t_{\mathrm{f}})$, Equation 7.35 yields

$$bt_{\mathrm{f}} = A_{\mathrm{f}} = \frac{M}{f_{\mathrm{b}}h} - \frac{A_{\mathrm{w}}}{6}. \tag{7.36a}$$

* May occur at transverse floorbeams with horizontal longitudinal live load forces applied due to traction and/or braking (see Chapter 4).

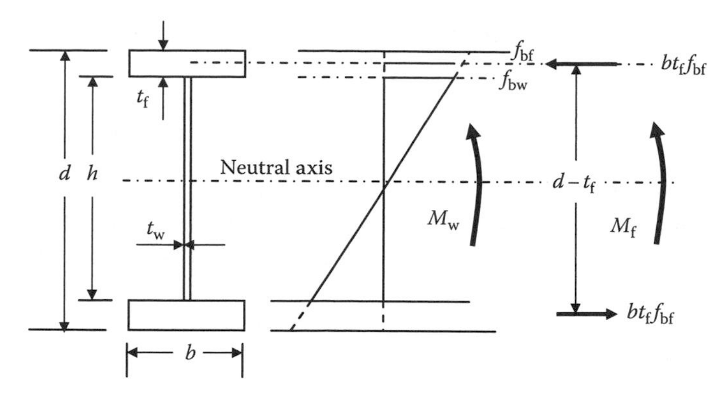

FIGURE 7.7 Preliminary proportioning of girder flanges.

An estimate of flange size can be made based on a preliminary web height and web thickness, $t_w \geq V/0.35F_y h$, where V is the maximum applied shear force, as

$$A_f = \frac{M}{f_b h} - \frac{V}{0.35F_y(6)} = \frac{M}{f_b h} - \frac{V}{2.10F_y}. \tag{7.36b}$$

The preliminary web height is typically estimated from typical L/d ratios for economical design (see Chapter 3), available plate and section sizes, plate slenderness, underclearance requirements, and/or aesthetic considerations.

7.2.6 Plate Girder Design

Modern plate girders typically consist of welded flange and stiffened web plates (Figure 7.8). Railway bridge girders are generally of large size, and ensuring the stability of plate elements in compression zones and consideration of stress concentration effects in tension zones are critical components of the design. Simple span plate girders of about 150 ft or less are generally economical for railway bridge construction. Longer spans are feasible for continuous construction. However, continuous spans are

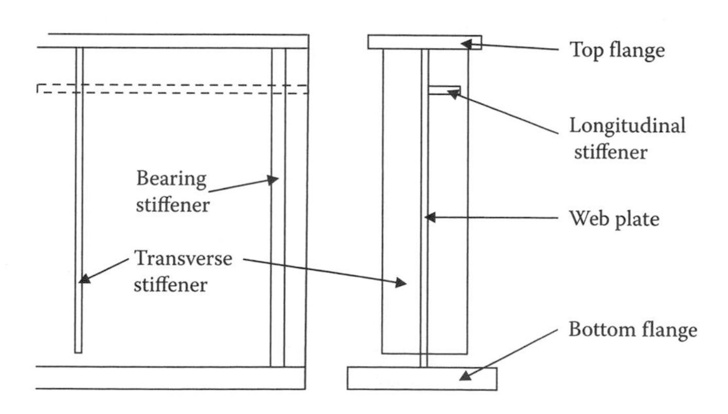

FIGURE 7.8 Cross section of the modern plate girder.

less frequently used due to uplift considerations (related to the large live load to dead load ratio of steel railway bridges) and stresses imposed by foundation movements that may not be readily detected.

Box girders are also relatively rare in railway bridge construction due to welded stiffener fatigue issues but may be used where large torsional stiffness is required (e.g., curved bridges).

Hybrid girders using both HPS and HSLA steels (see Chapter 2) may be economical for some long span applications (particularly over supports of continuous spans where flexural and shear stresses may be relatively high).

However, most ordinary steel railway span designs will be governed by deflection (which is stiffness related) and fatigue requirements (see Chapter 5), which are both material independent for structural steel. These criteria and fabrication issues must be carefully considered when assessing the use of hybrid girders with plate elements of differing steel type and grade.

The main elements of a plate girder (flange plates, web plates, main element splices, bearing stiffeners, and their respective connections) are designed to resist tensile and compressive normal axial and bending stresses, and shear stresses in the cross section. In ASD, secondary elements (stiffener plates) are designed to provide stability to the main elements of the girder cross section.

7.2.6.1 Main Girder Elements

Flanges of modern welded plate girders are made using a single plate in the cross section. The required thickness and width of the flange plates will be governed by strength, stability, fatigue, and serviceability criteria. Cover plates should not be used as flange plate dimensions may be varied along the length of the girder as required.* The thickness of flange plates may be limited by issues relating to steel-making and subsequent fabrication effects on lamellar tearing† and toughness.‡

Modern steel-making processes such as TMCP may alleviate many of the metallurgical concerns relating to thick plates but designers should carefully review typical welding details, materials, and procedures used for fabrication when considering maximum plate thickness. Most modern railway girder bridges of span less than 150′ can be designed with flanges less than 3 in. thick, which should generally preclude concerns with respect to welded fabrication and through-thickness restraint.

Webs of plate girders are generally relatively thin plates§ when designed in accordance with shear strength criteria. However, because of slenderness, the required thickness of the web plate also depends on flexural and shear stability considerations. Therefore, in order to avoid thick webs, the plates are often stiffened longitudinally

* Designers should note that it is often less costly to fabricate flanges without changes in dimension due to butt welding and transitioning requirements. Designs that utilize varying flange plate widths and lengths may be desirable and economical for long span fabrication, but consultation with experienced bridge fabricators is often warranted.

† Lamellar tearing occurs in thick plates due to large through-thickness strains produced by fabrication process effects such as weld metal shrinkage at highly restrained locations (i.e., joints and connections).

‡ Ductility is affected by the triaxial strains created by weld shrinkage and restraint in thick plates (see Chapter 2).

§ Particularly for longer spans typical of railway plate girders.

(to resist flexural buckling) and/or transversely (to resist shear buckling). The height of web plates in long plate girders may be quite large and designers should carefully review available plate sizes from steel making, fabrication, and shipping perspectives in order to avoid costly longitudinal splices and limit, to within practical requirements, transverse splices.

Welded or bolted splices may be used when required due to plate length limitations. These splices are generally designed for the shear and bending moment at the spliced section and/or specific strength criteria recommended by AREMA (2008).

The connection of web and flange plates in modern plate girder fabrication is generally performed by high-quality automatic welding. These welds must be designed to resist the total longitudinal shear at the connection as well as other loads directly applied to the flanges and/or web (see Chapter 9).

7.2.6.1.1 Girder Tension Flanges and Splices

7.2.6.1.1.1 Tension Flanges Overall girder bending capacity at service loads is not reduced by the usual number and pattern of holes in a girder cross section (the same capacity as for gross section but the stresses are distributed differently because of localized stress concentrations). However, AREMA (2008) recommends that girder tension flanges be designed based on the moment of inertia of the entire net section, I_{xn}, (using the neutral axis determined from the gross section*) and the tensile yield stress. This is particularly appropriate for tension flange splices and as protection from the effects of possible occasional tensile overload stresses. Many designers proportion tension flanges based on net section properties being not greater than 85% of the gross section properties.

The plate girder net section modulus, S_{xn}, is determined as

$$S_{xn} = \frac{I_{xn}}{c_{\mathrm{t}}} \geq \frac{M_{\mathrm{t\,max}}}{F_{\mathrm{all}}} \geq \frac{M_{\mathrm{t\,max}}}{0.55F_{\mathrm{y}}} \tag{7.37}$$

or

$$S_{xn} \geq \frac{\Delta M}{F_{\mathrm{fat}}}, \tag{7.38}$$

where c_{t} is the distance from the neutral axis to the extreme fiber in tension, $M_{\mathrm{t\,max}}$ is the maximum tensile bending moment due to all load effects and combinations (Table 4.5), ΔM is the maximum bending moment range due to fatigue load (see Chapter 5) and F_{fat} is the allowable fatigue stress range for the appropriate fatigue detail category.

Residual stresses, which must be considered in the design of dynamically loaded axial and flexural tensile members, and all axial compression members, are of negligible value in the design of statically loaded bending members. This is because the presence of residual stresses may cause an initial inelastic behavior, but subsequent

---

* The neutral axis must consist of a smooth line that, because fiber stresses cannot suddenly change, will vary only slightly at the typically few cross sections with holes. Therefore, the neutral axis will be essentially at the location of the neutral axis of the gross section along the girder length.

statically applied loads of the same or smaller magnitude will result in elastic behavior (Brockenbrough, 2006). Therefore, except for fatigue considerations related to dynamic live loads, residual stresses need not be explicitly considered in the design of bending members such as plate girders.*

7.2.6.1.1.2 Tension Flange Splices AREMA (2008) recommends that splices in the main members have a strength not less than that of the member being spliced. It is also recommended that splices in girder flanges should comprise elements that are not lesser in section than the flange element being spliced. Two elements in the same flange cannot be spliced at the same location.†

Therefore, bolted‡ splice elements in girder flanges should

- Have a cross-sectional area that is at least equal to that of the flange element being spliced.
- Comprise splice elements of sufficient cross section and location such that the moment of inertia of the spliced member is no less than that of the member at the splice location.

Splice elements may be single or double plates. Single plate splices are generally used on the exterior surfaces of flange splices to ensure a greater moment of inertia at the splice. Two plates§ are often used for larger girder splices where single-shear bolted connections are too long and a double-shear connection is required.

The splice fasteners (see Chapter 9) should be designed to transfer the force in the element being spliced to the splice material. Welded splices are usually made with CJP (full penetration) groove welds with a strength that is at least equal to the base material being spliced.

7.2.6.1.2 Girder Compression Flanges and Splices

7.2.6.1.2.1 Compression Flanges AREMA (2008) recommends that girder compression flanges be designed based on the moment of inertia of the entire gross section, I_{xg}, and the tensile** yield stress.

Therefore the plate girder gross section modulus, S_{xg}, is

$$S_{xg} = \frac{I_{xg}}{c_c} \geq \frac{M_{c\,max}}{F_{call}}, \tag{7.39}$$

where c_c is the distance from the neutral axis to the extreme fiber in compression; $M_{c\,max}$ is the maximum compressive bending moment due to all load effects and

---

* Residual stresses are also not explicitly considered in fatigue design because fatigue strength is based on nominal stress tests on elements and members containing residual stresses from manufacture or fabrication.

† This is usually only applicable to built-up section flanges, which are not often used for modern plate girder fabrication.

‡ These may be shop or field splices.

§ It is good practice that the centroid of the splice plates each side of the flange plate be coincident with the centroid of the flange being spliced.

** Tensile yield stress is almost equal to compressive yield stress (Chapter 2).

combinations (Table 4.5); F_{call} is the allowable compressive stress, based on stability considerations since the girder compression flange is susceptible to lateral–torsional instability prior to yielding.

In addition to lateral–torsional buckling effects on the allowable compressive stress, vertical and torsional buckling effects must be considered to ensure compression flange stability.

7.2.6.1.2.1.1 Lateral–Torsional Buckling Compression flange lateral–torsional instability is controlled by limiting allowable stresses to those given by Equations 7.20 or 7.24. It should be noted that Equation 7.20 was developed assuming an I section with equal flanges. Therefore, the smallest flange area, $A_f = bt_f$, should be used in Equation 7.20 when establishing critical buckling stress. It should also be noted that Equation 7.24 precludes inelastic buckling by ensuring that the girder compression area L/r_y does not exceed the value of Equation 7.22, which is presented again as Equation 7.40. Therefore, the length, L, should be taken as the largest distance between compression flange lateral supports, L_p, and r_{cy} is determined as the minimum radius of gyration of the compression flange and that portion of the web in compression (from the neutral axis to the edge of the web plate).

The larger of either Equations 7.20 or 7.24, presented again as Equations 7.41 and 7.42, respectively, is adopted to determine the allowable compressive bending stress, F_{call}, for design of the compression flange. Therefore the compression flange design requirements are

$$\frac{L_p}{r_{cy}} \leq 5.55\sqrt{\frac{E}{F_y}} \tag{7.40}$$

and the larger of

$$F_{call} = \left(\frac{0.13\pi E}{L_p d(\sqrt{1+\upsilon})/A_f}\right) \tag{7.41}$$

or

$$F_{call} = 0.55F_y - \frac{0.55F_y^2}{6.3\pi^2 E}\left(\frac{L_p}{r_{cy}}\right)^2, \tag{7.42}$$

where L_p is the largest distance between compression flange lateral supports, $A_f = bt_f$ is the area of the smallest flange in the girder (even if tension flange), and r_{cy} is the minimum radius of gyration of the compression flange and that portion of the web in compression. However, as illustrated by Figure 7.3, F_{call} cannot exceed $0.55F_y$.

7.2.6.1.2.1.2 Vertical Flexural Buckling If the web plate buckled due to bending in the compression zone, it would be unable to provide support for the attached compression flange and the compression flange could then buckle vertically as shown in Figure 7.9. To avoid compression flange vertical buckling, flexural buckling of the web plate is precluded by limiting the ratio of web height, h, to thickness, t_w, or by including a longitudinal stiffener.

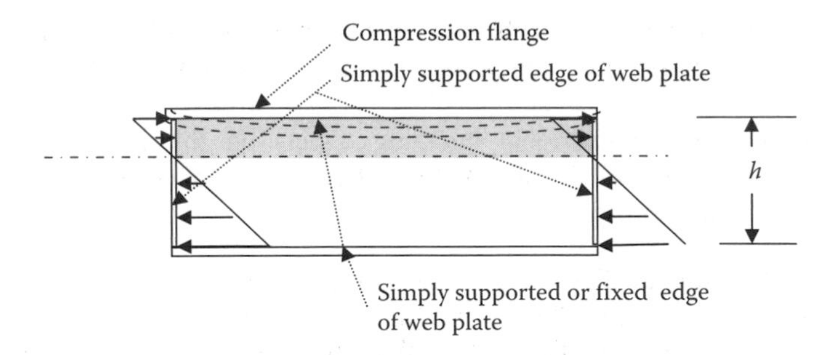

FIGURE 7.9 Pure flexural buckling of the web plate causing compression flange buckling.

The critical elastic buckling stress of a rectangular plate is

$$F_{cr} = \frac{k\pi^2 E t_w^2}{12(1-\upsilon^2)h^2}, \tag{7.43}$$

where F_{cr} is the critical buckling stress, k is the buckling coefficient depending on loading and plate edge conditions, and υ is Poisson's ratio (0.3 for steel). k ranges from 23.9 for simply supported edge conditions to 39.6 for fixed edge conditions assumed at the two edges (at the flanges) of a long plate in pure bending (Timoshenko and Woinowsky-Kreiger, 1959). AREMA (2008) conservatively uses $k = 23.9$ and reduces the ratio of web height to thickness to 90% of the theoretical values to account for web geometry imperfections. Substitution of $k = 23.9$ into Equation 7.43 yields

$$\frac{h}{t_w} \le 0.90\sqrt{\frac{23.9\pi^2 E}{12(1-0.3^2)(F_{cr})}} \le 4.18\sqrt{\frac{E}{F_{cr}}}, \tag{7.44}$$

which will preclude elastic buckling due to pure bending (Figure 7.10). Rearrangement of Equation 7.44 provides

$$t_w \ge 0.24h\sqrt{\frac{F_{cr}}{E}} \tag{7.45}$$

and if $F_{cr} = 0.55F_y$

$$t_w \ge 0.18h\sqrt{\frac{F_y}{E}}. \tag{7.46}$$

The allowable compressive bending stress, F_{cr}, is given by Equations 7.41 and 7.42. Therefore, when the actual calculated flexural stress at the compression flange, f_c, is less than F_{cr},

$$t_w \ge 0.18h\sqrt{\frac{F_y}{E}}\sqrt{\frac{f_c}{F_{cr}}} \tag{7.47}$$

and longitudinal stiffeners are not required for web flexural buckling stability.

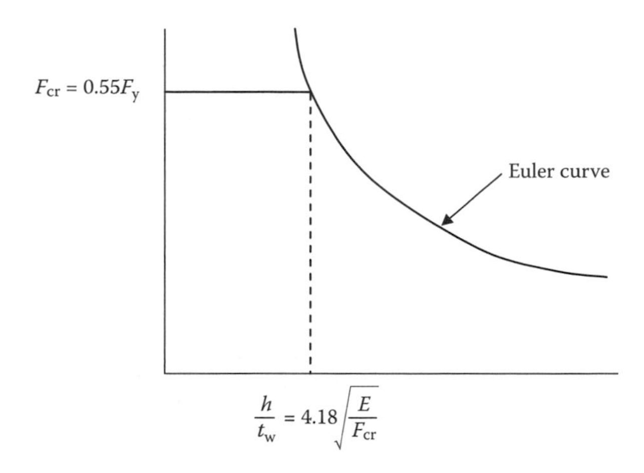

FIGURE 7.10 Elastic buckling curve for rectangular plate buckling under pure bending.

However, in cases where longitudinal stiffeners are provided, the minimum web thickness criteria to avoid flexural buckling (and thereby prevent vertical buckling of the compression flange) of Equation 7.47 is reduced. The optimum location for longitudinal web plate stiffeners is at $0.22h$ from the compression flange (Rockey and Leggett, 1962). When a longitudinal stiffener is placed at $h/5$ from the inside surface of the compression flange, as recommended by AREMA (2008), the critical elastic buckling stress of a rectangular plate (Equation 7.43) is

$$F_{cr} = \frac{129\pi^2 E t_w^2}{12(1-\upsilon^2)h^2}, \tag{7.48}$$

where the theoretical $k = 129$ (Galambos, 1988). Equation 7.48 with $F_{cr} = 0.55F_y$ can be expressed as

$$t_w \geq 0.08h\sqrt{\frac{F_y}{E}}. \tag{7.49}$$

Equation 7.49 indicates that the web thickness to preclude elastic critical flexural buckling of the web with a longitudinal stiffer can be 43% (23.9/129) of that required without a longitudinal stiffener (Equation 7.46). AREMA (2008) recommends that the web thickness with a longitudinal stiffener be no less than 50% of that required without a longitudinal stiffener.

7.2.6.1.2.1.3 Torsional Buckling Torsional buckling of the compression flange is essentially the buckling problem of uniform compression on a plate free at one side and partially restrained at the other (Salmon and Johnson, 1980). The critical elastic plate buckling stress is

$$F_{cr} = \frac{k\pi^2 E t_f^2}{12(1-\upsilon^2)b^2} \tag{7.50}$$

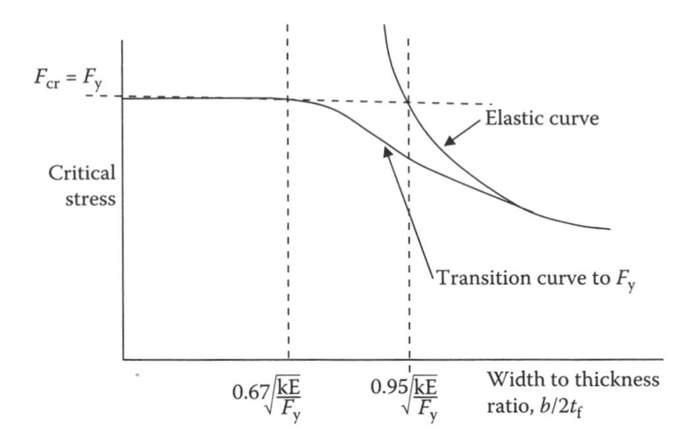

FIGURE 7.11 Plate buckling curve for uniform compression.

and the limiting width-to-thickness ratio* at $F_{cr} = F_y$ is

$$\frac{b}{2t_f} \leq \sqrt{\frac{k\pi^2 E}{12(1 - 0.3^2)F_y}} = 0.95\sqrt{\frac{kE}{F_y}}. \tag{7.51}$$

However, this is an elastic buckling curve and at $F_{cr} = F_y$ the plate axial strength is overestimated (above the transition curve as shown in Figure 7.11). To mitigate this, it is customary to use a limiting width-to-thickness ratio of

$$\frac{b}{2t_f} \leq 0.67\sqrt{\frac{kE}{F_y}}, \tag{7.52}$$

which is the approximate value corresponding to the transition curve at $F_{cr} = F_y$. For plates with a free edge, the buckling coefficient, k, is 0.425 with the other edge considered as simply supported and 1.277 with the other edge considered as fixed (Bleich, 1952). Tests have indicated that the lowest value of buckling coefficient, k, for partially restrained elements is about 0.70 (typical of a girder flange) (Tall, 1974). Therefore, substitution of $k = 0.7$ into Equation 7.52 yields

$$\frac{b}{2t_f} \leq 0.56\sqrt{\frac{E}{F_y}}. \tag{7.53}$$

AREMA (2008) recommends that this width to thickness ratio for local flange buckling be decreased further based on practical experience with local compression forces from ties,† fabrication tolerances, and other unaccounted effects.

* Based on yield strength of the plate.
† Particularly, if poorly framed.

The recommended compression flange width-to-thickness ratio is

$$\frac{b}{2t_f} \leq 0.43\sqrt{\frac{E}{F_y}}, \tag{7.54}$$

when there are no ties bearing directly on compression flanges and

$$\frac{b}{2t_f} \leq 0.35\sqrt{\frac{E}{F_y}}, \tag{7.55}$$

when ties bear directly on compression flanges.

7.2.6.1.2.2 Compression Flange Splices Splices in girder compression flanges are treated in a similar manner to those for tension flanges. The requirements are outlined above in the section on girder tension flange design.

7.2.6.1.3 Girder Web Plates and Splices

7.2.6.1.3.1 Web Plates Economical railway girders have relatively thin web plates. Therefore, in addition to designing the web plate to carry shear forces (Equation 7.32), it is also necessary to ensure stability of the web plates in girders. Figure 7.12 indicates the forces on the web plate that may create instability.

The stability criteria for shear, bending, and compression forces are separately developed and appropriately combined to investigate web plate stability. Inelastic buckling, due to residual stresses, load eccentricities, and geometric tolerances, is modeled by a buckling strength transition parabola formulation consistent with other structural instability criteria (e.g., axial and flexural compression).

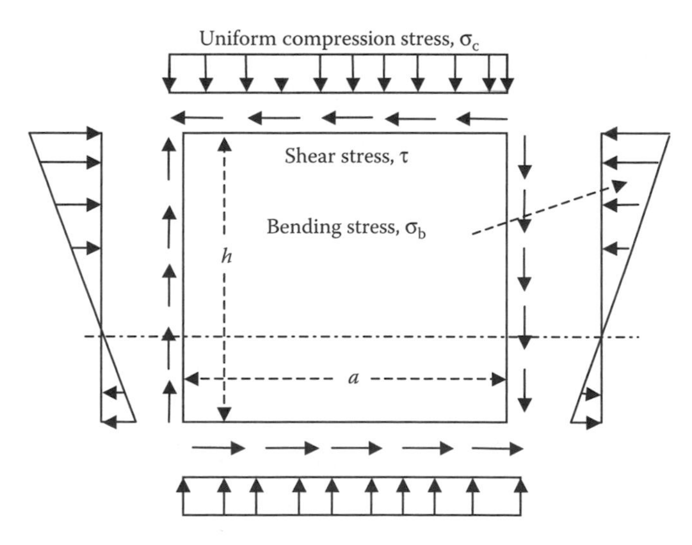

FIGURE 7.12 Stresses on girder web plates.

7.2.6.1.3.1.1 Elastic Buckling under Pure Bending Flexural buckling of the web plate is precluded by limiting the ratio of web height, h, to thickness, t_w, or by including a longitudinal stiffener. The elastic buckling of the web plate under bending was considered above in conjunction with the investigation of vertical buckling of the compression flange.

Web plate design without longitudinal stiffeners considering Equation 7.47, presented again as Equation 7.56, requires a minimum web plate thickness to preclude elastic flexural buckling of

$$t_w \geq 0.18h\sqrt{\frac{F_y}{E}}\sqrt{\frac{f_c}{F_{cr}}} \geq 0.24h\sqrt{\frac{f_c}{E}}. \tag{7.56}$$

Rearrangement of Equation 7.56 and substitution of $F_{cr} = 0.55F_y$ (preclude elastic buckling per Figure 7.10) yield the criteria that

$$\frac{h}{t_w} \leq 4.18\sqrt{\frac{E}{f_c}}. \tag{7.57}$$

Otherwise, longitudinal stiffeners are required for web flexural buckling stability.

Web plate design with a longitudinal stiffener at $0.20h$ from the compression flange considering Equation 7.49, presented again as Equation 7.58, requires a minimum web plate thickness to preclude elastic flexural buckling of

$$t_w \geq 0.08h\sqrt{\frac{F_y}{E}}\sqrt{\frac{f_c}{F_{cr}}} \geq 0.10h\sqrt{\frac{f_c}{E}}. \tag{7.58}$$

7.2.6.1.3.1.2 Elastic Buckling under Pure Shear The critical elastic plate buckling shear stress is

$$\tau_{cr} = \frac{k\pi^2 E t_w^2}{12(1-\upsilon^2)h^2}, \tag{7.59}$$

where $k = 5.35$ for infinitely long simply supported plate under pure shear (Timoshenko and Gere, 1961).

Shear yield stress, τ_y, is related to tensile yield stress, F_y, as (see Chapter 2)

$$\tau_{cr} = \tau_y = \frac{F_y}{\sqrt{3}}. \tag{7.60}$$

Therefore, from Equation 7.59,

$$\frac{h}{t_w} \leq 2.89\sqrt{\frac{E}{F_y}}. \tag{7.61}$$

7.2.6.1.3.1.3 Inelastic Buckling under Pure Shear In order to account for residual stresses (which, however, are generally not great in girder webs) and geometric eccentricities (such as out-of-flatness, which is typical in girder webs), Equation 7.61 is reduced to

$$\frac{h}{t_w} \leq 2.12\sqrt{\frac{E}{F_y}} \tag{7.62}$$

for design purposes. Therefore, when $h \geq 2.12 t_w \sqrt{E/F_y}$, transverse web stiffeners are required.

7.2.6.1.3.1.4 Combined Bending and Shear

7.2.6.1.3.1.4.1 Strength Criteria The web plate is subjected to a combination of shear forces and bending moment depending on location in the span. Since shear stress is greatest at the neutral axis where normal flexural stresses are zero and normal bending stress is greatest at the flange where shear stresses are less than average, it is generally sufficient to design for shear and flexural allowable stresses independently. Also, in ordinary steel railway girder design, the bending moment carried by the web plate is relatively small (see Equation 7.36a). However, the design engineer may need to review shear and flexure interaction in situations when

- Flexural stress is at a maximum allowable value and shear stress is greater than 55% of allowable shear stress or
- Shear stress is at a maximum allowable value and bending stress exceeds 70% of allowable flexural stress.

This interaction criterion is plotted in Figure 7.13. An interaction equation can be developed, using a factor of safety (FS) of 1.82, as

$$f_b \leq \left(0.75 - 1.05\frac{f_v}{F_y}\right) F_y \leq 0.55 F_y, \tag{7.63}$$

where f_v is the shear stress in the web; $F_v = 0.35 F_y$ is the allowable web shear stress; f_b is the flexural stress in the web and is equal to M_w/S_w, where M_w is the moment carried by the web plate (see Equation 7.35) and S_w is the web plate section modulus; $F_b = 0.55 F_y$ is the allowable web flexural stress.

7.2.6.1.3.1.4.2 Stability Criteria Shear and flexural buckling may have to be considered together when f_v/τ exceeds 0.40* (Timoshenko and Gere, 1961). For a

---

* When $f_v/\tau < 0.4$ the critical bending stress is negligibly affected by the presence of shear stresses.

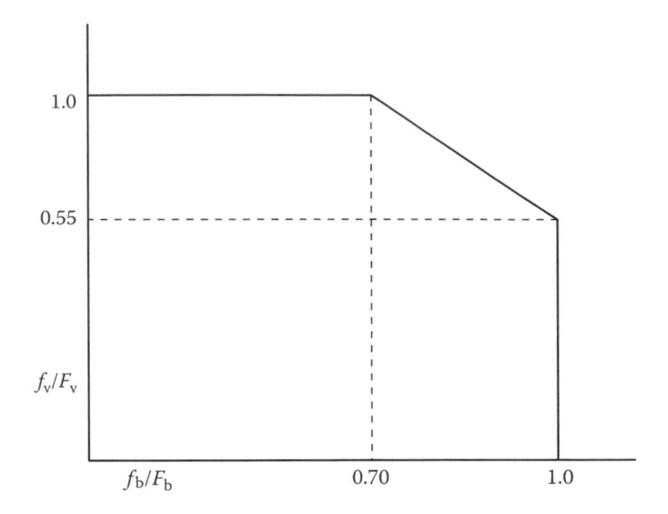

FIGURE 7.13 Web plate combined bending and shear.

simply supported plate a simple circular interaction formula (Equation 7.64) has been found to closely represent experimental data (Galambos, 1988).

$$\left(\frac{f_b}{F_{cr}}\right)^2 + \left(\frac{f_v}{\tau_{cr}}\right)^2 = 1. \tag{7.64}$$

7.2.6.1.3.2 Web Plate Splices AREMA (2008) recommends that splices in the web plates of girders be designed for the following criteria:

- A plate each side of the web with each plate designed for half the shear strength of the gross section of the web plate and having a minimum net moment of inertia of half the net moment of inertia of the web plate.
- The flexural strength of the net section of the web combined with the maximum shear force at the splice.

The web splice fasteners (see Chapter 9) should be designed to transfer the shear force, V, and moment, Ve, due to eccentricity, e, of the centroid of the bolt group from the shear force location. Welded splices are usually made with CJP groove welds with strength at least equal to the base material being spliced. The entire cross section should be welded.

7.2.6.1.4 Girder Flange-to-Web Plate Connection

In modern plate girder spans, the flange-to-web plate connection is made with welds. AREMA (2008) indicates that CJP, PJP, or fillet welds may be used for the flange-to-web connection.

However, PJP or fillet welds in deck plate girders (DPGs) with open decks or noncomposite concrete decks must be designed such that fatigue strength is controlled by weld toe cracking (to preclude cracking in the weld throat). Many design engineers also specify CJP flange-to-web welds for open DPG spans to ensure that vertically applied wheel loads can be safely resisted by the top flange-to-web weld.

7.2.6.1.4.1 Top Flange-to-Web Connection (Simply Supported Girder Spans) In addition to vertical wheel loads, the top flange-to-web weld connection must transmit horizontal shear due to the varying flange bending moment, $\mathrm{d}M$, along the girder length. The change in flange force, $\mathrm{d}P_\mathrm{f}$, due to bending along a length of girder, $\mathrm{d}x$, (Figure 7.14) is

$$\mathrm{d}P_\mathrm{f} = \frac{\mathrm{d}M}{I}\bar{y}(A_\mathrm{f}) = \frac{(V\,\mathrm{d}x)}{I}\bar{y}(A_\mathrm{f}) = \frac{VQ_\mathrm{f}}{I}\,\mathrm{d}x. \tag{7.65}$$

The horizontal shear flow, $q_\mathrm{f} = \mathrm{d}P_\mathrm{f}/\mathrm{d}x$, for which the top (compression) flange weld is designed, is

$$q_\mathrm{f} = \frac{\mathrm{d}P_\mathrm{f}}{\mathrm{d}x} = \frac{VQ_\mathrm{f}}{I}, \tag{7.66}$$

where V is the shear force, $\bar{y}$ is the distance from the top flange centroid to the neutral axis, and $Q_\mathrm{f} = A_\mathrm{f}\bar{y}$ (statical moment of the top flange area about the neutral axis).

The shear force from wheel live load, W, with 80% impact (AREMA, 2008), acting in a vertical direction along the top flange-to-web connection of DPG spans is

$$w = \frac{1.80(W)}{S_\mathrm{w}}, \tag{7.67}$$

where S_w is the wheel load longitudinal distribution ($S_\mathrm{w} = 3$ ft for open deck girders or $S_\mathrm{w} = 5$ ft for ballasted deck girders).

The resultant force per unit length of the weld is

$$q = \sqrt{q_\mathrm{f}^2 + w^2}. \tag{7.68}$$

The required effective area of the weld can then be established based on the allowable weld stresses recommended by AREMA (2008) as shown in Table 7.2 (see also Chapter 9).

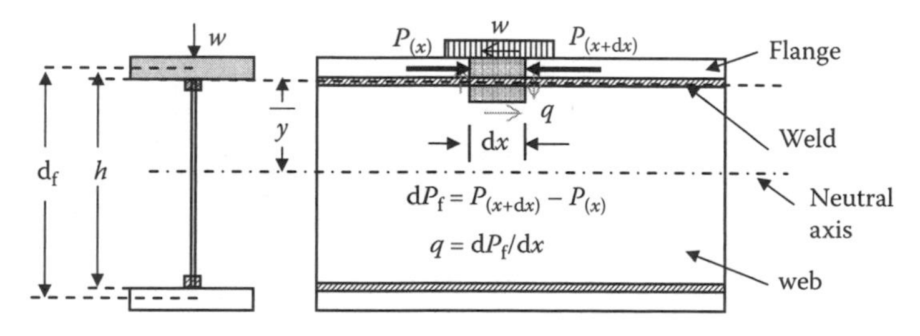

FIGURE 7.14 Forces transferred between the flange and web.

TABLE 7.2
Allowable Weld Stresses

Weld Type	Allowable Shear Stress (ksi)
CJP or PJP	$0.35F_y$
Fillet (60 ksi electrode)	16.5 but $<0.35F_y$ on base metal
Fillet (70 ksi electrode)	19.0 but $<0.35F_y$ on base metal
Fillet (80 ksi electrode)	22.0 but $<0.35F_y$ on base metal

7.2.6.1.4.2 Bottom Flange-to-Web Connection (Simply Supported Girder Spans) The horizontal shear flow for which the bottom (tension) flange weld is to be designed is

$$\Delta q_f = \frac{\Delta V Q_f}{I}, \tag{7.69}$$

where ΔV is the shear force range from live load plus impact, $\bar{y}$ is the distance from the bottom flange-to-web connection to the neutral axis, and $Q_f = A_f(d_f - \bar{y})$ (statical moment of the bottom flange area about the neutral axis). The required effective area of weld can then be established based on the allowable weld fatigue shear stress (typically Category B or B′ depending on weld backing bar usage) for the range of live load shear flow.

7.2.6.1.5 Girder Bearing Stiffeners

Concentrated loads (e.g., reactions at the ends of girders) create localized concentrated compressive stresses that may exceed yield stress. The localized yielding or web crippling may be resisted by web plates of sufficient thickness or by pairs of stiffeners. Web crippling can be conservatively analyzed as shown in Figure 7.15. The minimum

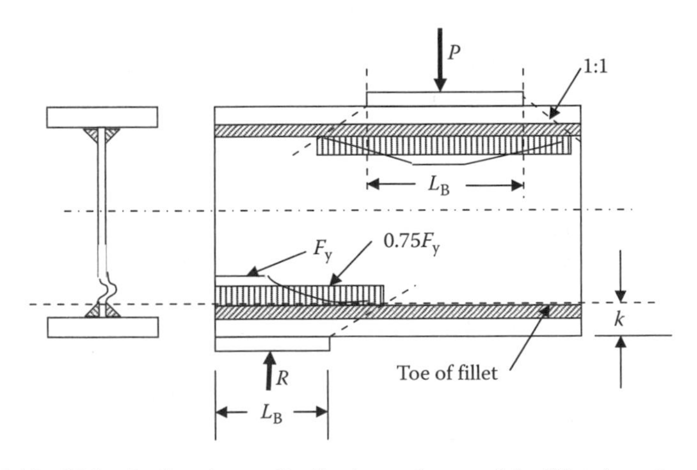

FIGURE 7.15 Web crippling (stress distribution at the toe of the fillet shown).

web plate thickness is

$$t_w \geq \frac{R}{0.75F_y(L_B + k)} \quad \text{for end reaction, } R, \tag{7.70a}$$

$$t_w \geq \frac{P}{0.75F_y(L_B + 2k)} \quad \text{for interior concentrated load, } P, \tag{7.70b}$$

where L_B is the length of bearing.

Steel railway girders will generally require bearing stiffeners due to the high magnitude loads. However, in situations where concentrated loads may not cause web crippling in accordance with Equations 7.70a or 7.70b, it is often advisable to install at least nominal stiffeners, in any case (an example is given in Akesson, 2008). Bearing stiffeners must be connected to both flanges and extend to near the edge of the flange. Bearing stiffeners are designed for the following criteria:

- Compression member behavior (yield and stability).
- Bearing stress.
- Local plate buckling.

7.2.6.1.5.1 Compression Member Behavior of Bearing Stiffeners The bearing stiffener is designed as a compression member with an effective cross section comprising the area of the stiffener elements, A_{bs}, and a portion of the web, A_{wbs}. The effective area, A_{ebs}, and effective moment of inertia, I_{ebs}, of the bearing stiffener cross sections shown in Figure 7.16 are

$$A_{ebs} = 2A_{bs} + A_{wbs} = 2A_{bs} + 12t_w \quad \text{for end reaction, } R, \tag{7.71a}$$

$$I_{ebs} = 2I_{bs} + 2A_{bs}\bar{y}^2 + t_w^4 \quad \text{for end reaction, } R, \tag{7.71b}$$

$$A_{ebs} = 2A_{bs} + 25t_w \quad \text{for interior concentrated load, } P, \tag{7.72a}$$

$$I_{ebs} = 2I_{bs} + 2A_{bs}\bar{y}^2 + 2.08t_w^4 \quad \text{for interior concentrated load, } P. \tag{7.72b}$$

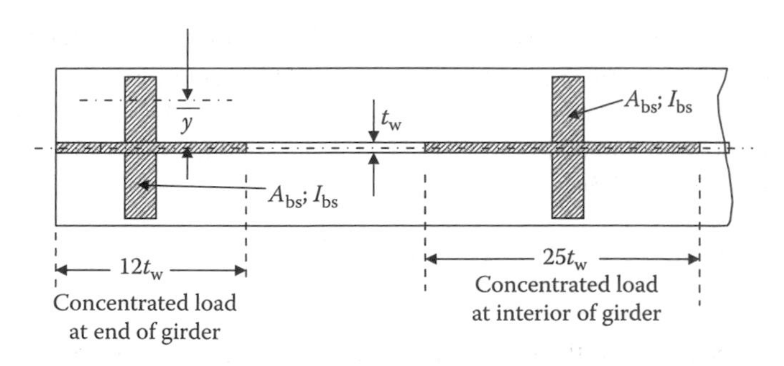

FIGURE 7.16 Bearing stiffener effective cross section.

The bearing stiffener may be designed as a compression member using $K = 0.75$ (AREMA, 2008) (see also Chapter 6) for an allowable compressive stress, F_{call}, of

$$F_{\text{call}} = 0.55F_{\text{y}}; \quad \text{when } \frac{h}{r_{\text{ebs}}} \prec 0.839\sqrt{\frac{E}{F_{\text{y}}}}, \tag{7.73}$$

$$F_{\text{call}} = 0.60F_{\text{y}} - \left(17500\frac{F_{\text{y}}}{E}\right)^{3/2}\left(\frac{0.75h}{r_{\text{ebs}}}\right); \quad 0.839\sqrt{\frac{E}{F_{\text{y}}}} \leq \frac{h}{r_{\text{ebs}}} \leq 6.712\sqrt{\frac{E}{F_{\text{y}}}}, \tag{7.74}$$

$$F_{\text{call}} = \frac{0.685\pi^2 E}{(h/r_{\text{ebs}})^2}; \quad \text{when } \frac{h}{r_{\text{ebs}}} \succ 6.712\sqrt{\frac{E}{F_{\text{y}}}}, \tag{7.75}$$

where h is the height of the bearing stiffener (clear distance between girder top and bottom flanges)

$$r_{\text{esb}} = \sqrt{\frac{I_{\text{esb}}}{A_{\text{esb}}}}.$$

The allowable load on the bearing stiffener is

$$P_{\text{call}} = F_{\text{call}}(A_{\text{ebs}}), \tag{7.76}$$

which should not exceed the maximum reaction, R, or concentrated load, P.

7.2.6.1.5.2 Bearing Stresses Since a part of the bearing stiffener area, A_{bs}, is removed at the top and bottom to clear the beam or girder rolling or weld fillets, the reduced bearing area, A'_{bs}, must be considered. AREMA (2008) recommends an allowable bearing stress for milled stiffeners and parts in contact of $0.83F_{\text{y}}$. Based on this, the reduced bearing stiffener area (area of the bearing stiffener in contact with the flange plate), A'_{bs}, is

$$A'_{\text{bs}} \geq \frac{R}{0.83F_{\text{y}}}. \tag{7.77}$$

7.2.6.1.5.3 Local Plate Buckling Local buckling of bearing stiffeners is essentially the problem of uniform compression on a plate free at one side and partially restrained at the other. The maximum permissible width-to-thickness ratio is, therefore, the same as that established previously for local buckling of a girder compression flange plate (Equation 7.54)

$$\frac{b_{\text{bs}}}{t_{\text{bs}}} \leq 0.43\sqrt{\frac{E}{F_{\text{y}}}}, \tag{7.78}$$

where b_{bs} is the width of the outstanding leg of the bearing stiffener and t_{bs} is the thickness of the outstanding leg of the bearing stiffener.

7.2.6.2 Secondary Girder Elements

Stiffeners are secondary elements, but are of utmost importance to ensure the stability of some main load carrying elements of plate girders. Specifically, the web plate is usually stiffened by transverse, and sometimes, longitudinal stiffeners. The web stiffeners generally consist of welded and/or bolted plates or angles, which may also have an effect on the main member allowable fatigue stress range if the stiffener attachments are within tensile regions of the plate girder* (see Chapter 5).

7.2.6.2.1 Longitudinal Web Plate Stiffeners

Equation 7.57 indicates that if $h/t_w \succ 4.18\sqrt{E/f_c}$, longitudinal stiffeners are required to preclude web flexural buckling instability. The minimum recommended (AREMA, 2008) web plate thickness with longitudinal stiffeners is 50% of Equation 7.56 or

$$t_w \geq 0.09h\sqrt{\frac{F_y}{E}}\sqrt{\frac{f_c}{F_{cr}}} \geq 0.12\sqrt{\frac{f_c}{E}}. \tag{7.79}$$

The longitudinal stiffeners should be proportioned such that they have a flexural rigidity, EI_{ls}, which creates straight nodes in the buckled plate. The plate buckling coefficient, k, for critical buckling stress with a longitudinal stiffener at 25% of web depth is $k = 101$. Using energy methods, it can be shown that (Bleich, 1952)

$$I_{ls} = \frac{ht_w^3}{12(1-\upsilon^2)}\left[\left(12.6 + 50\left(\frac{A_{ls}}{ht_w}\right)\right)\left(\frac{a}{h}\right)^2 - 3.4\left(\frac{a}{h}\right)^3\right]. \tag{7.80}$$

AREMA (2008)† uses a similar formula

$$I_{ls} = ht_w^3\left(2.4\left(\frac{a}{h}\right)^2 - 0.13\right), \tag{7.81}$$

where I_{ls} is the moment of inertia for a single longitudinal stiffener about the face of the web plate (if longitudinal stiffeners are used both sides of the web, the moment of inertia is taken about the centerline of the web‡), $A_{ls} = b_{ls}t_{ls}$ is the cross-sectional area of the longitudinal stiffener, and a is the distance between intermediate transverse stiffeners (Figure 7.17). Equation 7.81 also fits experimental data for pure bending with longitudinal stiffeners at $h/5$ and small values of A_{ls}/ht_w (0.05–0.25) (Salmon and Johnson, 1980).

The thickness of the longitudinal stiffener, t_{ls}, to avoid local buckling is essentially the buckling problem of uniform compression on a plate free at one side and partially

* For example, the bottom of transverse web stiffeners attached near the tension flange in simply supported plate girders. Such attachments, particularly if welded, should be reviewed with respect to their effect on allowable fatigue stress range.

† Many other ASD design codes, recommendations, guidelines, and manuals use the same, or similar, equation.

‡ It is usually not necessary and, therefore, unusual to use longitudinal stiffeners on both sides of the web.

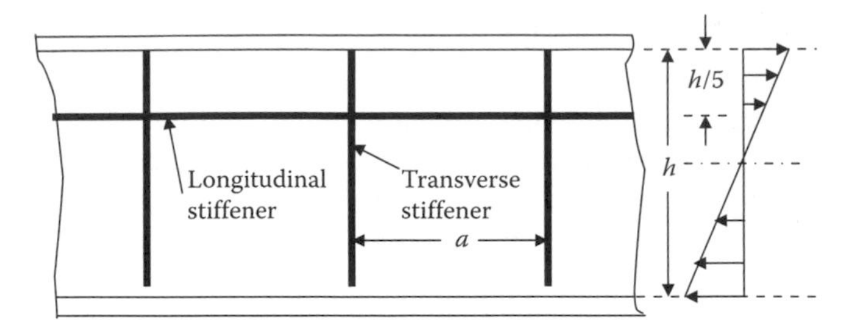

FIGURE 7.17 Web plate under pure bending.

restrained at the other. From Equation 7.52, the width-to-thickness ratio is

$$\frac{b_{ls}}{t_{ls}} \leq 0.67\sqrt{\frac{kE}{F_y}}, \tag{7.82}$$

which is the approximate value corresponding to the transition curve at $F_{cr} = F_y$ (Figure 7.11). The buckling coefficient, k, is 1.277 for plates with one edge fixed and the other edge free (typical of a longitudinal stiffener) (Bleich, 1952). Substitution of $k = 1.277$ into Equation 7.82 yields

$$\frac{b_{ls}}{t_{ls}} \leq 0.76\sqrt{\frac{E}{F_y}}. \tag{7.83}$$

Assuming the actual calculated stress $f = F_y$ and applying a safety factor of 1.82

$$\frac{b_{ls}}{t_{ls}} \leq 0.42\sqrt{\frac{E}{f}}, \tag{7.84}$$

$$t_{ls} \geq 2.39 b_{ls}\sqrt{\frac{f}{E}}, \tag{7.85}$$

from which the minimum longitudinal stiffener dimensions for stability can be determined.

7.2.6.2.2 Transverse Web Plate Stiffeners

Recent investigations have determined that it is appropriate to consider the design of transverse web stiffeners as flexural members resisting bending forces created by the restraint that the transverse stiffener imposes on lateral deflections of the web plate at the shear strength limit state (Kim et al., 2007). Therefore, if transverse stiffeners are required, the necessary spacing, a, to provide adequate rigidity through creation of nodal lines is determined by considering that AREMA (2008) restricts $a/h \leq 1$ so that the shear buckling coefficient is

$$k = 4.0 + \frac{5.34}{(a/h)^2}. \tag{7.86}$$

Therefore, the critical shear buckling stress is

$$\tau_{cr} = \frac{\left(4.0 + [5.34/(a/h)^2]\right)\pi^2 E t_w^2}{12(1-\upsilon^2)h^2}, \tag{7.87}$$

which may be rearranged to provide

$$\frac{a}{t_w} = \sqrt{\frac{5.34\pi^2}{12(1-\upsilon^2)(\tau_{cr}/E) - 4\pi^2(t_w/h)^2}}. \tag{7.88}$$

Using an FS of 1.5 Equation 7.88 is*

$$\frac{a}{t_w} = \sqrt{\frac{1.336}{1.5(0.277)(\tau_{cr}/E) - (t_w/h)^2}}. \tag{7.89}$$

Flexural buckling is precluded (Equation 7.44) where

$$\frac{h}{t_w} \leq 4.18\sqrt{\frac{E}{F_{cr}}} \leq 5.64\sqrt{\frac{E}{F_y}} \tag{7.90}$$

and considering $\tau_{cr} \leq 0.35F_y$ Equation 7.89 becomes

$$\frac{a}{t_w} = 3.42\sqrt{\frac{E}{F_y}} \tag{7.91}$$

or (with $\tau = 0.35F_y$)

$$\frac{a}{t_w} = 2.02\sqrt{\frac{E}{\tau}} \tag{7.92}$$

and AREMA (2008) recommends

$$a \leq 1.95 t_w \sqrt{\frac{E}{\tau}} \tag{7.93}$$

to establish transverse stiffener spacing. Based on web plate imperfections AREMA (2008) also provides a practical recommendation for maximum stiffener spacing of 96 in.

An equation for the required moment of inertia of a transverse web stiffener was developed from analytical and experimental tests as (Bleich, 1952)

$$I_{ts} = \frac{4 a t_w^3}{12(1-\upsilon^2)}\left(7\left(\frac{h}{a}\right)^2 - 5\right), \tag{7.94}$$

* AREMA (2008) uses the relatively lower factor of safety of 1.5 for shear buckling in recognition of the postbuckling strength of web plates in shear.

which may be simplified to

$$I_{ts} = 2.5a_0t_w^3\left(\left(\frac{h}{a}\right)^2 - 0.7\right), \quad (7.95)$$

where a_0 is the actual stiffener spacing used in design which must be $\leq h$ (see development of Equation 7.86) and ≤ 96 in.; a is the required stiffener spacing (Equation 7.93).

In elastic design the stiffeners are not required to carry a force* and, therefore, there is no need to design them, nor their connections, for strength. The size of the transverse stiffener is determined from Equation 7.95, which is based solely on rigidity considerations, and nominal welded or bolted connection to the web is typically used. AREMA (2008) recommends connection to the compression flange to provide additional stability to both the stiffener itself and against torsional buckling of the compression flange.† This connection may be accomplished with bolts or fillet welds or by careful grinding to ensure a uniform and tight fit against the flange. Wrap around fillet welds must not be used for welding transverse stiffeners to either the web or the compression flange.

However, if lateral bracing is attached to a stiffener, the connection at the top flange must be designed to transmit 2.5% of the compression flange force and other lateral forces from wind, centrifugal, or nosing (see Chapters 4 and 5). From a lateral bracing perspective, the connection at the bottom flange is less important as it resists only forces from wind. The stiffener connection to the web must also be designed to resist the forces from out-of-plane bending of the beams or girders they are connected to and forces due to lateral distribution of the live load.‡ AREMA (2008) recommends that transverse web stiffeners be adequately attached to both the top and bottom flanges when bracing is connected to the stiffeners (although not required for rolled beams on single track spans without skew or curvature).

The web plate stiffeners must not be welded to the tension flange, as such a transverse weld is a very poor fatigue detail. Furthermore, welds connecting transverse stiffeners to web plates should not be made close to the tension flanges because of stress concentration effects. Extensive testing and analytical work have established that the stiffener weld should be a minimum of 4–6 times the web plate thickness from the near toe of the tension flange-to-web weld (Basler and Thurlimann, 1959). AREMA (2008) recommends this distance as $6t_w$. Careful consideration of details is required (such as provision of bolted angles at the bottom of the stiffeners (D'Andrea et al., 2001) or peening pretreatments§) where brace frames are attached to transverse stiffeners that may precipitate out-of-plane distortion in the web gap. Even though fabrication cost is increased, some design engineers will provide bolted transverse

---

* Such as a compressive force if tension field action (postbuckling) is assumed in the web (analogous to truss post with web behavior like truss diagonal).

† Necessary, in particular, for relatively thin top flanges with open deck construction where lateral flexure of the top flange may precipitate weld cracking at the transverse stiffener-to-compression flange connection.

‡ Typically present in superstructures with relatively large skew, curvature, or track eccentricity.

§ Ultrasonic impact treatment is a modern pretreatment to improve these poor fatigue details at the base of welded transverse stiffeners (Roy and Fisher, 2006).

stiffener connections, particularly when they serve as bracing connection plates, in order to preclude detrimental out-of-plane web gap weld fatigue effects.

7.2.7 Box Girder Design

Box girders have a high flexural capacity and torsional rigidity. The design of box girders is generally analogous to the design of plate girders. However, the large compression flange makes it necessary to utilize stiffened steel plates or concrete slabs. Steel plate decks are often used when span lengths are large enough that the dead load from concrete slab decks becomes a disproportionate portion of the total load on the span. The top flange typically also serves as the ballasted deck.

7.2.7.1 Steel Box Girders

Steel box girders typically employ an orthotropic steel deck plate. The strength (yield and stability), fatigue, and serviceability design of orthotropic deck plates require careful consideration of fabrication and details. The design of orthotropic plate deck bridges is beyond the scope of this book and the reader is referred to books by Wolchuk (1963), Cusens and Pama (1979), Troitsky (1987), and others that provide definitive information regarding the analysis and design of orthotropic steel deck plate bridges.

7.2.7.2 Steel–Concrete Composite Box Girders

The design of steel–concrete composite box girder spans is also generally analogous to that for steel–concrete composite plate girder spans. The latter are discussed in greater detail in this chapter.

7.3 SERVICEABILITY DESIGN OF NONCOMPOSITE FLEXURAL MEMBERS

AREMA (2008) recommends that mid-span deflection of simply supported spans due to live load plus impact, $\Delta LL + I$, not exceed $L/640$, where L is the span length. Some engineers or bridge owners recommend even more severe live load plus impact deflection criteria to attain stiffer spans, which offer improved performance from a track–train dynamics perspective.

It is also recommended that camber be provided for dead load deflections in spans exceeding 90 ft. Camber of truss spans is recommended to account for deflections from dead load plus a live load of 3000 lb per track foot.

The serviceability criteria for steel railway spans are also discussed in Chapter 5.

Example 7.1

A 90 ft simple span steel ($F_y = 50$ ksi) DPG is to be designed for the forces shown in Table E7.1.

TABLE E7.1

Design Force	Shear Force, V (kips)	Bending Moment, M (ft-kips)
Dead load, DL	110	2500
Live load + 37% impact (maximum)	376	7314
Maximum (DL + LL)	486	9814
Live load + 13% impact (fatigue)	310	6032

Preliminary girder design:

Use a girder depth of 90/12 = 7.5′ = 90″ (see Chapter 3) and assume 2-1/2″ thick flanges (reasonable practical thickness*);

$$t_w \geq \frac{486}{0.35(50)(90-5)} \geq 0.33'',$$

use 3/8″ (a minimum of 0.335″ as per AREMA, 2008).

However, a minimum web slenderness of 134 is recommended for 50 ksi steel and $t_w \geq 90 - 5/134 \geq 0.63''$ without longitudinal stiffeners (0.32″ with longitudinal stiffeners). Use a 5/8″ thick web without longitudinal stiffeners. The designer should confirm that a 5/8″ thick plate is available with a minimum width of 86″ (allowing for trimming) in sufficient lengths to avoid excessive or poorly located vertical web plate splices.

$$A_f = \frac{7314(12)}{16(90-5)} - \frac{53.1}{6} = 55.7\ \text{in.}^2$$

($f_b = 16$ ksi is the allowable stress for fatigue Category B with no welded attachments, which means that, if transverse web stiffeners are required, a bolted connection to the web will be necessary)

$$A_f = \frac{9814(12)}{0.55(50)(90-5)} - \frac{53.1}{6} = 41.5\ \text{in.}^2$$

($f_b = 27.5$ ksi is the allowable bending stress).

Use 2-1/2″ × 20″ ($A_f = 50.0\ \text{in.}^2$); the designer should confirm that a 2-1/2″ thick plate is available in sufficient lengths to avoid excessive or poorly located transverse flange plate splices.

The girder section properties are shown in Figure E7.1 and Table E7.2.

$$\frac{\sum Ay_b}{\sum A} = \frac{6890.6}{153.13} = 45.00\ \text{in.},$$

* This is a reasonable plate thickness from an economic and technical (lamellar tearing, grain size) perspective.

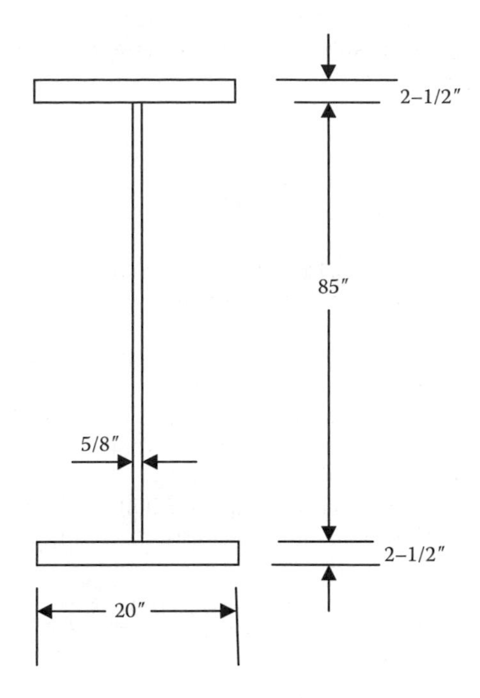

FIGURE E7.1

$$I_g = \sum A(y_b - y)^2 + \sum I_o = 191{,}406 + 32{,}038 = 223{,}444\ \text{in.}^4,$$

$$r_{cy} = \sqrt{\frac{I_{cy}}{A}} = \sqrt{\frac{[42.5(5/8)^3 + 2.5(20)^3]}{12[20(2.50) + 42.5(5/8)]}} = 4.67\ \text{in.},$$

$$S_g = \frac{223{,}444}{45.0} = 4965\ \text{in.}^3,$$

$$S_n = \frac{n_{ng}(223{,}444)}{45.0} = 4965 n_{ng}\ \text{in.}^3,$$

$n_{ng} = I_n/I_g$ (net-to-gross area ratio).
Weight of the girder = 153.1(490/144) = 521 lb/ft (46,900 lb total per girder).

TABLE E7.2

Element	A (in.^{2})	y_b (in.)	Ay_b (in.^{3})	$y_b - y$ (in.)	$A(y_b - y)^2$ (in.^{4})	I_o (in.^{4})
Top flange	50.00	88.75	4437.5	43.75	95,703	26.0
Web	53.13	45.00	2390.6	0	0	31,986
Bottom flange	50.00	1.25	62.5	−43.75	95,703	26.0
$\sum$	153.13		6890.6		191,406	32,038

Assume that $n_{ng} = 0.90$ (subsequent check after detailing bolted connections should be made),

$$I_n \sim 0.90(223{,}444) \sim 201{,}100\ \text{in.}^4$$

$$S_n \approx \frac{201{,}100}{45} \sim 4469\ \text{in.}^3$$

$\sigma_{tmax} \sim 9814(12)/4469 = 26.4$ ksi$< F_{tall} < 0.55(50) < 27.5$ ksi OK (a subsequent check of assumed dead load after final proportioning and detailing should be made).

For a 90 ft long plate girder, design for $> 2{,}000{,}000$ cycles with a mean impact percentage of 35% of the maximum impact.

$\Delta\sigma_{max} \sim 6032(12)/4469 = 16.2$ ksi. Therefore, details with a fatigue detail less than Category B (allowable fatigue stress range $= 16$ ksi) should not be used near the bottom flange area of the span.

$\tau_{max} = 486/(85)(5/8) = 9.15$ ksi$< F_{rail} < 0.35(50) < 17.5$ ksi OK.

Equation 7.24 yields the maximum unsupported length as

$$\left(\frac{L}{r_y}\right)_{max} \leq \sqrt{\frac{2(6.3)\pi^2(29{,}000)}{(50)^2}} \leq 38.0$$

$L \leq (38.0)(4.67) = 177$ in. $= 14.75$ ft (AREMA, 2008 recommends a maximum of 12 ft),

$$F_{call} = 0.55(50) - \frac{0.55(50)^2}{6.3\pi^2(29{,}000)}\left(\frac{144}{4.67}\right)^2 = 27.5 - 0.73 = 26.8\ \text{ksi}$$

$> \sigma_{cmax} > 9814(12)/4965 = 23.7$ ksi OK.

The girder will be laterally supported (after erection) by brace frames at a maximum spacing of 12 ft.

Girder design (for fabrication and erection loads):

During fabrication and erection the entire girder may be laterally unsupported. It is assumed that erection lifts will not be done coincident with windy conditions.

girder self-weight ~1.15(521)~600 lb/ft (with 15% contingency load),

$$\sigma_{girder} = [0.6(90)^2/8](12)/4469 = 1.6\ \text{ksi}$$

using an allowable bending stress of $1.25(0.55)F_y = 0.69F_y$ (see Table 4.5),

$$\left(\frac{L}{r_y}\right)_{girder} = \sqrt{\frac{0.69(50) - 1.6}{0.69(50)^2/6.3\pi^2(29{,}000)}} = 185,$$

and the maximum unsupported length for fabrication and erection is $L = 4.67(185)/12 = 72.1$ ft (80% of girder length) using a basic allowable stress of $1.25[0.55F_y]$.

Detailed design of a tension flange:

$$S_{xn} = \frac{I_{xn}}{c_t} \geq \frac{M_{t\,max}}{F_{all}} \geq \frac{M_{t\,max}}{0.55F_y} \geq \frac{9814(12)}{0.55(50)} \geq 4282\ \text{in.}^3\ \text{(OK, 4469 in.}^3\ \text{provided)}$$

for Category B weld;

$$S_{xn} \geq \frac{M_{max\text{-}range}}{F_{fat}} \geq \frac{(6032)(12)}{16} \geq 4524\ \text{in.}^3\ \text{(OK, \sim1\% overstress with 4469 in.}^3\ \text{provided).}$$

A subsequent check on the net section is required after the detailed design is completed.

Detailed design of a compression flange:
Brace frames will be placed at equal intervals of 11.25 ft

$$\frac{L_p}{r_{cy}} = \frac{11.25(12)}{4.67} = 28.9 \leq 5.55\sqrt{\frac{E}{F_y}} \leq 134\ \text{OK},$$

$$F_{call} = 0.55(50) - \frac{0.55(50)^2}{6.3\pi^2(29{,}000)}\left(\frac{(11.25)(12)}{4.67}\right)^2 = 27.5 - 0.64 = 26.9\ \text{ksi},$$

or

$$F_{call} = \left(\frac{0.13\pi(29{,}000)(50.0)}{(11.25)(12)(90)(\sqrt{1+0.3})}\right) = 42.8\ \text{ksi}.$$

Since $F_{call} \leq 0.55\,F_y \leq 27.5$; $F_{call} = 26.9$ ksi

$$S_{xg} = \frac{I_{xg}}{c_c} \geq \frac{M_{c\,max}}{F_{call}} \geq \frac{9814(12)}{26.9} \geq 4378\ \text{in.}^3\ \text{(OK, 4965 in.}^3\ \text{provided).}$$

Vertical buckling of the compression flange is avoided by precluding flexural buckling of the web when

$$t_w \geq 0.18h\sqrt{\frac{F_y}{E}}\sqrt{\frac{f_c}{F_{cr}}} \geq 0.18(85)\sqrt{\frac{50}{29{,}000}}\sqrt{\frac{9814(12)/4965}{26.9}} \geq 0.64\sqrt{\frac{23.7}{26.9}}$$
$$\geq 0.60\ \text{in. (OK, 0.625 in. provided).}$$

Local buckling of the compression flange:
The plate girder will have ties directly supported on the top (compression) flange; therefore

$$\frac{b}{2t_f} = \frac{20}{2(2.50)} = 4.0 \leq 0.35\sqrt{\frac{E}{F_y}} \leq 8.4\ \text{OK}$$

Detailed design of a web plate:

$$A_w \geq \frac{V}{0.35F_y} \geq \frac{486}{0.35(50)} \geq 27.8\,\text{in.}^2 \text{ (OK, } 53.13\,\text{in.}^2 \text{ provided).}$$

Flexural buckling:

$$\frac{h}{t_w} = \frac{85}{0.625} = 136 \leq 4.18\sqrt{\frac{E}{f_c}} \leq 4.18\sqrt{\frac{29{,}000}{22.6}} \leq 150.$$

Therefore, no longitudinal stiffeners are required for web flexural buckling stability. This was also shown in the calculation relating to the compression flange vertical buckling above.

Shear buckling:

$$h = 85 \geq 2.12t_w\sqrt{\frac{E}{F_y}} \geq 2.12(0.625)\sqrt{\frac{29{,}000}{50}} \geq 31.9\,\text{in.}$$

Therefore, transverse web stiffeners are required.

Combined bending and shear:

$$f_b = \frac{9814(12)}{4469} = 26.3 \leq \left(0.75 - 1.05\frac{f_v}{F_y}\right)F_y$$

$$\leq \left(0.75 - 1.05\frac{486/53.13}{50}\right)50\,\text{ksi} = 27.9.$$

However, $f_b = 0.55\,F_y = 27.5$ ksi OK.

This interaction criteria does not generally require checking, and in the case where $f_v/(0.35F_y) = 0.52 < 0.55$ (see Figure 7.13) and the moment carried by the web plate is approximately $((53.1/6)/(53.1/6 + (20)(2.5)))100 = 15.0\,\%$ of the total moment, combined bending and shear need not be considered.

Flange-to-web connection:
Top weld:

$$q = \sqrt{\left(\frac{VQ_f}{I}\right)^2 + \left(\frac{1.80W}{S_w}\right)^2} = \sqrt{\left(\frac{486(50.0)(43.75)}{223{,}444}\right)^2 + \left(\frac{1.80(40)}{3(12)}\right)^2}$$

$$= \sqrt{4.62^2 + 2.00^2} = 5.04\,\text{k/in.}$$

For a CJP weld, the weld size must be $\geq 0.5\sqrt{2}(5.04)/(0.35(50)) \geq 0.20$ in. OK since the web thickness is 0.625 in.

Bottom weld:

$$\Delta q = \frac{\Delta VQ_f}{I} = \frac{(310)(50.0)(43.75)}{201{,}100} = 3.37\,\text{k/in.}$$

For a CJP weld with backing bar removed, the weld size must be $\geq 0.5\sqrt{2}(3.37)/16 \geq 0.15$ in. OK since the web thickness is 0.625 in.

In general, CJP weld design does not need to be considered.

Design of Web Plate Stiffeners:

Use angles bolted to the web in order to preclude fatigue issues related to welding at the base of the transverse stiffeners.

The spacing of the stiffeners is

$$a_0 < h = 85\text{ in.}$$

$$< 96\text{ in.}$$

$$< a \leq 1.95t_w\sqrt{\frac{E}{\tau}} \leq 1.95(0.625)\sqrt{\frac{29{,}000}{486/53.13}} \leq 68.6\text{ in.}$$

Use a stiffener spacing of $(11.25)(12)/2 = 67.5$ in. (1/2 the distance between brace frames)

$$I_{ts} = 2.5a_0t_w^3\left(\left(\frac{h}{a}\right)^2 - 0.7\right) = 2.5(67.5)(0.625)^3\left(\left(\frac{85}{68.6}\right)^2 - 0.7\right) = 34.4\text{ in.}^4.$$

As shown in Figure E7.2a, a single $6 \times 4 \times \frac{1}{2}$ angle on one side of the web plate provides $I_{ts} = 17.4 + 4.75(1.99)^2 = 36.2\text{ in.}^4$.

Design of Bearing Stiffeners:

As shown in Figure E7.2b, the bearing stiffeners are to consist of $4 - L8 \times 4 \times \frac{1}{2}$.

$$A_{ebs} = 2A_{bs} + A_{wbs} = 2A_{bs} + 12t_w = 2(2)(5.75) + 12(0.625) = 30.50\text{ in.}^2,$$

$$I_{ebs} = 2I_{bs} + 2A_{bs}\bar{y}^2 + t_w^4 = 2(2)(38.5) + 2(2)(5.75)(3.17)^2 + (0.625)^4$$

$$= 385.6\text{ in.}^4,$$

$$r_{ebs} = \sqrt{\frac{385.6}{30.50}} = 3.56\text{ in.},$$

$$\frac{h}{r_{ebs}} = \frac{85}{3.56} = 23.9,$$

$0.839\sqrt{E/F_y} = 20.2$ and, therefore, the governing expression for allowable compressive stress is Equation 7.74

$$F_{call} = 0.60F_y - \left(17{,}500\frac{F_y}{E}\right)^{3/2}\left(\frac{0.75h}{r_{ebs}}\right) = 30 - 0.124(17.9) = 27.8\text{ ksi},$$

$$P_{call} = (27.8)(30.50) = 847\text{ kips} > 486\text{ kips OK.}$$

The reduced bearing stiffener area assuming a 1/2″ clip is made to clear the fillet at the web-to-flange junction (from double bevel CJP groove welds), $A'_{bs} = 2(2)(8 - 0.50)(0.5) = 15.0\text{ in.}^2$.

$$R = 486 \leq A'_{bs}(0.83F_y) \leq (15.0)(0.83)(50) \leq 622.5\text{ kips OK}$$

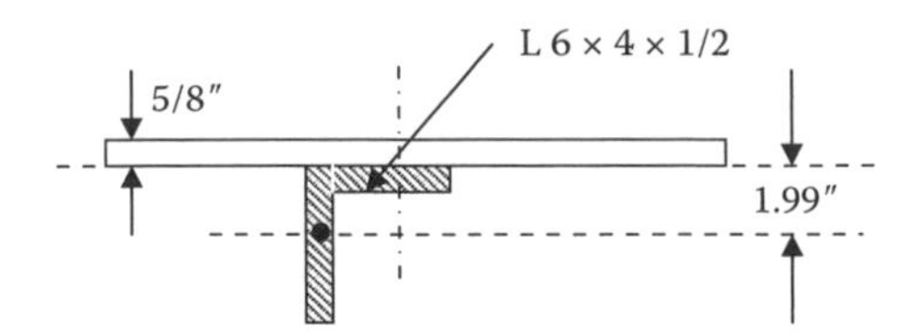

FIGURE E7.2a

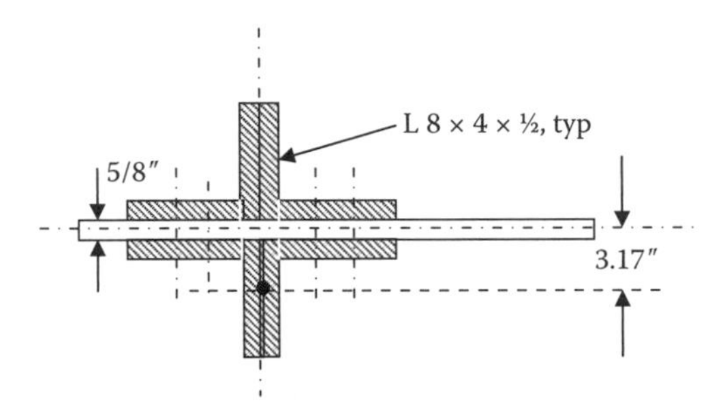

FIGURE E7.2b

Local buckling of outstanding compression elements:

$$\frac{b_{bs}}{t_{bs}} = \frac{8}{0.5} = 4.0 \leq 0.43\sqrt{\frac{E}{F_y}} \leq 10.4 \text{ OK}$$

Serviceability–deflection criteria

The required gross moment of inertia for a LL + I deflection criterion of L/f_Δ is (see Chapter 5)

$$I \geq \frac{M_{LL+I} L f_\Delta}{1934} \geq \frac{7314(90) f_\Delta}{1934} \geq 340.4 f_\Delta \text{ in.}^4.$$

The required section gross moment of inertia for various deflection criteria, f_Δ, is shown in Table E7.3.

The section $I_g = 223{,}444$ in.^{4} provides a deflection that is 2.5% less than the AREMA (2008) criteria.

TABLE E7.3

Deflection Criteria, L/Δ_{LL+I}	Required Gross Moment of Inertia, I_g (in.^{4})
500	170,180
640 (AREMA, 2008)	217,832
800	272,290
1000	340,362

The deflections are estimated as (see Chapter 5)

$$\Delta_{LL+I} = \frac{0.104 M_{LL+I} L^2}{EI} = \frac{0.104(7314)(12)[90(12)]^2}{29{,}000(223{,}444)} = 1.64\text{ in.},$$

$$\Delta_{DL} = \frac{0.104 M_{DL} L^2}{EI} = \frac{0.104(2500)(12)[90(12)]^2}{29{,}000(223{,}444)} = 0.56\text{ in.},$$

therefore consider a camber of 1/2″.

In practice, dimensions of the various main and secondary elements may be revised further to attain greater economy of material.

7.4 STRENGTH DESIGN OF STEEL AND CONCRETE COMPOSITE FLEXURAL MEMBERS

Railway bridges with ballasted decks are beneficial from operational, structural, and maintenance perspectives (see Chapter 3). The ballast may be placed on timber, steel, or concrete decks.

Timber decks are generally not effective from structural and maintenance perspectives. Steel plate decks have only the strength or stiffness to span small lengths under railway live loads. Therefore, steel plate decks are generally not appropriate on deck type bridges with girders or trusses spaced at wide distances. Floor systems are needed to support stiffened or unstiffened steel plate decks unless the longitudinal members are closely spaced.* The use of steel plate decks† is often appropriate for long span construction to reduce the superstructure dead load stresses. Steel quantity, fabrication (welding), and shipping/erection (deck plate size) considerations need to be carefully reviewed for steel deck plate bridges.

Reinforced and/or prestressed concrete decks have the strength and stiffness required for use as ballasted decks in ordinary steel railway bridge construction (Figure 7.18). Concrete decks may be noncomposite (not positively connected to the steel bridge span) or made composite. Relative slip between the concrete deck and steel span will occur with noncomposite construction and, even with the substantial dead load, the deck may translate under the action of modern train braking and locomotive traction forces (see Chapter 4). Composite steel and concrete construction has the following benefits for steel railway bridge construction:

- Ease of site access for the materials used in railway bridge construction (provided that concrete transport or site batching is available, there is reduced shipping and erection of large steel sections and/or plate decks requiring field bolting).
- Improved train ride and reduced track and deck maintenance (given that adequate deck drainage and waterproofing are provided).

* The fabrication and erection difficulties associated with steel decks on closely spaced longitudinal members may make them impractical except for use in short spans.

† In particular, orthotropic steel plate decks.

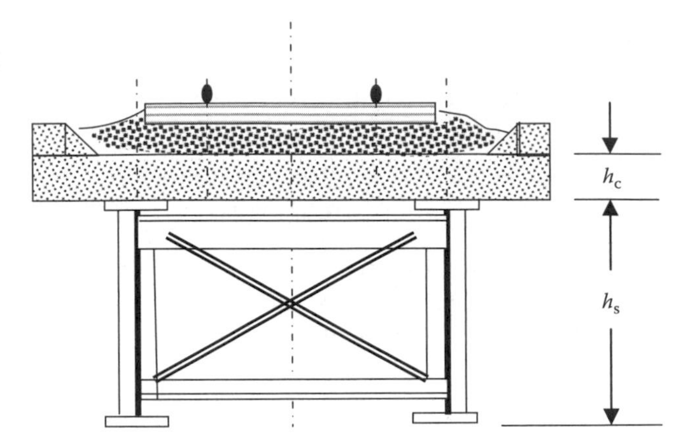

FIGURE 7.18 Cross section of a typical composite steel and concrete DPG span.

- Reduction in weight of fabricated steel (typically between 10% and 20%).
- Reduction of superstructure depth (may be required for clearances, see Chapter 3).
- Increased superstructure stiffness (improved performance under live load).
- Increased capacity against overloads.
- Deck acts to resist lateral forces at the top flange of girders (no need for horizontal bracing and reduced requirements relating to vertical bracing, see Chapter 5). (AREMA (2008) also recommends that noncomposite decks with at least 1 in. of steel flange embedded in concrete be considered as sufficient lateral resistance to preclude the need for top lateral bracing.)

Therefore, while composite construction is often utilized, its effectiveness depends on the mechanical interaction between the concrete deck and steel superstructure. This mechanical connection between steel and concrete is usually accomplished with proprietary shear studs developed specifically for this purpose. However, other mechanical shear transfer connectors have been used to attain composite action. AREMA (2008) recommends the use of shear studs of 3/4 or 7/8 in. diameter, d, with a minimum length of $4d$ or channels with a minimum height of 3 in. Mechanical connectors are typically welded to the steel superstructure and embedded in the concrete deck when cast (for cast-in-place construction) or within grouted recesses (for precast concrete construction*).

Steel girders, beams, floorbeams, and stringers are built as composite flexural members in railway bridge spans and must be designed to resist the internal normal and shear stresses created by combinations of external actions at various limit states (Table 4.5). However, in addition to the usual strength (yielding and stability), fatigue, fracture, and serviceability design criteria, structures of composite materials require consideration of stiffness and strain compatibility at the interface between the steel and concrete materials.

* Care must be exercised in precast concrete deck slab construction to ensure that grouted and mortared recesses, joints, and slab bedding are properly designed and constructed.

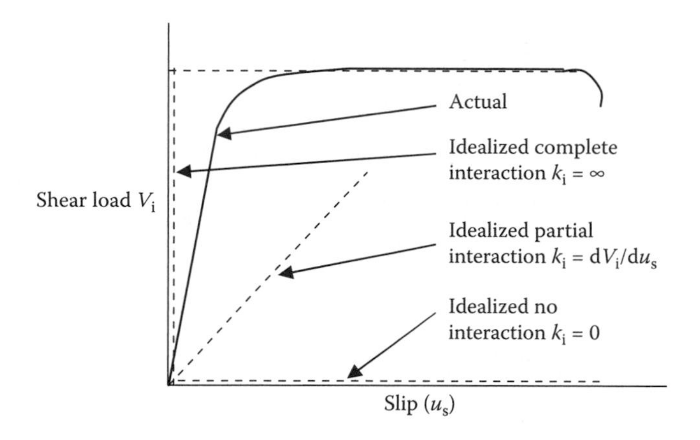

FIGURE 7.19 Stiffness of the interface connection.

7.4.1 Flexure in Composite Steel and Concrete Spans

When a composite steel and concrete span bends, the horizontal shear at the steel to concrete interface must be resisted in order that the materials act integrally. The number and size of mechanical connectors governs the interface behavior in terms of shear strength and stiffness of the connection. The allowable stress design method of AREMA (2008) provides for a linear elastic analysis at service loads. Linear elastic analysis may be used provided complete (or near complete) interaction (i.e., adequate horizontal stiffness) occurs at the interface. The flexural stress is then

$$\sigma = \frac{Mc}{I}, \tag{7.96}$$

where M is the bending moment, σ is the normal (flexural) stress, c is the distance from the neutral axis to the extreme fiber, and I is the moment of inertia.

The horizontal stiffness of the interface connection, k_i, is determined from the horizontal shear load versus interface slip relationship (Figure 7.19). All connections between concrete decks and steel superstructures will exhibit some degree of slip* or partial interaction. Partial interaction analysis requires consideration of nonlinear behavior, although "equivalent" linear elastic analyses of partial interaction have been developed (Newmark et al., 1951). Complete interaction enables a linear elastic analysis to be performed with a connection stiffness, $k_i = \infty$.

The relationship between interface connection stiffness, slip, and slip strain in a simply supported composite steel and concrete span is shown in Figure 7.20. With $k_i = 0$ no interaction occurs and with $k_i = \infty$ complete interaction occurs. As indicated above, practical structures will behave between these extremes and exhibit partial interaction. However, because of the relatively large number of shear connectors required for strength in railway girders, the connection stiffness will be very large and may be idealized as infinitely stiff with complete interaction. Complete interaction strain compatibility indicates that, slip, $u_s = 0$, and slip strain, $du_s/dx = \varepsilon_c - \varepsilon_s = 0$,

* Slippage must occur in order to mobilize shear connector resistance.

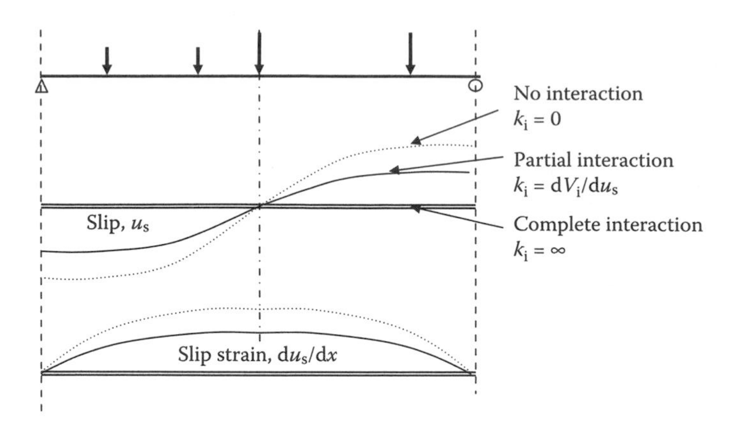

FIGURE 7.20 Slip and slip strain distribution in a simply supported composite beam.

need not be considered in the flexural analysis. The strain distribution through a composite steel and concrete beam is shown in Figure 7.21 for no, partial, and complete interaction.

Transformed section methods may be used to determine cross section stresses at service load levels since complete interaction allows for a linear elastic analysis (elastic E_s and E_c), with the same stress and strain profile (Gere and Timoshenko, 1984). The modular ratio, $n = E_s/E_c$, can be established and used as the transformation ratio for steel and concrete elements. However, long-term dead load stresses do not remain constant with time because of creep and shrinkage of the concrete deck. The dead load stresses will increase in the steel elements. Long-term effects are considered by a simplified approach to shrinkage and creep that uses a plastic modulus $n_{cr} = 3n$ (Viest, Fountain and Singleton, 1958).

The elastic stress distribution will also depend on how load is transferred to the composite steel concrete span (i.e., dependent on construction scheme). If the steel

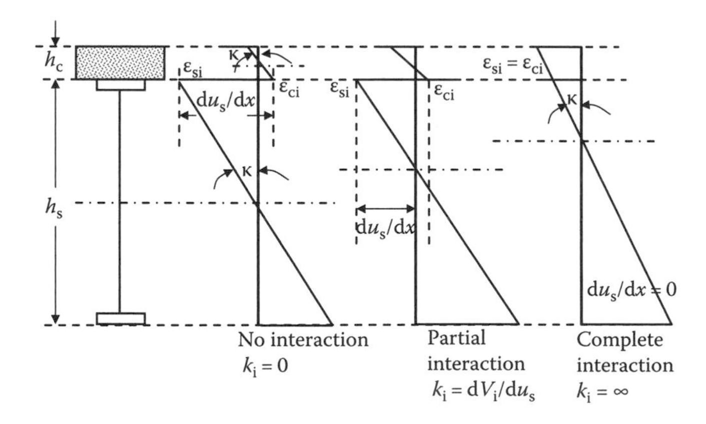

FIGURE 7.21 Strain profile through composite beam at various connection stiffness (no, partial or complete interaction) (κ is the curvature of the beam.

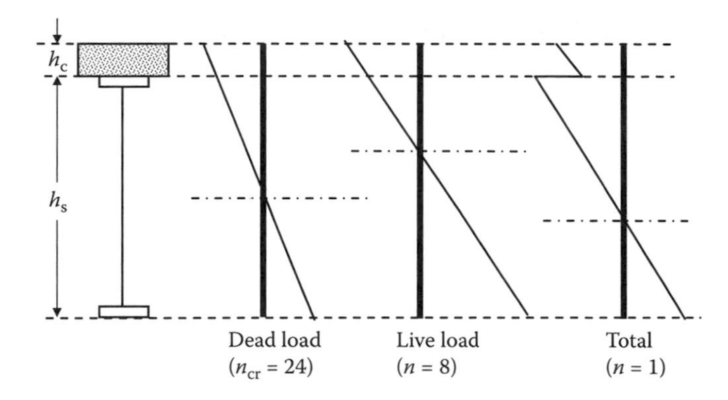

FIGURE 7.22a Stress profile through a composite beam using shored or falsework supported construction.

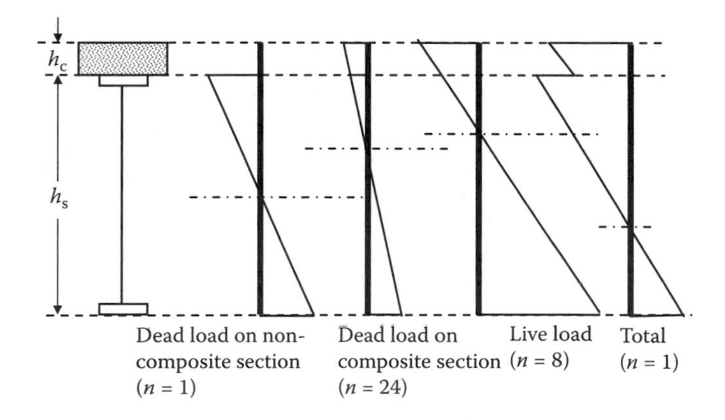

FIGURE 7.22b Stress profile through a composite beam using unshored or unsupported construction.

beams are supported by shoring or falsework until after the concrete deck hardens,* the entire composite section resists load. If, however, the construction is unshored (typical of railway bridges constructed over waterways, highways, or other railways) the composite section does not resist load until after the concrete slab hardens and the steel beams or girders must resist a portion of the dead load (concrete deck, steel beam, and construction equipment). The stress distributions for complete interaction through the composite section for these two construction methods are shown in Figures 7.22a and 7.22b.

---

* AREMA (2008) indicates that this occurs when the concrete has attained 75% of its specified 28-day compressive strength.

7.4.2 Shearing of Composite Beams and Girders

The linear elastic shear stress on the composite section is

$$\tau = \frac{VQ}{It} \tag{7.97}$$

and the shear flow at any section is

$$q = t\tau = \frac{VQ}{I}, \tag{7.98}$$

where V is the shear force, τ is the shear stress, q is the shear flow (along the length of girder), I is the moment of inertia, $Q = Ay_{\text{na}}$ is the statical moment of area about the neutral axis, and t is the thickness of the element.

7.4.2.1 Web Plate Shear

AREMA (2008) recommends that shear should be resisted by the steel girder web only. With maximum shear occurring at the neutral axis of the web plate, the minimum gross cross-sectional area of the steel girder web plate, A_{w}, is (Equation 7.32)*

$$A_{\text{w}} \geq \frac{V}{0.35F_{\text{y}}}. \tag{7.99}$$

Pure flexural, pure shear, and combined flexural and shear buckling of the web plate must also be considered in the same manner required for noncomposite girders.

7.4.2.2 Shear Connection between Steel and Concrete

The shear connection strength is also affected by the method of construction. Shored or supported construction requires that the shear flow at the steel to concrete interface be determined based on composite section properties for short- and long-term dead load, and short-term live load effects. In unshored or unsupported construction, shear flow at the steel-to-concrete interface must be determined based on composite section properties for the long-term effects of dead load and short-term live load effects. The shear flow, q_{i}, at the steel-to-concrete interface is

$$q_{\text{i}} = \tau b_{\text{f}} = \frac{VQ_{\text{c}}}{I_{\text{cp}}}, \tag{7.100}$$

---

* This is based on an "average" shear stress instead of the calculation of the shear stress through the cross section using Equation 7.97. For some wide flange (I-beam) sections the "average" shear stress may be about 75% of the maximum shear stress through the cross section calculated using Equation 7.97.

where V is the shear force, $Q_c = A_c y_{na}$ is the statical moment of the concrete slab about the neutral axis, A_c is the transformed area of the concrete slab, b_f is the width of the girder flange, y_{na} is the distance from the centroid of the concrete slab to the neutral axis, and I_{cp} is the moment of inertia of the composite section.

Shear transfer connectors must be designed to resist the shear flow at the steel to concrete interface. AREMA (2008) recommends that shear studs be at least 3″ long (4″ long shear studs are commonly used) and have the following strength:

$$S_r = \frac{C_{sr}\pi(d_s)^2}{4}\ \text{kips},$$

$$S_m = \frac{20.0\pi(d_s)^2}{4}\ \text{kips}.$$

For channels the recommended strength is

$$S_r = D_{sr} w_c\ \text{kips},$$

$$S_m = 3600 w_c\ \text{kips},$$

where S_r is the allowable horizontal design force for fatigue per connector, kips; S_m is the allowable maximum horizontal design force per connector, kips; d_s is the diameter of the shear stud (3/4 or 7/8 in.); C_{sr} is 7.0 for fatigue design cycles $N \geq$ 2,000,000 cycles and 10.0 for fatigue design cycles $N =$ 2,000,000 cycles; D_{sr} is 2100 for fatigue design cycles $N \geq$ 2,000,000 cycles and 2400 for fatigue design cycles $N =$ 2,000,000 cycles; w_c is the length of the channel perpendicular to the shear flow (transverse to the flange).

The distribution of shear connectors along the span is made based on the magnitude of the shear flow along the span length. Since live load shear flow varies along the span length, L, the shear connector spacing may also vary. The form of a typical shear flow influence line for the determination of live load maximum and range of shear flow is shown in Figure 7.23. Based on the maximum shear flow and live load shear flow range, a practical spacing over a length, s_i, with some acceptable overstress (usually about 10%) can be made as illustrated in Figure 7.24.

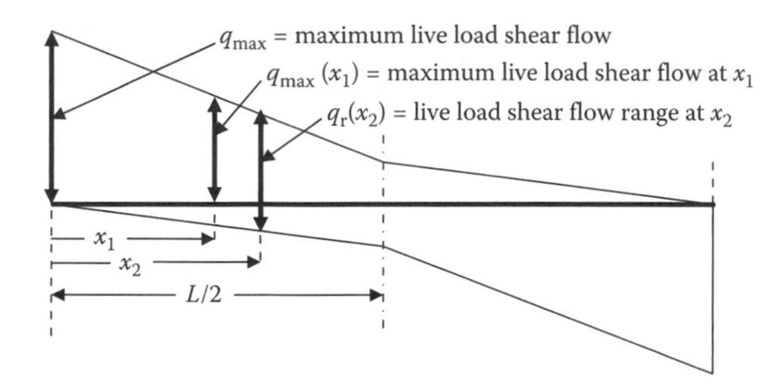

FIGURE 7.23 Distribution of shear flow along the span length.

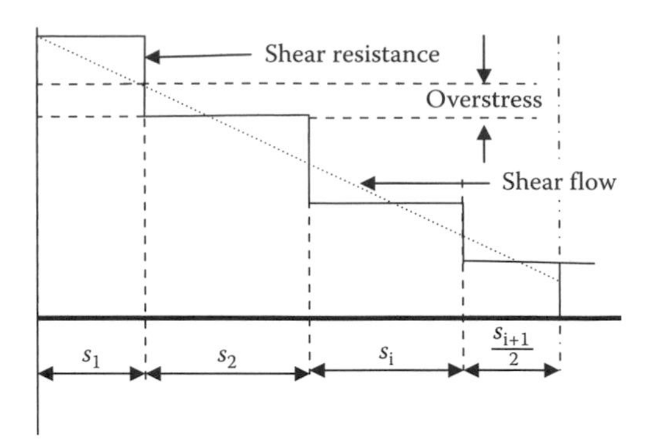

FIGURE 7.24 Distribution of shear resistance (studs or channels) along the span length.

7.5 SERVICEABILITY DESIGN OF COMPOSITE FLEXURAL MEMBERS

AREMA (2008) recommends that mid-span deflection of simply supported spans due to live load plus impact, $\Delta LL + I$, should not exceed $L/640$, where L is the span length. Some engineers or bridge owners recommend even more severe live load plus impact deflection criteria to attain stiffer spans, which offer improved performance from a structural* and track–train dynamics perspective.

It is also recommended that camber be provided for dead load deflections exceeding 1 in. For composite spans, the dead load deflections depend on the construction method employed (shored or unshored).

The serviceability criteria for steel railway spans are also discussed in Chapter 5.

Example 7.2 outlines the design of a composite steel–concrete span for Cooper's E80 live load considering both shored and unshored construction for the flexural design.

Example 7.2

A 90 ft simple span steel ($F_y = 50$ ksi) ballasted deck plate girder (BDPG) is to be designed for the forces shown in Table E7.4a.

Section properties of the span

The steel girder section properties are shown in Figure E7.1 (see Example E7.1) and Table E7.4b. The composite steel and concrete girder section properties are shown in Figure E7.3 and Tables E7.5 and E7.6 for short-term loads and in Tables E7.7 and E7.8 for long-term loads.

Steel section properties: see Example 7.1

* Concrete bridge decks generally exhibit better behavior on stiffer spans.

TABLE E7.4a
Design Forces

Design Force	Shear Force, V (kips)	Bending Moment (ft-kips)
Dead load on unshored steel section, DL1	70	1430
Dead load on composite section (unshored), DL2	40	1070
Total DL (DL1 + DL2)	110	2500
Live load + 37% impact (maximum)	376	7314
Maximum (DL1 + DL2 + LL)	486	9814
Live load + 13% impact (fatigue)	310	6032

TABLE E7.4b
Steel Section Properties

Element	A (in.^{2})	y_b (in.)	Ay_b (in.^{3})	$y_b - y$ (in.)	$A(y_b - y)^2$ (in.^{4})	I_o (in.^{4})
Top flange	50.00	88.75	4437.5	43.75	95,703	26.0
Web	53.13	45.00	2390.6	0	0	31,986
Bottom flange	50.00	1.25	62.5	−43.75	95,703	26.0
$\sum$	153.13		6890.6		191,406	32,038

Composite steel–concrete section properties:

Short-term loads:

$$\frac{\sum Ay_b}{\sum A} = \frac{17578}{265.6} = 66.18 \text{ in.,}$$

$$I_g = \sum A(y_b - y)^2 + \sum I_o = 162{,}134 + 224{,}382 = 386{,}516 \text{ in.}^4.$$

Assuming I_n steel section $= 0.90I_o = 0.90(223{,}444) = 201{,}100$ in.^{4},
I_n composite section $= 162{,}134 + 201{,}100 = 363{,}234$ in.^{4}.

Long-term loads:

$$\frac{\sum Ay_b}{\sum A} = \frac{10454}{190.63} = 54.84 \text{ in.,}$$

$$I_g = \sum A(y_b - y)^2 + \sum I_o = 75{,}308 + 223{,}757 = 299{,}065 \text{ in.}^4.$$

Assuming I_n steel section $= 0.90I_o = 0.90(223{,}444) = 201{,}100$ in.^{4},
I_n composite section $= 75{,}308 + 201{,}100 = 276{,}408$ in.^{4}.

Flexure and shear design

Flexural stresses are summarized in Figure E7.4 and Table E7.9 or Figure E7.5 and Table E7.10 for unshored and supported deck construction, respectively.

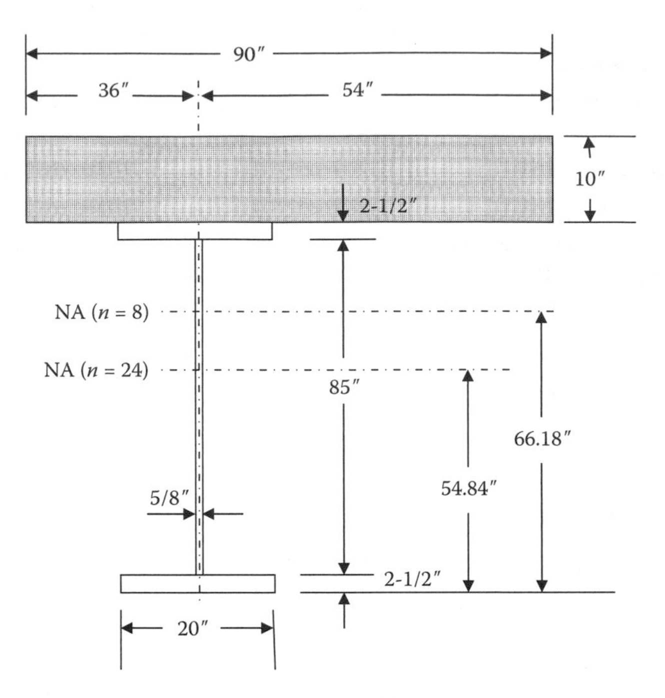

FIGURE E7.3 Composite section with $n = 8$ and $n = 24$ (elastic modular ratio and plastic modular ratio).

TABLE E7.5
Composite Steel and Concrete Section Properties—Short-Term Loads (Includes Live Load)

Element	A (in.^{2})	y_b (in.)	Ay_b (in.^{3})	$y_b - y$ (in.)	$A(y_b - y)^2$ (in.^{4})	I_o (in.^{4})
Steel section	153.13	45.00	6891	−21.18	68,667	223,444
Concrete slab ($n = 8$)	112.50	95.00	10,688	28.82	93,467	938
$\sum$	265.6		17,578		162,134	224,382

TABLE E7.6
Composite Steel and Concrete Section Properties with $n = 8$

Location (Figure E7.2)	n	c (in.)	I (Gross or Net Depending on Location of NA) (in.^{4})	nS (Gross or Net) (in.^{3})
Top concrete	8	33.82	386,516	91,429
Bottom concrete	8	23.82	386,516	129,812
Top steel	1	23.82	386,516	16,227
Bottom steel	1	66.18	363,234	5489

Note: NA = neutral axis.

TABLE E7.7
Composite Steel and Concrete Section Properties—Long-Term Loads (for Shored Construction Includes all Dead Loads and for Unshored Construction Includes Dead Load Supported by Steel Section only and Dead Load Supported by Composite Section (e.g., Track, Ballast, Walkways, and Conduits))

Element	A (in.^{2})	y_b (in.)	Ay_b (in.^{3})	$y_b - y$ (in.)	$A(y_b - y)^2$ (in.^{4})	I_o (in.^{4})
Steel section	153.13	45.00	6891	−9.84	14,825	223,444
Concrete slab ($n = 24$)	37.50	95.00	3563	40.16	60,483	313
$\sum$	190.63		10,454		75,308	223,757

TABLE E7.8
Composite Steel and Concrete Section Properties with $n = 24$

Location (Figure E7.2)	n	c (in.)	I (Gross or Net Depending on Location of NA) (in.^{4})	nS (Gross or Net) (in.^{3})
Top concrete	24	45.16	299,065	158,936
Bottom concrete	24	35.16	299,065	204,140
Top steel	1	35.16	299,065	8,506
Bottom steel	1	54.84	276,408	5,040

In this example there is not a great difference in unshored and shored flexural stresses due to the relatively small dead load stress on the noncomposite (steel only) section during unshored construction.

Shear stresses:

$f_v = (486)/[(85)(0.625)] = 9.15$ ksi (the shear is resisted entirely by the steel girder web)

Allowable stresses:

F_{call} concrete $= 0.40f'_c = 0.40(3) = 1.2$ ksi (minimum 28 day compressive strength of 3,000 psi) ≥ 1.15 ksi OK

F_{tall} steel $= F_{call}$ steel $= 0.55F_y = 0.55(50) = 27.5$ ksi ≥ 22.4 ksi OK

$F_{fat} = 16$ ksi (Category B with loaded length of 90 ft) ≥ 13.2 ksi OK

$F_{vall} = 0.35(50) = 17.50$ ksi ≥ 9.15 ksi OK

Girder design for fabrication and erection loads – see Example 7.1

Detailed design of the girder

Detailed design of the web plate

Flexural buckling:

$$\frac{h}{t_w} = \frac{85}{0.625} = 136 \leq 4.18\sqrt{\frac{E}{f_c}} \leq 4.18\sqrt{\frac{29{,}000}{22.6}} \leq 150.$$

Therefore, no longitudinal stiffeners are required for web flexural buckling stability.

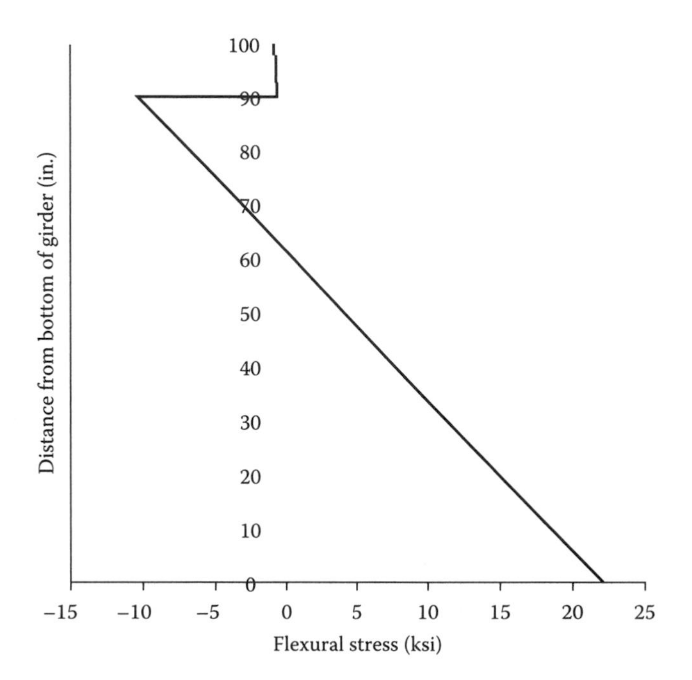

FIGURE E7.4 Composite steel and concrete section flexural stresses—unshored construction.

Shear buckling:

$$h = 85 \geq 2.12 t_w \sqrt{\frac{E}{F_y}} \geq 2.12(0.625)\sqrt{\frac{29{,}000}{50}} \geq 31.9\,\text{in.}$$

Therefore, transverse web stiffeners are required.

TABLE E7.9
Composite Steel and Concrete Section Flexural Stresses—Unshored Construction

Location (Figure E7.3)	DL1 Flexural Stress on Noncomposite Section ($n = 1$) (ksi)	DL2 Flexural Stress on Composite Section ($n = 24$) (ksi)	Maximum LL + I Flexural Stress on Composite Section ($n = 8$) (ksi)	Range of LL + I Flexural Stress on Composite Section ($n = 8$) (ksi)	Maximum Flexural Stress (ksi)
Top concrete	—	0.08	0.96	0.79	1.04
Bottom concrete	—	0.06	0.68	0.56	0.74
Top steel	3.46	1.51	5.41	4.47	10.38
Bottom steel	3.84	2.55	15.99	13.19	22.38

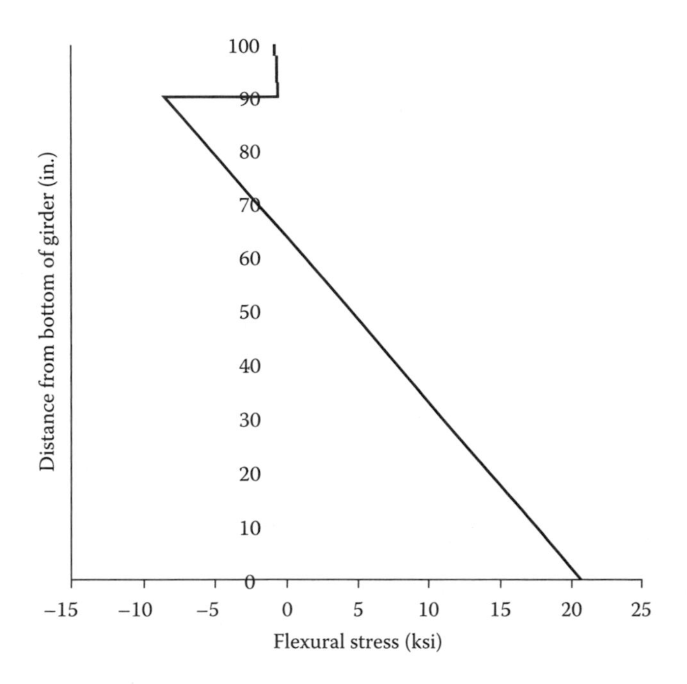

FIGURE E7.5 Composite steel and concrete section flexural stresses—shored construction.

Combined bending and shear:

$$f_b = 22.38 \leq \left(0.75 - 1.05\frac{f_v}{F_y}\right)F_y \leq \left(0.75 - 1.05\frac{486/53.13}{50}\right)50 \leq 27.9\ \text{ksi},$$

but no greater than 27.5 ksi. OK

Flange-to-web connection

Only unshored construction is considered in the flange-to-web connection design for brevity. Similar calculations may be performed if a shored construction method is utilized.

TABLE E7.10
Composite Steel and Concrete Section Flexural Stresses—Shored Construction

Location (Figure E7.3)	DL Flexural Stress on Composite Section ($n = 24$) (ksi)	Maximum LL + I Flexural Stress on Composite Section ($n = 8$) (ksi)	Range of LL + I Flexural Stress on Composite Section ($n = 8$) (ksi)	Maximum Flexural Stress (ksi)
Top concrete	0.19	0.96	0.79	1.15
Bottom concrete	0.15	0.68	0.56	0.83
Top steel	3.53	5.41	4.47	8.94
Bottom steel	5.95	15.99	13.19	21.94

The section properties at the top and bottom weld are shown in Tables E7.11 and E7.12, respectively. Shear flow at the weld, calculated based on these section properties, is shown in Table E7.13.

TABLE E7.11
Top Weld of Flange-to-Web Connection

Section	A (in.^{2})	y_{na} (in.)	Ay_{na} (in.^{3})	I_g (in.^{4})	Ay_{na}/I_g (in.$^{-1}$)
Noncomposite (steel only)	50.00	43.75	2188	223,444	9.79×10^{-3}
Composite—long term ($n = 24$)	37.50	40.16	1506	299,065	10.71×10^{-3}
	50.00	33.91	1696		
Composite—short term ($n = 8$)	112.50	28.82	3242	386,516	11.31×10^{-3}
	50.00	22.57	1129		

TABLE E7.12
Bottom Weld of Flange-to-Web Connection

Section	A (in.^{2})	y_{na} (in.)	Ay_{na} (in.^{3})	I_n (in.^{4})	Ay_{na}/I_n (in.$^{-1}$)
Noncomposite (steel only)	50.00	43.75	2188	201,100	10.88×10^{-3}
Composite—long term ($n = 24$)	50.00	53.59	2680	276,408	9.70×10^{-3}
Composite—short term ($n = 8$)	50.00	64.93	3247	363,234	8.94×10^{-3}

Maximum shear flow = 5.37 k/in.

Maximum LL + I range shear flow = 3.51 k/in. (AREMA recommends that even stress ranges in compression areas be considered due to the relatively high "effective mean stress" caused by residual stresses)

Allowable weld shear stress = 17.5 ksi (fillet)

Allowable fatigue stress range = 16 ksi (fatigue detail Category B)

Weld size = 5.37/[2(0.71)17.5] = 0.21 in.

Weld size = 3.52/[2(0.71)16] = 0.16 in.

Use min 5/16″ fillet welds or CJP welds (some designers will specify CJP welds because of the vertical load on the top flange-to-web weld).

TABLE E7.13
Shear Flow at Flange-to-Web Connection

Location (Figure E7.3)	DL1 Shear Flow on Noncomposite Section ($n = 1$) (k/in.)	DL2 Shear Flow on Composite Section ($n = 24$) (k/in.)	Maximum LL+I Shear Flow on Composite Section ($n = 8$) (k/in.)	Range of LL+I Shear Flow on Composite Section ($n = 8$) (k/in.)	Maximum Shear Flow (k/in.)
Top weld	0.69	0.43	4.25	3.51	5.37
Bottom weld	0.76	0.39	3.37	2.77	4.52

TABLE E7.14
Shear Forces along the Span

Location (Figure E7.6)	Distance, x, From End a (in.)	$+V_{LL}$ (kips)	$-V_{LL}$(kips) (Linear Interpolation)	V_{DL2} (kips)	V_r (kips)	V_{max} (kips)
A	0	275	0	40	310	417
B	270	158	37	20	220	287
C	540 (center)	74	74	0	167	203

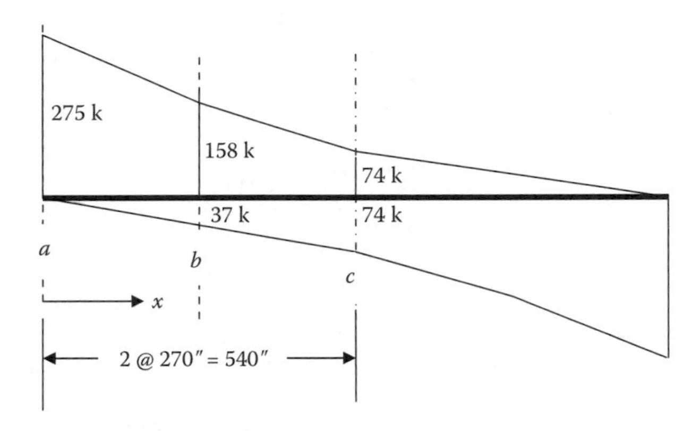

FIGURE E7.6 Shear forces from live load along the span.

Shear Stud design

Only unshored construction is considered in the shear stud design for brevity. Similar calculations may be performed if a shored construction method is utilized.

Also for brevity, shear stud spacing will be calculated at only three locations on the girder as indicated in Table E7.14 (in the design of practical girders, a smaller interval is recommended*).

The negative live load shear flow (the shear reverses when a wheel passes over a stud) is estimated to be a linear interpolation of the center span live load shear as shown in Figure E7.6.

The shear flow at the steel–concrete interface (Table E7.15) is

$$q_{LL+I} = \frac{V_{LL+I}Q_{cp}(n=8)}{I_{cp}(n=8)} = \frac{3242(V_{LL+I})}{386{,}516} = \frac{V_{LL+I}}{119},$$

$$q_{max} = q_{LL+I} + q_{DL2} = q_{LL+I} + \frac{V_{DL2}Q_{cp}(n=24)}{I_{cp}(n=24)} = \frac{V_{LL+I}}{119} + \frac{1506V_{DL2}}{299{,}065}$$

$$= \frac{V_{LL+I}}{119} + \frac{V_{DL2}}{199}.$$

* Equivalent uniform load charts (such as Steinman charts) are useful in determining live load shear forces at various locations along the span.

TABLE E7.15
Shear Flow along the Span

Location (Figure E7.6)	Distance, x, From End a (in.)	V_r (kips)	V_{DL2} (kips)	q_r (k/in.)	q_{DL2} (k/in.)	q_{max} (k/in.)
A	0	310	40	2.61	0.20	3.70
B	270	220	20	1.85	0.10	2.51
C	540	167	0	1.40	0	1.71

The shear flow due to dead load on the composite section is small and ignored in the practical design of composite steel and concrete girders. This is also evident from comparison of the allowable design load for the shear stress range, S_r, and the allowable design load for the maximum shear stress, S_m.

$$S_r = \frac{7.0\pi(0.875)^2}{4} = 4.21 \text{ kips},$$

$$S_m = \frac{20.0\pi(0.875)^2}{4} = 12.0 \text{ kips}.$$

Use three shear studs across the flange width;

s is the spacing required (Table E7.16) and is equal to $3(4.21)/q_{LL+I} = 12.63/q_{LL+I}$.

The actual shear stud spacing can be arranged in order that the maximum overstress is, for example, 10%, as shown in Figure E7.7.

Design of web plate stiffeners:

Use angles bolted to the web in order to preclude issues related to welding at the base of transverse stiffeners. Use a single 6 × 4 × 1/2 angle at 67.5 in. centers as shown in Example 7.1.

Design of bearing stiffeners: Use four 8 × 4 × 1/2 angles as shown in Example 7.1.

Serviceability—deflection criteria

The required gross moment of inertia for a LL+I deflection criteria of L/f_Δ is (see Chapter 5)

$$I \geq \frac{M_{LL+I} L f_\Delta}{1934} \geq \frac{7314(90)f_\Delta}{1934} \geq 340.4 f_\Delta \text{ in.}^4$$

The required section gross moment of inertia for various deflection criteria, f_Δ, is shown in Table E7.17.

TABLE E7.16
Shear Stud Spacing along the Span

Location	q_r (k/in.)	s (in.)
A	2.61	4.9
B	1.85	6.8
C	1.40	9.0

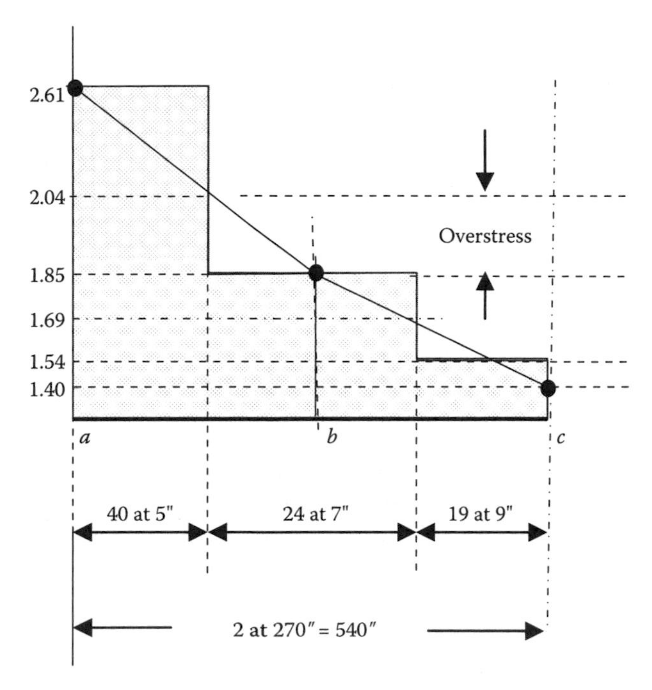

FIGURE E7.7 Shear flow and resistance along the span.

The section $I_g = 386,516$ in.^{4} provides a very stiff structure. The deflections are estimated as (see Chapter 5)

$$\Delta_{LL+I} = \frac{0.104 M_{LL+I} L^2}{EI} = \frac{0.104(7314)(12)[90(12)]^2}{29{,}000(386{,}516)} = 0.95 \text{ in.}$$

$$\Delta_{DL} = \frac{0.104 M_{DL1} L^2}{EI_{g1}} + \frac{0.104 M_{DL2} L^2}{EI_{g2}}$$

$$= \frac{0.104(1430)(12)[90(12)]^2}{29{,}000(223{,}444)} + \frac{0.104(1070)(12)[90(12)]^2}{29{,}000(299{,}065)}$$

$$= 0.32 + 0.18 = 0.50'',$$

no camber is required.

TABLE E7.17

Deflection Criteria, L/Δ_{LL+I}	Required Gross Moment of Inertia, I_g, (in.^{4})
500	170,180
640 (AREMA, 2008)	217,832
800	272,290
1000	340,362

TABLE E7.18

Location (Figures E7.1 and E7.3)	Noncomposite Section Stress (ksi) (Figure E7.1)	Composite Section Stress (ksi) (Figure E7.3)
Top concrete (maximum flexure)	—	1.04
Bottom concrete (maximum flexure)	—	0.74
Top steel (maximum flexure)	23.7	10.38
Web (maximum shear)	9.15	9.15
Bottom steel (maximum flexure)	26.4	22.4
Bottom steel (LL+I flexure)	16.2	13.2

Summary of the design

Examples 7.1 and 7.2 are not intended to be examples of optimum design but to provide numerical examples of noncomposite and composite steel and concrete girder designs. The summary of stresses and deflections of the two span designs [noncomposite (Example 7.1) and composite (Example 7.2)] shown in Table E7.18 reveals where reductions in element sizes can be made and the advantages of composite girder design may be exploited.

REFERENCES

Akesson, B., 2008, *Understanding Bridge Collapses*, Taylor & Francis, London, UK.

American Railway Engineering and Maintenance-of-Way Association (AREMA), 2008, Steel Structures, in *Manual for Railway Engineering*, Chapter 15, Lanham, MD.

Basler, K. and Thurlimann, B., 1959, *Plate Girder Research*, National Engineering Conference Proceedings, AISC, Chicago, IL.

Bleich, F., 1952, *Buckling Strength of Metal Structures*, McGraw-Hill, New York.

Brockenbrough, R.L., 2006, Properties of structural steels and effects of steelmaking and fabrication, in *Structural Steel Designer's Handbook*, Chapter 1, R.L. Brockenbrough (ed.), McGraw-Hill, New York.

Cusens, A.R. and Pama, R.P., 1979, *Bridge Deck Analysis*, Wiley, New York.

D'Andrea, M., Grondin, G.Y., and Kulak, G.L., 2001, *Behaviour and Rehabilitation of Distortion-Induced Fatigue Cracks in Bridge Girders*, University of Alberta Structural Engineering Report No. 240, Edmonton, Canada.

Galambos, T.V. (ed.), 1988, *Guide to Stability Design Criteria for Metal Structures*, Wiley, New York.

Gere, J.M. and Timoshenko, S.P., 1984, *Mechanics of Materials*, Wadsworth, Belmont, CA.

Kim, Y.D., Jung, S.-K., and White, D.W., 2007, Transverse stiffener requirements in straight and horizontally curved steel i-girders, *Journal of Bridge Engineering*, 12(2), ASCE, Reston, VA.

Newmark, N.M., Seiss, C.P., and Viest, I.M., 1951, Tests and analysis of composite beams with incomplete interaction, *Proceedings of the Society for Experimental Stress Analysis*, 9(1), 75–92.

Roark, R.J. and Young, W.C., 1982, *Formulas for Stress and Strain*, McGraw-Hill, New York.

Rockey, K.C. and Leggett, D.M.A., 1962, The buckling of a plate girder web under pure bending when reinforced by a single longitudinal stiffener, *Proceedings of the Institute of Civil Engineers*, 21, 161.

Roy, S. and Fisher, J.W., 2006, Modified AASHTO design S–N curves for post-weld treated welded details, *Journal of Bridge Engineering*, 2(4), Taylor & Francis, Abingdon, Oxford, UK.

Salmon, C.G. and Johnson, J.E., 1980, *Steel Structures Design and Behavior*, Harper and Row, New York.

Seaburg, P.A. and Carter, C.J., 1997, *Torsional Analysis of Structural Steel Members*, AISC, Chicago, IL.

Tall, L. (ed.), 1974, *Structural Steel Design*, Wiley, New York.

Timoshenko, S.P. and Gere, J.M., *Theory of Elastic Stability*, 2nd Edition, McGraw-Hill, New York.

Timoshenko, S. and Woinowsky-Krieger, S., 1959, *Theory of Plates and Shells*, McGraw-Hill, New York.

Trahair, N.S., 1993, *Flexural-Torsional Buckling of Structures*, CRC Press, Boca Raton, FL.

Troitsky, M.S., 1987, *Orthotropic Bridges Theory and Design*, 2nd Edition, James F. Lincoln Arc Welding Foundation, Cleveland, OH.

Viest, I.M., Fountain, R.S., and Singleton, R.C., 1958, *Composite Construction in Steel and Concrete*, McGraw-Hill, New York.

Wang, C.M., Reddy, J.N., and Lee, K.H., 2000, *Shear Deformable Beams and Plates*, Elsevier, Kidlington, Oxford, UK.

Wolchuk, R., 1963, *Design Manual for Orthotropic Steel Plate Deck Bridges*, AISC, Chicago, IL.

8 Design of Steel Members for Combined Forces

8.1 INTRODUCTION

Structural steel members in railway superstructures are usually designed to resist only axial or transverse loads as outlined in Chapters 6 and 7, respectively. These external loads create internal normal and shear stresses in members of the superstructure. However, in some situations, it is necessary to consider members subjected to combinations of forces.

Combined stresses in railway bridges typically arise from biaxial bending of unsymmetrical cross sections, unsymmetrical bending from transverse force eccentricities and combined axial and bending forces caused by eccentricities, member out-of-straightness, self-weight,* and applied lateral loads such as wind. For linear elastic materials and small deformations, superposition of stresses from combined loads is appropriate.

8.2 BIAXIAL BENDING

If bending moments, M_x and M_y, are applied at the centroid of an unsymmetrical section as shown in Figure 8.1, bending will occur in both the yz and xz planes.

However, on unsymmetrical cross sections these planes are not principal planes, and each moment contributes to a portion of the total bending about each axis. The flexural stress, σ_p, at a location, p, with coordinates x and y is

$$\sigma_p = E\varepsilon_x + E\varepsilon_y = -E(\kappa_x x + \kappa_y y), \tag{8.1}$$

where ε_x is the strain on the xz plane, ε_y is the strain on the yz plane, κ_x is the curvature on the xz plane, κ_y is the curvature on the yz plane, and E is the modulus of elasticity of steel and is equal to 29,000 ksi.

* This is the case for members that are not in a vertical plane.

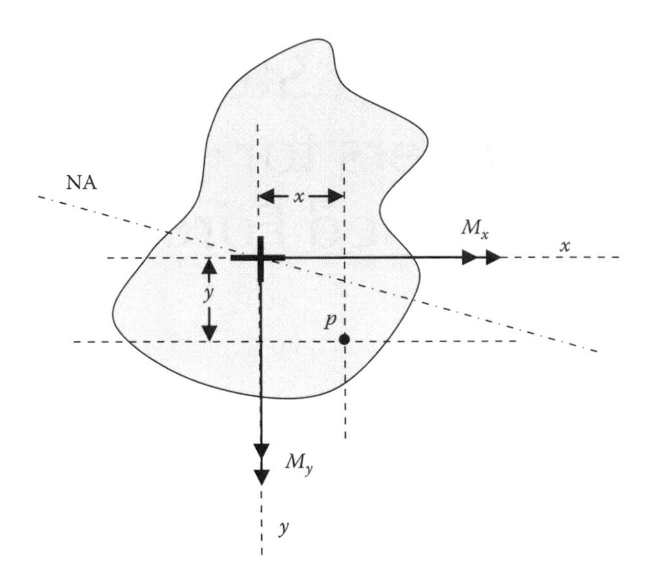

FIGURE 8.1 Biaxial bending of unsymmetrical cross section.

The bending moments, M_x and M_y, can then be written as

$$M_y = \int \sigma x \, dA = -E(\kappa_y \int xy \, dA + k_x \int x^2 \, dA) = -E(\kappa_y I_{xy} + \kappa_x I_y), \quad (8.2)$$

$$M_x = \int \sigma y \, dA = -E(\kappa_y \int y^2 \, dA + k_x \int xy \, dA) = -E(\kappa_y I_x + \kappa_x I_{xy}), \quad (8.3)$$

where I_x is the moment of inertia about the x axis and I_y is the moment of inertia about the y axis.

Equations 8.2 and 8.3 may be solved simultaneously for κ_x and κ_y and substituted into Equation 8.1 to obtain

$$\sigma = \frac{M_x I_y - M_y I_{xy}}{I_x I_y - I_{xy}^2} y + \frac{M_y I_x - M_x I_{xy}}{I_x I_y - I_{xy}^2} x. \quad (8.4)$$

Steel members in railway superstructures subjected to biaxial bending usually have two axes of symmetry. Therefore,

$$I_{xy} = 0 \quad (8.5)$$

and Equation 8.4 becomes

$$\sigma = \frac{M_x}{I_x} y + \frac{M_y}{I_y} x = f_{bx} + f_{by} \leq F_b, \quad (8.6)$$

where f_{bx} is the normal stress from bending moment, M_x, about the x axis, f_{by} is the normal stress from bending moment, M_y, about the y axis, and F_b is the allowable bending stress.

However, since the allowable bending stress may not be the same in each plane of bending, Equation 8.6 may be expressed as

$$\frac{f_{bx}}{F_{bx}} + \frac{f_{by}}{F_{by}} = \frac{(M_x/I_x)y}{F_{bx}} + \frac{(M_y/I_y)x}{F_{by}} \leq 1, \quad (8.7)$$

where F_{bx} is the allowable bending stress in the direction of the x axis and F_{by} is the allowable bending stress in the direction of the y axis.

The interaction Equation 8.7 may be used for design considering both tensile and compressive flexural stresses by using the appropriate allowable bending stress for flexural tension ($F_{bx}, F_{by} = F_{tall} = 0.55F_y$) or compression ($F_{bx}, F_{by} = F_{call}$).

8.3 UNSYMMETRICAL BENDING (COMBINED BENDING AND TORSION)

The best design strategy is to avoid torsion. However, in some cases it is unavoidable. Torsion is combined with bending when transverse loads are not applied through the shear center of the member.

When a torsional moment is applied, pure torsion always exists. Pure torsion creates shearing stresses in the flanges and webs of structural shapes such as channels and I-shaped beams. However, warping torsion also exists when cross sections do not remain plane due to some form of restraint. Warping torsion creates shearing and normal stresses in the flanges of I shapes and normal stresses in the flanges and web of channels. These torsional shear and normal stresses must be superimposed on the shear and normal stresses in flanges and webs due to flexure.

The torsional moment resistance, T, of a cross section to a constant torsional moment is

$$T = T_t + T_w, \quad (8.8)$$

where T_t is the pure torsional (or St. Venant) moment resistance and is given by

$$T_t = GJ\frac{d\theta}{dz}, \quad (8.9)$$

T_w is the warping torsional moment resistance and is given by

$$T_w = -EC_w\frac{d^3\theta}{dz^3}, \quad (8.10)$$

z is the longitudinal axis of the beam or girder, G is the shear modulus of elasticity of steel (~11,200 ksi), E is the tensile modulus of elasticity of steel (~29,000 ksi), J is the torsional constant of the cross section, and C_w is the warping constant of the cross section (see Table 7.1).

The shear stresses from pure torsion effects of the applied torsional moments are

$$\tau_t = \frac{T_t t}{J} \quad (8.11)$$

and substitution of Equation 8.9 into Equation 8.11 yields

$$\tau_t = \frac{T_t t}{J} = Gt\left(\frac{d\theta}{dz}\right), \tag{8.12}$$

where t is the thickness of the element.

The shear stresses from warping effects of the applied torsional moments on an I-shaped section are

$$\tau_w = \frac{3}{2}\frac{T_w}{A_f h} \tag{8.13}$$

and substitution of Equation 8.10 into Equation 8.13 yields

$$\tau_w = -\frac{3}{2}\frac{EC_w}{A_f h}\left(\frac{d^3\theta}{dz^3}\right) = -\frac{Eb_f^2 h}{16}\left(\frac{d^3\theta}{dz^3}\right), \tag{8.14}$$

where A_f is the area of the flange and is equal to $b_f t_f$,

$$C_w = \frac{I_y h^2}{4} = \frac{(I_f)h^2}{2} = \frac{t_f b_f^3 h^2}{24},$$

h is the distance between centroids of flanges for I-shaped members.

The normal stresses from warping effects of the applied torsional moments on an I-shaped section are determined by considering the normal stress from the lateral bending of the flanges

$$\sigma_w = \frac{M_l x}{I_f} = \frac{EI_f h}{2}\left(\frac{d^2\theta}{dz^2}\right) = \frac{EC_w}{h}\left(\frac{d^2\theta}{dz^2}\right), \tag{8.15}$$

where x is the distance on the flange from the neutral axis of flexural stress distribution in the flange (maximum at $x = b/2$), M_l is the lateral bending moment on one flange, I_f is the moment of inertia of the flange.

The differential equation of torsion is (from Equations 8.8 through 8.10)

$$T = GJ\frac{d\theta}{dz} - EC_w\frac{d^3\theta}{dz^3}. \tag{8.16}$$

For torsional moments that vary uniformly along the length (z axis) of a member, dT/dz is

$$\frac{dT}{dz} = t' = GJ\frac{d^2\theta}{dz^2} - EC_w\frac{d^4\theta}{dz^4}. \tag{8.17}$$

For torsional moments that vary linearly along the length (z axis) of a member, dT/dz is

$$\frac{dT}{dz} = \frac{t'z'}{L} = GJ\frac{d^2\theta}{dz^2} - EC_w\frac{d^4\theta}{dz^4}, \tag{8.18}$$

where t' is the maximum torsional moment applied at the end support, $L - z'$ is the distance from the end support with maximum torsional moment, L is the length of the span.

The angle of twist, θ, is provided by the solution of Equations 8.16, 8.17, or 8.18 and depends on both loading and boundary conditions. The general form of the angle of twist, θ, can be expressed as (Kuzmanovic and Willems, 1983)

$$\theta = A\sinh\left(\frac{z}{a}\right) + B\cosh\left(\frac{z}{a}\right) + C + D(z), \tag{8.19}$$

where A, B, C, and D are constants depending on boundary conditions and loading, $D(z)$ is an expression in terms of z, depending on loading, and $a = \sqrt{EC_w/GJ}$ (a characteristic length).

The equations for the angle of twist and its derivatives for a concentrated torsional moment, T', applied at the center of a simply supported span are (Salmon and Johnson, 1980)

$$\theta = A\sinh\left(\frac{z}{a}\right) + B\cosh\left(\frac{z}{a}\right) + C + \frac{T'z}{2GJ} = \frac{T'a}{2GJ}\left(\frac{z}{a} - \frac{\sinh(z/a)}{\cosh(L/2a)}\right), \tag{8.20a}$$

$$\frac{d\theta}{dz} = \frac{T'}{2GJ}\left(1 - \frac{\cosh(z/a)}{\cosh(L/2a)}\right), \tag{8.20b}$$

$$\frac{d^2\theta}{dz^2} = \frac{T'}{2GJa}\left(-\frac{\sinh(z/a)}{\cosh(L/2a)}\right), \tag{8.20c}$$

$$\frac{d^3\theta}{dz^3} = \frac{T'}{2GJa^2}\left(-\frac{\cosh(z/a)}{\cosh(L/2a)}\right). \tag{8.20d}$$

Example 8.1 illustrates the use of Equations 8.20a–d for the determination of combined stresses due to torsion and flexure.

The equations for the angle of twist and its derivatives for a uniformly distributed torsional moment, t', applied at the center of a simply supported span are (Kuzmanovic and Willems, 1983)

$$\theta = \frac{t'a^2}{GJ}\left(-\tanh\left(\frac{L}{2a}\right)\sinh\left(\frac{z}{a}\right) + \cosh\left(\frac{z}{a}\right) - \frac{z^2}{2a^2} + \frac{zL}{2a^2} - 1\right), \tag{8.21a}$$

$$\frac{d\theta}{dz} = \frac{t'a}{GJ}\left(-\tanh\left(\frac{L}{2a}\right)\cosh\left(\frac{z}{a}\right) + \sinh\left(\frac{z}{a}\right) - \frac{z}{a} + \frac{L}{2a}\right), \tag{8.21b}$$

$$\frac{d^2\theta}{dz^2} = \frac{t'}{GJ}\left(-\tanh\left(\frac{L}{2a}\right)\sinh\left(\frac{z}{a}\right) + \cosh\left(\frac{z}{a}\right) - 1\right), \tag{8.21c}$$

$$\frac{d^3\theta}{dz^3} = \frac{t'}{GJa}\left(-\tanh\left(\frac{L}{2a}\right)\cosh\left(\frac{z}{a}\right) + \sinh\left(\frac{z}{a}\right)\right). \tag{8.21d}$$

Example 8.3 illustrates the use of Equations 8.21a–d for the determination of combined stresses due to torsion and flexure.

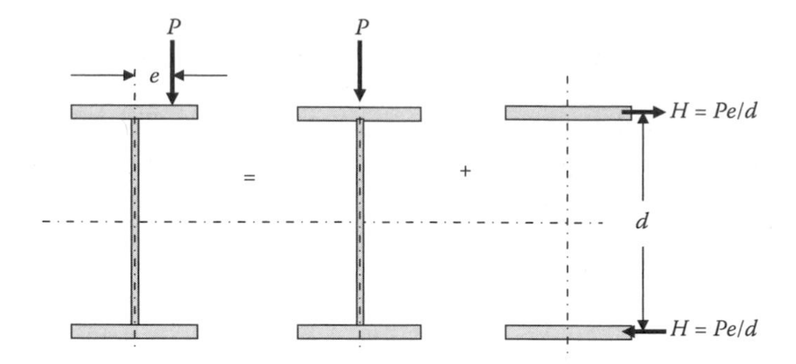

FIGURE 8.2 Equivalent static system for eccentrically applied vertical load.

Equation 8.19 and its derivatives have also been solved for other typical boundary and loading conditions and are provided for design use in equations and charts (Seaburg and Carter, 1997). However, even with such design aids, the solution of Equations 8.12, 8.14, and 8.15 is generally too cumbersome for routine design work and an approximate method, based on a flexure analogy, is often employed.

In this method, it is assumed that the torsional moment acts as a horizontal force couple in the plane of the flanges. The horizontal forces create bending moments in the flanges and the problem is solved by using a simplified analysis involving flexure only. If the torsion is created by eccentric vertical loads, an equivalent static system consisting of a vertical load applied at the shear center and horizontal forces applied at each flange is appropriate (Figure 8.2). Examples 8.2 and 8.4 illustrate the use of the flexure analogy for torsional stresses created by an eccentric vertical load. If the torsion is created by an applied horizontal force, an equivalent static system consisting of vertical and horizontal loads applied at the shear center (creating biaxial bending) and horizontal forces applied at each flange is appropriate (Figure 8.3a). For the latter case, an equivalent static system as shown in Figure 8.3b may also be used. The method is conservative as it ignores pure torsion and assumes torsional moment is resisted entirely by warping torsion. Therefore, normal stresses due to warping are

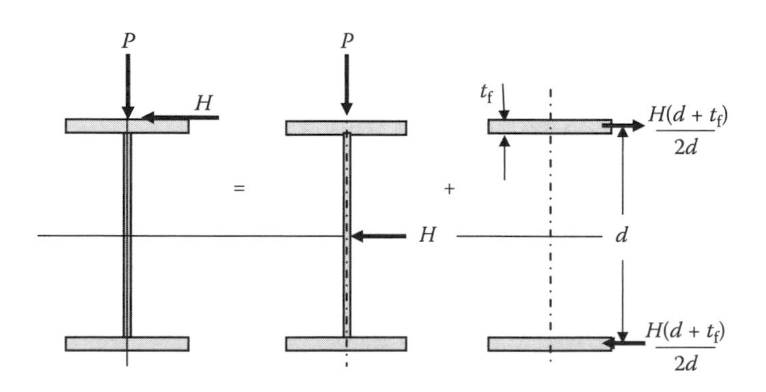

FIGURE 8.3a Equivalent static system for applied vertical and horizontal loads.

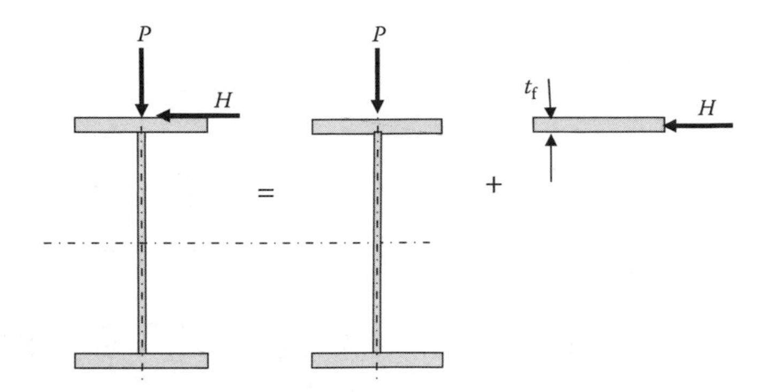

FIGURE 8.3b Alternative equivalent static system for applied vertical and horizontal loads.

overestimated. Modification factors that reduce the normal lateral bending stress have been developed to mitigate this conservativeness.

However, designers should use the flexure analogy with caution, particularly in cases where torsional effects are relatively large or for members with unusual loading or support conditions. Examples 8.1 (concentrated torsional moment) and 8.3 (uniformly distributed torsional moment) outline solutions developed from Equation 8.16. Examples 8.2 and 8.4 outline the solution of the same problems using the flexure analogy.

Example 8.1

A machinery girder in a movable bridge is to support a concentrated load from equipment with an eccentricity of 6 in. as shown in Figure E8.1. Note that only the stresses from the 35 kip equipment load are considered in this example. Dead loads and other loads on the machinery girder are not considered.

Section properties are

$$I_x = 5430\ \text{in.}^4,$$

$$S_x = 339\ \text{in.}^3,$$

$$I_y = 216\ \text{in.}^3,$$

$$J = \sum \frac{bt^3}{3} = \frac{2(12(0.75)^3) + (30.5(0.44)^3)}{3} = 4.23\ \text{in.}^4,$$

$$C_w = \frac{I_y h^2}{4} = \frac{216(31.25)^2}{4} = 52{,}734\ \text{in.}^6,$$

$$a = \sqrt{\frac{EC_w}{GJ}} = 180\ \text{in.},$$

$$L/a = 1.20,$$

$$\frac{T'}{2GJ} = \frac{Pe}{2GJ} = \frac{35(6)}{2(11200)(4.23)} = 2.22 \times 10^{-3}\ \text{in.}^{-1}.$$

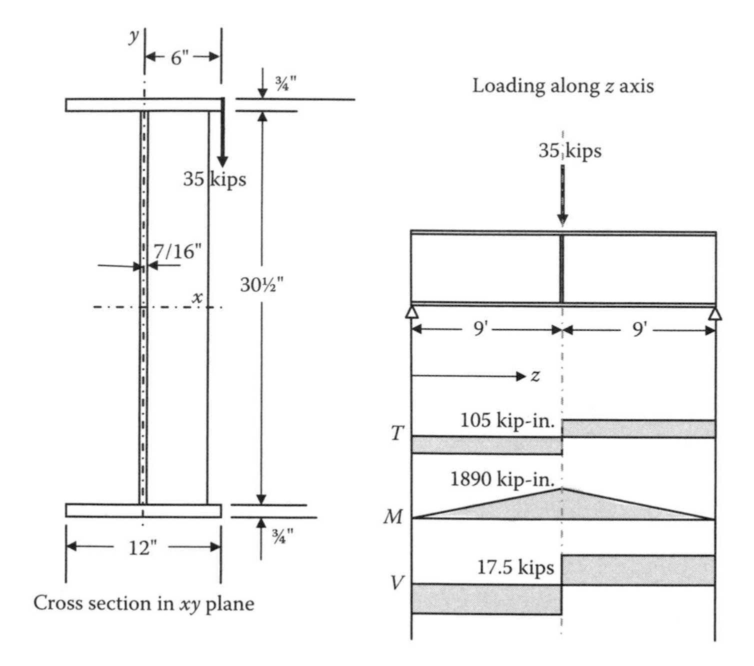

FIGURE E8.1

The solution of the differential Equation 8.16 is

$$\theta = A\sinh\left(\frac{z}{a}\right) + B\cosh\left(\frac{z}{a}\right) + C + \frac{T}{2GJ}z.$$

For boundary conditions $\theta = d^2\theta/dz^2 = 0$ at $z = 0$ and $z = L$,

$$\theta = \frac{T'a}{2GJ}\left(\frac{z}{a} - \frac{\sinh(z/a)}{\cosh(L/2a)}\right) = 0.40\left(\frac{z}{180} - \frac{\sinh(z/180)}{1.186}\right),$$

$$\frac{d\theta}{dz} = \frac{T'}{2GJ}\left(1 - \frac{\cosh(z/a)}{\cosh(L/2a)}\right) = 2.22 \times 10^{-3}\left(1 - \frac{\cosh(z/180)}{1.186}\right),$$

$$\frac{d^2\theta}{dz^2} = \frac{T'}{2GJa}\left(-\frac{\sinh(z/a)}{\cosh(L/2a)}\right) = 1.23 \times 10^{-5}\left(-\frac{\sinh(z/180)}{1.186}\right),$$

$$\frac{d^3\theta}{dz^3} = \frac{T'}{2GJa^2}\left(-\frac{\cosh(z/a)}{\cosh(L/2a)}\right) = 6.87 \times 10^{-8}\left(-\frac{\cosh(z/180)}{1.186}\right).$$

Pure (St. Venant) torsion (Table E8.1):

$$\tau_t = Gt\left(\frac{d\theta}{dz}\right) = 11{,}200(t)2.22 \times 10^{-3}\left(1 - \frac{\cosh(z/180)}{1.186}\right)$$

$$= 24.82t\left(1 - \frac{\cosh(z/180)}{1.186}\right).$$

TABLE E8.1

Element of the Girder	Pure Torsion Shear Stress, τ_t (ksi)	
	$z = 0$ (end)	$z = L/2 = 108$ in. (center)
Flange, $t = 0.75$ in.	2.92	0
Web, $t = 0.44$ in.	1.70	0

Warping torsion (lateral bending of the flanges) (Tables E8.2 and E8.3):

$$\tau_w = \frac{-Eb_f^2 h}{16}\left(\frac{d^3\theta}{dz^3}\right) = \frac{-29{,}000(12)^2(31.25)}{16}6.87 \times 10^{-8}\left(-\frac{\cosh(z/180)}{1.186}\right)$$

$$= 0.472\cosh\left(\frac{z}{180}\right),$$

$$\sigma_w = \frac{Eb_f h}{4}\left(\frac{d^2\theta}{dz^2}\right) = \frac{29{,}000(12)(31.25)}{4}1.23 \times 10^{-5}\left(-\frac{\sinh(z/180)}{1.186}\right)$$

$$= -28.20\sinh\left(\frac{z}{180}\right).$$

Flexure of the girder (Tables E8.4 and E8.5):

$$M = 35(216)/4 = 1890\ \text{kip-in.},$$

$$\sigma_b(0) = 0,$$

$$\sigma_b\left(\frac{L}{2}\right) = \pm\frac{1890}{339} = \pm 5.58\ \text{ksi},$$

$$V = 35/2 = 17.5\ \text{k},$$

$$Q_{\text{flange}} = [(12 - 0.44)(0.75)]\left(\frac{32 - 0.75}{2}\right) = 135.5\ \text{in.}^3,$$

TABLE E8.2

Element of the Girder	Warping Torsion Shear Stress, τ_w (ksi)	
	$z = 0$ (end)	$z = L/2 = 108$ in. (center)
Flange, $t = 0.75$ in.	0.47	0.56

TABLE E8.3

Element of the Girder	Warping Torsion Normal Stress, σ_w (ksi)	
	$z = 0$ (end)	$z = L/2 = 108$ in. (center)
Flanges, $t = 0.75$ in.	0	−17.95

TABLE E8.4

Element of the Girder	Flexural Normal Stress, σ_b (ksi)	
	$z = 0$ (end)	$z = L/2 = 108$ in. (center)
Top Flange, $t = 0.75$ in.	0	−5.58

$$Q_{web} = [(12)(0.75)]\left(\frac{32 - 0.75}{2}\right) + \frac{30.5}{2}(0.44)\frac{30.5}{4} = 191.8\,\text{in.}^3,$$

$$\tau_{b\ flange}(0) = \frac{17.5(135.5)}{5430(0.75)} = 0.58\,\text{ksi},$$

$$\tau_{b\ web}(0) = \frac{17.5(191.8)}{5430(0.44)} = 1.40\,\text{ksi},$$

Combined stresses (Table E8.6):
Both pure, τ_t, and warping, τ_w, shear stresses due to torsion are relatively small, but warping normal stress (18.0 ksi) is large.

Example 8.2

Use the flexure analogy (Figure E8.2) for torsion to find the shear and normal stresses due to combined flexure and torsion of Example 8.1.

$$H = \frac{35(6)}{32 - 0.75} = 6.72\,\text{kips},$$

$$M_H = \frac{6.72(216)}{4} = 362.9\,\text{kip-in.},$$

$$\sigma_{bH} = \frac{362.9}{[0.75(12)^2/6]} = 20.2\,\text{ksi}.$$

TABLE E8.5

Element of the Girder	Flexural Shear Stress, τ_b (ksi)	
	$z = 0$ (end)	$z = L/2 = 108$ in. (center)
Flange, $t = 0.75$ in.	0.58	0.58
Web, $t = 0.44$ in.	1.40	1.40

TABLE E8.6

Element of the Girder	Shear Stress (ksi)	
	$z = 0$ (end)	$z = L/2 = 108$ in. (center)
Flange, $t = 0.75$ in.	3.97	1.14
Web, $t = 0.44$ in.	3.10	1.40

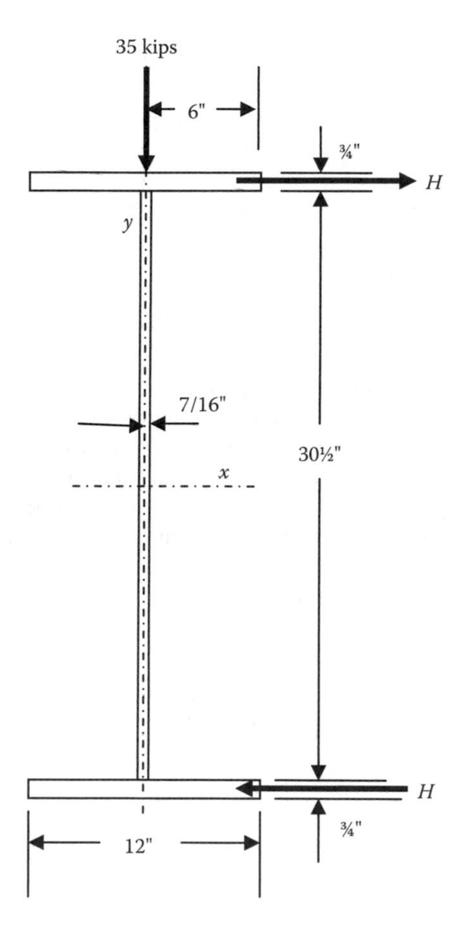

FIGURE E8.2

Modification factors, β, that reduce the normal lateral bending stress have been developed as a corrective measure since the flexure analogy overestimates normal flange stresses due to warping. For the case of $L/a = 1.20$ and torsional moment, T, applied at the center of the span, $\beta \sim 0.94$ (Salmon and Johnson, 1980; Kulak and Grondin, 2002).

$$\sigma_{bH} = \frac{362.9\beta}{[0.75(12)^2/6]} = 20.2\beta = 20.2(0.94) = 19.0\,\text{ksi},$$

which is close to the value of 18.0 ksi obtained in Example 8.1. The flexure analogy models flange normal warping stresses well for many typical steel railway bridge elements.

$$V_H = \frac{6.72}{2} = 3.36\,\text{kips},$$

$$\tau_{bH} = \frac{3(3.36)}{2[12(0.75)]} = 0.56\,\text{ksi},$$

Combined stresses:

$$\sigma_{\text{flange}} = \sigma_{\text{b}}\left(\frac{L}{2}\right) + \sigma_{\text{w}}\left(\frac{L}{2}\right) = 5.58 + 19.0 = 24.6\,\text{ksi},$$

which is close to the value of 23.5 ksi obtained in Example 8.1.

$$\tau_{\text{web}} = \tau_{\text{b}} = 1.40\,\text{ksi},$$

$$\tau_{\text{flange}} = \tau_{\text{b}} + \tau_{\text{w}} = 0.58 + 0.56 = 1.14\,\text{ksi}.$$

The shear stress in the flange and web is correct at $z = L/2$, where pure torsional shear stresses, τ_{t}, are zero but are underestimated at $z = 0$ because pure torsion is not considered in the flexure analogy. However, for many typical steel railway bridge elements, torsional shear stresses are of much less concern than flange normal stresses due to warping torsion. In such cases, the flexure analogy is appropriate for ordinary torsion design problems.

Example 8.3

The end floorbeam in a ballasted through plate girder bridge is subjected to a uniformly distributed load, w, at an eccentricity of 2.5 in. as shown in Figure E8.3.

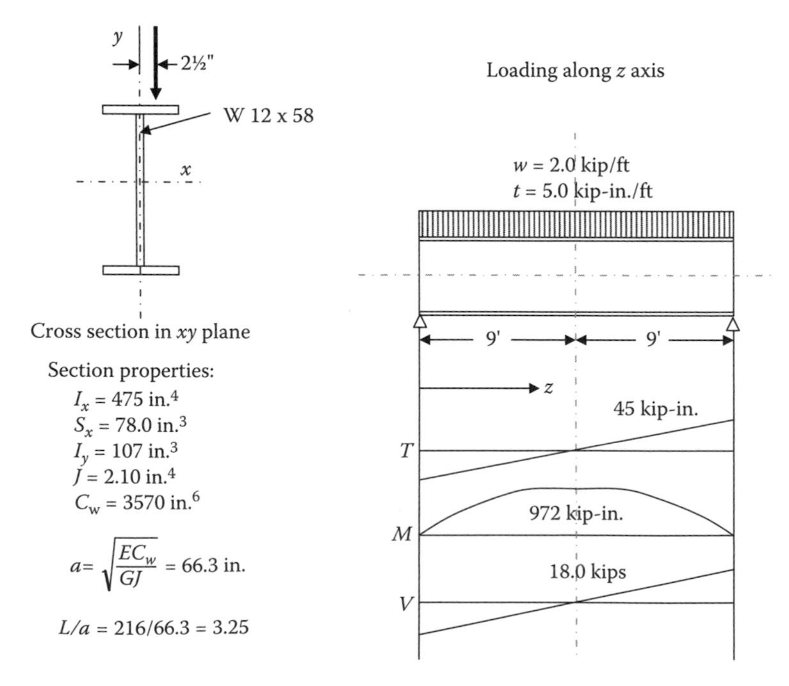

FIGURE E8.3

TABLE E8.7

Element of the Girder	Pure Torsion Shear Stress, τ_t (ksi)	
	$z = 0$ (end)	$z = L/2 = 108$ in. (center)
Flange, $t = 0.625$ in.	5.79	0
Web, $t = 0.375$ in.	3.48	0

The solution of the differential equation is (Equation 8.21a)

$$\theta = 0.078\left(-0.926\sinh\left(\frac{z}{66.3}\right) + \cosh\left(\frac{z}{66.3}\right) - \frac{z^2}{8791} + \frac{z}{40.70} - 1\right)$$

and its differentials are

$$\frac{d\theta}{dz} = 1.177 \times 10^{-3}\left(-0.926\cosh\left(\frac{z}{66.3}\right) + \sinh\left(\frac{z}{66.3}\right) - \frac{z}{66.3} + 1.629\right),$$

$$\frac{d^2\theta}{dz^2} = 1.775 \times 10^{-5}\left(-0.926\sinh\left(\frac{z}{66.3}\right) + \cosh\left(\frac{z}{66.3}\right) - 1\right),$$

$$\frac{d^3\theta}{dz^3} = 2.677 \times 10^{-7}\left(-0.926\cosh\left(\frac{z}{a}\right) + \sinh\left(\frac{z}{a}\right)\right).$$

Pure (St. Venant) torsion (Table E8.7):

$$\tau_t = Gt\left(\frac{d\theta}{dz}\right) = 13.18(t)\left(-0.926\cosh\left(\frac{z}{66.3}\right) + \sinh\left(\frac{z}{66.3}\right) - \frac{z}{66.3} + 1.629\right).$$

Warping torsion (lateral bending of flanges) (Tables E8.8 and E8.9):

$$\tau_w = \frac{-Eb_f^2h}{16}\left(\frac{d^3\theta}{dz^3}\right)$$

$$= \frac{-29{,}000(10)^2(12.25)}{16}(2.677 \times 10^{-7})\left(-0.926\cosh\left(\frac{z}{a}\right) + \sinh\left(\frac{z}{a}\right)\right),$$

$$\tau_w = 0.60\left(-0.926\cosh\left(\frac{z}{66.3}\right) + \sinh\left(\frac{z}{66.3}\right)\right),$$

TABLE E8.8

Element of the Girder	Warping Torsion Shear Stress, τ_w (ksi)	
	$z = 0$ (end)	$z = L/2 = 108$ in. (center)
Flange, $t = 0.625$ in.	−0.55	0

TABLE E8.9

Element of the Girder	Warping Torsion Normal Stress, σ_w (ksi)	
	$z = 0$ (end)	$z = L/2 = 108$ in. (center)
Flange, $t = 0.625$ in.	0	−9.81

$$\sigma_w = \frac{Eb_f h}{4}\left(\frac{d^2\theta}{dz^2}\right)$$

$$= \frac{29{,}000(10)(12.25)}{4}(1.775 \times 10^{-5})\left(-0.926\sinh\left(\frac{z}{66.3}\right) + \cosh\left(\frac{z}{66.3}\right) - 1\right),$$

$$\sigma_w = 15.76\left(-0.926\sinh\left(\frac{z}{66.3}\right) + \cosh\left(\frac{z}{66.3}\right) - 1\right).$$

Flexure of the girder (Tables E8.10 and E8.11):

$$M = 972\text{ kip-in.},$$

$$\sigma_b(0) = 0,$$

$$\sigma_b\left(\frac{L}{2}\right) = \pm\frac{972}{78} = \pm 12.46\text{ ksi},$$

$$V = 18.0\text{ k},$$

$$Q_{flange} = 18.2\text{ in.}^3,$$

$$Q_{web} = 43.2\text{ in.}^3,$$

$$\tau_{b\ flange}(0) = \frac{18.0(18.2)}{475(0.625)} = 1.11\text{ ksi},$$

$$\tau_{b\ web}(0) = \frac{18.0(43.2)}{475(0.375)} = 4.37\text{ ksi}.$$

TABLE E8.10

Element of the Girder	Flexural Normal Stress, σ_b (ksi)	
	$z = 0$ (end)	$z = L/2 = 108$ in. (center)
Top Flange, $t = 0.625$ in.	0	−12.46

TABLE E8.11

Element of the Girder	Flexural Shear Stress, τ_b (ksi)	
	$z = 0$ (end)	$z = L/2 = 108$ in. (center)
Flange, $t = 0.625$ in.	1.11	0
Web, $t = 0.375$ in.	4.37	0

TABLE E8.12

Element of the Girder	Shear Stress (ksi) $z = 0$ (end)	Shear Stress (ksi) $z = L/2 = 108$ in. (center)
Flange, $t = 0.625$ in.	7.45	0
Web, $t = 0.375$ in.	7.84	0

TABLE E8.13

Element of the Girder	Normal Stress (ksi) $z = 0$ (end)	Normal Stress (ksi) $z = L/2 = 108$ in. (center)
Flange, $t = 0.625$ in.	0	22.27

Combined stresses (Tables E8.12 and E8.13):

In this case, where there are no bearing stiffeners to transfer loads between flanges, local flexural normal stresses in the flange from the eccentric load must also be considered and superimposed on the top flange stresses.

Example 8.4

Use the flexure analogy for torsion to find the shear and normal stresses due to combined flexure and torsion of Example 8.3.

$$H = \frac{2.0(2.5)}{12.25 - 0.625} = 0.43 \text{ kips/ft},$$

$$M_H = \frac{0.43(216)^2}{(12)8} = 209.0 \text{ kip-in.},$$

$$\sigma_{bH} = \frac{209.0}{(0.625(10)^2/6)} = 20.1 \text{ ksi}.$$

Modification factors, β, that reduce the normal lateral bending stress have been developed as a corrective measure since the flexure analogy overestimates normal flange stresses due to warping. For the case of $L/a = 3.25$ and uniform torsional moment, t', $\beta = 0.48$ (Salmon and Johnson, 1980).

$$\sigma_{bH} = 20.1\beta = 20.1(0.48) = 9.65 \text{ ksi},$$

which is very close to the value of 9.81 ksi obtained in Example 8.3.

$$V_H = \frac{0.43(18)}{2} = 3.87 \text{ kips},$$

$$\tau_{bH} = \frac{3(3.87)}{2[10(0.625)]} = 0.93 \text{ ksi}.$$

Combined stresses:

$$\sigma_{\text{flange}} = \sigma_b\left(\frac{L}{2}\right) + \sigma_w\left(\frac{L}{2}\right) = 12.46 + 9.65 = 22.1\,\text{ksi},$$

which is very close to the value of 22.3 ksi obtained in Example 8.3.

$$\tau_{\text{web}} = \tau_b = 4.37\,\text{ksi},$$

$$\tau_{\text{flange}} = \tau_b + \tau_w = 1.11 + 0.93 = 2.04\,\text{ksi}.$$

The shear stress in the flange and web are underestimated at $z = 0$ because pure torsion is not considered in the flexural analogy.

8.4 COMBINED AXIAL FORCES AND BENDING OF MEMBERS

Members are subjected to axial forces and bending moments due to axial force eccentricities (often unintentional and related to connection eccentricities, member out-of-straightness and/or secondary deflection effects) and when axial members are laterally loaded (typically self-weight and/or wind). These normal axial and flexural stresses must be combined. Axial tension combined with bending is generally of lesser concern than axial compression combined with bending due to the potential for instability of slender compression members.

8.4.1 Axial Tension and Uniaxial Bending

The tensile load reduces the bending effects on the member when tensile axial loads act simultaneously with bending. For the beam shown in Figure 8.4

$$M = \frac{wL^2}{8} - T\Delta. \tag{8.22}$$

However, the deflection, Δ, is dependent on the bending moment, M, which is itself dependent on the deflection Δ. The deflection

$$\Delta = \frac{5wL^4}{384EI} - \frac{T\Delta L^2}{8EI} \tag{8.23}$$

may be solved iteratively.* The bending moment, M, is

$$M = \frac{wL^2}{8} - \frac{T}{EI}\left(\frac{5wL^4}{384} - \frac{T\Delta L^2}{8}\right). \tag{8.24}$$

However, since the effect of the tensile force on the deflection, Δ, can conservatively be neglected in the analysis† of linear elastic members, the principle of superposition

* A digital computer algorithm is generally required.

† In usual structures the effect is small (Bresler, Lin and Scalzi, 1968).

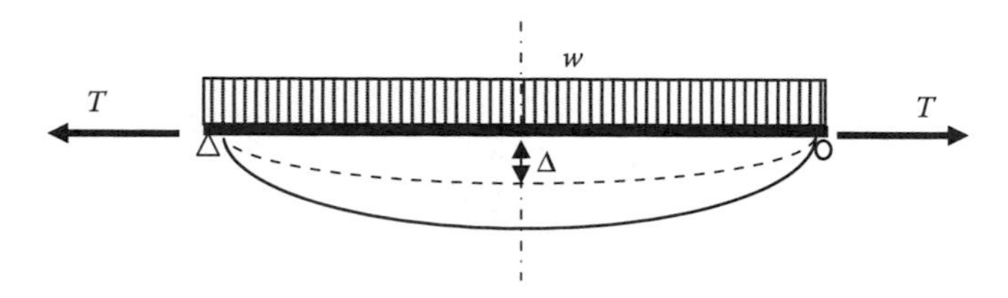

FIGURE 8.4 Member subjected to combined tensile axial force and bending.

may be applied to ensure that failure by yielding does not occur. AREMA (2008) recommends that, if bending (even with superimposed axial tensile stresses) causes compression in some parts of the cross section, the flexural compressive stress and stability criteria should be considered. Therefore, the allowable flexural stress may differ from the allowable axial tensile stress, which provides the interaction equation

$$\pm\frac{\sigma_b}{F_b} + \frac{\sigma_t}{F_t} \leq 1, \tag{8.25}$$

where σ_b is the maximum tensile or compressive bending stress, σ_t is the maximum tensile axial stress, F_b is the allowable tensile or compressive bending stress, F_t is the allowable axial tensile stress on the gross section and is equal to $0.55F_y$.

When flexural compression with axial tension results in tensile stresses, Equation 8.25 is

$$\sigma_b + \sigma_t \leq 0.55F_y, \tag{8.26}$$

which is the AREMA (2008) recommendation. When flexural compression with axial tension results in compressive stresses, AREMA (2008) recommends

$$-\sigma_b + \sigma_t \leq F_{call}, \tag{8.27}$$

where F_{call} is the allowable compressive bending stress.

8.4.2 Axial Compression and Uniaxial Bending

The compressive load increases the bending effects on the member when compressive axial loads act simultaneously with bending (Figure 8.5). For the beam shown in Figure 8.5

$$M = \frac{wL^2}{8} + P\Delta. \tag{8.28}$$

Again, the deflection, Δ, is dependent on the bending moment, M, which is itself dependent on the deflection Δ. The deflection is

$$\Delta = \frac{5wL^4}{384EI} + \frac{P\Delta L^2}{8EI} \tag{8.29}$$

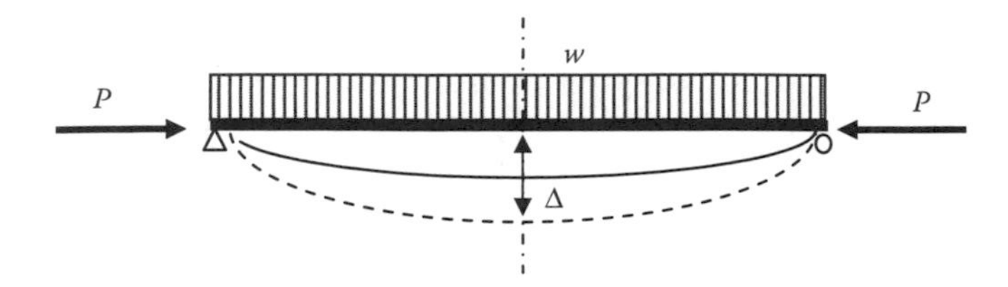

FIGURE 8.5 Member subjected to combined compressive axial force and bending.

and the bending moment, M, is determined as

$$M = \frac{wL^2}{8} - \frac{P}{EI}\left(\frac{5wL^4}{384} + \frac{P\Delta L^2}{8}\right). \tag{8.30}$$

Equation 8.29 indicates that the deflection builds upon itself (a deflection causes more deflection) and instability may occur due to this P–Δ effect. Therefore, an iterative solution to Equation 8.30 is required, which is not efficient for routine design work. Alternately, for some boundary conditions and loads, the differential equation for axial compression and flexure may be solved. However, for routine design work, limitations on combined stresses or semiempirical interaction equations have been developed. AREMA (2008) uses interaction equations for both the yielding and stability criteria.

8.4.2.1 Differential Equation for Axial Compression and Bending on a Simply Supported Beam

Consider a member loaded with a general uniform lateral load, $w(z)$, a concentrated load, Q, at a location, a, end moments, M_A and M_B, and compressive axial force, P, as shown in Figure 8.6 . The bending moments at z due to loads M_A, M_B, Q, and $w(z)$ are combined into a collective bending moment, M_p, such that

$$M_p(z) = M_{w(z)} + M_{MA} + M_{MB} + M_{Q(z)}, \tag{8.31}$$

where $M_{w(z)}$ is the bending moment at z due to $w(z)$, M_{MA} and M_{MB} are the bending moments due to M_A and M_B, $M_{Q(z)}$ is the bending moment at z due to $Q(z)$.

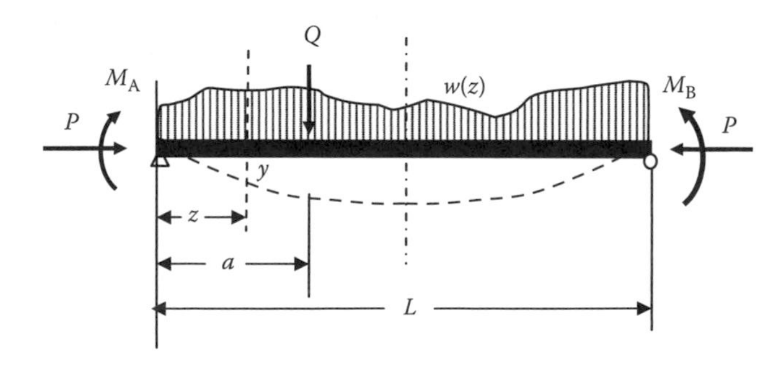

FIGURE 8.6 General loading of combined axial compression and bending member.

The moment, M_z, at z is then

$$M_Z = M_p(z) + Py. \tag{8.32}$$

Substitution of $M_Z = -EI\, d^2y/dz^2$ into Equation 8.32 yields

$$\frac{d^2y}{dz^2} + k^2y = -\frac{M_p(z)}{EI}, \tag{8.33}$$

where $k^2 = \frac{P}{EI}$.

Differentiating twice yields

$$\frac{d^4y}{dz^4} + k^2\frac{d^2y}{dz^2} = -\frac{1}{EI}\left(\frac{d^2M_p(z)}{dz^2}\right). \tag{8.34}$$

Equation 8.34 is the differential equation for axial compression and bending.

8.4.2.1.1 Axial Compression and Bending from a Uniformly Distributed Transverse Load

$$M_p(z) = M_{w(z)} = \frac{wz(L-z)}{2}, \tag{8.35}$$

$$\frac{d^2M_p(z)}{dz^2} = -w, \tag{8.36}$$

and the differential equation for axial compression and flexure (Equation 8.34) is

$$\frac{d^4y}{dz^4} + k^2\frac{d^2y}{dz^2} = \frac{w}{EI}. \tag{8.37}$$

The solution of Equation 8.37 is (Chen and Lui, 1987)

$$y(z) = \frac{w}{EIk^4}\left(\tan\frac{kL}{2}\sin kz + \cos kz - 1\right) - \frac{w}{2EIk^2}z(L-z), \tag{8.38}$$

$$\frac{d^2y(z)}{dz^2} = -\frac{w}{EIk^2}\left(\tan\frac{kL}{2}\sin kz + \cos kz - 1\right), \tag{8.39}$$

$$M_Z = -EI\frac{d^2y}{dz^2} = \frac{w}{EIk^2}\left(\tan\frac{kL}{2}\sin kz + \cos kz - 1\right). \tag{8.40}$$

The maximum moment at the center span is

$$M_{Z=L/2} = \frac{w}{k^2}\left(\sec\frac{kL}{2} - 1\right) = \frac{wL^2}{8}\left(\frac{8\,(\sec(kL/2) - 1)}{k^2L^2}\right), \tag{8.41}$$

where $(8(\sec kL/2 - 1)/k^2L^2)$ is a moment magnification factor accounting for the effects of the axial compressive force, P. The secant function can be expanded in a power series as (Beyer, 1984)

$$\sec\frac{kL}{2} = 1 + \frac{1}{2}\left(\frac{kL}{2}\right)^2 + \frac{5}{24}\left(\frac{kL}{2}\right)^4 + \frac{61}{720}\left(\frac{kL}{2}\right)^6 + \cdots. \tag{8.42}$$

Since $P_e = \pi^2EI/L^2$ (Euler buckling load) and $k = \sqrt{P/EI}$,

$$\frac{kL}{2} = \frac{\pi}{2}\sqrt{\frac{P}{P_e}}. \tag{8.43}$$

Substitution of Equation 8.43 into Equations 8.41 and 8.42 provides

$$M_{z=L/2} = \frac{wL^2}{8}\left(1 + 1.028\left(\frac{P}{P_e}\right) + 1.031\left(\frac{P}{P_e}\right)^2 + 1.032\left(\frac{P}{P_e}\right)^3 + \cdots\right) \tag{8.44a}$$

$$= \frac{wL^2}{8}\left(1 + 1.028\left(\frac{P}{P_e}\right)\left(1 + 1.003\left(\frac{P}{P_e}\right) + 1.004\left(\frac{P}{P_e}\right)^2 + \cdots\right)\right). \tag{8.44b}$$

Equation 8.44b may be approximated as

$$M_{z=L/2} \approx \frac{wL^2}{8}\left(1 + 1.028\left(\frac{P}{P_e}\right)\left(1 + \left(\frac{P}{P_e}\right) + \left(\frac{P}{P_e}\right)^2 + \left(\frac{P}{P_e}\right)^3 + \cdots\right)\right) \tag{8.45}$$

$$\approx \frac{wL^2}{8}\left(1 + 1.028\left(\frac{P}{P_e}\right)\left(\frac{1}{1-(P/P_e)}\right)\right) \tag{8.46}$$

$$\approx \frac{wL^2}{8}\left(\frac{1}{1-(P/P_e)}\right), \tag{8.47}$$

where $1/(1 - (P/P_e))$ is an approximate moment magnification factor appropriate for use in design.

8.4.2.1.2 Axial Compression and Bending from a Concentrated Transverse Load

$$M_p(z) = M_{Q(z)} = \frac{Qz(L-a)}{L} \quad \text{for } 0 \le z \le a, \tag{8.48a}$$

$$M_p(z) = M_{Q(z)} = \frac{Qa(L-z)}{L} \quad \text{for } a \le z \le L. \tag{8.48b}$$

Substitution of Equations 8.48a and 8.48b into Equation 8.33 yields

$$\frac{d^2y}{dz^2} + k^2y = -\frac{Qz(L-a)}{EIL} \quad \text{for } 0 \leq z \leq a, \tag{8.49a}$$

$$\frac{d^2y}{dz^2} + k^2y = -\frac{Qa(L-z)}{EIL} \quad \text{for } a \leq z \leq L. \tag{8.49b}$$

The general solutions to Equations 8.49a and 8.49b are

$$y = A\sin kz + B\cos kz - \frac{Qz(L-a)}{EILk^2} \quad \text{for } 0 \leq z \leq a, \tag{8.50a}$$

$$y = C\sin kz + D\cos kz - \frac{Qa(L-z)}{EILk^2} \quad \text{for } a \leq z \leq L. \tag{8.50b}$$

Differentiating Equations 8.50a and 8.50b with boundary conditions of $y(0) = y(L) = 0$ and noting that displacement, $y(a)$, and slope, $dy(a)/dz$, are continuous at $z = a$ provides

$$M_Z = -EI\frac{d^2y}{dz^2} = -\frac{Q}{k}\frac{\sin k(L-a)}{\sin kL}\sin kz \quad \text{for } 0 \leq z \leq a, \tag{8.51a}$$

$$M_Z = -EI\frac{d^2y}{dz^2} = \frac{Q\sin ka}{k}\left(\frac{\sin kz}{\tan kL} - \cos kz\right) \quad \text{for } a \leq z \leq L. \tag{8.51b}$$

The maximum moment at the center span (by substitution of $z = L/2$ into Equations 8.51a or 8.51b) is

$$M_{z=L/2} = \frac{QL}{4}\left(\frac{2\tan kL/2}{kL}\right). \tag{8.52}$$

The tangent function in Equation 8.52 can be expanded in a power series as (Beyer, 1984)

$$\tan\frac{kL}{2} = \frac{kL}{2} + \frac{1}{3}\left(\frac{kL}{2}\right)^3 + \frac{2}{15}\left(\frac{kL}{2}\right)^5 + \frac{17}{315}\left(\frac{kL}{2}\right)^7 + \cdots, \tag{8.53}$$

which when substituted into Equation 8.52, and after making further simplifications similar to those outlined in Section 8.4.2.1.1, yields

$$M_{z=L/2} \approx \frac{QL}{4}\left(\frac{1-0.2\,(P/P_e)}{1-(P/P_e)}\right), \tag{8.54}$$

where $[(1-0.2(P/P_e))/(1-(P/P_e))]$ is an approximate moment magnification factor appropriate for use in design.

8.4.2.1.3 Axial Compression and Bending from End Moments

$$M_\mathrm{p}(z) = M_\mathrm{MA} + M_\mathrm{MB} = M_\mathrm{A} - \left(\frac{M_\mathrm{A} + M_\mathrm{B}}{L}\right)z. \tag{8.55}$$

Substitution of Equation 8.55 into Equation 8.33 yields

$$\frac{\mathrm{d}^2y}{\mathrm{d}z^2} + k^2y = -\frac{M_\mathrm{A}}{EI} + \left(\frac{M_\mathrm{A} + M_\mathrm{B}}{EIL}\right)z. \tag{8.56}$$

The general solution to Equation 8.56 is

$$y = A\sin kz + B\cos kz + \frac{M_\mathrm{A} + M_\mathrm{B}}{EILk^2}z - \frac{M_\mathrm{A}}{EIk^2} \tag{8.57}$$

Considering boundary conditions of $y(0) = y(L) = 0$

$$y = -\frac{(M_\mathrm{A}\cos kL + M_\mathrm{B})}{EIk^2\sin kL}\sin kz + \frac{M_\mathrm{A}}{EIk^2}\cos kz + \frac{M_\mathrm{A} + M_\mathrm{B}}{EILk^2}z - \frac{M_\mathrm{A}}{EIk^2}. \tag{8.58}$$

Differentiation of Equation 8.58 yields

$$\frac{\mathrm{d}y}{\mathrm{d}z} = -\frac{(M_\mathrm{A}\cos kL + M_\mathrm{B})}{EIk\sin kL}\cos kz - \frac{M_\mathrm{A}}{EIk}\sin kz + \frac{M_\mathrm{A} + M_\mathrm{B}}{EILk^2}, \tag{8.59}$$

$$\frac{\mathrm{d}^2y}{\mathrm{d}z^2} = \frac{(M_\mathrm{A}\cos kL + M_\mathrm{B})}{EI\sin kL}\sin kz - \frac{M_\mathrm{A}}{EI}\cos kz, \tag{8.60}$$

$$\frac{\mathrm{d}^3y}{\mathrm{d}z^3} = \frac{k(M_\mathrm{A}\cos kL + M_\mathrm{B})}{EI\sin kL}\cos kz + \frac{kM_\mathrm{A}}{EI}\sin kz. \tag{8.61}$$

The bending moment and shear forces are

$$M_\mathrm{Z} = -EI\frac{\mathrm{d}^2y}{\mathrm{d}z^2} = -\frac{(M_\mathrm{A}\cos kL + M_\mathrm{B})}{\sin kL}\sin kz + M_\mathrm{A}\cos kz, \tag{8.62}$$

$$V_\mathrm{z} = -EI\frac{\mathrm{d}^3y}{\mathrm{d}z^3} = -\frac{k(M_\mathrm{A}\cos kL + M_\mathrm{B})}{\sin kL}\cos kz - kM_\mathrm{A}\sin kz. \tag{8.63}$$

The maximum moment occurs where the shear force is zero. If Equation 8.63 is equated to zero, we obtain

$$\tan kz_\mathrm{M} = -\frac{M_\mathrm{A}\cos kL + M_\mathrm{B}}{M_\mathrm{A}\sin kL}, \tag{8.64}$$

where z_M is the location of the maximum moment along the z axis. Expressions for $\sin(kz_\mathrm{M})$ and $\cos(kz_\mathrm{M})$ may be obtained from Equation 8.64 for substitution into Equation 8.62 to obtain the maximum bending moment, M_max, as

$$M_\mathrm{max} = -M_\mathrm{B}\left(\sqrt{\frac{(M_\mathrm{A}/M_\mathrm{B})^2 - 2\,(M_\mathrm{A}/M_\mathrm{B})\cos kL + 1}{\sin^2 kL}}\right). \tag{8.65}$$

If the end moments are equal but opposite in direction (i.e., beam bent in single curvature)

$$M = M_A = -M_B, \tag{8.66}$$

then Equation 8.65 is

$$M_{max} = M\sqrt{\frac{2(1-\cos kL)}{\sin^2 kl}} = M \sec \frac{kL}{2}, \tag{8.67}$$

which is the secant formula. In this case, the maximum moment occurs at $z_M = L/2$.

8.4.2.1.4 Combined Axial Compression and Flexural Loading

The bending moments from combined transverse uniformly distributed and axial compression loads, transverse concentrated and axial compression loads, and end moment and axial compression loads have been developed from differential Equations 8.33 and 8.34. For combined loads, such as those shown in Figure 8.6, the bending moments from each load case may be superimposed, provided that the axial compression force is the same for each load case.

A general way of superimposing the effects from each load case is to superpose the deflected shapes by summation of Equations 8.38, 8.50, and/or 8.58, depending on the applicable load cases. The location, z_M, of the maximum bending moment, M_{max}, may be obtained by setting $d^3y/dz^3 = 0$, and the maximum bending moment may be obtained by substitution of z_M into the equation for bending moment, $-EI(d^2y/dz^2)$.

However, it is evident that the design of members subjected to simultaneous bending and axial compression by methods involving the solution of differential Equations 8.33 and 8.34 is relatively complex and not well suited to routine design work. Because of this, interaction equations have been developed based on bending moment–curvature–axial compression relationships.

8.4.2.2 Interaction Equations for Axial Compression and Uniaxial Bending

Interaction equations used in ASD are determined from interaction equations developed for ultimate loads. When members remain elastic, the axial compressive force, P, has no effect on the moment–curvature–axial compression relationship. However, in the inelastic range, where partial yielding has occurred, the moment–curvature–axial compression relationship is dependent on the axial compression. The nonlinear relationship between bending moment, M, and axial compressive force, P, depends on the curvature, ϕ, of the member. The ultimate axial compressive load is attained when yielding under combined bending and axial compression reduces the member stiffness, due to partial yielding of the cross section, and creates instability. Also, in the range of inelastic behavior, the superposition of effects due to bending, M, and axial compression, P, is not applicable.

The moment–curvature–axial compression relationships for members are determined analytically for various degrees of member yielding.* A typical moment–curvature–axial compression relationship is shown in Figure 8.7.

* Indicated by depth of yielded material from the fibers with the largest compressive strain.

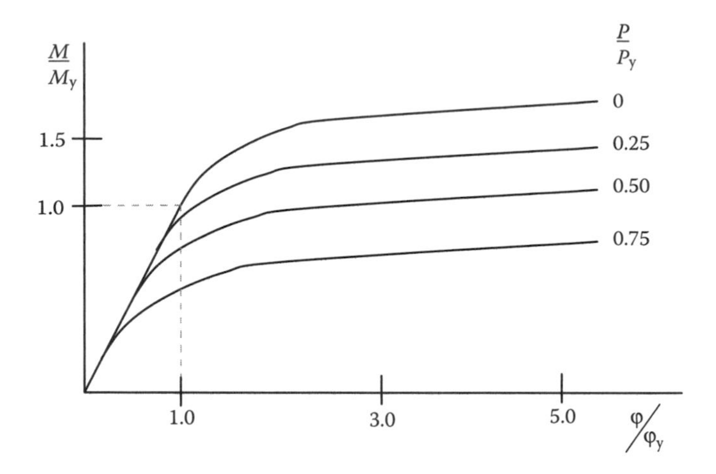

FIGURE 8.7 Typical plot of moment–curvature relationship for a member subjected to bending and axial compression.

Once the moment–curvature–axial compression relationship is established, each value of P/P_y with a slenderness ratio, *KL*/*r* (where *r* is the radius of gyration in the plane of bending), is combined with various values of M/M_y until instability occurs (at M_u). Since M/M_y creates deflection, Δ, which creates an additional bending moment $(P/P_y)\Delta$, an iterative analysis, such as numerical integration by Newmark's method or other step-by-step numerical integration techniques, is often used. Once the appropriate values of P/P_y and M/M_u are determined (when deflections calculated in two successive iterations are in sufficient agreement) for various values of *KL*/*r*, interaction diagrams, such as that shown in Figure 8.8, may be made. These interaction curves may be approximated by interaction equations for various values of L/r.

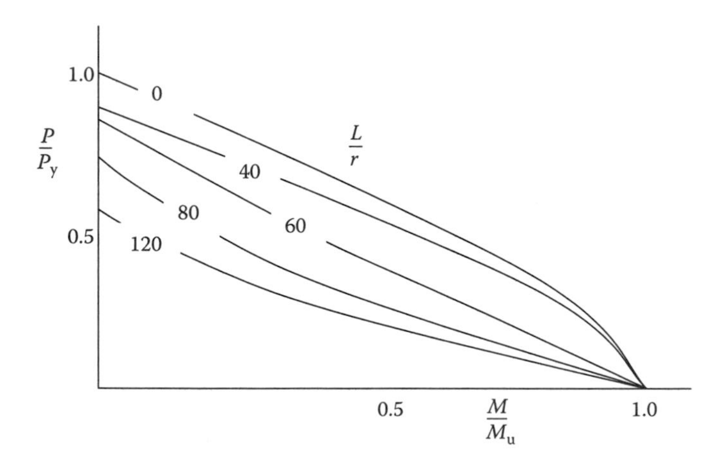

FIGURE 8.8 Typical interaction curves for a member subjected to bending and axial compression.

The curve for $L/r = 0$ may be approximated as

$$\frac{P}{P_y} + \frac{M}{1.18M_u} = 1.0 \tag{8.68}$$

or

$$\frac{\sigma_a}{0.55F_y} + \frac{\sigma_b}{1.18F_u} = 1.0. \tag{8.69}$$

For service load design, Equation 8.69 may be conservatively expressed as

$$\frac{\sigma_a}{0.55F_y} + \frac{\sigma_b}{F_b} = 1.0, \tag{8.70}$$

where σ_a is the normal stress due to applied axial compression, σ_b is the normal stress due to applied bending moment, F_u is the ultimate bending stress, F_b is the allowable compressive stress for bending alone. This interaction equation is applicable to members with low slenderness, such as locations that are braced in the plane of bending, where yielding will be the failure criterion. However, for members with larger slenderness, stability must also be investigated in the yielding criterion.

The yield criterion for members of larger slenderness is established as Equation 8.70 but by considering F_a instead of F_y due to the potential for allowable axial compressive stresses to be controlled by instability when $KL/r \geq 0.629\sqrt{E/F_y}$ (see Chapter 6). Equation 8.70 is then

$$\frac{\sigma_a}{F_a} + \frac{\sigma_b}{F_b} = 1.0. \tag{8.71}$$

The interaction curves may be approximated for various values of slenderness, L/r, by interaction equations as

$$\frac{P}{P_{cr}} + \frac{M}{M_u(1 - P/P_e)} = 1.0 \tag{8.72}$$

or

$$\frac{\sigma_a}{F_{cr}} + \frac{\sigma_b}{F_u(1 - \sigma_a/\sigma_e)} = 1.0, \tag{8.73}$$

where P_{cr} is the critical axial buckling load, P_e is the Euler buckling load and is equal to $\pi^2 EI/L^2$, F_{cr} is the critical axial buckling stress, σ_e is the Euler buckling stress and is equal to $\pi^2 E/(KL/r)^2$.

Equation 8.73, using an $FS = 1.95$ for service load design for axial buckling, may be conservatively expressed as

$$\frac{\sigma_a}{F_a} + \frac{\sigma_b}{F_b[1 - \sigma_a/(\sigma_e/1.95)]} = 1.0 \tag{8.74}$$

or

$$\frac{\sigma_a}{F_a} + \frac{\sigma_b}{F_b\left(1 - (\sigma_a/0.514\pi^2 E)\,(KL/r)^2\right)} = 1.0, \tag{8.75}$$

where F_a is the allowable stress for axial compression alone and K is the effective length factor and is equal to $\pi/L\sqrt{EI/P_{cr}}$ (see Chapter 6).

The yield criterion (Equation 8.71) and the stability criteria (Equation 8.75) should be investigated for all members subjected to simultaneous bending and axial compression.

8.4.3 Axial Compression and Biaxial Bending

The strength of members subjected to axial compression and biaxial bending is complex. Theoretical procedures have been developed for short members and longer members to produce interaction curves (Chen and Astuta, 1977; Culver, 1966) and confirmed as reasonable by experiment and in computer studies for typical members by Birnstiel (1968), Pillai (1980), and others (see Galambos, 1988).

Since a design methodology for axial compression and biaxial bending must include the case of axial compression and uniaxial bending, it would appear reasonable to extend the interaction Equations 8.71 and 8.75 to the case of biaxial bending with axial compression. Therefore, the interaction formula relating to the stability criterion is

$$\frac{\sigma_a}{F_a} + \frac{\sigma_{bx}}{F_{bx}\left(1 - \sigma_a/0.514\pi^2 E\,(K_x L_x/r_x)^2\right)} + \frac{\sigma_{by}}{F_{by}\left(1 - \sigma_a/0.514\pi^2 E\,\left(K_y L_y/r_y\right)^2\right)} = 1.0. \tag{8.76}$$

For members with low slenderness (where yielding controls) or at locations of supports or where braced in the plane of bending,

$$\frac{\sigma_a}{F_a} + \frac{\sigma_{bx}}{F_{bx}} + \frac{\sigma_{by}}{F_{by}} = 1.0, \tag{8.77}$$

where $F_a = 0.55F_y$ when $L/r = 0$ (points of bracing or supports), σ_{bx} is the normal bending stress about the x axis, σ_{by} is the normal bending stress about the y axis, F_{bx} is the allowable bending stress about the x axis, F_{by} is the allowable bending stress about the y axis, and $K_x L_x/r_x$ and $K_y L_y/r_y$ are the effective slenderness ratios of the member about the axes x and y, respectively (see Chapter 6).

8.4.4 AREMA Recommendations for Combined Axial Compression and Biaxial Bending

AREMA (2008) recommends that members subjected to axial compression and biaxial bending be designed in accordance with Equations 8.76, 8.77, and 8.70 extended for biaxial bending.

However, AREMA (2008) recognizes that, for members with relatively small axial compressive forces, the secondary effects are negligible. Therefore, when $\sigma_a/F_a \leq 0.15$, Equation 8.76 may be expressed as

$$\frac{\sigma_a}{F_a} + \frac{\sigma_{bx}}{F_{bx}} + \frac{\sigma_{by}}{F_{by}} \leq 1.0. \tag{8.78}$$

Both yielding and stability effects must be considered when $\sigma_a/F_a \succ 0.15$. AREMA (2008) recommends that, when $\sigma_a/F_a \succ 0.15$, Equation 8.70 (yield criterion) extended for biaxial bending and Equation 8.76 (stability criterion) should be used for design as

$$\frac{\sigma_a}{0.55F_y} + \frac{\sigma_{bx}}{F_{bx}} + \frac{\sigma_{by}}{F_{by}} \leq 1.0 \quad (8.79)$$

and

$$\frac{\sigma_a}{F_a} + \frac{\sigma_{bx}}{F_{bx}\left(1 - (\sigma_a/0.514\pi^2 E)(K_x L_x/r_x)^2\right)} + \frac{\sigma_{by}}{F_{by}\left(1 - (\sigma_a/0.514\pi^2 E)\left(K_y L_y/r_y\right)^2\right)} \leq 1.0, \quad (8.80)$$

where σ_a is the normal axial compressive stress, σ_{bx} is the normal flexural stress about the x axis, σ_{by} is the normal flexural stress about the y axis, F_a is the allowable axial compressive stress for axial compression only (Chapter 6), F_{bx} is the allowable flexural compressive stress for bending only (Chapter 7) about the x axis, and F_{by} is the allowable flexural compressive stress for bending only (Chapter 7) about the y axis.

Equation 8.79 relates to the yield criterion, which is appropriate to consider at support locations, members with very low slenderness ratios, and locations braced in the planes(s) of bending. The design of members subjected to axial compression and bending is usually governed by the stability criterion of Equation 8.80.

8.5 COMBINED BENDING AND SHEAR OF PLATES

Combined bending and shear may be significant in the webs of plate girders as outlined in Chapter 7, Section 7.2.6, concerning the plate girder design.

REFERENCES

American Railway Engineering and Maintenance-of-Way Association (AREMA), 2008, Steel structures, in *Manual for Railway Engineering*, Chapter 15, Lanham, MD.

Beyer, W.H. (Ed.), 1984, *Standard Mathematical Tables*, 27th Edition, CRC Press, Boca Raton, FL.

Birnstiel, C., 1968, Experiments on H-columns under biaxial bending, *Journal of Structural Engineering*, ASCE, (94), ST10, 2429–2450.

Bresler, B., Lin, T.Y., and Scalzi, J.B., 1968, *Design of Steel Structures*, 2nd Edition, Wiley, New York.

Chen, W.F. and Astuta, T., 1977, *Theory of Beam Columns*, Vols 1 and 2, McGraw-Hill, New York.

Chen, W.F. and Lui, E.M., 1987, *Structural Stability*, Elsevier, New York.

Culver, C.G., 1966, Exact solution of the biaxial bending equations, *Journal of Structural Engineering*, ASCE, (92), ST2, 63–84.

Galambos, T.V., 1988, *Guide to Stability Design Criteria for Metal Structures*, Wiley, New York.

Kulak, G.L. and Grondin, G.Y., 2002, *Limit States Design in Structural Steel*, 7th Edition, CISC, Toronto, ON.

Kuzmanovic, B.O. and Willems, N., 1983, *Steel Design for Structural Engineers*, 2nd Edition, Prentice-Hall, New Jersey.

Pillai, U.S., 1980, *Comparison of Test Results with Design Equations for Biaxially Loaded Steel Beam Columns*, Civil Engineering Research Report No. 80-2, Royal Military College of Canada, Kingston, Canada.

Salmon, C.G. and Johnson, J.E., 1980, *Steel Structures Design and Behavior*, 2nd Edition, Harper & Row, New York.

Seaburg, P.A. and Carter, C.J., 1997, *Torsional Analysis of Structural Steel Members*, AISC, Chicago, IL.

9 Design of Connections for Steel Members

9.1 INTRODUCTION

The design of connections is of equal importance to the safety and reliability of steel railway bridges as the design of the axial and flexural members that are connected to form the superstructure (see Chapters 6 and 7, respectively). Connections in modern steel railway superstructures are made with welds, bolts,* and/or pins.† Typically, these connections transmit axial shear (e.g., truss member connections and beam flange splices), combined axial tension and shear (e.g., semirigid and rigid beam framing connections that transmit shear and moment), or eccentric shear (e.g., welded flexible beam framing connections and beam web plate splices). Connection behavior is often complex but may be represented for routine design with relatively simple mathematical models. AREMA (2008) recommends design forces for axial and flexural member connection design.

Truss member end connections at the top chord of deck trusses or the bottom chord of through trusses should be designed for the allowable strength of the member. Vertical post end connections in deck trusses without diagonals in adjacent panels at the top chord and hangers in through trusses should be designed for 125% of the calculated maximum force in the member. The connections must also be designed considering the allowable fatigue stress ranges‡ and the calculated live load stress range magnitude (with reduced impact) for members subjected to tensile cyclical stress ranges from live load (see Chapter 4).

Beam framing connections generally behave as rigid (fixed or with substantial rotational restraint), semirigid (intermediate level of rotational restraint), or flexible (little or no rotational restraint) at service loads. The connections transfer only shear forces if considered as flexible (i.e., as simply supported beams). AREMA (2008) recommends

* Rivets may be used in the design of some historical structures. However, riveting is not a modern or often used fastening method and the engineer should confirm that expertise in riveting is available for both design and installation.

† Pins are generally only used in special circumstances in modern steel railway superstructures such as at suspended spans of cantilever structures (Chapter 5) or as components of support bearings in long spans.

‡ For example, for slip-resistant bolted connections without presence of stresses from out-of-plane bending, the allowable fatigue stress range is 18 ksi for less than 2 million stress range cycles and 16 ksi for greater than 2 million stress range cycles (Fatigue Detail Category B).

that flexible beam framing end connections in beams (typically assumed in the design of stringers and floorbeams) and girders be designed for 125% of the calculated shear force. Connections considered as semirigid or rigid may be designed for the combined bending moment and shear force applied at the joint. Rotational end restraint may be modeled for analysis using rotational springs with spring stiffness, $k_{\phi} = M_r/\phi$, at the ends of the beam. Knowledge of k_{ϕ}, from analytical and experimental research, enables the determination of the end bending moment, M_r, for connection design.

Secondary and bracing member connections must be designed for the lesser of the allowable strength of, or 150% of the calculated maximum force in, the member. The connections in members used as struts and ties to reduce the unsupported member length of other members should be designed for 2.5% of the force in the member being braced.

9.2 WELDED CONNECTIONS

Welding is the metallurgical fusion of steel components or members through an atomic bond. The steel must be melted to effect the coalescence and, therefore, a relatively large quantity and concentration of heat energy is required. The heat energy is usually supplied by an electric arc created between a metal electrode and the base metal during the welding of structural steel components and members.

There are many electric arc welding processes in use. SMAW, SAW, and FCAW are the most commonly used processes for railway superstructure fabrication. Several passes are often required during the welding process to ensure fine-grain metallurgy of the deposited weld metal. The maximum size of weld placed in a single pass depends on position of the weld and is specified in the American Welding Society Bridge Welding Code (ANSI/AASHTO/AWS D1.5, 2005).

Residual stresses (already introduced into rolled plates and members at steel mills) combined with welding heat cycles can cause distortions to occur. Residual (or locked-in) stresses may be increased if the distortions are restrained. The distortions and/or residual stresses occur when welds contract more than the base metal along the weld longitudinal axis and/or due to the transverse contraction of weld metal (which tends to pull plates together and may involve a transverse angular distortion). Weld balancing and multipass welding procedures can often mitigate distortion or excessive residual stress. However, when residual stresses are inevitable, such as in thick butt welded plates, supplementary heat treatments are often required to "stress relieve" the weld and base metal adjacent to the weld.

Lamellar tearing can occur in thick plates with welded joints where tensile stresses are directed through the plate thickness by weld shrinkage (usually not of concern for plates less than about $1\frac{1}{2}$ in. thick). Joint preparation and welding sequence are important to mitigate lamellar tearing.

Therefore, it is of critical importance that welding processes and procedures produce quality welds with proper profiles, good penetration, complete contact with the base metal at all surfaces, no cracks, porosity, and/or inclusions.

There are many weld and joint types used for welded connections in railway superstructure steel fabrication (see Sections 9.2.2 and 9.2.3, respectively). Fillet and groove welds are the most prevalent weld types for steel railway superstructure fabrication (slot and plug welds are generally not used). Stud welding is used in composite span fabrication to attach studs to the top flange of beam and girder spans (see Chapter 7). Butt and "T" joints are the most common welded joints used in steel railway superstructures. However, corner, lap, and edge joints might be used in secondary and nonstructural steel elements.

Welded connection design must consider the force path and strain compatibility between weld and base material. Important aspects of welded connection design are base metal weldability (see Chapter 2), deposited weld metal quality, element thickness, restraint conditions, and other details such as preparation and weld quantity. In general, the engineer should design the smallest welds possible to mitigate distortion and residual stresses in welded joints. However, when welding does cause slight distortion, it may be corrected by mechanical and heat straightening of the component or member. Steel fabricators have procedures for mechanical and heat straightening. Many of these procedures are based on Federal Highway Administration guidelines (FHWA, 1998).

Welds are continuous and rigid. Therefore, they facilitate crack propagation. As a consequence, welds must not be susceptible to fatigue crack initiation and weld designs must avoid conditions that create stress concentrations (such as excessive weld reinforcement, concentrations at intersecting welds,* highly constrained joints, discontinuous backing bars, plug, and slot welds). Weld fracture in nonredundant FCM must be carefully considered in weld design. Requirements for base metal materials, welding process, consumables (including weld metal toughness), joint preparation, preheat, and interpass temperatures recommended by AREMA (2008) and AWS (2005) will provide for welds that are not susceptible to fracture.

Welding inspection is critical for quality control and assurance and is an economical method of ensuring that welding is properly performed to produce connections in conformance with the design requirements. Welds are typically inspected by magnetic particle, ultrasonic, and radiographic methods.

9.2.1 Welding Processes for Steel Railway Bridges

The arc welding processes most commonly in use for railway superstructure steel fabrication are SMAW, SAW, and FCAW.†

9.2.1.1 Shielded Metal Arc Welding

The SMAW process is a manual welding process where the consumable electrode (electrode metal is transferred to the base metal) is coated with powdered materials in

* This might occur, for example, where horizontal gusset plates and vertical transverse or intermediate stiffeners are connected near the tension flange of a girder.

† For nonredundant FCM only SMAW, SAW, or FCAW processes are generally allowed. Therefore, other arc welding processes such as gas metal arc welding (GMAW) and electroslag welding (ESW) are not commonly used in steel railway superstructure fabrication.

a silicate binder. The heated coating converts into a shielding gas to ensure that there is no oxidation or atmospheric contamination* of the weld material in the arc stream and pool. The coating residue forms as a slag on the weld surface (provided enough time exists for the slag to float to the weld surface) and some is absorbed. The coating also assists in stabilizing the electric arc during the welding process.

9.2.1.2 Submerged Arc Welding

The SAW process is an automatic welding process where the bare metal consumable electrode is covered with a granular fusible flux material. This flux shields the arc stream and pool. Economical and uniform welds with good mechanical properties, ductility and corrosion resistance are produced by the modern SAW process.

9.2.1.3 Flux Cored Arc Welding

The FCAW process is often semiautomatic and uses a continuous wire as the consumable electrode. The wire is annular with the core filled with the flux material.† In this manner the flux material at the core behaves similarly to the coating in the SMAW or granular flux in the SAW process.

9.2.1.4 Stud Welding

Stud welding is essentially an SMAW process made automatic. The stud acts as the electrode as it is driven into a molten pool of metal. A ceramic ferrule placed at the base of the stud provides the molten pool of metal and also serves as the electrode coating for protection of the weld. A complete penetration weld (with a small exterior annular fillet weld) is achieved across the entire stud shank or body. This welding process is used extensively in the fabrication of composite steel and concrete beams and girders (Chapter 7).

9.2.1.5 Welding Electrodes

Electrodes commonly used for structural steel welding are designated as E60XX or E70XX. These electrodes have a consumable metal tensile strength of 60 and 70 ksi, respectively. The "XX" in the electrode designation refers to numbers that describe other requirements related to use, such as welding position, power supply, coating type, and arc type/characteristics. Low hydrogen electrodes are often specified to provide superior weld properties and preclude possible embrittlement of the weld from absorbed hydrogen.

9.2.2 Weld Types

Welds are groove, fillet, slot, or plug welds. Slot and plug welds exhibit poor fatigue behavior and are not recommended for use on steel railway superstructures.

* Generally, in the form of nitrides and oxides, which promote brittle weld behavior.

† Exterior coatings on the continuous wire would be removed while being fed through the electrode holder to reach the arc.

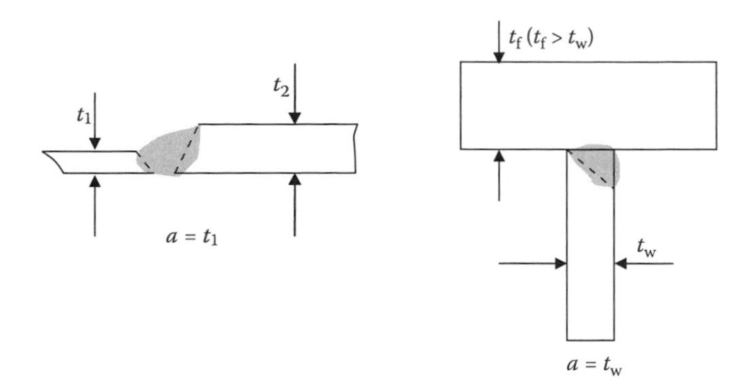

FIGURE 9.1 Size of CJP groove welds.

9.2.2.1 Groove Welds

Groove welds generally require joint preparation and may be CJP or PJP welds. CJP welds are made through the thickness of the pieces being joined. PJP welds are made without weld penetration through the thickness of the pieces being joined. Groove welds are single bevel or double bevel,* square, V, U, or J welds. Descriptions of these welds and prequalification requirements† for both CJP and PJP welds are shown in AWS (2005). The size, a, of a CJP groove weld is the thinner of the plates joined (Figure 9.1). The size, a, of a PJP weld is usually the depth of the preparation chamfer‡ (Figure 9.2).

Minimum PJP groove weld sizes that ensure fusion are recommended in AWS (2005). These minimum recommended PJP groove weld sizes depend on the thickness of the thickest plate or element in the joint. The minimum PJP groove weld size is recommended as 1/4 in. except for joints with plates or elements of base metal thickness greater than 3/4 in. where minimum recommended weld size is 5/16 in. However, the weld size need not exceed the thickness of the thinnest part in the joint. PJP welds should not be used for members loaded such that there is tensile stress normal to the effective throat of the PJP weld.

AREMA (2008) recommends only CJP groove welds be used for connections with the exception that PJP groove welds may be used to connect plate girder flange and web plates (T-joint in Figure 9.2).

9.2.2.2 Fillet Welds

Fillet welds do not require any joint preparation and are readily made by the SMAW, SAW, and FCAW processes. The size of a fillet weld, a or b, is determined based on

* Double bevel, V, U, or J welds are generally required for plates greater than about 5/8 in. thick to avoid excessive weld material consumption, distortion, and/or residual stresses.

† Prequalification requirements relate to the welding process, base metal thickness, groove preparation, welding position, and supplemental gas shielding if FCAW is used.

‡ For some welds the entire depth of preparation cannot be used and the PJP weld size may be taken as 1/8 or 1/4 in. less than the preparation depth. In the case of PJP square butt joints, the weld size should not exceed 75% of the plate thickness.

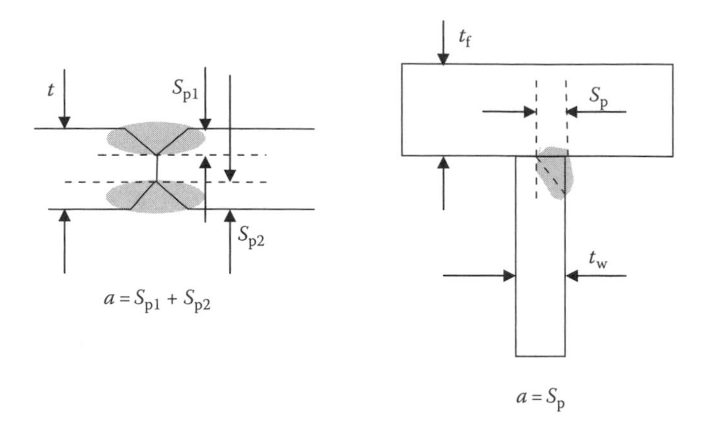

FIGURE 9.2 Size of PJP groove welds.

the thickness of the weld throat, t_e, required to resist shear (Figure 9.3). The throat depth $= 0.707a$ for fillet welds with equal leg length, $a = b$ (the usual case). Minimum fillet weld sizes that ensure fusion are recommended in AWS (2005). Minimum fillet weld size, a_{min}, is a function of the thickness of the thickest plate or element in the joint. Minimum size for single pass fillet welds is generally recommended as 1/4 in. except for plates or elements with base metal with thickness greater than 3/4 in. where minimum fillet weld size is 5/16 in. However, the weld size need not exceed

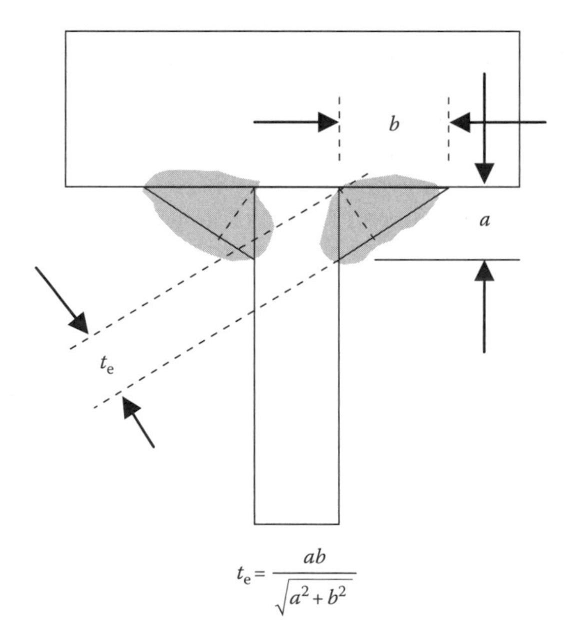

FIGURE 9.3 Size of fillet welds.

the thickness of the thinnest part in the joint, in which case care must be taken to provide adequate preheating of the weld area.

The minimum connection plate thickness recommendations of AREMA (2008) should preclude cutting of an element, plate, or component by fillet weld penetration. Maximum filet weld size, a_{max}, is recommended in AWS (2005) in order to avoid excessive base metal melting and creation of potential stress concentrations. The maximum fillet weld size is the thickness of the thinnest of the plates or elements in a joint for elements or plates with thickness less than 1/4 in. The maximum fillet weld size is the thickness of the thinnest plate or element less 5/64 in. for joints with thickness of the thinnest plates or elements greater than 1/4 in.

The minimum effective length of fillet welds is generally recommended as $4a$. AREMA (2008) recommends that fillet welds used to resist axial tension that is eccentric to the weld line or cyclical tensile stresses must be returned continuously around any corner for a minimum of $2a$. AREMA (2008) also recommends that wrap-around fillet welds not be used when welding intermediate transverse stiffeners to girder webs.

9.2.3 Joint Types

Welds are used in lap, edge, "T," corner, and butt joints. Welded lap joints are generally used only in secondary members and edge joints are used only in nonstructural members. However, "T," corner, and butt joints are commonly used for main girder fabrication and splicing of steel railway superstructure elements.

Welded lap joints (Figure 9.4a–d) are simple joints sometimes used in secondary members of steel railway bridges. The joints in Figure 9.4a–c are typically subjected to eccentric loads. Figure 9.4d shows a type of lap joint used to connect attachments, such as stiffeners, to girder web plates. Lap joints typically use fillet welds.

Welded "T" and corner joints (Figure 9.4e and f) are typically used to connect web plates and flange plates of plate and box girder spans, respectively. "T" and corner joints may use fillet or groove welds and are typically subjected to horizontal shear from bending along the longitudinal weld axis.

Welded butt joints (Figure 9.4g) often join plate ends (such as at girder flange and web plate splices) with complete penetration groove welds. Butt joints are also used in welded splices of entire elements or sections. There is no force eccentricity in typical butt joints, but, particularly in tension zones, butt joints require careful consideration of residual stresses.* Weld and connection element transitions for butt welded plates of different thickness and/or width are recommended by AREMA (2008). Butt joints should not be used to join plates with a difference in both thickness and width unless the element resists only axial compression.† Edge preparation and careful alignment during welding are critical for good quality butt joints.

* Stress relieving is often required.

† Stress concentrations may be large for butt welded connections subject to tension.

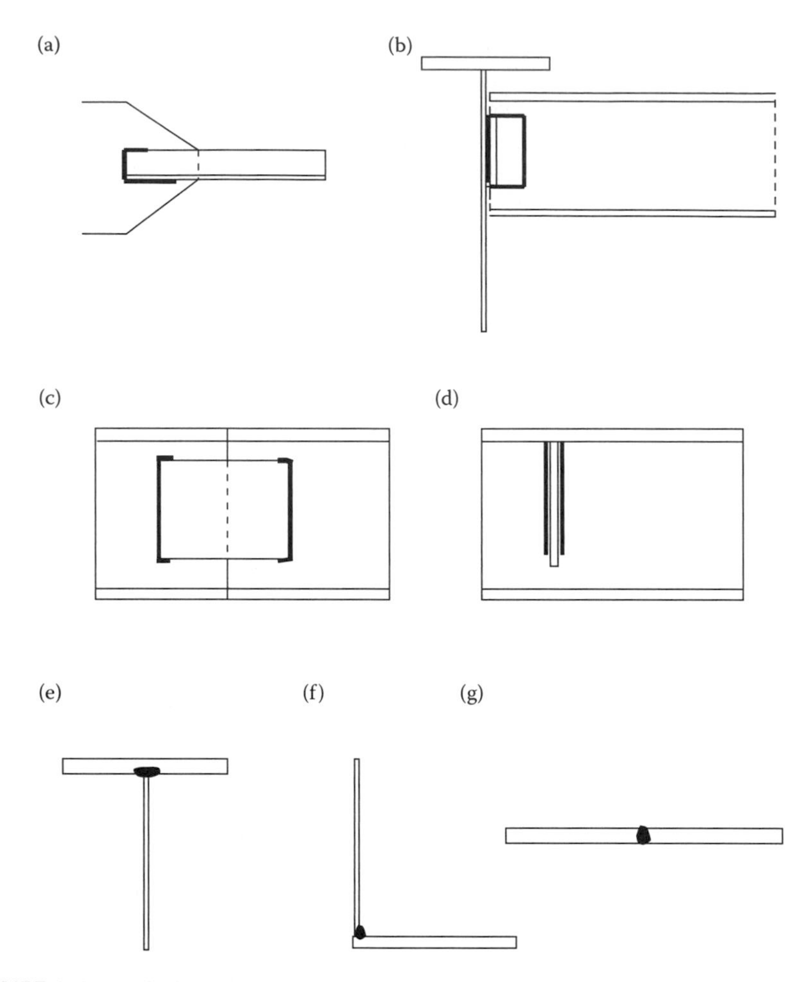

FIGURE 9.4 Typical welded joints in steel railway superstructures.

9.2.4 Welded Joint Design

9.2.4.1 Allowable Weld Stresses

Fillet welds transmit forces by shear stress in the weld throat and groove welds transmit loads in the same manner as the elements that are joined (e.g., by shear, axial, and/or bending stresses).

Therefore, for fillet welds, the allowable shear stress is the smaller of $0.28F_u$ through the weld throat based on electrode strength or $0.35F_y$ at the weld leg based on base metal strength.

For groove welds the allowable shear stress is $0.35F_y$ and the allowable tension or compression stress is $0.55F_y$ based only on base metal strength. This is because complete joint penetration welds using matching electrodes (as specified by AWS, 2005) are at least as strong as the base metal under static load conditions. In addition,

CJP welds made in accordance with the AREMA (2008) FCP will be of the same or greater fatigue strength than the base metal.

However, for PJP groove or fillet welds subjected to shear, axial, and/or flexural tensile stresses due to live load, the allowable fatigue stress range for the appropriate Fatigue Detail Category and equivalent number of constant stress cycles (Chapter 5) must be considered. The allowable fatigue stress ranges may be small and govern the required weld size for weak fatigue details such as transversely loaded fillet or PJP welds and fillet or groove welds used on attachments with poor transition details.*

9.2.4.2 Fatigue Strength of Welds

Stress concentrations are created by welding processes. These processes may introduce discontinuities within the weld, distortion of members, residual stresses, and stress raisers due to poor weld profiles. Stress concentration factors for butt joint welds typically range from 1.0 to 1.6 and from between 1.0 and 2.8 (or more) for other joints (Kuzmanovic and Willems, 1983). These stress concentration effects are included in the nominal stress range fatigue testing of many different weld types, joint configurations, and loading directions. This provides the design criteria, in terms of the allowable fatigue stress ranges, for the various Fatigue Detail Categories recommended in AREMA (2008). Further discussion of allowable fatigue stresses for design is contained in Chapter 5.

9.2.4.3 Weld Line Properties

It is intuitive and convenient to design welds as line elements. The effective weld area, A_e (on which allowable stresses are assumed to act), is

$$A_e = t_e L_w, \tag{9.1}$$

where t_e is the thickness of thinner element for CJP groove welds, or depth of the preparation chamfer less 1/8 in. for PJP groove weld root angles between 45° and 60° (for SWAW and SAW welds), or depth of the preparation chamfer for PJP groove weld root angles greater than or equal to 60° (for SWAW and SAW welds), or the throat length equal to $0.707a$ (for fillet welds with equal legs, a)† and L_w is the length of the weld.

If we consider the weld as a line, the allowable force per unit length, F_w, on the weld is

$$F_w = t_e(f_{all}), \tag{9.2}$$

where f_{all} is the allowable weld stress, and, from Equation 9.2,

$$t_e = \frac{F_w}{f_{all}}. \tag{9.3}$$

* Generally, such details should be avoided to preclude low allowable fatigue stress ranges, which may render the superstructure design uneconomical.

† Fillet weld throat length is sometimes increased by a small amount to recognize the inherent strength of fillet welds.

Therefore, considering the weld as a line provides a direct method of designing welds for bending and/or torsion of the weld line. The moment of inertia in the direction parallel to and perpendicular to the longitudinal weld axis is required for situations where welds are subjected to bending and torsion. Example 9.1 illustrates the calculation of weld line properties for a particular weld configuration.

Example 9.1

Determine the weld line properties for the weld configuration shown in Figure E9.1.

$$x_1 = \frac{2t_e b(b/2)}{2bt_e + dt_e} = \frac{b^2}{2b + d},$$

$$I_x = t_e\left(\frac{d^3}{12} + 2b\left(\frac{d}{2}\right)^2\right),$$

$$I_y = t_e\left(\frac{2b^3}{12} + 2b\left(\frac{b}{2} - x_1\right)^2 + dx_1^2\right),$$

$$I_p = I_x + I_y = t_e\left(\frac{d^3}{12} + 2b\left(\frac{d}{2}\right)^2 + \frac{2b^3}{12} + 2b\left(\frac{b}{2} - x_1\right)^2 + dx_1^2\right)$$

$$= t_e\left(\frac{8b^3 + 6bd^2 + d^3}{12}\right),$$

$$S_x = \frac{I_x}{d/2} = t_e\left(\frac{d^2}{6} + bd\right).$$

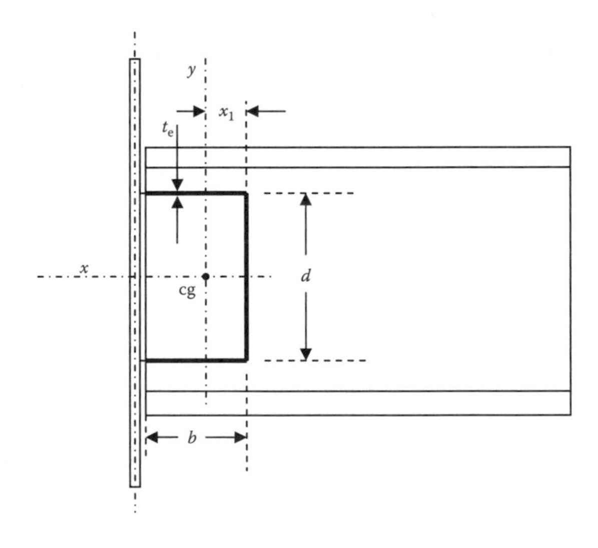

FIGURE E9.1

Weld line properties for any weld configuration may be determined as shown in Example 9.1. Table 9.1 provides weld line properties for various commonly used weld configurations.

9.2.4.4 Direct Axial Loads on Welded Connections

Axial weld connections should have at least the strength of the members being connected and be designed to avoid large eccentricities.

Groove welds are often used for butt welds between axial tension or compression members (Figure 9.4g). Eccentricities are avoided and, with electrodes properly chosen to match the base metal (see AWS, 2005), CJP groove welds are designed in accordance with the base metal strength and thickness.

Fillet welds are designed to resist shear stress on the effective area, A_e. The size of fillets welds is often governed by the thickness of the elements being joined and it is necessary to determine the length of fillet weld to transmit the axial force without eccentricity at the connection. Example 9.2 outlines the design of an axially loaded full strength fillet weld connection that eliminates eccentricities.

Example 9.2

Design the welded connection for some secondary wind bracing shown in Figure E9.2. The steel is Grade 50 ($F_y = 50$ ksi) and E70XX electrodes are used for the SMAW fillet weld.

Considering an estimated shear lag coefficient of 0.90 (see Chapter 6), the member strength is

$$T = 0.55(50)(5.75) = 158.1 \text{ kips}$$

or

$$T = (0.90)0.47(70)(5.75) = 170.3 \text{ kips.}$$

The minimum fillet weld size is 1/4 in. and maximum fillet weld size is 7/16 in. Try a 5/16 in. fillet weld.

$t_e = (5/16)(0.707) = 0.22$ in.

From Equation 9.2, the allowable strength of the weld line is $F_w = t_e(f_{all}) = 0.22(19) = 4.18$ kips/in.

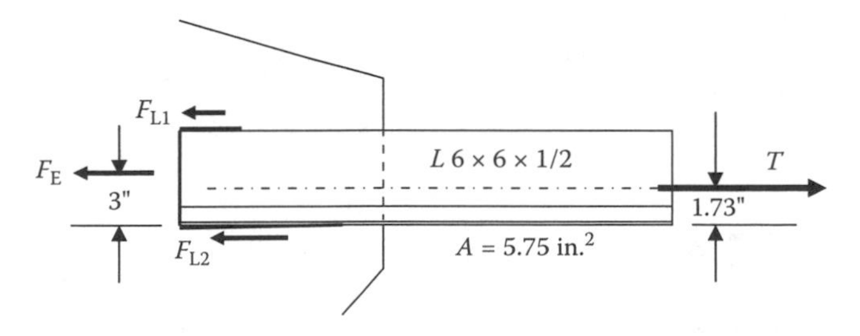

FIGURE E9.2

TABLE 9.1
Properties of Weld Lines

Weld Configuration	Location of center of gravity (cg) (x_1 and y_1)	Section Modulus/ Weld Size	Polar Moment of Inertia About cg/Weld Size
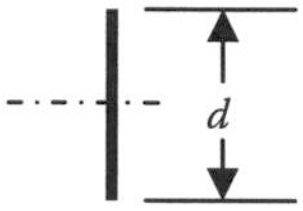	$x_1 = 0$ $y_1 = \frac{d}{2}$	$\frac{S_x}{t_e} = \frac{d^2}{6}$	$\frac{I_p}{t_e} = \frac{d^3}{12}$
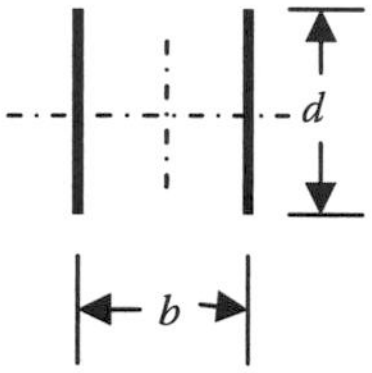	$x_1 = \frac{b}{2}$ $y_1 = \frac{d}{2}$	$\frac{S_x}{t_e} = \frac{d^2}{3}$	$\frac{I_p}{t_e} = \frac{d(3b^2 + d^2)}{6}$
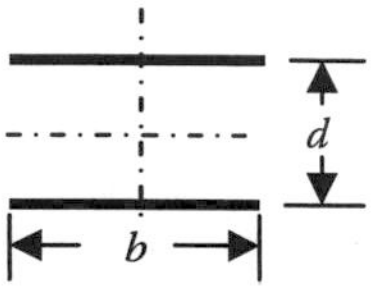	$x_1 = \frac{b}{2}$ $y_1 = \frac{d}{2}$	$\frac{S_x}{t_e} = bd$	$\frac{I_p}{t_e} = \frac{d(3d^2 + b^2)}{6}$
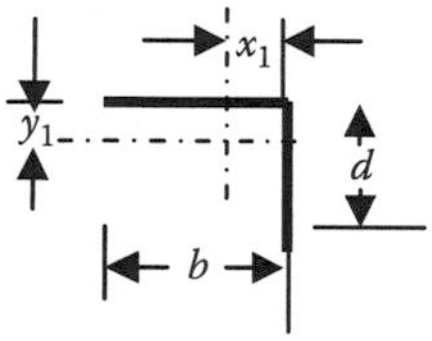	$x_1 = \frac{b^2}{2(b+d)}$ $y_1 = \frac{d^2}{2(b+d)}$	$\frac{S_x}{t_e} = \frac{d^2 + 4bd}{6}$	$\frac{I_p}{t_e} = \frac{(b+d)^4 - 6b^2d^2}{12(b+d)}$
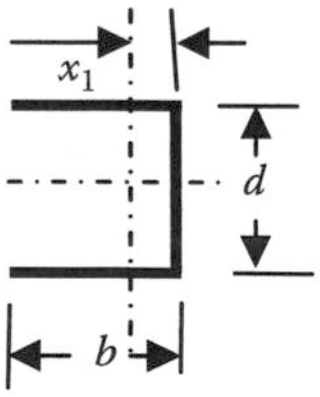	$x_1 = \frac{b^2}{2b+d}$ $y_1 = \frac{d}{2}$	$\frac{S_x}{t_e} = bd + \frac{d^2}{6}$	$\frac{I_p}{t_e} = \frac{8b^3 + 6bd^2 + d^3}{12}$
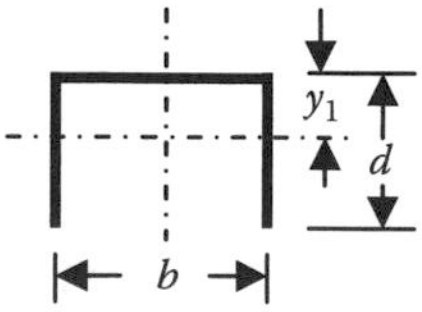	$x_1 = \frac{b}{2}$ $y_1 = \frac{d^2}{2d+b}$	$\frac{S_x}{t_e} = \frac{d^2 + 2bd}{3}$	$\frac{I_p}{t_e} = \frac{8d^3 + 6db^2 + b^3}{12}$
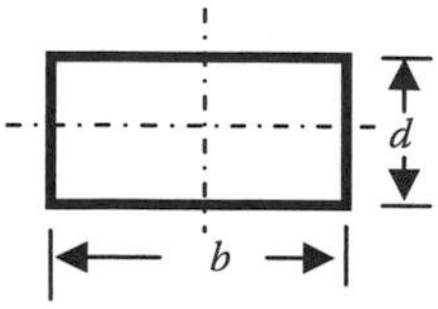	$x_1 = \frac{b}{2}$ $y_1 = \frac{d}{2}$	$\frac{S_x}{t_e} = bd + \frac{d^2}{3}$	$\frac{I_p}{t_e} = \frac{(b+d)^3}{6}$

Shear stress on base metal $= \dfrac{4.18}{(5/16)} = 13.4\,\text{ksi} \leq 0.35(50) \leq 17.5\,\text{ksi}$ OK

Shear stress on fillet weld throat $= \dfrac{4.18}{0.707(5/16)} = 18.9\,\text{ksi} \leq 0.28(70) \leq 19.6\,\text{ksi}$ OK

$F_E = 4.18(6) = 25.1$ kips

Taking moments about F_{L2} yields

$$F_{L1} = \frac{T(1.73) - F_E(3)}{6} = \frac{158.1(1.73) - 25.1(3)}{6} = 33.0\,\text{kips}$$

$$F_{L2} = T - F_{L1} - F_E = 158.1 - 33.0 - 25.1 = 100.0\,\text{kips}$$

Length of weld $L1 = 33.0/4.18 = 7.9$ in., say 8 in.

Length of weld $L2 = 100.0/4.18 = 23.9$ in., say 24 in.

The effect of the force eccentricity is to require that the fillet welds are balanced such that weld $L2$ is 16 in. longer than weld $L1$. The length of weld $L2$ may be reduced by reducing the eccentricity.

If the connection length is assumed to be $(8 + 24)/2 = 16$ in., the shear lag coefficient, U, is $U = (1 - x/L) = (1 - 1.73/16) = 0.89$, which is sufficiently close to $U = 0.90$ assumed.

9.2.4.5 Eccentrically Loaded Welded Connections

Even small load eccentricities must be considered in design since welded connections have no initial pretension (such as that achieved by the application of torque to bolts). Many welded connections are loaded eccentrically (e.g., the connections shown in Figures 9.4b–d). Eccentric loads will result in combined shear and torsional moments or combined shear and bending moments, depending on the direction of loading with respect to weld orientation in the connection.

9.2.4.5.1 Connections Subjected to Shear Forces and Bending Moments

A connection such as that of Figure 9.4d is shown in greater detail in Figure 9.5. The fillet welds each side of the stiffener resist both shear forces and bending moments.

The shear stress on the welds is

$$\tau = \frac{P}{A} = \frac{P}{2t_e d} \tag{9.4}$$

and the flexural stress (using S_x from Table 9.1) is

$$\sigma_b = \frac{M}{S_x} = \frac{3Pe}{t_e d^2}. \tag{9.5}$$

The stress resultant is

$$f = \sqrt{\tau^2 + \sigma_b^2} = \frac{P}{2t_e d}\sqrt{1 + \left(\frac{6e}{d}\right)^2}. \tag{9.6}$$

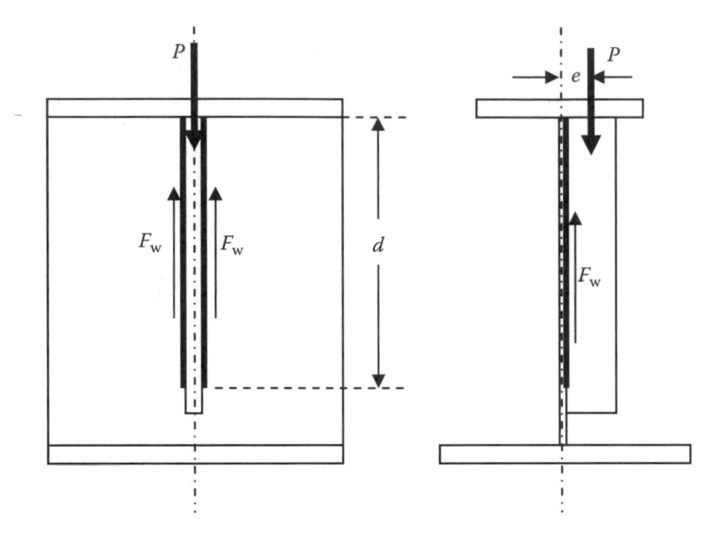

FIGURE 9.5 Bending and shear forces on fillet welds.

9.2.4.5.2 *Connections Subjected to Shear Forces and Torsional Moments*

A connection such as that of Figure 9.4b is shown in greater detail in Figure 9.6. The fillet welds each side of the leg of the connection angle (or plate) against the beam web resist both shear forces and torsional moments.

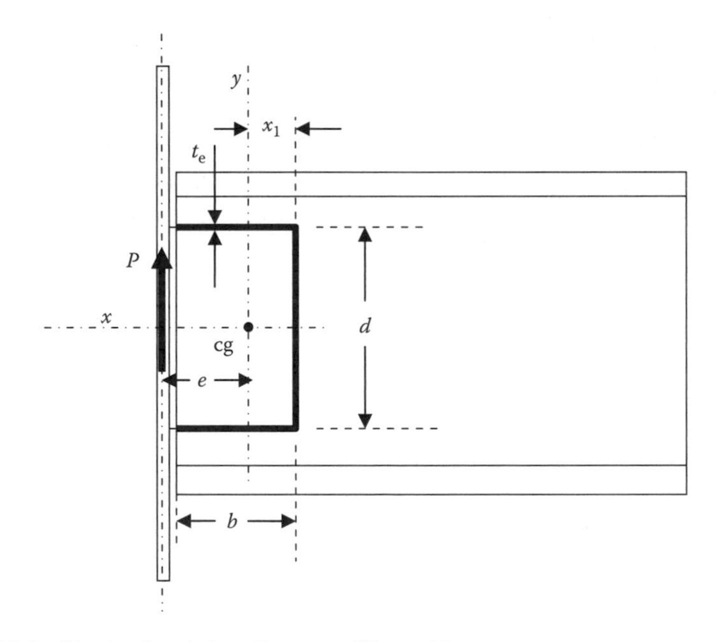

FIGURE 9.6 Torsional and shear forces on fillet welds.

The shear stress on the welds is

$$\tau = \frac{P}{A} = \frac{P}{2t_e(2b + d)} \tag{9.7}$$

and the torsional stress (using I_p from Table 9.1) is

$$\sigma_{tx} = \frac{Ty}{I_p} = \frac{6(Pe)y}{t_e\left(8b^3 + 6bd^2 + d^3\right)}, \tag{9.8a}$$

$$\sigma_{ty} = \frac{Tx}{I_p} = \frac{6(Pe)x}{t_e\left(8b^3 + 6bd^2 + d^3\right)}, \tag{9.8b}$$

where x is the distance from the centroid to the point of interest on the weld in the x-direction and y is the distance from the centroid to the point of interest on the weld in the y-direction.

The stress resultant at any location on the weld described by locations x and y is

$$f = \sqrt{(\tau + \sigma_{ty})^2 + \sigma_{tx}^2}. \tag{9.9}$$

9.2.4.5.3 *Beam Framing Connections*

Welded beam framing connections are not used in the main members of steel railway superstructures due to the cyclical load regime (see Example 9.3). However, when used on secondary members, such as walkway supports, welded beam framing connections are subjected to shear force, P, and end bending moment, M_e, on welds on the outstanding legs of the connection angles. The legs of the connection angles fastened to the web of the beam are also subject to an eccentric shear force, which creates a torsional moment, Pe (Figure 9.7).

Beam framing connections are often assumed to transfer shear only in usual design practice (i.e., it is assumed that the beam is simply supported and $M_e = 0$). However, in reality, due to end restraint, some proportion of the fixed end moment, δM_f, typically exists ($\delta = M_e/M_f$, where M_f is the fixed end beam moment). Welded connection behavior in structures is often semirigid with a resulting end moment (Blodgett, 2002). The magnitude of the end moment depends on the rigidity of the support. For example, a beam end connection to a heavy column flange may be quite rigid ($\delta \to 1$), while a beam end connection framing into the web of a girder or column may be quite flexible ($\delta \to 0$) (Figure 9.8).

A rigid connection may be designed for the end moment due to full fixity, M_f, and corresponding shear force, P. A semirigid connection will require an understanding of the end moment (M_e)–end rotation (ϕ_e) relationship (often nonlinear) to determine rotational stiffness and the end moment to be used in conjunction with shear force for design. Moment–rotation curves, developed from theory and experiment, for welded joint configurations* are available in the technical literature (e.g., Faella et al., 2000).

* Mainly for beam to column flange connections.

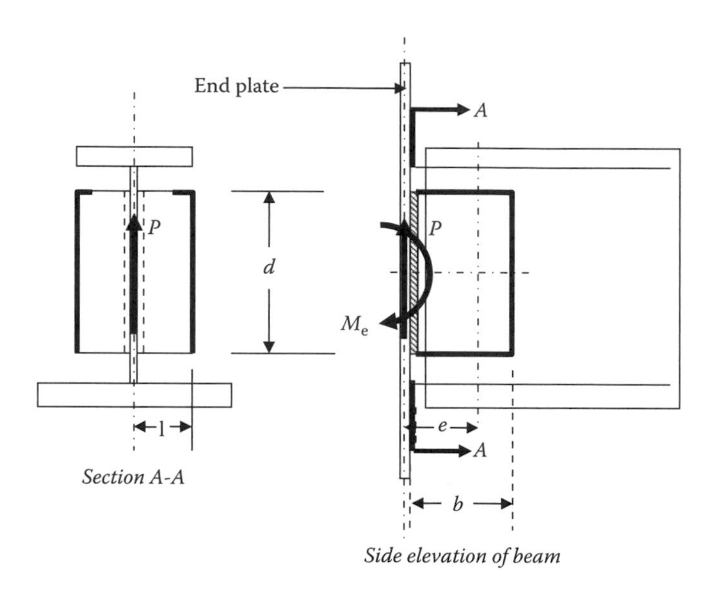

FIGURE 9.7 Simple welded beam framing connection.

A flexible end connection will deform and resist very little bending moment. Simple beam framing connections (Figure 9.7) that exhibit the characteristics of a flexible connection may be designed for shear force, P, only ($M_e = 0$). AREMA (2008) recognizes that most connections actually exhibit some degree of semirigid behavior and allows flexible connection design (with angle thickness that allows for deformation*) provided the design shear force is increased by 25%. Therefore, flexible welded beam framing connections may be designed considering shear on the outstanding legs of the connection and, due to the eccentricity of the shear force, the combined shear and torsion on the leg of the connection angles fastened to the web of the beam. Otherwise, a semirigid connection design considering both beam end moment and shear is required.

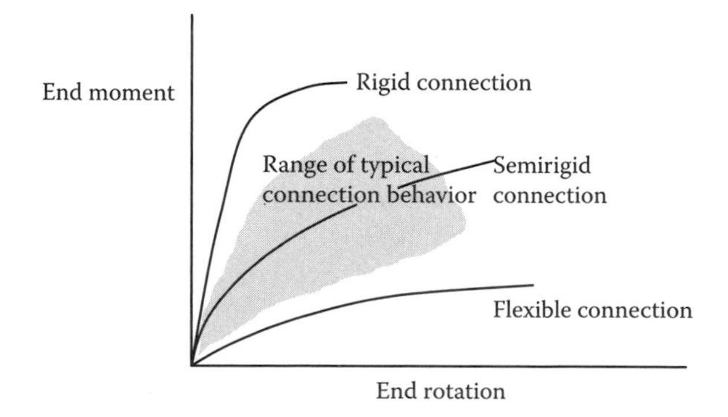

FIGURE 9.8 Typical moment–rotation curves for welded beam end connections.

* In particular, the outstanding legs must be sufficiently flexible to disregard any beam end bending effects.

The outstanding legs of angles in the simple beam framing connection (often referred to as clip angles) must deform sufficiently to allow for flexible connection behavior. An approximate solution for the maximum thickness of angle to allow sufficient deformation in a welded beam framing connection, over the depth, d, can be developed assuming the shear force, P, is applied at a distance, e (see Figures 9.6 and 9.7). The bending stress, f_{wa}, in the leg of the angle connected to the end plate (e.g., a beam, girder, or column web plate) from a load, P, applied at an eccentricity, e, is (from Equation 9.5)

$$f_{wa} = \frac{M}{S} = \frac{3Pe}{t_a d^2}, \quad (9.10)$$

where P is the shear force applied at eccentricity, e, t_a is the thickness of angle, and d is the depth of the connected angle.

The tensile force, T_P, on the connection angle (pulling the angle away from the end plate connection) is

$$T_P = f_{wa}(2)t_a = \frac{6Pe}{d^2}. \quad (9.11a)$$

This tensile force, T_P, creates a bending moment in the angle legs (assuming the legs behave as simply supported beams of length $2l$) of

$$M_{wa} = \frac{T_P(2l)}{4} \quad (9.11b)$$

and the stress in the angle leg is

$$f_{wa} = \frac{6M_{wa}}{t^2} = \frac{3T_P l}{t^2} = \frac{18Pel}{t^2 d^2}. \quad (9.12)$$

The deformation of the connection angles is

$$\Delta = \frac{T_P(2l)^3}{48EI} = \frac{2T_P l^3}{Et^3} = \frac{f_{wa} l^2}{1.5Et}. \quad (9.13)$$

Example 9.3 illustrates the design of a welded beam framing connection.

Example 9.3

Design the welded simple beam framing connection shown in Figure E9.3 for a shear force of $P = 52$ kips. The uniformly loaded beam is 20 ft long, has a strong axis moment of inertia of 1200 in.^{4} and frames into the web of a plate girder. The allowable shear stress on the fillet welds is 17.5 ksi. Electrodes are E70XX. The return weld in Section X-X will have a minimum length of twice the weld size and may be neglected in the design.

$P' = 1.25(52) = 65$ kips (AREMA recommendation for flexible connection design)

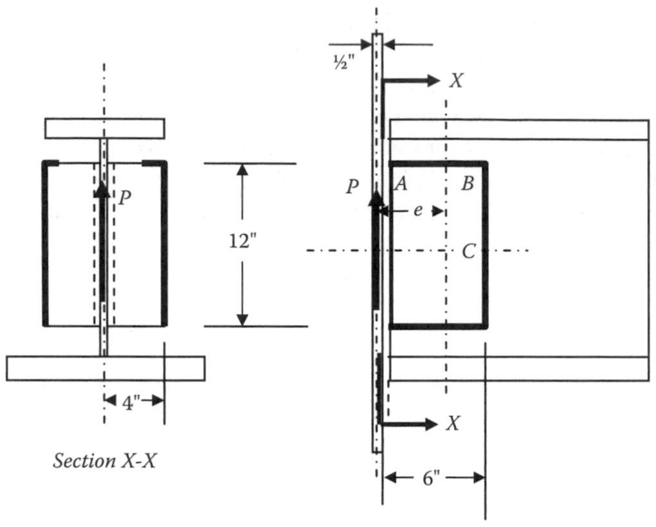

FIGURE E9.3

From Table 9.1:

$$e = 6 - x_1 = 6 - \frac{b^2}{2b+d} = 6 - \frac{36}{24} = 4.5\,\text{in.}$$

Shear and bending on welds in Section X-X:
Bending stress on the welds:
From Table 9.1, the section modulus of the welds, S_w, is

$$S_w = \frac{t_e d^2}{3} = 48t_e\,\text{in.}^3$$

$$\sigma_w = \pm\frac{M_e}{S_w} = \frac{3M_e}{t_e d^2} = \pm\frac{M_e}{48t_e}.$$

Since, for a flexible connection, $M_e = 0$ there is no flexural stress in the weld

Shear stress on the weld:

$$\tau = \frac{P'}{2t_e d} = \frac{2.71}{t_e}\,\text{ksi}$$

The stress resultant is

$$f_1 = \sqrt{\tau^2 + \sigma_b^2} = \frac{1}{t_e}\sqrt{2.71^2 + 0^2} = \frac{2.71}{t_e}\,\text{ksi}$$

Therefore, the required weld thickness is

$$t_e \geq \frac{2.71}{0.28(70)} \geq 0.14\,\text{in for shear in the throat of welds or}$$

$$t_e \geq \frac{2.71}{(17.5)(0.707)} \geq 0.22\,\text{in for shear of equal leg welds on the end plate.}$$

Angle thickness to allow deformation:

From Equations 9.12 and 9.13 the deformation of the connection angles is

$$\Delta = \frac{12Pel^3}{d^2Et^3} = \frac{12(52)(4.5)(4)^3}{(12)^2(29000)t^3} = \frac{0.0430}{t^3}\text{ in.}$$

If the beam is uniformly loaded with distributed load, w,

$$w = \frac{2P}{L} = \frac{2(52)}{20(12)} = 0.433\text{ k/in.},$$

the end rotation, Y_b, is

$$\theta_b = \frac{wL^3}{24EI} = \frac{0.433(240)^3}{24(29000)I} = \frac{8.60}{I}\text{ rad.}$$

The rotation occurs about the bottom of the angle so that the deformation at the top of the angle, Δ, is

$$\Delta = 12\theta_b = \frac{103.2}{I}$$

so that

$$t \le 0.0747(I)^{1/3}.$$

For a beam with $I = 1200\text{ in.}^4$,

$t \le 0.0747(1200)^{1/3} \le 0.79\text{ in.}$

The angle thickness should be based on the requirement for transmitting shear or the minimum element thickness recommended by AREMA (2008), but should not be greater than about 3/4 in. thick (note the calculation of 0.79 in is approximate) to ensure adequate flexibility for consideration as a flexible beam framing connection.

Shear and torsion on welds in side elevation:

The shear stress on the welds is

$$\tau = \frac{P'}{2t_e(2b+d)} = \frac{1.35}{t_e}\text{ ksi.}$$

The torsional stress (using I_p from Table 9.1) is

$$\sigma_{tx} = \frac{6(P'e)y}{t_e\left(8b^3 + 6bd^2 + d^3\right)} = \frac{0.203y}{t_e}\text{ ksi}$$

$$\sigma_{ty} = \frac{0.203x}{t_e}\text{ ksi.}$$

The stress resultant, f_2, at any location on the weld described by locations x and y is

$$f_2 = \sqrt{(\tau + \sigma_{ty})^2 + \sigma_{tx}^2} = \frac{1}{t_e}\sqrt{(1.35 + 0.203x)^2 + (0.203y)^2}.$$

TABLE E9.1

Location	x (in.)	y (in.)	$t_e f_2$ (k/in.)
A	4.5	6	2.57
B	1.5	6	2.06
C	1.5	0	1.66

The weld stresses are computed in Table E9.1 for various locations on the welds on the beam web.

Therefore, the required weld thickness is

$$t_e \geq \frac{2.57}{0.28(70)} \geq 0.13 \text{ in.}$$

for shear and torsion in the throat of welds on the beam web

$$t_e \geq \frac{2.57}{(17.5)(0.707)} \geq 0.21 \text{ in.}$$

for shear and torsion of equal leg welds on the beam web.

Fatigue must be considered if the applied load is cyclical. A connection such as that shown in Figure E9.3 is a very poor connection from a fatigue perspective with an allowable fatigue stress of 8 ksi. In that case, the required weld thickness for shear and bending of the welds on the end plate will greatly exceed the maximum allowable fillet weld size based on the thickness of the thinnest plate or element in the connection.

Eccentrically loaded welded connections should not be used to join members subjected to cyclical live loads. The very low fatigue strength of these joints (e.g., transversely loaded fillet welds subject to tension or stress reversal) makes them unacceptable for joints in main carrying members. For such joints, bolted connections are more appropriate.

9.2.4.5.4 Girder Flange-to-Web "T" Joints

A connection such as those shown in Figure 9.4e and f transmits horizontal shear forces from bending and, if present, direct transverse loads. This connection between girder flanges and web plates may be made with CJP groove welds, PJP groove welds, or fillet welds using the SAW process. Some engineers specify fillet or PJP groove welds when the connection is subjected to only horizontal shear from bending. CJP groove welds may be required when weld shear due to direct loading is combined with horizontal shear from bending.

The horizontal shear flow, q_f, for which the flange-to-web weld is designed, is (see Chapter 7)

$$q_f = \frac{dP_f}{dx} = \frac{VQ_f}{I}, \tag{9.14}$$

where V is the maximum shear force, $q_f = A_f\bar{y}$ (statical moment of the flange area about the neutral axis), $\bar{y}$ is the distance from the flange centroid to the neutral axis.

TABLE 9.2
Allowable Weld Stresses

Weld Type	Stress State	Allowable Stress (ksi)
CJP or PJP	Tension or compression	$0.55F_y$
CJP or PJP	Shear	$0.35F_y$
Fillet (60 ksi electrode)	Shear	16.5 but $<0.35F_y$ on base metal
Fillet (70 ksi electrode)	Shear	19.0 but $<0.35F_y$ on base metal
Fillet (80 ksi electrode)	Shear	22.0 but $<0.35F_y$ on base metal

If present,* the shear force, acting in a vertical direction, is (see Chapter 7)

$$w = \frac{1.80(W)}{S_w}, \tag{9.15}$$

where S_w is the wheel load longitudinal distribution ($S_w = 3$ ft for open deck girders or $S_w = 5$ ft for ballasted deck girders).

The resulting force per unit length of the weld (from Equations 9.14 and 9.15) is

$$q = \sqrt{q_f^2 + w^2}. \tag{9.16}$$

The required effective area of the weld can then be established based on the allowable weld stresses recommended by AREMA (2008) as shown in Table 9.2.

The required effective area of welds subject to horizontal shear from cyclical flexure ranges must also be established based on the allowable weld fatigue stresses recommended by AREMA (2008) (typically Category B or B′ depending on weld backing bar usage).

9.3 BOLTED CONNECTIONS

Bolting is the connection of steel components or members by mechanical means. Bolted connections are relatively easy to make and inspect (in contrast to the equipment and skills required for welding and inspection of welds). The strength of the mechanical connection is affected by the bolt installation process.

9.3.1 Bolting Processes for Steel Railway Bridges

9.3.1.1 Snug-Tight Bolt Installation

Connection strength depends on the bearing, shear, and tensile strength of the bolts and connected material for bolts installed without pretension (snug-tight). The bolts are made snug-tight manually with a wrench or power tool applied to the nut. The full effort of a person installing a bolt with a wrench or a nut installed with a power tool until wrench impact will generally provide the small pretension required to retain the nut on

* For example, vertical loads are transferred through the flange-to-web weld in open and noncomposite deck plate girder spans.

the bolt in statically loaded structures. These are bearing-type connections. Bearing-type connections are generally not used in steel railway superstructures because of live load stress reversals and cyclical stresses in main members and vibration in both main and secondary members.

9.3.1.2 Pretensioned Bolt Installation

Connections made with pretensioned bolts rely on friction between plates or element surfaces (faying surfaces) for strength. These are slip-resistant connections and are used extensively in modern steel railway superstructures. Pretensioned bolted joints are made by snug-tight bolt installation followed by increasing the torque applied to the bolt. The applied torque creates tension in the bolt (and corresponding compression of, and friction between, the connection elements). The minimum required bolt pretension, T_{bP}, is

$$T_{bP} \geq 0.70P_{bU} \geq 0.70F_{bU}A_{st}, \tag{9.17}$$

where P_{bU} is the minimum specified tensile strength of the bolt; F_{bU} is the minimum specified tensile stress of the bolt material; A_{st} is the tensile stress area of the bolt and is equal to the cross-sectional area through the threaded portion of the bolt.

To attain this minimum bolt tension, AREMA (2008) recommends that nuts be rotated between 1/3 and 1 turn from the snug-tight condition, depending on bolt length and angle of connection plates with respect to the bolt axis.* This will establish a pretension in the bolt, P_{bP}, which is greater than the minimum required bolt pretension, T_{bP}, for the bolt, as shown in Figure 9.9. Alternatively, slip-resistant bolted joints may be made using specialized twist-off-type bolts or direct tension indicators.

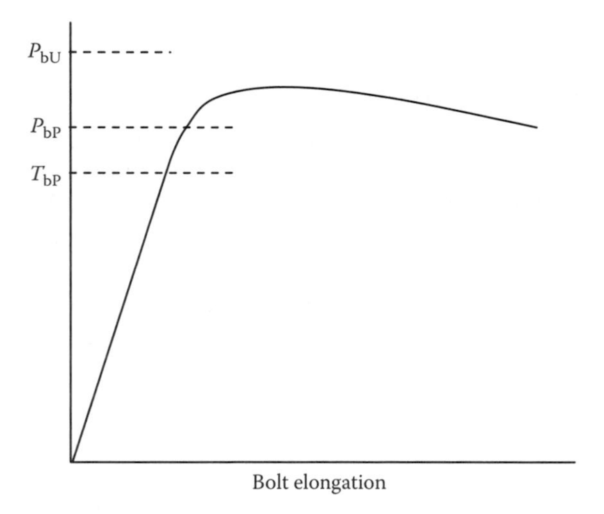

FIGURE 9.9 Bolt tension forces and elongation during application of bolt torque.

* Bolts will generally not fail until nut rotation exceeds about 1.75 times from the snug-tight condition (Kulak, 2002).

TABLE 9.3
Minimum Tensile Strength of High-Strength Steel Bolts

Bolt Type	F_{bU} (ksi)
A325 with bolt diameter, d_b, ≤ 1 in.	120
A325 with bolt diameter, d_b, >1 in.	105
A490	150

Bolts in connections should be installed first at the stiffest locations of the connection and retightened, if required, following the installation of other bolts in the connection (due to possible relaxation of the previously tightened bolts). A tension measuring device should be used to measure the tension in a representative number of bolts in the connection.

9.3.2 Bolt Types

Fasteners used in modern steel structures are either common or high-strength bolts.

9.3.2.1 Common Steel Bolts

Common* bolts are specified by ASTM Standard A307. A307 bolts are generally not used in applications involving live load stress reversals, cyclical stresses, and/or vibration. A307 bolts are also not used in steel railway superstructure fabrication due to their low strength.

9.3.2.2 High-Strength Steel Bolts

High-strength steel bolts are specified by ASTM Standards A325 and A490. The minimum tensile strength, P_{bU}, for A325 bolts and A490 bolts is shown in Table 9.3. A325 Type 3 high-strength steel bolts are available in atmospheric corrosion-resistant steel (see Chapter 2).

Equation 9.17 can be used to establish the minimum required bolt pretension, T_{bP}, as shown in Table 9.4. In order to account for the threaded portion of bolts, an effective bolt area, $A_{st} = 0.75(A_b)$, is used, such that

$$T_{bP} \geq 0.70 F_{bU} A_{st} \geq 0.53 F_{bU} A_b, \tag{9.18}$$

where A_b is the cross-sectional area of the bolt based on nominal bolt diameter.

9.3.3 Joint Types

Bolts are used in lap, "T," corner, and butt joints. Bolted lap joints (Figure 9.10a–d) are often used in members of steel railway bridges. The joints in Figure 9.10b and c may

* Also called machine, ordinary, unfinished, or rough bolts.

TABLE 9.4
Minimum Required Bolt Pretension for High-Strength Steel Bolts in Slip-Resistant Connections

Bolt Diameter, d_b, (in.)	Minimum Required Bolt Pretension, T_{bP} (kips)	
	A325 Bolts	A490 Bolts
1/2	12	16 (AREMA uses 15)
5/8	19	24
3/4	28	35
7/8	38 (AREMA uses 39)	48 (AREMA uses 49)
1	50 (AREMA uses 51)	62 (AREMA uses 64)
$1\frac{1}{2}$	98 (AREMA uses 103)	140 (AREMA uses 148)

be subjected to eccentric loads. Figure 9.10c shows a beam splice arrangement using bolted lap joints. Figure 9.10d shows a type of lap joint used to connect attachments such as stiffeners to girder web plates.

Bolted "T" and corner joints (Figure 9.10e and f) are rarely used to connect web plates and flange plates of plate and box girder bending members in modern steel railway superstructures. However, "T" and corner joints may be used in the fabrication of built-up axial members.

Bolted butt joints (Figure 9.10g) are typically used in join plate ends in a similar fashion to the flange splice joints shown in Figure 9.10c.

9.3.4 Bolted Joint Design

9.3.4.1 Allowable Bolt Stresses

Forces in a connection are transmitted through the effective shear, bearing, and tensile strength of the bolts. Bearing-type and slip-resistant connections exhibit different behavior in effective shear, but similar bolt bearing and tension behavior.

9.3.4.1.1 Allowable Effective Shear Stress

The allowable effective shear force on bearing-type connections is based on the allowable shear strength of the bolt shanks in the joint. Slip-resistant connections have an effective shear strength based on the magnitude of the prestress force and the shear slip coefficient of the steel connection elements. Following the failure of slip-resistant connections, the connection will behave as a bearing-type connection.

9.3.4.1.1.1 Allowable Effective Shear Stress in Bearing-Type Connections
Figure 9.11 illustrates that the behavior of a bolt under shear load is inelastic and without a well-defined yield stress. Therefore, bolt strength is determined based on ultimate shear strength. Experimentation has shown that the ultimate shear strength, F_{bv}, is in direct proportion to the ultimate tensile strength, P_{bU}, and is not affected by bolt prestress (Kulak et al., 1987). Therefore, the allowable shear stress of a bolt

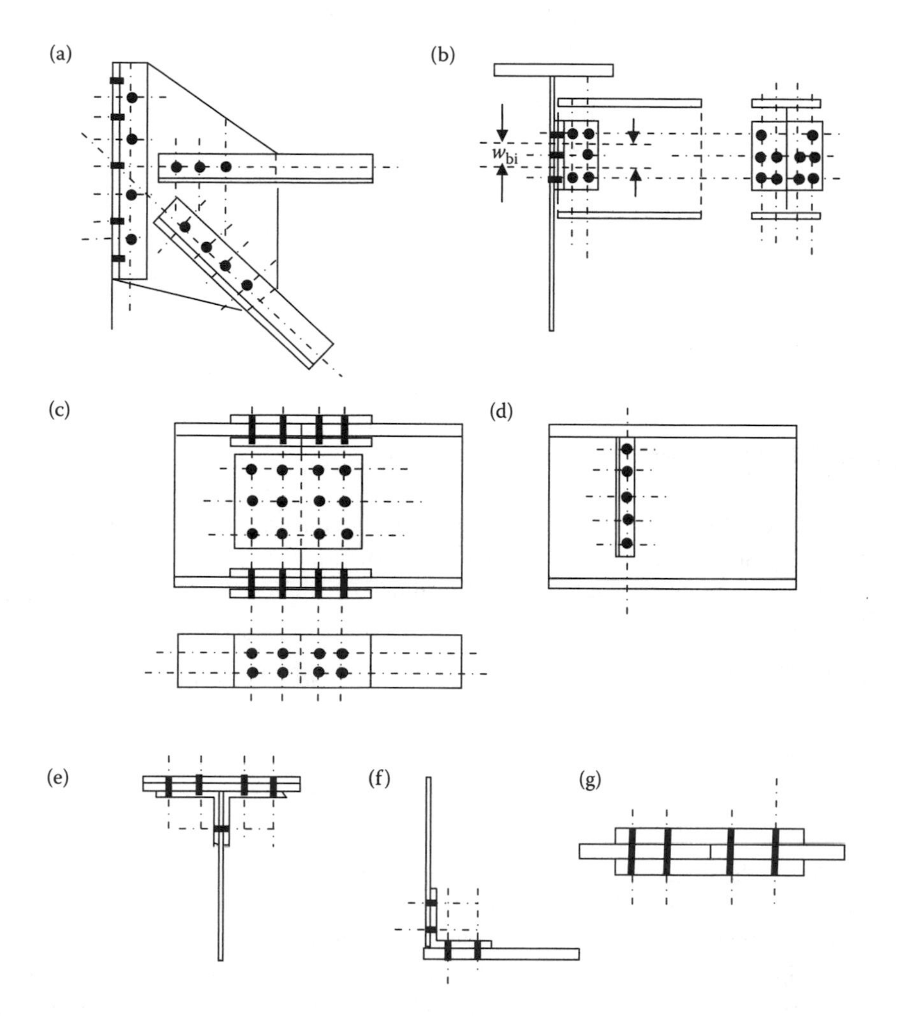

FIGURE 9.10 Typical bolted joints in steel railway superstructures.

(including a reduction of 0.80 due to the approximation) is

$$f''_{bv} \approx \frac{(0.80)0.62F_{bU}}{FS} \approx \frac{0.50F_{bU}}{FS}. \tag{9.19}$$

The allowable shear stress, f''_{bv}, for A325 bolts is 30 ksi (considering the nominal bolt diameter) if it is assumed that FS = 2.0 and F_{bU} = 120 ksi.* The allowable shear stress, f''_{bv}, for A325 bolts is 0.70(30) = 21 ksi if shear is assumed through the threaded portion of the bolt.

* Typically the bolt ultimate tensile strength is taken as 120 ksi for the development of allowable bolt stresses. This ignores the 12.5% reduction in ultimate tensile strength for A325 bolts with diameter exceeding 1 in.

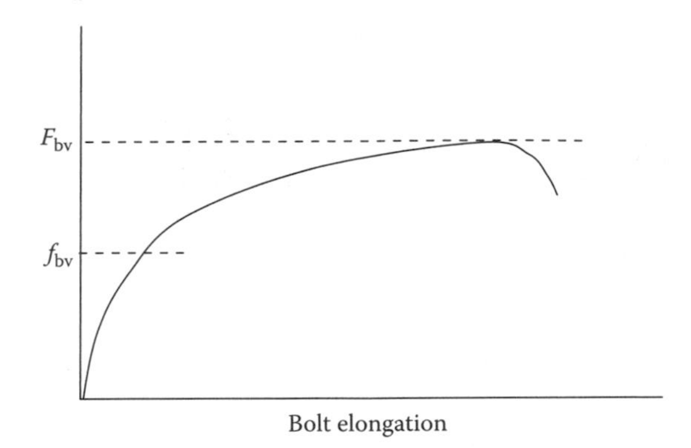

FIGURE 9.11 Bolt shear stress and elongation.

AREMA (2008) recommends only slip-resistant connections and, therefore, does not provide an allowable stress, f''_{bv}, based on shearing of the bolt shank. In slip-resistant connections, service loads are transmitted by friction and bolt shank shearing will not govern the design.

9.3.4.1.1.2 Allowable Effective Shear Stress in Slip-Resistant Connections The shear slip force, P_{bv}, is

$$P_{bv} = mnf'_{bv}(A_b) = k_s m\alpha \sum_{i=1}^{n} T_{bPi}, \tag{9.20}$$

where f'_{bv} is the effective allowable bolt shear stress; k_s is the shear slip coefficient of steel connection; m is the number of slip planes (faying surfaces); n is the number of bolts in connection; T_{bPi} is the specified pretension in bolt i; $\alpha = T_{bi}/T_{bPi}$; T_{bi} is the actual pretension in bolt i. Therefore, the effective allowable shear stress (which is based on the magnitude of the prestress force and the shear slip coefficient) is

$$f'_{bv} = \frac{k_s\alpha \sum_{i=1}^{n} T_{bPi}}{nA_b}. \tag{9.21a}$$

Using $A_{st} = 0.75(A_b)$, Equation 9.21a for A325 bolts, when the specified pretension in each bolt, T_{bPi}, is equal, becomes

$$f'_{bv} = \frac{k_s\alpha T_{bPi}}{A_b} = \frac{k_s\alpha(0.70F_{bU}A_{st})}{A_b} = 63\alpha k_s. \tag{9.21b}$$

In tests done to establish an empirical relationship for the effective allowable bolt shear stress, the slip probability level, mean slip coefficient, k_{sm}, and bolt pretension are not explicitly determined. They are combined into a slip factor, D, that incorporates the k_s and k_{sm} relationship, and α of Equation 9.21b as

$$f'_{bv} = \frac{k_s\alpha T_{bPi}}{A_b} = (0.53)Dk_{sm}F_{bU} = 63Dk_{sm}. \tag{9.22}$$

TABLE 9.5
Mean Slip Coefficients for Steel Surfaces

Class	Surface Description	Mean Slip Coefficient, k_{sm}
A	Clean mill scale and blast cleaned surface before coating	0.33
B	Blast cleaned surface with or without coating	0.50
C	Galvanized and roughened surfaces	0.40

AREMA (2008) outlines three slip critical connection faying surface conditions for design (Table 9.5). Tests done with turn-of-nut and calibrated wrench bolt installations will yield different results for the slip factor, D. The "Specification for Structural Joints Using ASTM A325 or A490 Bolts" (RCSC, 2000) provides values of slip factor, D, based on a 5% slip probability* and method of installation as shown in Table 9.6.†

Substitution of the mean slip coefficient, k_{sm} (Table 9.5), and slip factor, D (Table 9.6), into Equation 9.22 provides the effective allowable shear stress for a 5% slip probability as shown in Table 9.7. It is usual practice to specify turn-of-nut bolt installation and use $f'_{bv} = 17.0$ ksi for the design of slip-resistant connections. This provides an allowable shear force per bolt of 10.2 kips for a 7/8 in diameter bolt with a nominal cross-sectional area of 0.60 in.^{2}.

TABLE 9.6
Slip Coefficient for A325 Bolts with 5% Slip Probability

	Slip Coefficient, D	
Mean Slip Coefficient, k_{sm}	Turn-of-Nut Installation	Calibrated Wrench Installation
0.33	0.82	0.72
0.50	0.90	0.79
0.40	0.90	0.78

TABLE 9.7
Effective Allowable Shear Stress for A325 Bolts Based on 5% Slip Probability

	Effective Allowable Shear Stress, f'_{bv} (ksi)	
Mean Slip Coefficient, k_{sm}	Turn-of-Nut Installation	Calibrated Wrench Installation
0.33	17.0	15.0
0.50	28.4 (AREMA uses 28.0)	24.9
0.40	22.7 (AREMA uses 22.0)	19.7

* A slip probability of 5% (corresponds to a 95% confidence level for the test data) is appropriate for usual steel railway superstructure design.

† RCSC (2000) also provides slip factors for 1% and 10% slip probability.

9.3.4.1.1.3 Allowable Bearing Stress in Connections Bearing failures are manifested as either yielding due to bearing of the connection elements against the bolt and/or block shearing of the connection elements near an edge.

The ultimate bearing strength, F_B, of the connection element material bearing on the bolt shank is related to the ultimate tensile strength, F_U, by the following linear relationship (Kulak et al., 1987):

$$F_B = \left(\frac{l_e}{d_b}\right) F_U, \tag{9.23}$$

where l_e is the distance from the centerline of the bolt to the nearest edge in the direction of the force and d_b is the diameter of the bolt. Using FS = 2.5 against bearing on the plate material and the AREMA (2008) recommendation that $l_e \geq 3d_b$, results in an allowable bolt bearing stress, f_B, of

$$f_B = \frac{F_B}{\text{FS}} = 1.2F_U. \tag{9.24}$$

The yield strength in pure shear, F_v (see Chapter 2), is

$$F_v = \frac{F_y}{\sqrt{3}} \tag{9.25}$$

and the yield strength, P_y, of the shear block failure shown in Figure 9.12 is

$$P_y = 2t_p\left(l_e - \frac{d_b}{2}\right)\frac{F_y}{\sqrt{3}} = 1.15t_p\left(l_e - \frac{d_b}{2}\right)F_y, \tag{9.26}$$

where t_p is the plate thickness.

The bearing strength of the bolt, F_{bB}, is

$$F_{bB} = f_B d_b t_p, \tag{9.27}$$

which must not exceed the yield strength of the shear block given by Equation 9.26. Therefore,

$$P_y = 1.15t_p\left(l_e - \frac{d_b}{2}\right)F_y \geq F_{bB} = f_B d_b t_p. \tag{9.28a}$$

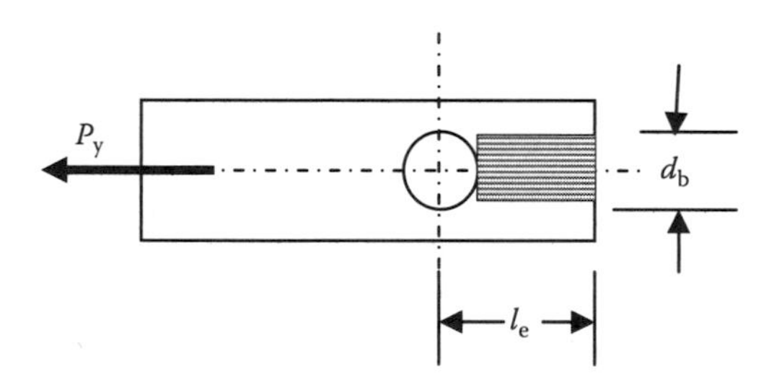

FIGURE 9.12 Shear block failure due to bolt bearing.

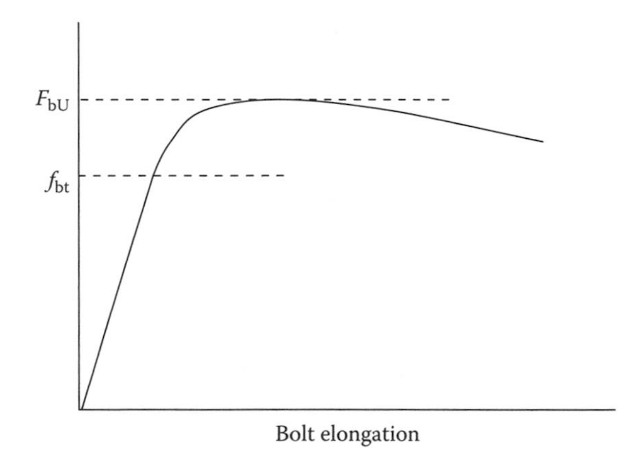

FIGURE 9.13 Bolt direct tension forces and elongation.

Rearrangement of Equation 9.28a yields

$$\frac{l_e}{d_b} \geq \frac{0.87 f_B}{F_y} + 0.5, \tag{9.28b}$$

which may be conservatively simplified (for $l_e/d_b \geq 1.4$*) to

$$\frac{l_e}{d_b} \geq \frac{f_B}{F_y}. \tag{9.29}$$

Rearrangement of Equation 9.29 and using FS = 2.0† provides the allowable bearing stress as

$$f_B \leq \frac{l_e F_y}{d_b} \leq \frac{l_e F_u}{d_b(\mathrm{FS})} = \frac{l_e F_u}{2 d_b}. \tag{9.30}$$

Equations 9.24 and 9.30 are the allowable bearing stresses on bolts recommended by AREMA (2008).

9.3.4.1.1.4 Allowable Tension Stress in Connections Figure 9.13 illustrates that the behavior of a bolt under tensile load is elastic for small elongations. The strength of a bolt loaded in direct tension is not affected by pretension stresses from installation by a method that applies the pretension to the bolt from torquing (Kulak, 2002). This is because the pretension load is readily dissipated as the direct tensile load is applied to a connection. Therefore, bolt strength is determined based on ultimate tensile strength. The allowable tensile stress, f_{bt}, in an A325 bolt, using FS = 2.0 and $A_{st} = 0.75(A_b)$, is

$$f_{bt} = \frac{0.75 F_{bU}}{\mathrm{FS}} = 45\,\mathrm{ksi}. \tag{9.31}$$

---

* This is the case in practical structures.

† A lower FS is used due to the conservative nature of the assumptions made to develop Equation 9.29.

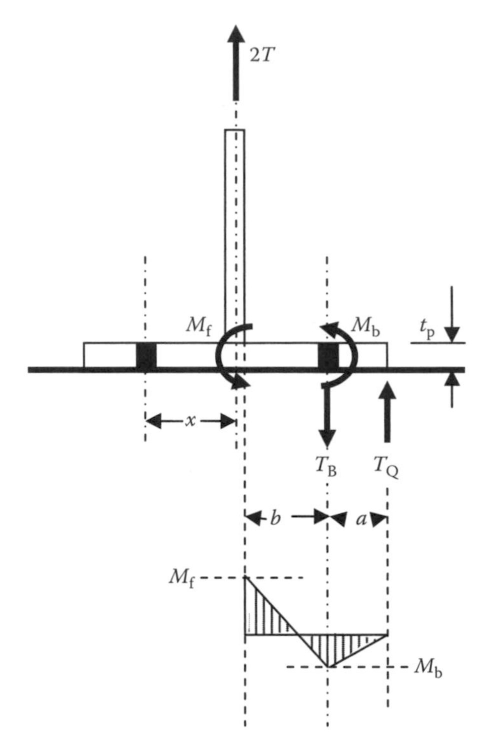

FIGURE 9.14 Prying action on a bolted tension joint.

However, as shown in Figure 9.14, the bolts in tension connections are subjected to additional tensile forces, T_Q, created by the prying action resulting from the flexibility of the connection leg. From Figure 9.14, the bending moment at the bolt line, M_b, is

$$T(b) = M_f + \left(\frac{A_{nb}}{A_{gf}}\right)M_b = M_f + \left(\frac{A_{nb}}{A_{gf}}\right)\alpha M_f = M_f(1 + \alpha\eta) \tag{9.32}$$

and with $M_f = \frac{w_b t_p^2}{4}F_y$,

$$T_Q(a) = M_b = \alpha\eta M_f = \frac{\alpha\eta w_b t_p^2 F_y}{4}, \tag{9.33}$$

where T is the applied tensile force per bolt; T_Q is the prying tensile force per bolt; $T_B = T + T_Q$ is the total tensile force per bolt; $\eta = A_{nb}/A_{gf}$; A_{nb} is the net area of the flange at the bolt line; A_{gf} is the gross area of the flange at the intersection with the web plate; $\alpha = M_b/M_f$ and depends on T_Q/T; and w_{bi} is the tributary area for prying of each bolt i (see Figure 9.10b).

The bolt tension, $T_B = T + T_Q$, with substitution of Equations 9.32 and 9.33, is

$$T_B = T + T_Q = \frac{M_f(1+\eta\alpha)}{b} + \frac{\alpha\eta w_b t_p^2 F_y}{4a} = \frac{M_f(1+\eta\alpha)}{b} + \frac{\alpha\eta M_f}{a}$$
$$= T\left(1 + \frac{\alpha\eta b}{(1+\alpha\eta)a}\right). \tag{9.34}$$

Therefore,

$$T_Q = T\left(\frac{\alpha\eta b}{(1+\alpha\eta)a}\right). \tag{9.35}$$

Further manipulation of these equations provides the thickness, t_p, as

$$t_p \leq \sqrt{\frac{4T_B ab}{w_b F_y(a + \alpha\eta(a+b))}}. \tag{9.36}$$

However, Equations 9.34 through 9.36 are difficult to use in routine design. Analytical and experimental studies have provided a semiempirical equation as (Douty and McGuire, 1965)

$$T_Q = T\left(\frac{0.50 - \left(w_b t_p^4/30ab^2 A_b\right)}{a/b\,((a/3b)+1) + \left(w_b t_p^4/6ab^2 A_b\right)}\right). \tag{9.37}$$

Equation 9.37 may be further simplified as (Kulak et al., 1987)

$$T_Q = T\left(\frac{3b}{8a} - \frac{t_p^3}{20}\right). \tag{9.38}$$

Further analytical and empirical studies (Nair et al., 1974) have provided other empirical equations for the prying force, but Equation 9.38 is simple and conservative for use in routine design of bolted connections subjected to tension.

Connections with bolts subjected to direct tension should generally be avoided in the main members of steel railway superstructures. Bolt tension and prying may occur combined with shear in connections such as those shown in Figures 9.10a and 9.10b. AREMA (2008) recommends the allowable tensile stress on fasteners, including the effects of prying, as 44 and 54 ksi for A325 and A490 bolts, respectively.

9.3.4.1.1.5 Allowable Combined Tension and Shear Stress in Connections Connections in steel railway superstructures may be subjected to combined shear and tension forces (e.g., the beam connection of Figure 9.10b). An ultimate strength interaction equation developed from tests (Chesson et al., 1965) is shown in Figure 9.15

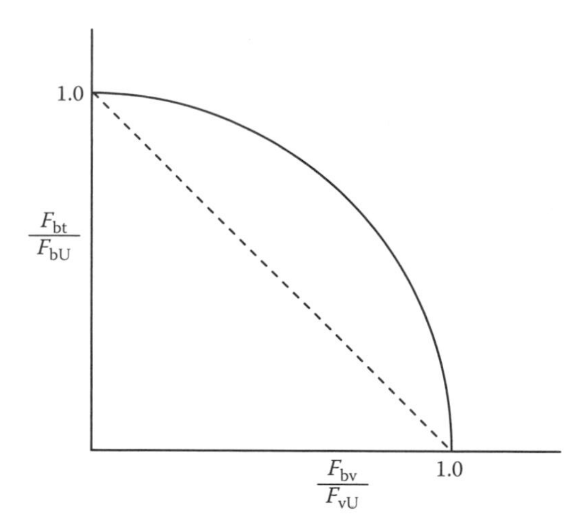

FIGURE 9.15 Bolt shear and tension interaction.

(solid line) and Equation 9.39

$$\left(\frac{F_{bt}}{F_{bU}}\right)^2 + \left(\frac{F_{bv}}{F_{vU}}\right)^2 = 1.0, \tag{9.39}$$

where F_{bt} is the ultimate tensile stress under combined shear and tension; F_{bU} is the ultimate tensile stress under tension only; F_{bv} is the ultimate shear stress under combined shear and tension; F_{vU} is the ultimate shear stress under shear only and is equal to $0.62F_{bU}$ (see Equation 9.19).

Therefore, Equation 9.39 may be expressed as

$$\left(\frac{F_{bt}}{F_{bU}}\right)^2 + 2.60\left(\frac{F_{bv}}{F_{bU}}\right)^2 = 1.0. \tag{9.40}$$

Equation 9.39, in terms of allowable stresses, is

$$\left(\frac{\sigma_{bt}}{f_{bt}}\right)^2 + \left(\frac{\tau_{bv}}{f_{bv}}\right)^2 = 1.0, \tag{9.41}$$

where σ_{bt} is the tensile stress in the bolt (including prying action effects). f_{bt} is the allowable bolt tensile stress for bearing-type connections or the nominal tensile stress from pretension for slip-resistant connections and is equal to T_{bP}/A_b. τ_{bv} is the shear stress in the bolt; f_{bv} is the allowable effective bolt shear stress and is equal to either f'_{bv} for slip-resistant connections (Equation 9.22) or f''_{bv} for bearing-type connections (Equation 9.19); T_{bP} is the bolt pretension (see Equation 7.17 and Table 9.3); A_b is the nominal area of the bolt ($1.33A_{st}$); and A_{st} is the effective bolt area through the threaded portion of the bolt.

Therefore, for slip-resistant connections,

$$\left(\frac{\sigma_{\mathrm{bt}}}{(T_{\mathrm{bP}}/A_{\mathrm{b}})}\right)^2 + \left(\frac{\tau_{\mathrm{bv}}}{f'_{\mathrm{bv}}}\right)^2 = 1.0. \tag{9.42}$$

The elliptical Equation 9.42 may be simplified by a straight-line approximation (dashed line for ultimate stress values in Figure 9.15) as

$$\left(\frac{\sigma_{\mathrm{bt}}}{(T_{\mathrm{bP}}/A_{\mathrm{b}})}\right) + \left(\frac{\tau_{\mathrm{bv}}}{f'_{\mathrm{bv}}}\right) = 1.0 \tag{9.43}$$

and may be rearranged as

$$\tau_{\mathrm{bv}} \le f'_{\mathrm{bv}}\left(1 - \frac{\sigma_{\mathrm{bt}}}{(T_{\mathrm{bP}}/A_{\mathrm{b}})}\right). \tag{9.44}$$

If τ_{bv} is taken as the allowable shear stress, f_{bv}, when combined with tension, Equation 9.44 becomes

$$f_{\mathrm{bv}} \le f'_{\mathrm{bv}}\left(1 - \frac{\sigma_{\mathrm{bt}}}{(T_{\mathrm{bP}}/A_{\mathrm{b}})}\right), \tag{9.45}$$

which is the allowable shear stress for combined shear and tension recommended by AREMA (2008).

9.3.4.1.1.6 Allowable Fatigue Stresses in Bolted Connections The allowable fatigue stress of bolted joints depends on whether the bolts are loaded primarily in shear, such as in lap and butt joints (Figure 9.10a–g) or tension. AREMA (2008) recommends that all joints subject to fatigue by cyclical stresses must be pretensioned slip-resistant connections.

9.3.4.1.1.6.1 Allowable Shear Fatigue Stress in Bolted Connections Bearing-type connections are subject to fatigue damage accumulation and crack initiation at the edge of, or within, holes due to localized tensile stress concentrations. Slip-resistant connections are subject to fretting fatigue.

AREMA (2008) recommends the allowable stress range, based on Fatigue Detail Category B, of 18 ksi for the base metal of slip-critical connections subjected to 2 million cycles or less or 16 ksi for the base metal of slip-critical connections subjected to greater than 2 million stress range cycles (see Chapter 5). Bolts will generally not experience fatigue failure prior to the base metal and, therefore, AREMA (2008) contains no recommendations concerning allowable shear stress ranges for bolts.

9.3.4.1.1.6.2 Allowable Tensile Fatigue Stress in Bolted Connections The stress range in a bolt of a slip-resistant connection is affected by the pretension applied to the bolt and the rigidity of the connection joint.

The stress range is typically considerably less than the nominal tensile stress in the bolt in relatively rigid slip-resistant connections with small prying forces.

The prying force should be limited to a maximum of 30% of the external load on the bolt for connections subjected to cyclical stresses in steel railway superstructures. An allowable tensile stress range of 31.0 ksi for A325* bolts is recommended where prying forces do not exceed 30% of the external load on the bolt.

The AREMA (2008) criteria are appropriate given the need to discourage the use of bolted connections subject to direct cyclical tensile stress regimes in structures such as steel railway spans.

9.3.4.2 Axially Loaded Members with Bolts in Shear

These are typically truss member end connections with forces transferred from the axial members through connections consisting of bolts in shear and gusset plates (Figure 9.10a). Axial tension member end connections and gusset plates must be designed considering yielding, fracture, and block shear (tear-out) failure. Axial compression member end connections and gusset plates must be designed considering buckling and block shear.

9.3.4.2.1 Axial Member End Connection

The number of bolts required in the axial force connection may be determined by considering the allowable bolt shear stress, f_{bv}, or, if cyclically loaded, the allowable bolt shear fatigue stress (16 or 18 ksi depending on number of design stress range cycles). The allowable bearing stress, f_B, should be used for bearing-type connections. Bearing stress may also be considered for slip-critical connections as a precaution following slip failure. Determination of the number of bolts required to transfer forces from the member in the connection will determine the net section area required for the member and the basic dimensions of the connection gusset plates.

The design of axial members for strength (yield and fracture), fatigue, stability, and serviceability is discussed in Chapter 6. The effects of the connection in terms of net area and shear lag effects were considered in the design criteria for axial tension members. However, the design of axial member end connections requires attention to localized effects of bolt bearing stresses (in regard to member element thickness) and fracture or rupture by block shear.

Axial bolted connections should conform to the recommended minimum bolt spacing and edge distance criteria of AREMA (2008). The minimum bolt spacing (center-to-center bolts) is three times the bolt diameter. The minimum bolt spacing, s_b, along a line of force, based on bearing considerations, from Equation 9.30 is

$$s_b \geq \frac{2d_b\sigma_{bc}}{F_u} + \frac{d_b}{2}, \tag{9.46}$$

where σ_{bc} is the bearing stress on connection element of area d_bt_p and t_p is the thickness of connection element. The recommended minimum edge distance, e_b, is

* An allowable tensile stress range of 38.0 ksi is recommended for A490 bolts.

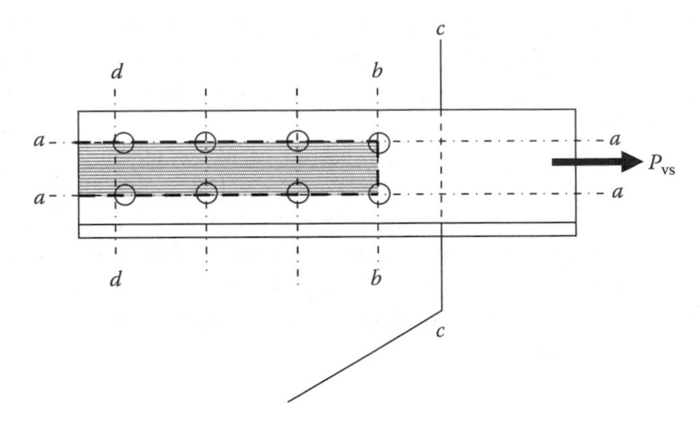

FIGURE 9.16 Block shear failure in a member at an axial tension connection.

$1.5 + 4t_p \leq 6$ in. and, based on bearing stress considerations,

$$e_b \geq \frac{2d_b\sigma_{bc}}{F_u}. \tag{9.47}$$

Block shear failure occurs in members at ultimate shear stress on planes along the bolt lines (lines a–a in Figure 9.16) and ultimate tensile stress on planes between bolt lines (line b–b in Figure 9.16). The failure ultimately results in the tear-out of a section (shaded area of the member in Figure 9.16). If the allowable shear stress is limited to $F_{vU} = 0.60F_u$/FS (see Equation 9.19) and allowable tensile stress is F_u/FS, the allowable block shear strength, P_{vs}, using FS = 2.0, is

$$P_{vs} = 0.30F_U A_{nv} + 0.50F_u A_{nt}. \tag{9.48}$$

However, a combination of yielding on one plane and fracture on the other plane is likely, depending on the connection configuration. When the net fracture strength in tension is greater than the net fracture strength in shear, $F_u A_{nt} \geq F_{vU} A_{nv} \geq 0.60F_u A_{nv}$, yielding will occur on the gross shear plane and the allowable block shear strength, P_{vs}, becomes

$$P_{vs} = 0.35F_y A_{gv} + 0.50F_u A_{nt}. \tag{9.49a}$$

Conversely, when the net fracture strength in tension is less than the net fracture strength in shear, $F_u A_{nt} \prec F_{vU} A_{nv} \prec 0.60F_u A_{nv}$, yielding will occur on the gross tension plane and the allowable block shear strength, P_{vs}, is

$$P_{vs} = 0.30F_u A_{nv} + 0.55F_y A_{gt}, \tag{9.49b}$$

where F_u is the ultimate tensile stress of connection element, F_y is the tensile yield stress of connection element, A_{gv} is the gross area subject to shear stress (thickness times gross length along lines a–a in Figure 9.16), A_{nv} is the net area subject to shear stress (thickness times net length along lines a–a in Figure 9.16), A_{gt} is the gross area subject to tension stress (thickness times gross length along lines b–b in Figure 9.16), and A_{nt} is the net area subject to tension stress (thickness times net length along line b–b in Figure 9.16).

AREMA (2008) recommends determination of allowable block shear strength using Equations 9.48 and either Equation 9.49a or 9.49b, depending on whether the net fracture strength in tension is greater or less than the net fracture strength in shear.

Connection shear lag effects that require consideration for axial tension member design are considered in Chapter 6. Shear lag is taken into account for design by determination of a reduced cross-sectional area or effective net area, A_e, which is based on the connection efficiency.

9.3.4.2.2 Gusset Plates

In general, gusset plates should be designed to be as compact as possible. This not only reduces material consumption but reduces slenderness ratios and free edge distances for greater buckling strength. Gusset plates have been traditionally designed using beam theory to determine axial, bending, and shear stresses at various critical sections of a gusset plate. However, the slender beam model is not an accurate model,* and consideration of the limit states of block shear (tear-out) and axial stress, based on an appropriate area, is used for the design of ordinary gusset plates.

Block shear in a gusset plate is analogous to the situation shown in Figure 9.16 but with the tear-out section extending from the edge of the gusset plate (line c–c in Figure 9.16) to the furthest line of bolts (line d–d in Figure 9.16). Equations 9.48 and either Equation 9.49a or 9.49b are also used to determine the allowable block shear strength of the gusset plate at each member end connection.

The axial stress in the gusset plate is required for comparison to allowable tensile and compressive axial stresses. Testing has shown that an effective length, w_e, perpendicular to the last bolt line (line d–d in Figure 9.17), on which axial stresses act, may be based on lines 30° to the bolt row lines (lines a–a in Figure 9.17) from the first perpendicular bolt line (line b–b in Figure 9.17) (Whitmore, 1952). The Whitmore effective length, w_e, is

$$w_e = 2l_c \tan(30°) + s_r = 1.15l_c + s_r. \tag{9.50}$$

The effective length, w_e, must often be reduced if it intersects other members or contains elements with different strengths (e.g., $\bar{w}_e$ in Figure 9.18). The axial tensile design of the gusset plate is then based on

$$\sigma_{at} = \frac{P}{w_e t_p} \leq 0.55F_y \tag{9.51a}$$

or

$$\sigma_{at} = \frac{P}{w_{ne} t_p} \leq 0.47F_U. \tag{9.51b}$$

* For example, tests show shear stresses are closer to V/A than 1.5 V/A as predicted by beam theory.

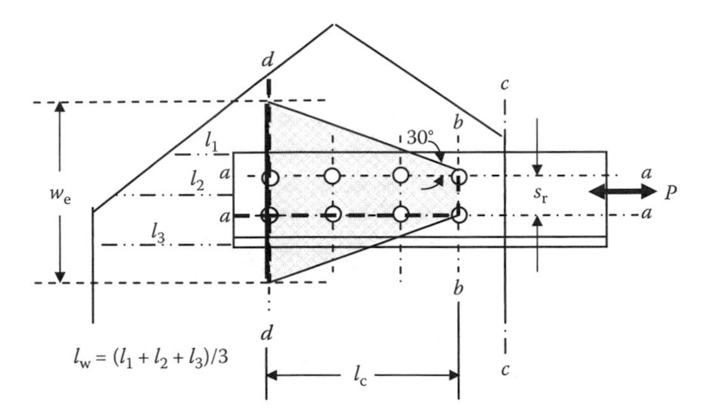

FIGURE 9.17 Whitmore stress block in a gusset plate at an axial member connection.

The limit state of yielding on the gross section, $w_e t_p$, immediately below the last line of bolts is represented by Equation 9.51a* and that of tensile failure on the net section, $w_{ne} t_p$, through the last line of bolts by Equation 9.51b. Compressive design (considering stability) is based on

$$\sigma_{ac} = \frac{P}{w_e t_p} \le F_{call}, \tag{9.52}$$

where w_{ne} is the net effective length, σ_{at} is the axial tension stress on effective area, $w_e t_p$ or $w_{ne} t_p$, σ_{ac} is the axial compression stress on effective area, $w_e t_p$, F_{call} is the allowable axial compression stress based on the effective slenderness ratio, Kl_w/r_w (many engineers restrict $Kl_w/r_w \le 100$–120), Kl_w is the effective buckling length of the gusset plate, l_w is the average distance, $(l_1 + l_2 + l_3)/3$, from last line of bolts (line d–d in Figure 9.17) to the edge of the gusset plate. l_w may extend to the first row of bolts at an interface member such as a bottom chord element (line f–f in Figure 9.18).

$$r_w = t_p/\sqrt{12}$$

t_p is the thickness of the gusset plate, K is the effective length factor typically taken as between 0.50 and 0.65 for properly braced gusset plates (Thornton and Kane, 1999).†

The use of block shear rupture and the Whitmore section analysis may be sufficient for the design of ordinary gusset plates. However, for heavily loaded railway truss members it is often appropriate to also check beam theory‡ shear forces, bending moments, and axial forces at critical sections (e.g., lines f–f, g–g, h–h and i–i in Figure 9.18). The critical sections such a g–g and h–h in Figure 9.18 should be

* Equation 9.51a is slightly conservative as the Whitmore section is taken through the center of the last line of bolts.

† If gusset plates are not braced against lateral movement, k may be greater than 1 (see Chapter 6, Figures 6.4 and 6.5, Table 6.4).

‡ An alternative to slender beam theory, the uniform force method, which is strongly dependent on connection geometry, has been used for building design (Thornton and Kane, 1999).

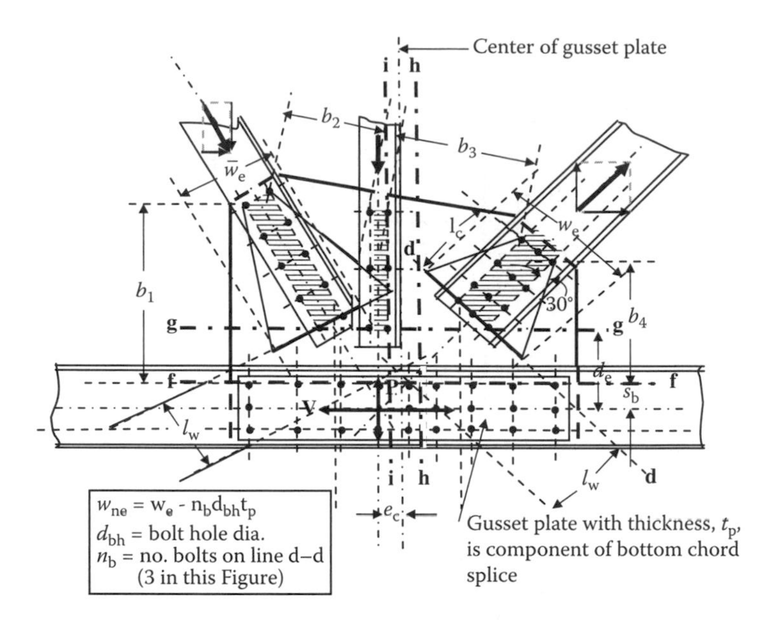

FIGURE 9.18 Typical truss gusset plate connection.

reviewed for combined stresses (see Chapter 8) from the following:

- Shear field, V, on the gross section of the gusset plate from resultant horizontal and vertical forces in the members
- Axial tension or compression, P, on the gross section of the gusset plate from resultant horizontal and vertical forces in the members
- Bending moment, for example $M = \pm V(d_e) \pm P(e_c)$ at section g–g in Figure 9.18.

Critical sections such as f–f and i–i in Figure 9.18 should also be reviewed for combined stresses from the following:

- Shear fracture, V, on the net section of the gusset plate from resultant horizontal forces in the members
- Axial tension or compression, P, on the net section of the gusset plate from resultant vertical forces in the members
- Bending moment

For gusset plates in very complex connections or in long span trusses, the detailed analysis of gusset plate connections by finite element analysis is often warranted.

In addition, free edge lengths on the gusset plate should be minimized to preclude localized buckling effects. Many engineers restrict b_i/t_p ratios to less than $2.06\sqrt{E/F_y}$ ($i = 1, 2, 3$, and 4 in Figure 9.18). Edge stiffening angles should be used when the free edge distance is large and be proportioned such that $b_i/\sqrt{w} < 120$.

Example 9.4 outlines the design of an axially loaded bolted connection using block shear and Whitmore stress block analyses.

Example 9.4

Design the slip-resistant bolted connection for some secondary wind bracing shown in Figure E9.4. The steel is Grade 50 ($F_y = 50$ ksi and $F_U = 65$ ksi), and 7/8 in. diameter A325 high-strength steel bolts are used in the connection for this 7 ft long member.

$T = 50$ kips, $f_{bv} = 17$ ksi (slip-resistant connection allowable bolt shear)

Secondary and bracing member connections must be designed for the lesser of the allowable strength of the member or 150% of the calculated maximum forces in the member.

$$T' = 1.5(50) = 75\ \text{kips}$$

Allowable tensile strength of member (see Chapter 6)

Shear lag effect, $U = 1 - (\overline{x}/L) = 1 - (1.68/9) = 0.81$. However, AREMA recommends the use of 0.60 for single angle connections.

Effective net area $= A_{ne} = 0.60(5.75 - 2(1)(0.5)) = 2.85$ in.^{2}

Allowable strength $= 0.47(65)(2.85) = 87.0$ K or $= 0.55(50)(5.75) = 158.1$ K

Design connection for 75 kips axial tension.

Member

The number of bolts in single shear (single shear plane or faying surface):

$$n = \frac{75}{17\left(\pi d_b^2/4\right)} = \frac{75}{17(0.60)} = \frac{75}{10.2} = 7.4, \text{ use minimum 8 bolts.}$$

check bearing stress

$$\sigma_{bc} = \frac{75}{8(0.375)(1.00)} = 25.0\ \text{ksi for a minimum 3/8 in. thick gusset plate,}$$

$$f_B \le \frac{l_e F_u}{2d_b} \le \frac{(3.5)(65)}{2(7/8)} \le 130\ \text{ksi}$$

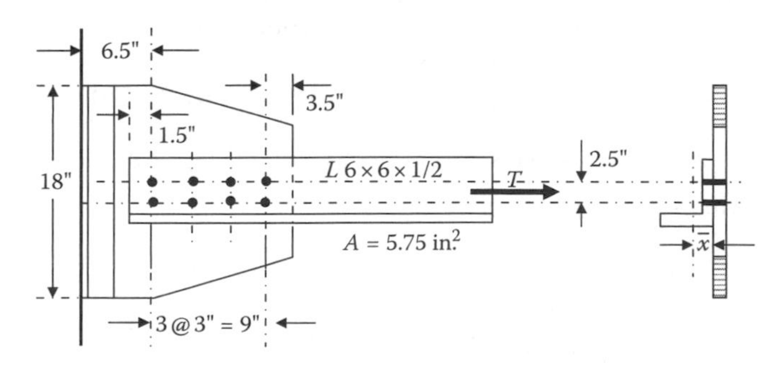

FIGURE E9.4

or

$$f_B \leq 1.2F_u \leq 1.2(65) \leq 78\,\text{ksi OK.}$$

Tensile stress in the member:

$\bar{x} = 1.68$ in.

$A_e = 2.85$ in.^{2}

Tensile axial stress in the angle net section $= \sigma_a = \frac{75}{2.85} = 26.3$ ksi $\leq 0.47(65) \leq 30.6$ ksi OK

Tensile axial stress in the angle gross section $= \sigma_a = \frac{75}{5.75} = 13.0$ ksi $*$ $0.55(50) * 27.5$ ksi OK

Block shear failure in angle:

$A_{gt} = 2.5(0.5) = 1.25$ in.^{2}

$A_{nt} = 1.25 - 0.5(1) = 0.75$ in.^{2}

$A_{gv} = 2(9 + 1.5)(0.5) = 10.50$ in.^{2}

$A_{nv} = 10.50 - 8(0.5)(1.0) = 6.50$ in.^{2}

$P_{vs} = 0.30F_U A_{nv} + 0.50F_U A_{nt} = 0.30(65)(6.50) + 0.50(65)(0.75) = 126.8 + 24.4 = 151.2$ kips

Tensile ultimate strength $= F_u A_{nt} = (65)(0.75) = 48.8$ kips

Shear ultimate strength $= 0.60F_u A_{nv} = 0.60(65)(6.50) = 253.4$ kips

Therefore, tensile yielding on the gross section and shear fracture on the net section is appropriate to consider.

$P_{vs} = 0.30F_U A_{nv} + 0.55F_y A_{gt} = 0.30(65)(6.50) + 0.55(50)(1.25) = 126.8 + 34.4 = 161.2$ kips

The allowable block shear stress is 151.2 kips $\geq$ 75 kips OK

The member design is governed by the tensile fracture criterion due to the considerable shear lag effect associated with the single angle connection used for this secondary member.

Gusset plate

Block shear failure in the gusset plate:

$A_{gt} = 2.5(0.375) = 0.94$ in.^{2}

$A_{nt} = 0.94 - 0.375(1) = 0.56$ in.^{2}

$A_{gv} = 2(9 + 3.5)(0.375) = 9.38$ in.^{2}

$A_{nv} = 9.38 - 8(0.375)(1.0) = 6.38$ in.^{2}

$P_{vs} = 0.30F_U A_{nv} + 0.50F_U A_{nt} = 0.30(65)(6.38) + 0.50(65)(0.56) = 124.3 + 18.2 = 142.5$ kips

Tensile ultimate strength $= F_u A_{nt} = (65)(0.56) = 36.4$ kips

Shear ultimate strength $= 0.60F_u A_{nv} = 0.60(65)(6.38) = 248.6$ kips

Therefore, tensile yielding on the gross section and shear fracture on the net section is appropriate to consider.

$P_{vs} = 0.30F_U A_{nv} + 0.55F_y A_{gt} = 0.30(65)(6.38) + 0.55(50)(0.94) = 124.3 + 25.9 = 150.2$ kips

The allowable block shear stress is 142.5 kips $\geq$ 75 kips OK

Axial tension in the gusset plate:

$$w_e = 2l_c \tan(30°) + s_r = 1.15(9) + 2.5 = 12.85\text{ in.}$$

$$\sigma_{at} = \frac{75}{(12.85)(0.375)} = 15.6 \leq 0.55F_y \leq 27.5\,\text{ksi OK}$$

or

$$\sigma_{at} = \frac{75}{(12.85 - 2(0.375)(1))(0.375)} = 16.5 \leq 0.47F_U \leq 30.6 \text{ ksi OK}$$

Axial compression in the gusset plate:

In this example $T = 50$ kips. However, investigate the allowable compression stresses in the gusset plate for the case of complete stress reversal $T = -C$ (typical of wind bracing).

$C = -50$ kips, $f_{bv} = 17$ ksi (slip-resistant connection allowable bolt shear), $C' = 1.5(-50) = -75$ kips

Allowable Compressive Strength of Member (see Chapter 6)
$r_{min} = r_{xy} = 1.18$ in.

$$\frac{kL}{r_{min}} = \frac{0.75(7)(12)}{1.18} = 53.4$$

$F_{call} = 0.60(50) - 0.165(53.4) = 21.2$ ksi

Allowable strength $= (21.2)(5.75) = 121.6$ kips compression

Design connection for 75 kips axial compression.

Compressive stress in the gusset plate:

$$\sigma_{ac} = \frac{C'}{w_e t_p} = \frac{75}{12.85(0.375)} = 15.6 \text{ ksi}$$

$$\frac{Kl_w}{r_w} = \frac{0.65(6.5)\sqrt{12}}{0.375} = 39.0 \geq 0.629\sqrt{\frac{E}{F_y}} \geq 15.2$$

$$F_c = 0.60F_y - \left(\frac{17500F_y}{E}\right)^{3/2}\left(\frac{Kl_w}{r_w}\right) = 0.60F_y - 165.7\left(\frac{Kl_w}{r_w}\right)$$

$$= 30{,}000 - 165.7(39.0) = 23{,}533 \text{ psi} = 23.5 \text{ ksi} \geq 15.6 \text{ ksi OK}$$

If the connection shown in Figure E9.4 has a compression diagonal creating a vertical force of 35 kips and horizontal force of 25 kips in addition to the 50 kip tensile force;

$P = T = 50 - 25 = 25$ kips

$V = 35$ kips

$M = 35(6.5) = 227.5$ in-kips (on section through the last line of bolts)

$A_g = 18(0.375) = 6.75 \text{ in.}^2$

$A_n = 6.75 - 2(1)(0.375) = 6.00 \text{ in.}^2$

$S_n = 0.375(18)^2/6 - 2(1)(0.375)(1.25) = 19.3 \text{ in.}^3$

$\tau_v = 35/6.75 = 5.2$ ksi (1.5V/A not used since slender beam theory not theoretically valid)

$\sigma_a = 25/6.00 = 4.2$ ksi

$\sigma_b = 227.5/19.3 = 11.8$ ksi

Use a linear interaction formula to examine combined stress effects

$$\frac{5.3}{0.35(50)} + \frac{4.2}{0.55(50)} + \frac{11.8}{0.55(50)} = 0.30 + 0.15 + 0.43 = 0.88 \leq 1.00 \text{ OK}$$

9.3.4.3 Eccentrically Loaded Connections with Bolts in Shear and Tension

Small load eccentricities may often be ignored in slip-resistant bolted connections, but large eccentricities should be considered in the design. Many bolted connections are loaded eccentrically (e.g., the connections shown in Figure 9.10b and c). Eccentric loads will result in combined shear and torsional moments or combined shear and bending moments, depending on the direction of loading with respect to the bolts in the connection.

9.3.4.3.1 Connections Subjected to Shear Forces and Bending Moments

A connection similar to that shown in Figure 9.10d is shown in Figure 9.19. The bolts each side of the stiffener bracket resist both shear forces and bending moments.

9.3.4.3.1.1 Bolt Shear Stress The shear stress on the bolts is

$$\tau_b = \frac{P}{n_s n_b A_b}, \tag{9.53}$$

where n_b is the number of bolts, n_s is the number of shear planes, and A_b is the nominal cross sectional area of the bolt.

9.3.4.3.1.2 Bolt Tensile Stress The tension, σ_{ti}, on bolt i from bending moment, $M = Pe$ is

$$\sigma_{ti} = \frac{M}{A_b S_{bi}} = \frac{Pe}{A_b S_{bi}} = \frac{Peh_{bi}}{A_b I_b}, \tag{9.54}$$

where A_b is the nominal cross sectional area of bolt i, S_{bi} is the "effective section modulus" of bolt $i = I_b/h_{bi}$, I_b is the "effective moment of inertia" of the bolt group, and h_{bi} is the distance from bolt i to the neutral axis of the bolt group.

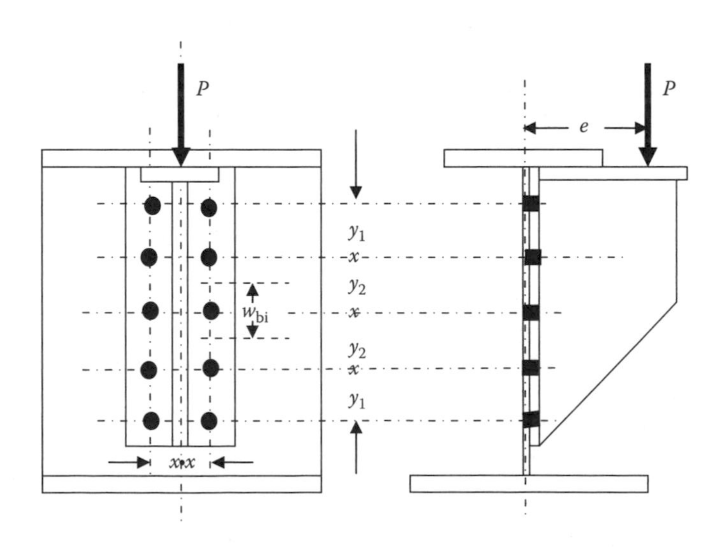

FIGURE 9.19 Bending and shear forces on a bolted connection.

For the connection in Figure 9.19, $I_b = 4(y_1 + y_2)^2 + 4(y_2)^2$ and the bolt tension on the most highly stressed bolt, σ_t (bolt farthest from the neutral axis of the bolt group), is

$$\sigma_t = \frac{Peh_{bi}}{A_b I_b} = \frac{Pe(y_1 + y_2)}{4\left((y_1 + y_2)^2 + (y_2)^2\right) A_b}.$$

If $y = y_1 = y_2$, $\sigma_t = Pe/10yA_b$. Prying action effects, which will increase the bolt tension, must also be considered in the connection design (e.g., by using Equation 9.38).

9.3.4.3.1.3 Combined Shear and Tension The bolts in Figure 9.19 are subject to shear force, $F_{bv} = \tau_b A_b$, and tensile force, $T_B = T + T_Q$, which must be combined to determine the allowable stress in the bolts of the connection. The allowable shear stress for combined shear and tension in a slip-resistant connection is (from Equation 9.45)

$$f_{bv} = \frac{F_{bv}}{A_b} \leq f'_{bv} \left(1 - \frac{(T_B)}{(T_{bP})}\right), \tag{9.55}$$

where f'_{bv} is the allowable bolt shear stress for slip-resistant connections, $T_B = T + T_Q$ is the total bolt tensile force, T_{bP} is the bolt pretension (see Equation 9.17 and Table 9.3), and A_b is the nominal area or bolt.

Example 9.5

Review the design of the slip-resistant single shear plane connection shown in Figure 9.19 using 7/8 in. diameter A325 bolts for a load, $P = 55$ kips with eccentricity, $e = 6$ in. The steel is ASTM A709 Grade 50 with $F_y = 50$ ksi and $F_u = 65$ ksi. The connection geometry is similar to Figure 9.19 with:

$y = y_1 = y_2 = 4.0$ in.
$x = 4.5$ in.
$w_{bi} = 4.0$ in.
$a = 1.5$ in. (see Figure 9.14)
$b = 4.25$ in.
$t_p = 0.5$ in.
$n_b = 10$ (number of bolts)
$n_s = 1$

Shear

$$\tau'_{bv} = \frac{P}{n_s n_b A_b} = \frac{55}{(1)10(0.60)} = 9.2\,\text{ksi}$$

Tension

$$T = \frac{Pe}{10y} = \frac{55(6)}{10(4.0)} = 8.3\,\text{kips}$$

$$T_Q = T\left(\frac{3b}{8a} - \frac{t_p^3}{20}\right) = 8.3\left(\frac{3(4.25)}{8(1.5)} - \frac{(0.5)^2}{20}\right) = 8.7\,\text{kips}$$

$$\sigma_{bt} = \frac{(8.3 + 8.7)}{0.60} = 28.3\,\text{ksi} \le f_{bt} \le 44.0\,\text{ksi OK}$$

Combined shear and tension

$$f_{bv} = f'_{bv}\left(1 - \frac{(T)}{(T_{bP})}\right) = 17\left(1 - \frac{(8.3 + 8.7)}{39}\right) = 9.6\,\text{ksi} \ge \tau'_{bv} \ge 9.2\,\text{ksi OK}$$

check bearing stress

$\sigma_{bc} = \frac{55}{10(0.50)(1.00)} = 11.0$ ksi, assuming that the minimum thickness of the connection plate and the web is 0.5 in.

$f_B \le \frac{l_e F_u}{2d_b} \le \frac{(1.5)(65)}{2(7/8)} \le 55.7$ ksi, assuming that the minimum edge distance is 1.5 in, OK

or $f_B \le 1.2F_u \le 1.2(65) \le 78$ ksi

9.3.4.3.2 *Connections Subjected to Eccentric Shear Forces (Combined Shear and Torsion)*

A connection subjected to eccentric shear is shown in Figure 9.20. The bolts in the connection resist direct shear forces from, P, and torsional shear forces from the moment, Pe. The direct shear stress, τ, on the bolts (all bolts with the same A_b), is

$$\tau = \frac{P}{n_s \sum A_b} = \frac{P}{n_s n_b A_b} \tag{9.56}$$

and the torsional shear stress, τ_T, on the bolts is

$$\tau_T = \frac{Per_T}{n_s J_b} = \frac{Per_T}{n_s \sum n_b A_b r_T^2} = \frac{Per_T}{n_s A_b \sum n_b (x_T^2 + y_T^2)}. \tag{9.57}$$

Equation 9.57 can be developed in the x- and y-directions as

$$\tau_{Tx} = \frac{Pey_T}{n_s A_b \sum n_b (x_T^2 + y_T^2)}, \tag{9.58a}$$

$$\tau_{Ty} = \frac{Pex_T}{n_s A_b \sum n_b (x_T^2 + y_T^2)}, \tag{9.58b}$$

where n_b is the total number of bolts in the connection; n_s is the number of shear planes; $J_b = \Sigma n_b A_b r_T^2$ is the polar moment of inertia of connection; $r_T = \sqrt{x_T^2 + y_T^2}$ is the distance from the bolt to the centroid of the bolt group, x_T is the distance from

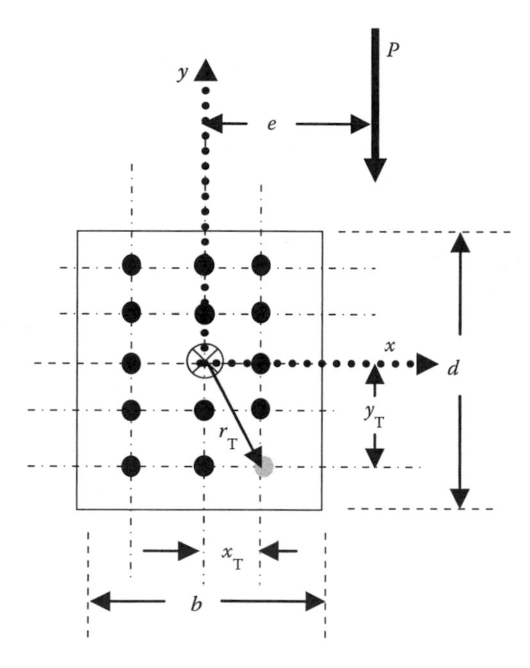

FIGURE 9.20 Eccentric shear forces on a bolted connection.

the centroid to the bolt in the x-direction, and y_T is the distance from the centroid to the bolt in the y-direction.

The resultant shear stress on any bolt described by locations x and y is

$$f = \sqrt{(\tau + \tau_{Ty})^2 + \tau_{Tx}^2}. \tag{9.59}$$

9.3.4.3.3 *Beam Framing Connections*

Bolted beam framing connections are often used in the main members of steel railway superstructures (Figure 9.21a).* These framing connections are subject to shear forces, P, and member end bending moments, M_e. Furthermore, the legs of the connection angles fastened to the web of the beam (a double-shear connection) may also be subject to a torsional moment, Pe, due to the eccentric application of shear force† (Figure 9.21a, side elevation).

Beam framing connections are often assumed to transfer shear only (i.e., it is assumed the beam is simply supported and $M_e = 0$), provided that adequate connection flexibility exists. However, due to some degree of end restraint, a corresponding proportion of fixed end moment, δM_e, typically exists ($\delta = M_e/M_f$ where M_f is the fixed end beam moment). The magnitude of the end moment depends on the rigidity of the support and can be of considerable magnitude (Al-Emrani, 2005).

* These connections can be single-shear or double-shear connections depending on configuration. For example, in the floor systems of many steel railway superstructures, a double-shear connection exists at interior floorbeams and, typically, a single-shear connection at end floorbeams.

† Depending on whether these effects are accounted for in the structural analysis.

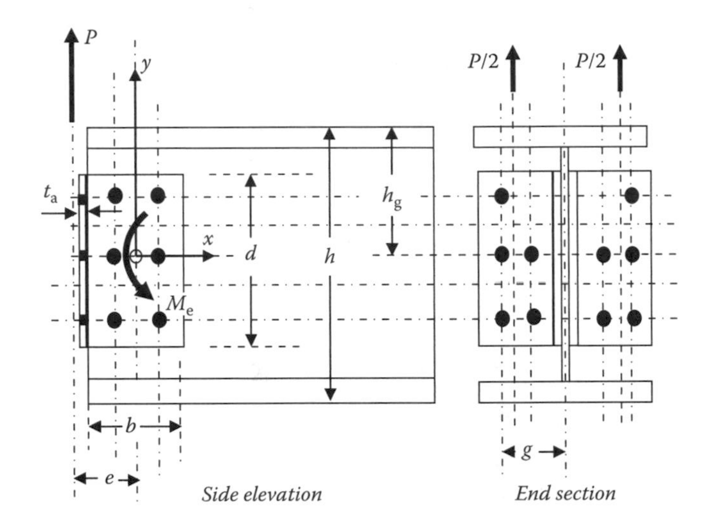

FIGURE 9.21a Bolted beam framing moment connection.

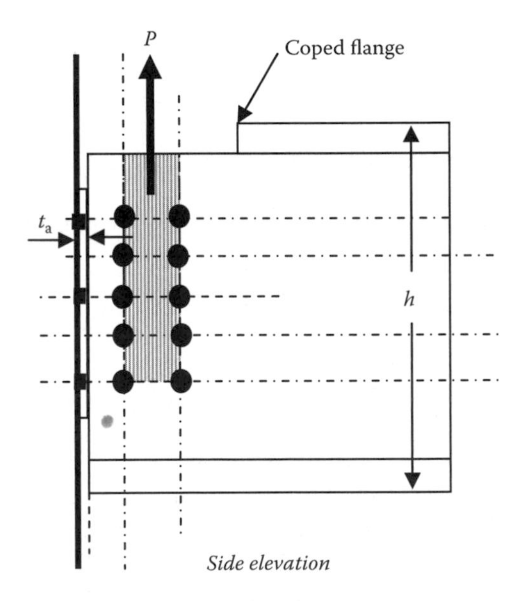

FIGURE 9.21b Bolted beam framing connection subject to block shear.

A rigid connection may be designed for the shear force, P, and the corresponding end moment due to full fixity, M_f. A semirigid connection will require consideration of the end moment (M_e)–end rotation (ϕ_e) relationship (often nonlinear) to determine rotational stiffness and the end moment for design. Moment–rotation curves, developed from theory and experiment, for bolted joint configurations* are available in

* Typically for beam to column flange connections.

the technical literature on connection design (e.g., Faella et al., 2000; Leon, 1999). A flexible end connection will deform and resist very little bending moment. Therefore, simple beam framing connections that exhibit the characteristics of a flexible connection (see Figure 9.8) may be designed for shear force, P, only ($M_e = 0$). AREMA (2008) recognizes that all connections actually exhibit some degree of semirigid behavior, but allows flexible connection design (with a bolt configuration that allows for adequate deformation and flexibility) provided the design shear force is increased by 25%. Otherwise, a semirigid connection design considering both beam end moment and shear is required.

Therefore, flexible bolted beam framing connections must be designed considering shear on the outstanding legs of the connection and, due to the eccentricity of the shear force, combined shear and torsion on the leg of the connection angles fastened to the web of the beam. However, for flexible bolted connections,* it is usual practice to disregard the moment, Pe, due to the eccentricity, e.

The angles in the simple beam framing connection (often referred to as clip angles) must deform in order to allow an appropriate level of flexible connection behavior. Bolted connections are made more flexible by providing a minimum gage distance, g, over a distance, h_g, from the top of the beam (see Figure 9.21a). AREMA (2008) recommends $h_g \geq h/3$ and

$$g \geq \sqrt{\frac{Lt_a}{8}}, \tag{9.60}$$

where L is the length of the beam span in inches and t_a is the thickness of angle in inches.

Equation 9.60 is based on analytical and experimental work regarding the fatigue strength of typical stringer to floorbeam connections (Yen et al., 1991).

If the beam flanges are coped at the connection, the design must also consider block shear (the combination of shear or tension yielding on one plane and tension or shear fracture on the other that may cause tear-out of the shaded area shown in Figure 9.21b). AREMA (2008) recommends determination of allowable block shear strength using Equations 9.48 and either Equations 9.49a or 9.49b, depending on whether the net fracture strength in tension is greater or less than the net fracture strength in shear.

Examples 9.6 and 9.7 illustrate bolted beam end framing connection design assuming no beam end moment (flexible connection) and with a beam end moment (semirigid or rigid),† respectively.

Example 9.6

Design the bolted simple beam framing connection using $6 \times 4 \times 1/2$ in. angles as shown in Figure E9.5 for a shear force of $P = 80$ kips. The beam

---

* In contrast to more rigid welded beam end connections where the moment, Pe, due to eccentricity, e, should be considered in the design.

† The beam end moments are generally determined by relatively sophisticated structural analyses that, for example, consider flexural members with equivalent rotational springs at supports.

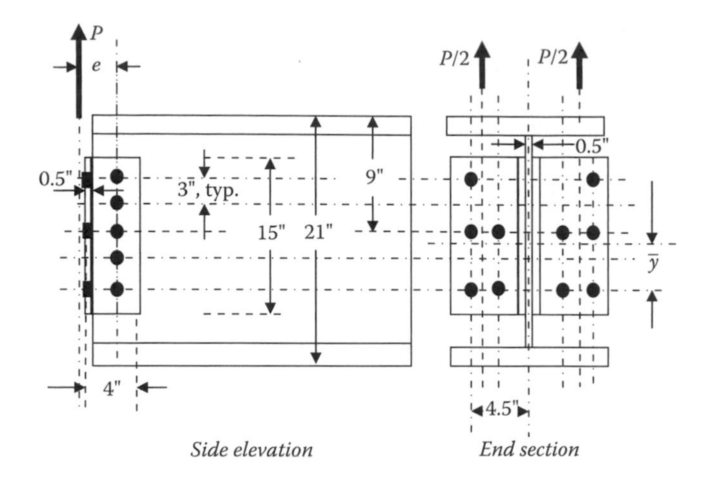

FIGURE E9.5

is 20 ft long and frames into the web of a plate girder with a single-shear connection. The allowable shear stress on the high-strength bolts is 17.0 ksi. The shear force is developed from a routine analysis considering a simply supported beam with complete connection flexibility.

$P' = 1.25(80) = 100$ kips (AREMA recommendation for flexible connection design)

The angle thickness should be based on the requirement for transmitting shear or minimum element thickness recommended by AREMA (2008). In this example the angle thickness is 1/2 in.

Bolt configuration to allow deformation:

$$g \geq \sqrt{\frac{Lt_a}{8}} \geq \sqrt{\frac{20(12)(0.5)}{8}} = 3.87 \leq 4.50 \text{ in. OK}$$

$h_g = 9$ in. $\geq h/3 \geq 7$ in. OK
therefore, flexible connection (neglect end bending effects).

Shear stress on the bolts:

$\tau = \frac{P'}{n_s n_b A_b} = \frac{100}{10(0.60)(1)} = 16.7$ ksi ≤ 17.0 ksi OK (single-shear connection, use five bolts in double-shear connection)

check bearing stress

$\sigma_{bc} = \frac{100}{10(0.50)(1.00)} = 20.0 \leq f_B$ ksi (both angles and beam web are 0.5 in. thick)

$f_B = \frac{l_e F_u}{2d_b} = \frac{(1.5)(65)}{2(7/8)} = 55.7$ ksi, assuming a minimum loaded edge distance of 1.5 in., OK
or

$f_B = 1.2F_u = 1.2(65) = 78$ ksi.

If the applied load is cyclical, fatigue must be considered. A connection such as that shown in Figure E9.5 has an allowable fatigue shear stress range of 16 ksi for connections subjected to greater than 2 million stress range cycles. The connection is, therefore, acceptable with respect to fatigue considerations.

Example 9.7

Design the bolted beam framing connection using 6 × 6 × 1/2 in. angles as shown in Figure E9.6 for a shear force, $P = 70$ kips and end moment, $M_e = 25$ ft-kip. The beam is 20 ft long and frames into the web of a plate girder with single-shear connections. The allowable shear stress on the high-strength bolts is 17.0 ksi. The shear force and bending moment were developed from an analysis considering partial rigidity of the connection, so the AREMA recommendation of a 25% increase in shear force is not used.

$P' = 70$ kips

$M_e = 25$ ft-kips = 300 in-kips in the direction creating tension at the top of the beam end

Shear and Bending on Bolts in the End Section:

Bolt forces due to bending:

The centroid of the connection is

$\bar{y} = \frac{2(12+9+6+3)}{10} = 6.0$ in.

The section modulus of the top (third row) bolts in the connection is

$S_{bt} = \frac{2((12-6.0)^2+(9-6.0)^2+(6-6.0)^2+(3-6.0)^2+(0-6.0)^2)}{(12-6.0)} = 30.0$ bolt-in for top bolts (in tension in the simple beam connection)

The section modulus of the second row bolts in the connection is

$S_{bb} = \frac{2((12-6.0)^2+(9-6.0)^2+(6-6.0)^2+(3-6.0)^2+(0-6.0)^2)}{(9-6.0)} = 60.0$ bolt-in for bottom bolts

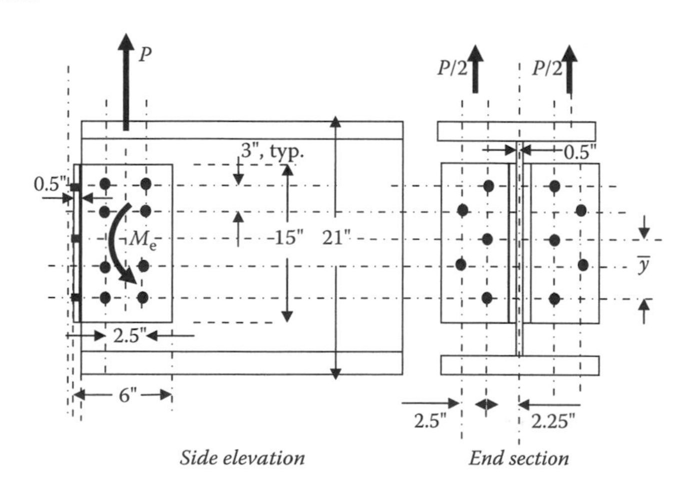

FIGURE E9.6

$$\sigma_{bt} = \frac{M_e}{S_{bt}A_b} = +\frac{300}{30.0(0.6)} = 16.67\text{ ksi for top bolts} \leq 44.0\text{ ksi OK}$$

$$\sigma_{bt} = \frac{M_e}{S_{bt}A_b} = +\frac{300}{60.0(0.6)} = 8.33\text{ for the second row bolts}$$

$$T_{bt} = \sigma_{bt}A_b = 16.67(0.6) = 10.0\text{ kips for top bolts}$$

$$T_{bt} = \sigma_{bt}A_b = 8.33(0.6) = 5.0\text{ kips for top bolts}$$

Prying action (Equation 9.38):

$$T_Q = 10.0\left(\frac{3b}{8a} - \frac{(t_p)^3}{20}\right) = 10.0\left(\frac{3(2.25 - 0.25 - .05)}{8(3.75)} - \frac{(0.5)^3}{20}\right)$$

$$= 0.14(10.0) = 1.4\text{ kips for the top row bolts}$$

$$T_Q = 5.0\left(\frac{3b}{8a} - \frac{(t_p)^3}{20}\right) = 5.0\left(\frac{3(4.00)}{8(1.25)} - \frac{(0.5)^3}{20}\right)$$

$$= 1.19(5.0) = 6.0\text{ kips for the second row bolts}$$

Tension in top bolts = $T_b = 10.0 + 1.4 = 11.4$ kips
Tension in the second row bolts = $T_b = 5.0 + 6.0 = 11.0$ kips

Shear stress on the bolts:

$$\tau = \frac{P'}{n_s n_b A_b} = \frac{70}{10(0.60)(1)} = 11.67\text{ ksi}$$

Combined shear and tension

$$f_{bv} = f'_{bv}\left(1 - \frac{(T_b)}{(T_{bP})}\right) = 17\left(1 - \frac{11.4}{39}\right) = 12.0\text{ ksi} \geq 11.67\text{ ksi OK}$$

check bearing stress

$$\sigma_{bc} = \frac{70}{10(0.50)(1.00)} = 14.0 \leq f_B\text{ ksi}$$

$$f_B = \frac{l_e F_u}{2d_b} = \frac{(1.5)(65)}{2(7/8)} \leq 55.7\text{ ksi,}$$

assuming a minimum loaded direction edge distance of 1.5 in., OK
or

$$f_B = 1.2F_u = 1.2(65) \leq 78\text{ ksi}$$

Shear and torsion on bolts in Side Elevation:
The direct shear stress, t, in the bolts is

$$\tau = \frac{P'}{n_s n_b A_b} = \frac{70}{(2)8(0.60)} = 7.3\text{ ksi}$$

and the torsional shear stress, τ_T, on the highest stressed bolts is

$$\tau_{Tx} = \frac{M_e y_T}{n_s A_b \sum n_b (x_T^2 + y_T^2)} = \frac{300(1.25)}{2(0.6)\left(4(6^2 + 1.25^2) + 4(3^2 + 1.25^2)\right)}$$

$$= \frac{300(1.25)}{2(0.6)(192.5)} = 1.6 \text{ ksi},$$

$$\tau_{Ty} = \frac{M_e x_T}{n_s A_b \sum n_b (x_T^2 + y_T^2)} = \frac{300(6)}{2(0.6)(192.5)} = 7.8 \text{ ksi}.$$

The maximum resultant shear stress in the bolts is

$$f = \sqrt{(\tau + \tau_{Ty})^2 + \tau_{Tx}^2} = \sqrt{(7.3 + 7.8)^2 + 1.6^2} = 15.2 \text{ ksi} \leq 17.0 \text{ ksi OK}$$

check bearing stress

$$\sigma_{bc} = \frac{(15.2)(0.60)}{1(0.50)(1.00)} = 18.2 \leq f_B \text{ ksi}$$

$$f_B = \frac{l_e F_u}{2 d_b} = \frac{(1.5)(65)}{2(7/8)} = 55.7 \text{ ksi}$$

assuming a minimum loaded edge distance of 1.5 in., OK
or

$$f_B = 1.2 F_u = 1.2(65) = 78 \text{ ksi}.$$

9.3.4.3.4 Axially Loaded Connections with Bolts in Direct Tension

Connections with bolts subject to direct tension should generally be avoided in the main members of steel railway superstructures. However, when bolts are subjected to direct tension, the additional bolt forces created by prying action of the connection leg must also be considered (e.g., by using Equation 9.38). The effects of the prying action on the allowable fatigue design stresses must also be considered as shown in Example 9.8.

Example 9.8

Design the bolted hanger-type connection shown in Figure E9.7 for an axial force consisting of

$P_{DL} = 20$ kips (dead load), $P_{LL+I} = 60$ kips (live load plus impact).
Use 7/8 in diameter A325 bolts.

$b = 2.0 - 0.5 = 1.5$ in.
$a = 3$ in.
$t_p = 0.5$ in.

$$T_Q = T\left(\frac{3b}{8a} - \frac{t_p^3}{20}\right) = \frac{(20 + 60)}{4}\left(\frac{3(1.5)}{8(3)} - \frac{(0.5)^3}{20}\right) = 20(0.18) = 3.60 \text{ ksi}$$

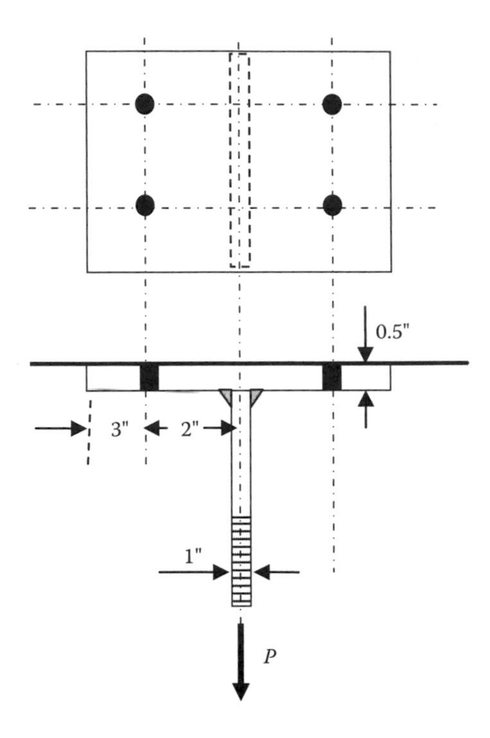

FIGURE E9.7

$T_B = T + T_Q = 20 + 3.60 = 23.6$ kips

$$\sigma_{bt} = \frac{T_B}{A_b} = \frac{23.6}{(0.6)} = 39.3 \leq 44.0 \text{ ksi OK.}$$

The tensile stress range, including prying stress, is

$$\Delta\sigma_{bt} = \frac{60(1.18)}{(4)0.6} = \frac{17.7}{0.6} = 29.5 \text{ ksi}$$

Since

$$\frac{T_Q}{T}\% = \frac{3.60}{20}(100) = 18\% \leq 30\%$$

the allowable tensile stress range is 31.0 ksi $\geq$ 29.5 ksi OK.

9.3.4.3.5 Axial Member Splices

A common axial member bolted splice involves the use of lap joints on the member elements as shown in Figure 9.10g. The bolted connection is designed as a slip-resistant connection, often including a review of bearing stresses in case of joint failure by slippage. Splices located in the center of truss members must also be sufficiently rigid to resist bending from self-weight and other lateral forces.

AREMA (2008) also recommends that truss web member axial splice connections be designed for 133% of allowable stress using the live load that will increase the maximum chord stress in the highest stresses chord by 33% (see Chapters 4, 5, and 6). The procedure that may be used for truss web member splice connection design is analogous to that outlined in Chapter 6, Example 6.4, for web members.

9.3.4.3.5.1 Axial Tension Member Splices Main member axial tension splices should be designed for the strength of member. For secondary members, AREMA (2008) recommends that the splice be designed for the lesser of the strength of the member or 150% of the maximum calculated tension.

Steel rods or bars may be spliced by turnbuckles and sleeve nuts. Rolled or built-up tension members are spliced by bolted plates and, therefore, designed as net area tension members (see Chapter 6) with due consideration of block shear at the connection. Generally, all elements of tension members are spliced on each side of the element to avoid eccentricities and shear lag effects. The connection bolts are designed for direct shear and bearing strength.

9.3.4.3.5.2 Axial Compression Member Splices Splice plates and bolts must transmit 50% of the force and be placed on four sides of the member in a manner that provides for the accurate and firm fit of the abutting elements in rolled or built-up compression members that are faced or finished to bear. This may result in compression members with only nominal splice plates and the designer may wish to ensure adequate bending rigidity by designing the splice for bending and shear from a minimum transverse force of 2.5% of the member axial compression. The connection bolts are designed for direct shear and bearing strength.

9.3.4.3.6 Beam and Girder Splices

Conventional beam and girder bolted splices involve the use of lap joints as shown in Figure 9.10c. The plates used in these splices are designed in accordance with AREMA (2008) as outlined in Chapter 7, Section 7.2.6, concerning plate girder design. The bolted connection is designed as a slip-resistant connection, often with a bearing check in the case of joint failure by slippage.

9.3.4.3.6.1 Beam and Girder Flange Splice Bolts Beam and girder flange splices should be designed for the strength of the member being spliced. Also, as outlined in Chapter 7, Section 7.2.6, bolted splice elements in girder flanges should

- Have a cross-sectional area that is at least equal to that of the flange element being spliced
- Comprise splice elements of sufficient cross section and location such that the moment of inertia of the member at the splice is no less than that of the member adjacent to the splice location.

The bolts in tension and compression flange lap joint splices are subjected to direct shear in transferring flange forces from bending, F_f, between the girder flange and splice plates. Depending on whether one or two plate splices are used, the bolts will be in single or double shear, respectively.

9.3.4.3.6.2 Beam and Girder Web Splice Bolts As outlined in Chapter 7, Section 7.2.6, bolted splice elements in girder web plates should be designed

- To transfer the shear force, V, including moment, Ve, due to eccentricity, e, of the centroid of the bolt group
- For the concurrent net flexural strength of the web plate.

The bolts in girder web lap joint splices are subjected to direct and torsional shear in transferring web plate forces from shear and bending between the girder web and splice plates. The web splice plates must be designed for the gross shear strength of the web plate and have a net moment of inertia not less than that of the web plate. Two web splice plates must always be used so that the bolts are in double shear.

The design of girder flange and web splice connection plates and bolts is outlined in Example 9.9.

Example 9.9

Design the bolted flange and web splices for the girder shown in Figure E9.8 with flange splices located where $M = 8800$ ft-kips and web splices where $M_V = 6000$ ft-kips and $V = 450$ kips.

The girder has the following properties (see Example 7.1):

$I_g = 223{,}444$ in.^{4}

$S_g = 4965$ in.^{3}

$A_w = 53.13$ in.^{2}

Allowable girder compressive bending capacity, $M_{call} = 0.55(F_y)(S_g) = 0.55(50)(4965)/12 = 11{,}378$ ft-kips (assuming fully laterally supported)

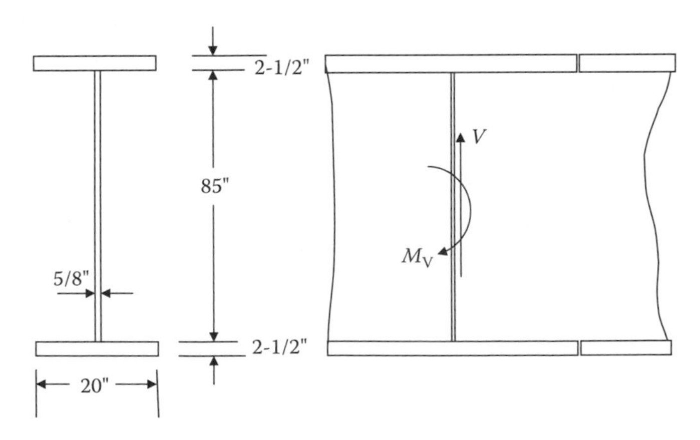

FIGURE E9.8

Allowable girder tensile bending capacity, $M_{tall} = 0.55(F_y)(S_n)$

Allowable girder shear capacity, $V_{all} = 0.35(F_y)(A_w) = 0.35(50)(53.13) = 929.8$ kips

Compression flange splices

Try a single plate flange splice as shown in Figure E9.9.

For the splice cross section to be at least equal to the flange cross section, $t_s = 2.5$ in.

Since the plates are spliced on the exterior of the flanges, the moment of inertia at the splice will be greater than the moment of inertia of the cross section being spliced. There must be enough bolts each side of the splice to develop the flange strength.

The maximum force in the compression flange plate at the splice location (average at the centroid) is

$$F_f = \frac{8800(12)}{4965}\frac{(1 + 42.5/45)}{2}(2.5)(20) = 21.3(0.97)(50) = 1031.5 \text{ kips.}$$

Note that a conservative approximation for the actual flange force, F_f', can be determined from the couple resisting the applied bending moment as

$$F_f' = \frac{8800(12)}{87.5} = 1206.9 \text{ kips}$$

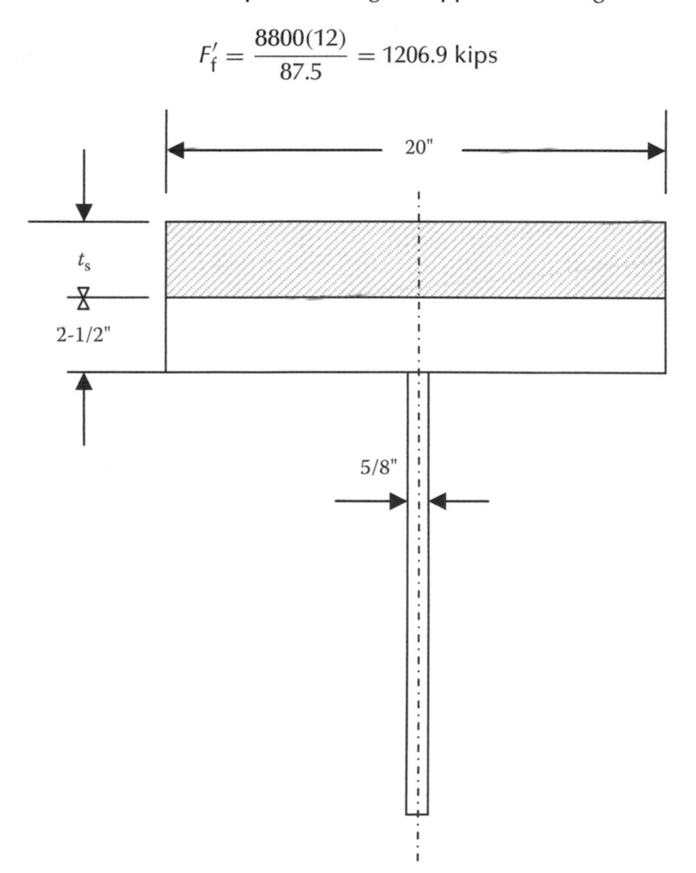

FIGURE E9.9

The allowable force in the compression flange plate at the splice location (average at the centroid) is

$$F_f = \frac{11{,}378(12)}{4965}\frac{(1 + 42.5/45)}{2}(2.5)(20) = 27.5(0.97)(50) = 1333.7 \text{ kips}.$$

Transfer of this force from girder flange to splice plate is by a slip-resistant lap joint using A325 bolts $f'_{bv} = 17.0$ ksi. The number of single shear bolts required is

$$n_{fs} = \frac{1333.7}{(1)(17.0)(0.6)} = 131 \text{ bolts}$$

each side of the splice. The joint will be too long. To reduce the amount of bolting required, a double shear splice using splice plates each side of the compression flange is designed as shown in Figure E9.10.

$$n_{fs} = \frac{1333.7}{(2)(17.0)(0.6)} = 65 \text{ bolts}$$

each side of the splice. Use 16 rows of four bolts.

For the splice cross section to be at least equal to the flange cross section (50 in.^{2}), try

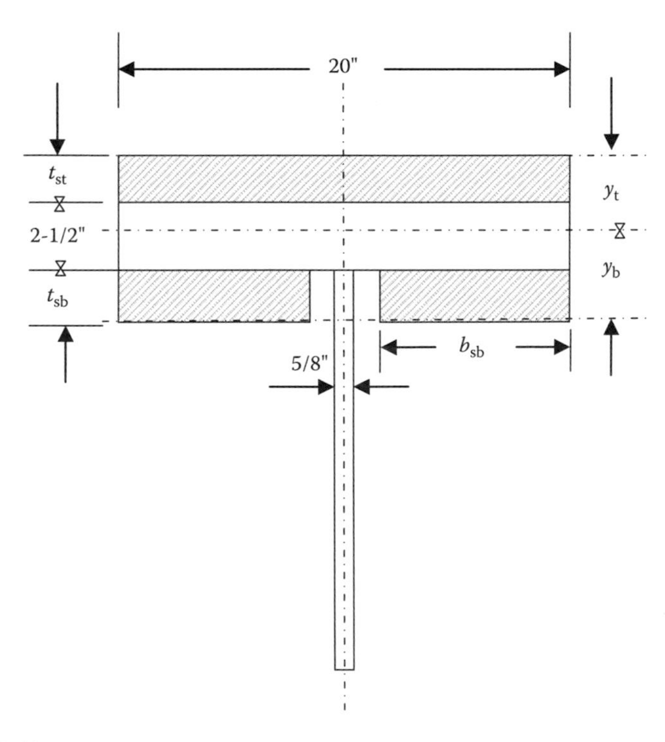

FIGURE E9.10

$t_{st} = 1.75$ in.
$t_{sb} = 1.00$ in.
$b_{sb} = 8$ in.
The area of the splice is $1.75(20) + 2(1.00)(8) = 35.0 + 16.0 = 51.0$ in.$^2 \geq$ 50 in.^{2} OK

$$y_t = 5.25 - \frac{(35.0)(4.375) + (50)(2.25) + (16.0)(0.5)}{51 + 50} = 5.25 - 2.71 = 2.54 \text{ in.}$$

$$y_b = 2.71 \text{ in.}$$

Since the top flange splice centroid is $2.54 - (1.75 + 2.5/2) = -0.46$ in. from the top flange centroid (0.46 in. farther from the girder neutral axis), the moment of inertia at the splice will be greater than the moment of inertia of the cross section being spliced.

The spliced flange eccentricity = 0.46 in. This eccentricity creates a maximum force bending moment of $1031.5(0.46)/12 = 39.5$ ft-kips, which is small (0.45% of maximum moment) and, therefore, acceptable without further consideration.

Check bearing stress

$$\sigma_{bc} = \frac{1333.7}{64(1.00)(1.00)} = 20.8 \leq f_B \text{ ksi}$$

$$f_B = \frac{l_e F_u}{2d_b} = \frac{(1.5)(65)}{2(7/8)} = 55.7 \text{ ksi}$$

assuming a minimum loaded edge distance of 1.5 in., OK
or

$$f_B = 1.2F_u = 1.2(65) = 78 \text{ ksi}$$

Tension flange splices

For the splice cross section to be at least equal to the flange net cross section ($50 - 4(1.00)(2.5) = 40$ in.^{2}), try
$t_{st} = 1.75$ in.
$t_{sb} = 1.00$ in.
$b_{sb} = 8$ in.
The net area of the splice is $1.75(20 - 4(1.00)) + 2(1.00)(8 - 2(1.00)) = 28.0 + 12 = 40.0$ in.$^2 \geq 40$ in.^{2} OK

The girder net section moment of inertia is

$$I_n = 223{,}444 - 2\left(4(1.00)(2.5)(43.75)^2\right) = 223{,}444 - 38{,}281 = 185{,}163 \text{ in.}^4$$

and the neutral axis is $50(88.75) + 53.13(45) + 40(1.25)/(50 + 53.13 + 40) = 48.1$ in from the underside of the bottom flange.

The girder allowable tensile bending capacity is $M_{tall} = 0.55(50)(185163/48.1)/12 = 27.5(3849.5)/12 = 8822$ ft-kips.

The net section moment of inertia of the spliced section is

$$I_g = 2(51)(43.75 + 0.46)^2 + 2(8.93 + 26.04 + 2(0.67)) + \frac{0.625(85)^3}{12} = 231{,}420 \text{ in.}^4$$

$$I_n = 231{,}420 - 2\left(4(1.00)(1.75)(45.875)^2 + 4(1.00)(1.00)(42.00)^2\right) = 231{,}420 - 43{,}575$$

$$= 187{,}845 \geq 185{,}163 \text{ in.}^4$$

OK, the moment of inertia at the splice is greater than the moment of inertia of the cross section being spliced.

The maximum force in the tension flange plate at the splice location (average at the centroid) is

$$F_f = \frac{8800(12)}{185{,}163/48.1} \frac{(1 + (42.5/45))}{2}(40) = 27.4(0.97)(40) = 1063.4 \text{ kips}$$

The allowable force in the tension flange plate at the splice location (average at the centroid) is

$$F_f = \frac{8822(12)}{185{,}163/48.1} \frac{(1 + (42.5/45))}{2}(40) = 27.5(0.97)(40) = 1066.0 \text{ kips}$$

The number of double-shear bolts required is

$$n_{fs} = \frac{1066.0}{(2)(17.0)(0.6)} = 52 \text{ bolts}$$

each side of the splice. Use 13 rows of four bolts as shown in Figure E9.11.

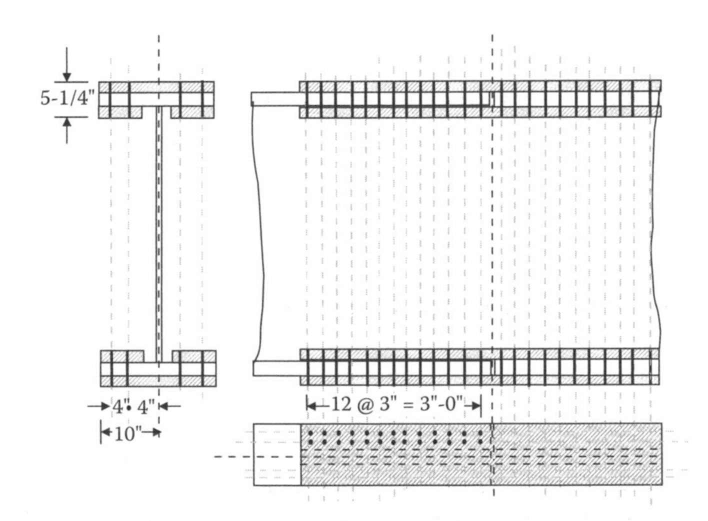

FIGURE E9.11

Check bearing stress

$$\sigma_{bc} = \frac{1066.0}{52(1.00)(1.00)} = 20.5 \leq f_B \text{ ksi}$$

$$f_B = \frac{l_e F_u}{2d_b} = \frac{(1.5)(65)}{2(7/8)} = 55.7 \text{ ksi}$$

assuming a minimum loaded edge distance of 1.5 in., OK
or

$$f_B = 1.2F_u = 1.2(65) = 78 \text{ ksi}$$

Since the weaker tension flange splice will govern at ultimate conditions, a stronger compression flange splice is not required. Therefore, both top and bottom flange splices will consist of 52 bolts each side of the double shear splice, provided the splice is strong enough to transmit the actual compression flange force of 1031.5 kips.

$$n_{fs} = \frac{1031.5}{(2)(17.0)(0.6)} = 51 \leq 52 \text{ provided, OK}$$

Long joints, particularly after slippage, do not provide for an equal distribution of bolt shear stress at gross section yielding. Therefore, the average bolt shear strength will be decreased in longer joints and, effectively, the joint has a lower factor of safety against yielding than bolts in shorter joints. However, theoretical and experimental investigations have shown that for joints less than about 50 in long, the FS remains at least 2.0, which is acceptable (Kulak et al., 1987). For long joints it is often recommended to consider reducing the allowable bolt shear stress by 20% to ensure FS $\geq$ 2.0.

Web plate splices

Try a web splice using 1/2 in. plates with 42 bolts each side of the splice in the 85 in web plate shown in Figure E9.12.

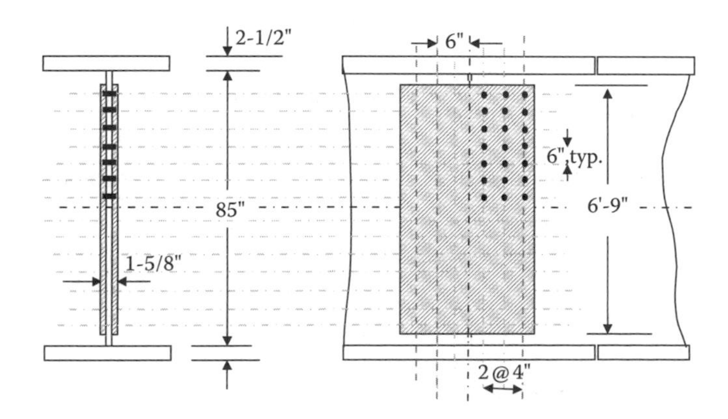

FIGURE E9.12

The gross section shear strength of the girder web plate is

$$V_{\text{all}} = (0.625)(85)(0.35)(50) = 929.8\,\text{kips.}$$

The gross section shear strength of the web splice is

$$V_{\text{sw}} = \frac{2((81)(0.50)(0.35)(50))}{1.5} = 945\ \text{kips} \geq 929.8\ \text{kips OK}$$

$$f_{\text{vs}} = \frac{1.5(929.8)}{2(81)(0.5)} = 17.2 \leq 17.5\ \text{ksi OK}$$

The net section moment of inertia of the web plate is

$$= \left(\frac{0.625(85)^3}{12} - (1.00)(0.625)(2)(39^2 + 33^2 + 27^2 + 21^2 + 15^2 + 9^2 + 3^2)\right)$$

$$= 31{,}986 - 5119 = 26{,}867\ \text{in.}^4$$

The net section flexural strength of the girder web is

$$M_{\text{sw}} = 0.55(50)\frac{(31{,}986 - 5{,}119)}{42.5(12)} = 1449\ \text{ft-kips}$$

The net section moment of inertia of the web splice is

$$= 2\left(\frac{0.5(81)^3}{12} - (1.00)(0.5)(2)(39^2 + 33^2 + 27^2 + 21^2 + 15^2 + 9^2 + 3^2)\right)$$

$$= 36{,}097\ \text{in.}^4$$

The web splice plates have a net moment of inertia greater than that of the web plate.

The maximum direct shear stress, t, on the bolts is

$$\tau = \frac{V}{n_{\text{s}} n_{\text{b}} A_{\text{b}}} = \frac{450}{2(42)(0.60)} = 8.9\,\text{ksi (all bolts with the same } A_b\text{)}$$

and the maximum torsional shear stress in the x- and y-directions is

$$\tau_{\text{T}x} = \frac{(Ve + M_{\text{sw}})y_{\text{T}}}{n_{\text{s}} A_{\text{b}} J_{\text{s}}}$$

$$\tau_{\text{T}y} = \frac{(Ve + M_{\text{sw}})x_{\text{T}}}{n_{\text{s}} A_{\text{b}} J_{\text{s}}}$$

$$Ve = 450(0.5) = 225.0\ \text{ft-kips}$$

$$M_{\text{sw}} = 1449\,\text{ft-kips}$$

$$J_s = \sum n_b(x_T + y_T)^2 = 4[(39^2 + 10^2) + (33^2 + 10^2) + (27^2 + 10^2) + (21^2 + 10^2)$$
$$+ (15^2 + 10^2) + (9^2 + 10^2) + (3^2 + 10^2) + (39^2 + 6^2) + (33^2 + 6^2)$$
$$+ (27^2 + 6^2) + (21^2 + 6^2) + (15^2 + 6^2) + (9^2 + 6^2) + (3^2 + 6^2) + (39^2 + 2^2)$$
$$+ (33^2 + 2^2) + (27^2 + 2^2) + (21^2 + 2^2) + (15^2 + 2^2) + (9^2 + 2^2) + (3^2 + 2^2)]$$
$$= 4(13{,}265) = 53{,}060 \text{ bolt-in.}^2$$

$$\tau_{Tx} = \frac{1674(12)(39)}{2(0.60)(53{,}060)} = 12.3 \text{ ksi},$$

$$\tau_{Ty} = \frac{1674(12)(10)}{2(0.60)(53{,}060)} = 3.2 \text{ ksi}.$$

The resultant shear stress on the most highly stressed bolt is

$$f = \sqrt{(\tau + \tau_{Ty})^2 + \tau_{Tx}^2} = \sqrt{(8.9 + 3.2)^2 + 12.3^2} = 17.2 \text{ ksi},$$

which is considered acceptable at 1.3% overstress OK
Check bearing stress

$$\sigma_{bc} = \frac{17.2(0.60)}{1(0.5)(1.00)} = 20.6 \le f_B \text{ ksi},$$

$$f_B = \frac{l_e F_u}{2d_b} = \frac{(1.5)(65)}{2(7/8)} = 55.7 \text{ ksi},$$

assuming a minimum loaded edge distance of 1.5 in., OK
or

$$f_B = 1.2F_u = 1.2(65) = 78 \text{ ksi}.$$

The web splice flexural stress, based on the girder web net bending strength, is

$$= \frac{1449(12)(40.5)}{36{,}097} = 19.5 \text{ ksi}.$$

The interaction between flexure and shear in the web splice plates is assessed, based on Equation 7.63 in Chapter 7, as

$$f_b = \left(0.75 - 1.05\frac{f_v}{F_y}\right)F_y = \left(0.75 - 1.05\frac{17.2}{50}\right)50 = 0.39(50) = 19.4 \text{ ksi},$$

which is considered acceptable at less than 0.5% overstress. OK

REFERENCES

Al-Emrani, M., 2005, Fatigue performance of stringer-to-floor-beam connections in riveted railway bridges, *Journal of Bridge Engineering*, 10(2), ASCE, Reston, VA.

American Railway Engineering and Maintenance-of-Way Association (AREMA), 2008, Steel structures, in *Manual for Railway Engineering*, Chapter 15, Lanham, MD.

American Welding Society (AWS), 2005, *Bridge Welding Code*, ANSI/AASHTO/AWS D1.5, Miami, FL.

Blodgett, O.W., 2002, *Design of Welded Structures*, Lincoln Arc Welding Foundation, Cleveland, OH.

Chesson, E., Faustino, N.L., and Munse, W.H., 1965, High-strength bolts subjected to tension and shear, *Journal of the Structural Division, ASCE*, 91(ST5), 155–180.

Douty, R.T. and McGuire, W., 1965, High strength bolted moment connections, *Journal of the Structural Division, ASCE*, 91(ST2), 101–128.

Faella, C., Piluso, V., and Rizzano, G., 2000, *Structural Steel Semirigid Connections*, CRC Press, Boca Raton, FL.

Federal Highway Administration (FHWA), 1998, *Heat-Straightening Repairs of Damaged Steel Bridges*, FHWA-IF-99-004, Washington, DC.

Kulak, G., 2002, *High Strength Bolts*, AISC Steel Design Guide 17, Chicago, IL.

Kulak, G., Fisher, J.W., and Struik, J.H.A., 1987, *Guide to Design Criteria for Bolted and Riveted Joints*, 2nd Edition, Wiley, New York.

Kuzmanovic, B.O. and Willems, N., 1983, *Steel Design for Structural Engineers*, 2nd Edition, Prentice-Hall, Englewood Cliffs, NJ.

Leon, R.T., 1999, Partially restrained connections, in *Handbook of Structural Steel Connection Design and Details*, A.R. Tamboli (Ed.), Chapter 4, McGraw-Hill, New York.

Nair, R.S., Birkemoe, P.C., and Munse, W.H., 1974, High strength bolts subjected to tension and prying, *Journal of the Structural Division, ASCE*, 100(ST2), 351–372.

Research Council on Structural Connections (RCSC), 2000, Specification for Structural Joints Using ASTM A325 or A490 Bolts, AISC, Chicago, IL.

Thornton, W.A. and Kane, T., 1999, Design of connections for axial, moment and shear forces, in *Handbook of Structural Steel Connection Design and Details*, A.R. Tamboli (Ed.), Chapter 2, McGraw-Hill, New York.

Whitmore, R.E., 1952, *Experimental Investigation of Stresses in Gusset Plates*, University of Tennessee Experiment Station Bulletin 16.

Yen, B.T., Zhou, Y., Wang, D., and Fisher, J.W., 1991, *Fatigue Behavior of Stringer-Floorbeam Connections*, Lehigh University Report 91-07, Bethlehem, PA.

Index

Note: n denotes footnote.

C

D

H

I

J

K

L

M

N

O

P

Y